T0327538

Structural Steel Design to Eurocode 3 and AISC Specifications

Structural Steel Design to Eurocode 3 and AISC Specifications

By

Claudio Bernuzzi

and

Benedetto Cordova

WILEY Blackwell

This edition first published 2016

Registered Office
John Wiley & Sons, Ltd, The Atrium, Southern Gate, Chichester, West Sussex, PO19 8SQ, United Kingdom

Editorial Offices
9600 Garsington Road, Oxford, OX4 2DQ, United Kingdom
The Atrium, Southern Gate, Chichester, West Sussex, PO19 8SQ, United Kingdom

For details of our global editorial offices, for customer services and for information about how to apply for permission to reuse the copyright material in this book please see our website at www.wiley.com/wiley-blackwell.

Based on *Progetto e verifica delle strutture in acciaio* by Claudio Bernuzzi.

Library of Congress Cataloging-in-Publication data applied for.

ISBN: 9781118631287

A catalogue record for this book is available from the British Library.

Wiley also publishes its books in a variety of electronic formats. Some content that appears in print may not be available in electronic books.

Cover image: photovideostock/Getty

Set in 10/12pt Minion by SPi Global, Pondicherry, India

1 2016

Contents

Preface

Over the last century, design of steel structures has developed from very simple approaches based on a few elementary properties of steel and essential mathematics to very sophisticated treatments demanding a thorough knowledge of structural and material behaviour. Nowadays, steel design utilizes refined concepts of mechanics of material and of theory of structures combined with probabilistic-based approaches that can be found in design specifications.

This book intends to be a guide to understanding the basic concepts of theory of steel structures as well as to provide practical guidelines for the design of steel structures in accordance with both European (EN 1993) and United States (ANSI/AISC 360-10) specifications. It is primarily intended for use by practicing engineers and engineering students, but it is also relevant to all different parties associated with steel design, fabrication and construction.

The book synthesizes the Authors' experience in teaching Structural Steel Design at the Technical University of Milan-Italy (Claudio Bernuzzi) and in design of steel structures for power plants (Benedetto Cordova), combining their expertise in comparing and contrasting both European and American approaches to the design of steel structures.

The book consists of 16 chapters, each structured independently of the other, in order to facilitate consultation by students and professionals alike. Chapter 1 introduces general aspects such as material properties and products, imperfection and tolerances, also focusing the attention on testing methods and approaches. The fundamentals of steel design are summarized in Chapter 2, where the principles of structural safety are discussed in brief to introduce the different reliability levels of the design. Framed systems and methods of analysis, including simplified methods, are discussed in Chapter 3. Cross-sectional classification is presented in Chapter 4, in which special attention has been paid to components under compression and bending. Design of single members is discussed in depth in Chapter 5 for tension members, in Chapter 6 for compression members, in Chapter 7 for members subjected to bending and shear, in Chapter 8 for members under torsion, and in Chapter 9 for members subjected to bending and compression. Chapter 10 deals with design accounting for the combination of compression, flexure, shear and torsion.

Chapter 11 addresses requirements for the web resistance design and Chapter 12 deals with the design approaches for frame analysis. Chapters 13 and 14 deal with bolted and welded connections, respectively, while the most common type of joints are described in Chapter 15, including a summary of the approach to their design. Finally, built-up members are discussed in Chapter 16. Several design examples provided in this book are directly chosen from real design situations. All examples are presented providing all the input data necessary to develop the design. The different calculations associated with European and United States specifications are provided in two separate text columns in order to allow a direct comparison of the associated procedures.

Last, but not least, the acknowledge of the Authors. A great debt of love and gratitude to our families: their patience was essential to the successful completion of the book.

We would like to express our deepest thanks to Dr. Giammaria Gabbianelli (University of Pavia-I) and Dr. Marco Simoncelli (Politecnico di Milano-I) for the continuous help in preparing figures and tables and checking text. We are also thankful to prof. Gian Andrea Rassati (University of Cincinnati-U.S.A.) for the great and precious help in preparation of chapters 1 and 13.

Finally, it should be said that, although every care has been taken to avoid errors, it would be sanguine to hope that none had escape detection. Authors will be grateful for any suggestion that readers may make concerning needed corrections.

Claudio Bernuzzi and Benedetto Cordova

CHAPTER 1

The Steel Material

1.1 General Points about the Steel Material

The term *steel* refers to a family of iron–carbon alloys characterized by well-defined percentage ratios of main individual components. Specifically, iron–carbon alloys are identified by the carbon (C) content, as follows:

- *wrought iron*, if the carbon content (i.e. the percentage content in terms of weight) is higher than 1.7% (some literature references have reported a value of 2%);
- *steel*, when the carbon content is lower than the previously mentioned limit. Furthermore, steel can be classified into extra-mild (C < 0.15%), mild (C = 0.15 ÷ 0.25%), semi-hard (C = 0.25 ÷ 0.50%), hard (C = 0.50 ÷ 0.75%) and extra-hard (C > 0.75%) materials.

Structural steel, also called *constructional steel* or sometimes *carpentry steel*, is characterized by a carbon content of between 0.1 and 0.25%. The presence of carbon increases the strength of the material, but at the same time reduces its ductility and weldability; for this reason structural steel is usually characterized by a low carbon content. Besides *iron* and *carbon*, structural steel usually contains small quantities of other elements. Some of them are already present in the iron ore and cannot be entirely eliminated during the production process, and others are purposely added to the alloy in order to obtain certain desired physical or mechanical properties.

Among the elements that cannot be completely eliminated during the production process, it is worth mentioning both *sulfur* (S) and *phosphorous* (P), which are undesirable because they decrease the material ductility and its weldability (their overall content should be limited to approximately 0.06%). Other undesirable elements that can reduce ductility are *nitrogen* (N), *oxygen* (O) and *hydrogen* (H). The first two also affect the strain-ageing properties of the material, increasing its fragility in regions in which permanent deformations have taken place.

The most important alloying elements that may be added to the materials are *manganese* (Mn) and *silica* (Si), which contribute significantly to the improvement of the weldability characteristics of the material, at the same time increasing its strength. In some instances, *chromium* (Cr) and *nickel* (Ni) can also be added to the alloy; the former increases the material strength and, if is present in sufficient quantity, improves the corrosion resistance (it is used for stainless steel), whereas the latter increases the strength while reduces the deformability of the material.

Structural Steel Design to Eurocode 3 and AISC Specifications, First Edition. Claudio Bernuzzi and Benedetto Cordova.

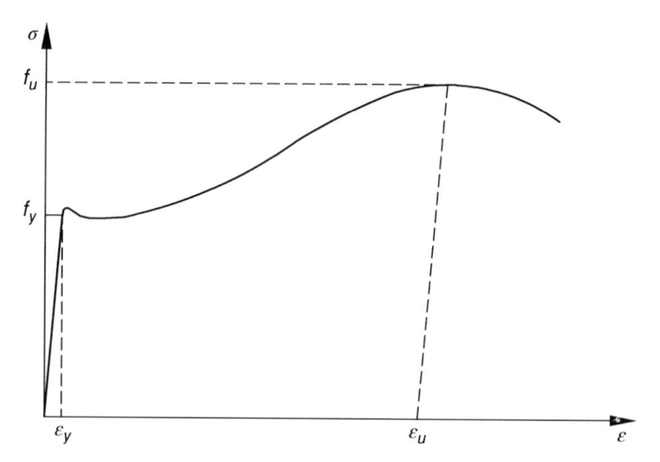

Figure 1.1 Typical constitutive law for structural steel.

Steel is characterized by a symmetric constitutive stress-strain law (σ–ε). Usually, this law is determined experimentally by means of a tensile test performed on coupons (samples) machined from plate material obtained from the sections of interest (Section 1.7). Figure 1.1 shows a typical stress-strain response to a uniaxial tensile force for a structural steel coupon. In particular, it is possible to distinguish the following regions:

- an initial branch that is mostly linear (*elastic phase*), in which the material shows a linear elastic behaviour approximately up to the yielding stress (f_y). The strain corresponding to f_y is usually indicated with ε_y (yielding strain). The slope of this initial branch corresponds to the modulus of elasticity of the material (also known as longitudinal modulus of elasticity or Young's modulus), usually indicated by E, with a value between 190 000 and 210 000 N/mm^2 (from 27 560 to 30 460 ksi, approximately);
- a *plastic phase*, which is characterized by a small or even zero slope in the σ–ε reference system;
- the ensuing branch is the *hardening phase*, in which the slope is considerably smaller when compared to the elastic phase, but still sufficient enough to cause an increase in stress when strain increases, up to the ultimate strength f_u. The hardening modulus has values between 4000 and 6000 N/mm^2 (from 580 to 870 ksi, approximately).

Usually, the uniaxial constitutive law for steel is schematized as a multi-linear relationship, as shown in Figure 1.2a, and for design purposes an elastic-perfectly plastic approximation is generally used; that is the hardening branch is considered to be horizontal, limiting the maximum strength to the yielding strength.

The yielding strength is the most influential parameter for design. Its value is obtained by means of a laboratory uniaxial tensile test, usually performed on coupons cut from the members of interest in suitable locations (see Section 1.7).

In many design situations though, the state of stress is biaxial. In this case, reference is made to the well-known Huber-Hencky–Von Mises criterion (Figure 1.2b) to relate the mono-axial yielding stress (f_y) to the state of plane stress with the following expression:

$$\sigma_1^2 - \sigma_1\sigma_2 + \sigma_2^2 + 3\sigma_{12}^2 = f_y^2 \quad (1.1)$$

where σ_1, σ_2 are the normal stresses and σ_{12} is the shear stress.

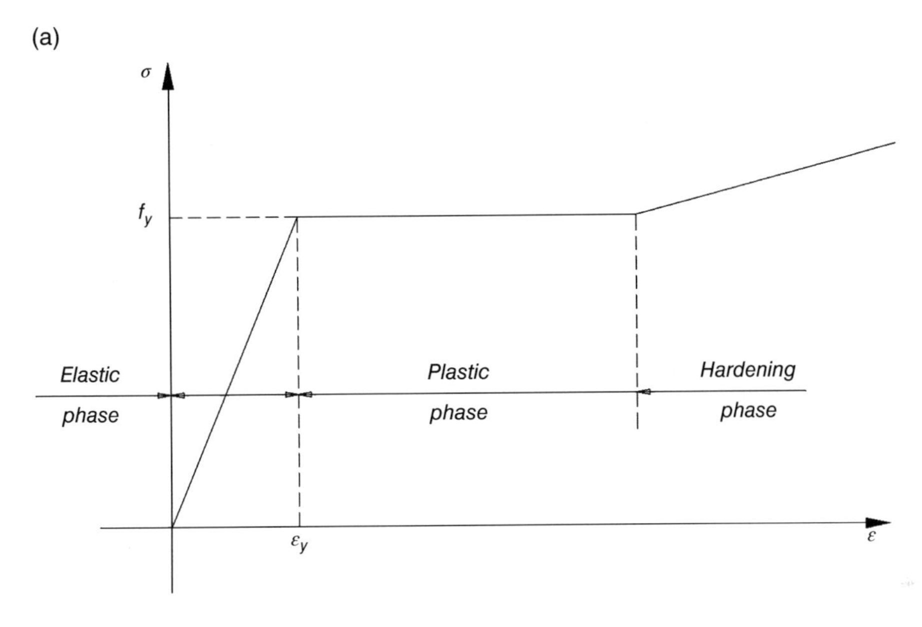

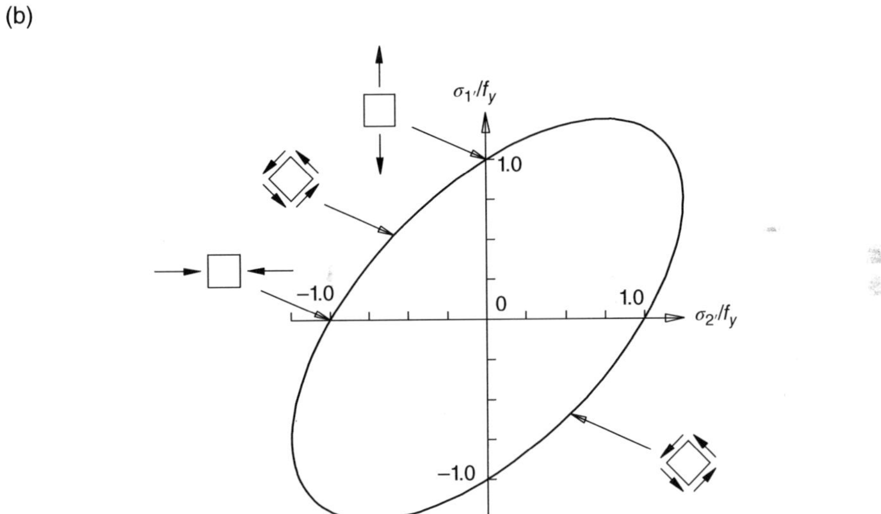

Figure 1.2 Structural steel: (a) schematization of the uniaxial constitutive law and (b) yield surface for biaxial stress states.

In the case of pure shear, the previous equation is reduced to:

$$\sigma_{12} = \tau_{12} = \frac{f_y}{\sqrt{3}} = \tau_y \tag{1.2}$$

With reference to the principal stress directions 1′ and 2′, the yield surface is represented by an ellipse and Eq. (1.1) becomes:

$$(\sigma_{1'})^2 + (\sigma_{2'})^2 - (\sigma_{1'})\cdot(\sigma_{2'}) = f_y^{\,2} \tag{1.3}$$

1.1.1 Materials in Accordance with European Provisions

The European provisions prescribe the following values for material properties concerning structural steel design:

Density:	$\rho = 7850$ kg/m^3 (= 490 lb/ft^3)
Poisson's coefficient:	$\nu = 0.3$
Longitudinal (Young's) modulus of elasticity:	$E = 210\,000$ N/mm^2 (= 30 460 ksi)
Shear modulus:	$G = \dfrac{E}{2(1+\nu)}$
Coefficient of linear thermal expansion:	$\alpha = 12 \times 10^{-6}$ per °C (=6.7 × 10^{-6} per °F)

The mechanical properties of the steel grades most used for construction are summarized in Tables 1.1a and 1.1b, for hot-rolled and hollow profiles, respectively, in terms of yield strength (f_y) and ultimate strength (f_u). Similarly, Table 1.2 refers to steel used for mechanical fasteners. With respect to the European nomenclature system for steel used in high strength fasteners, the generic tag ($j.k$) can be immediately associated to the mechanical characteristics of the material expressed in International System of units (I.S.), considering that:

- $j{\cdot}k{\cdot}10$ represents the yielding strength expressed in N/mm^2;
- $j{\cdot}100$ represents the failure strength expressed in N/mm^2.

Table 1.1a Mechanical characteristics of steels used for hot-rolled profiles.

	Nominal thickness t			
	$t \le 40$ mm		40 mm < $t \le 80$ mm	
EN norm and steel grade	f_y (N/mm^2)	f_u (N/mm^2)	f_y (N/mm^2)	f_u (N/mm^2)
EN 10025-2				
S 235	235	360	215	360
S 275	275	430	255	410
S 355	355	510	335	470
S 450	440	550	410	550
EN 10025-3				
S 275 N/NL	275	390	255	370
S 355 N/NL	355	490	335	470
S 420 N/NL	420	520	390	520
S 460 N/NL	460	540	430	540
EN 10025-3				
S 275 M/ML	275	370	255	360
S 355 M/ML	355	470	335	450
S 420 M/ML	420	520	390	500
S 460 M/ML	460	540	430	530
EN 10025-5				
S 235 W	235	360	215	340
S 355 W	355	510	335	490
EN 10025-6				
S 460 Q/QL/QL1	460	570	440	550

Table 1.1b Mechanical characteristics of steels used for hollow profiles.

EN norm and steel grade	Nominal thickness t			
	$t \leq 40$ mm		40 mm $< t \leq 65$ mm	
	f_y (N/mm^2)	f_u (N/mm^2)	f_y (N/mm^2)	f_u (N/mm^2)
EN 10210-1				
S 235 H	235	360	215	340
S 275 H	275	430	255	410
S 355 H	355	510	335	490
S 275 NH/NLH	275	390	255	370
S 355 NH/NLH	355	490	335	470
S 420 NH/NLH	420	540	390	520
S 460 NH/NLH	460	560	430	550
EN 10219-1				
S 235 H	235	360		
S 275 H	275	430		
S 355 H	355	510		
S 275 NH/NLH	275	370		
S 355 NH/NLH	355	470		
S 460 NH/NLH	460	550		
S 275 MH/MLH	275	360		
S 355 MH/MLH	355	470		
S420 MH/MLH	420	500		
S 460 NH/NLH	460	530		

Table 1.2 Nominal yielding strength values (f_{yb}) and nominal failure strength (f_{ub}) for bolts.

Bolt class	4.6	4.8	5.6	5.8	6.8	8.8	10.9
f_{yb} (N/mm^2)	240	320	300	400	480	640	900
f_{ub} (N/mm^2)	400	400	500	500	600	800	1000

The details concerning the designation of steels are covered in EN 10027 Part 1 (*Designation systems for steels – Steel names*) and Part 2 (*Numerical system*), which distinguish the following groups:

- *group 1*, in which the designation is based on the usage and on the mechanical or physical characteristics of the material;
- *group 2*, in which the designation is based on the chemical content: the first symbol may be a letter (e.g. C for non-alloy carbon steels or X for alloy steel, including stainless steel) or a number.

With reference to the group 1 designations, the first symbol is always a letter. For example:

- *B* for steels to be used in reinforced concrete;
- *D* for steel sheets for cold forming;
- *E* for mechanical construction steels;
- *H* for high strength steels;
- *S* for structural steels;
- *Y* for steels to be used in prestressing applications.

Focusing attention on the structural steels (starting with an *S*), there are then three digits *XXX* that provide the value of the minimum yielding strength. The following term is related to the technical conditions of delivery, defined in EN 10025 ('Hot rolled products of structural steel') that proposes the following five abbreviations, each associated to a different production process:

- the *AR* (*As Rolled*) term identifies rolled and otherwise unfinished steels;
- the *N* (*Normalized*) term identifies steels obtained through normalized rolling, that is a rolling process in which the final rolling pass is performed within a well-controlled temperature range, developing a material with mechanical characteristics similar to those obtained through a normalization heat treatment process (see Section 1.2);
- the *M* (*Mechanical*) term identifies steels obtained through a thermo-mechanical rolling process, that is a process in which the final rolling pass is performed within a well-controlled temperature range resulting in final material characteristics that cannot be obtained through heat treating alone;
- the *Q* (*Quenched and tempered*) term identifies high yield strength steels that are quenched and tempered after rolling;
- the *W* (*Weathering*) term identifies weathering steels that are characterized by a considerably improved resistance to atmospheric corrosion.

The YY code identifies various classes concerning material toughness as discussed in the following. Non-alloyed steels for structural use (EN 10025-2) are identified with a code after the yielding strength (XXX), for example:

- YY: alphanumeric code concerning toughness: S235 and S275 steels are provided in groups JR, J0 and J2. S355 steels are provided in groups JR, J0, J2 and K2. S450 steels are provided in group J0 only. The first part of the code is a letter, J or K, indicating a minimum value of toughness provided (27 and 40 J, respectively). The next symbol identifies the temperature at which such toughness must be guaranteed. Specifically, R indicates ambient temperature, 0 indicates a temperature not higher than 0°C and 2 indicates a temperature not higher than −20°C;
- C: an additional symbol indicating special uses for the steel;
- N, AR or M: indicates the production process.

Weldable fine grain structural steels that are normalized or subject to normalized rolling (EN 10025-3); that is, steels characterized by a granular structure with an equivalent ferriting grain size index greater than 6, determined in accordance with EN ISO 643 ('Micrographic determination of the apparent grain size'), are defined by the following codes:

- N: for the production process;
- YY: for the toughness class. The L letter identifies toughness temperatures not lower than −50°C; in the absence of the letter L, the reference temperature must be taken as −20°C.

Fine grain steels obtained through thermo-mechanical rolling processes (EN 10025-4) are identified by the following code:

- M: for the production process;
- YY: for the toughness class. The letter L, as discussed previously, identifies toughness temperatures no lower than −50°C; in the absence of the letter L, the reference temperature must be taken to be −20°C.

Weathering steels for structural use (EN 10025-5) are identified by the following code:

- the YY code indicates the toughness class: these steels are provided in classes J0, J2 and K2, indicating different toughness requirements at different temperatures.
- the W code indicates the weathering properties of the steel;
- P indicates an increased content of phosphorous;
- N or AR indicates the production process.

Quenched and tempered high-yield strength plate materials for structural use (EN 10025-6) are identified by the following codes:

- Q code indicates the production process;
- YY: identifies the toughness class. The letter L indicates a specified minimum toughness temperature of −40°C, while code L1 refers to temperatures not lower than −60°C. In the absence of these codes, the minimum toughness values refer to temperatures no lower than −20°C.

In Europe, it is mandatory to use steels bearing the CE marks, in accordance with the requirements reported in the Construction Products Regulation (CPR) No. 305/2011 of the European Community. The usage of different steels is allowed as long as the degree of safety (not lower than the one provided by the current specifications) can be guaranteed, accompanied by adequate theoretical and experimental documentation.

1.1.2 Materials in Accordance with United States Provisions

The properties of structural steel materials are standardized by ASTM International (formerly known as the *American Society for Testing and Materials*). Numerous standards are available for structural applications, generally dedicated to the most common product families. In the following, some details are reported.

1.1.2.1 General Standards

ASTM A6 (*Standard Specification for General Requirements for Rolled Structural Steel Bars, Plates, Shapes and Sheet Piling*) is the standard that covers the general requirements for rolled structural steel bars, plates, shapes and sheet piling.

1.1.2.2 Hot-Rolled Structural Steel Shapes

Table 1.3 summarizes key data for the most commonly used hot-rolled structural shapes.

- *W-Shapes*
  ASTM A992 is the most commonly used steel grade for all hot-rolled W-Shape members. This material has a minimum yield stress of 50 ksi (356 MPa) and a minimum tensile strength of 65 ksi (463 MPa). Higher values of the yield and tensile strength can be guarantee by ASTM A572 Grades 60 or 65 (Grades 42 and 50 are also available) or ASTM A913 Grades 60, 65 or 70 (Grace 50 is also available). If W-Shapes with atmospheric corrosion resistance characteristics are required, reference can be made to ASTM A588 or ASTM A242 selecting 42, 46 or 50 steel Grades. Finally, W-Shapes according to ASTM A36 are also available.
- *M-Shapes and S-Shapes*
  These shapes have been produced up to now in ASTM A36 steel grade. From some steel producers they are now available in ASTM A572 Grade 50. M-Shapes with atmospheric corrosion resistance characteristics can be obtained by using ASTM A588 or ASTM A242 Grade 50.

Table 1.3 ASTM specifications for various structural shapes (from Table 2-3 of the *AISC Manual*).

Steel type	ASTM designation		F_y minimum yield stress (ksi)	F_u tensile stress (ksi)	Applicable shape series									
					W	M	S	HP	C	MC	L	HSS rectangular	HSS round	Pipe
Carbon	A36		36	58–80										
	A53 Gr. B		35	60										
	A500	Gr. B	42	58										
			46	58										
		Gr. C	46	62										
			50	62										
	A501		36	58										
	A529	Gr. 50	50	65–100										
		Gr. 55	55	70–100										
High strength low	A572	Gr. 42	42	60										
alloy		Gr. 50	50	65										
		Gr. 55	55	70										
		Gr. 60	60	75										
		Gr. 65	65	80										
	A618	Gr. I and II	50	70										
		Gr. III	50	65										
	A913	50	50	60										
		60	60	75										
		65	65	80										
		70	70	90										
	A992		50–65	65										
Corrosion resistant	A242		42	63										
high strength			46	67										
low-alloy			50	70										
	A588		50	70										
	A847		50	70										

= Preferred material specification.
= Other applicable material specification.
= Material specification does not apply.

- *Channels*
  See what is stated about M- and S-Shapes.
- *HP-Shapes*
  ASTM A572 Grade 50 is the most commonly used steel grade for these cross-section shapes. If atmospheric corrosion resistance characteristics are required for HP-Shapes, ASTM A588 or ASTM A242 Grade 46 or 50 can be used. Other materials are available, such as ASTM A36, ASTM A529 Grades 50 or 55, ASTM A572 Grades 42, 55, 60 and 65, ASTM A913 Grades 50, 60, 65, 70 and ASTM A992.
- *Angles*
  ASTM A36 is the most commonly used steel grade for these cross-sections shapes. Atmospheric corrosion resistance characteristics of the angles can be guaranteed by using ASTM A588 or ASTM A242 Grades 46 or 50. Other available materials: ASTM A36, ASTM A529 Grades 50 or 55, ASTM A572 Grades 42, 50, 55 and 60, ASTM A913 Grades 50, 60, 65 and 70 and ASTM A992.
- *Structural Tees*
  Structural tees are produced cutting W-, M- and S-Shapes, to make WT-, MT- and ST-Shapes. Therefore, the same specifications for W-, M- and S-Shapes maintain their validity.

Table 1.4 Applicable ASTM specifications for plates and bars (from Table 2-4 of the *AISC Manual*).

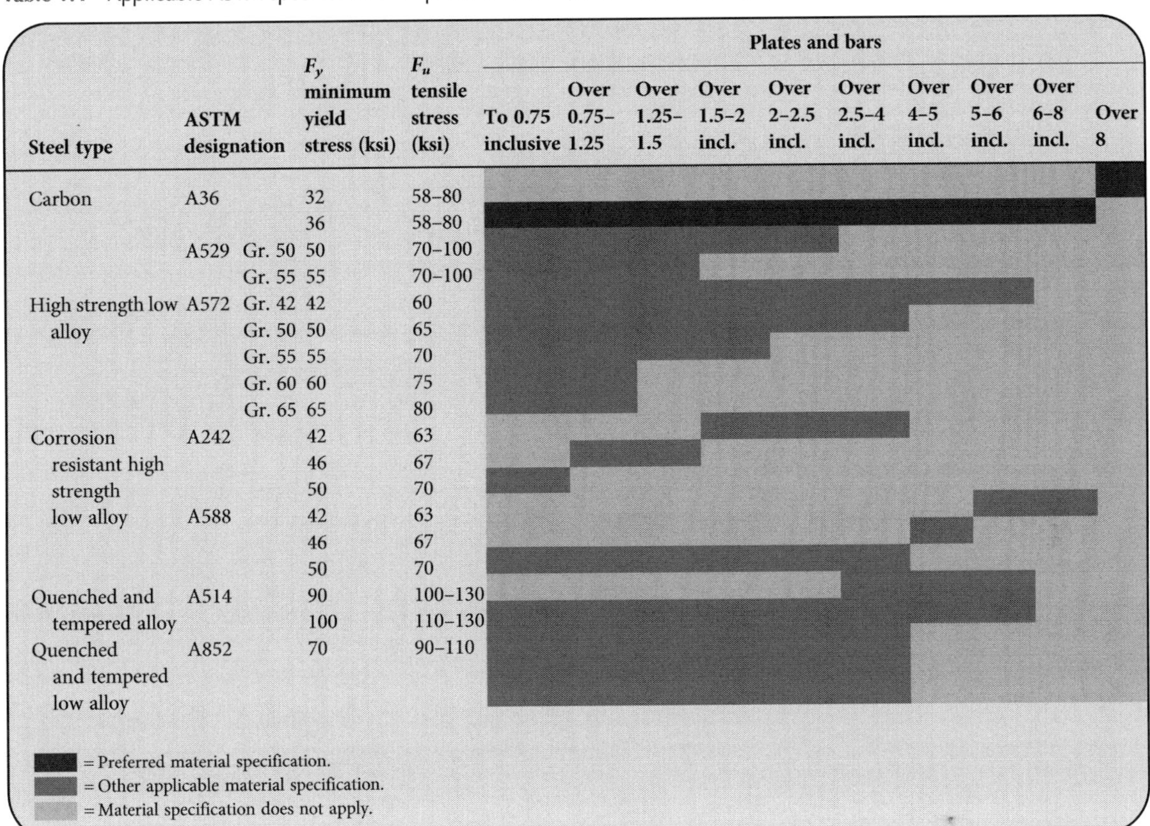

Steel type	ASTM designation		F_y minimum yield stress (ksi)	F_u tensile stress (ksi)	Plates and bars: To 0.75 inclusive	Over 0.75–1.25	Over 1.25–1.5	Over 1.5–2 incl.	Over 2–2.5 incl.	Over 2.5–4 incl.	Over 4–5 incl.	Over 5–6 incl.	Over 6–8 incl.	Over 8
Carbon	A36		32	58–80										
			36	58–80										
	A529	Gr. 50	50	70–100										
		Gr. 55	55	70–100										
High strength low alloy	A572	Gr. 42	42	60										
		Gr. 50	50	65										
		Gr. 55	55	70										
		Gr. 60	60	75										
		Gr. 65	65	80										
Corrosion resistant high strength low alloy	A242		42	63										
			46	67										
			50	70										
	A588		42	63										
			46	67										
			50	70										
Quenched and tempered alloy	A514		90	100–130										
			100	110–130										
Quenched and tempered low alloy	A852		70	90–110										

= Preferred material specification.
= Other applicable material specification.
= Material specification does not apply.

- *Square, Rectangular and Round HSS*
  ASTM A500 Grade B ($F_y = 46$ ksi and $F_u = 58$ ksi) is the most commonly used steel grade for these shapes. ASTM A550 Grade C ($F_y = 50$ ksi and $F_u = 62$ ksi) is also used. Rectangular HSS with atmospheric corrosion resistance characteristics can be obtained by using ASTM A847. Other available materials are ASTM A501 and ASTM A618.
- *Steel Pipes*
  ASTM A53 Grade B ($F_y = 35$ ksi and $F_u = 60$ ksi) is the only steel grade available for these shapes.

1.1.2.3 Plate Products

As to plate products, reference can be made to Table 1.4.

- *Structural plates*
  ASTM A36 $F_y = 36$ ksi (256 MPa) for plate thickness equal to or less then 8 in. (203 mm), $F_y = 32$ ksi (228 MPa) for higher thickness and $F_u = 58$ ksi (413 MPa) is the most commonly used steel grade for structural plates. For other materials, reference can be made to Table 1.4.
- *Structural bars*
  Data related to structural plates are valid also for bars with the exception that ASTM A514 and A852 are not admitted.

1.1.2.4 Sheets

ASTM A606 and ASTM A1011 are the two main standards for metal sheets. The former deals with weathering steel, the latter standardizes steels with improved formability that are typically used for the production of cold-formed profiles.

1.1.2.5 High-Strength Fasteners

ASTM A325 and A490 are the standards dealing with high-strength bolts used in structural steel connections. The nominal failure strength of A325 bolts is 120 ksi (854 MPa), without an upper limit, while the nominal failure strength of A490 bolts is 150 ksi (1034 MPa), with an upper limit of 172 ksi (1224 MPa) per ASTM, limited to 170 ksi (1210 MPa) by the structural steel provisions. ASTM F1852 and F2280 are standards for tension-control bolts, characterized by a splined end that shears off when the desired pretension is reached. Loosely, A325 (and F1852) bolts correspond to 8.8 bolts in European standards and A490 (and F2280) bolts correspond to 10.9 bolts.

ASTM F436 standardizes hardened steel washers for fastening applications. ASTM F959 is the standard for direct tension indicator washers, which are a special category of hardened washers with raised dimples that flatten upon reaching the minimum pretension force in the fastener.

ASTM A563 standardizes carbon and alloy steel nuts.

ASTM A307 is the standard for steel anchor rods; it is also used for large-diameter fasteners (above 1½-in.). ASTM F1554 is the preferred standard for anchor rods.

ASTM 354 standardizes quenched and tempered alloy steel bolts.

ASTM A502 is the standard of reference for structural rivets.

1.2 Production Processes

Steel can be obtained by converting wrought iron or directly by means of fusion of metal scrap and iron ore. Ingots are obtained from these processes, which then can be subject to hot- or cold-mechanical processes, eventually becoming final products (plates, bars, profiles, sheets, rods, bolts, etc.). These products, examined in detail in Section 1.5, can be obtained in various ways that can be practically summarized into the following techniques:

- *forming process by compression or tension* (e.g. forging, rolling, extrusion);
- *forming process by flexure and shear.*

Among these processes, the most important is the rolling process in both its hot- and cold-variations, by which most products used in structural applications (referred to as rolled products) are obtained. In the *hot-rolling* process, steel ingots are brought to a temperature sufficient to soften the material (approximately 1200°C or 2192°F), they first travel through a series of juxtaposed counter-rotating rollers (*primary rolling* – Figure 1.3) and are roughed into square or rectangular cross-section bars.

These semi-worked products are produced in different shapes that can be then further rolled to obtain plates, large- or medium-sized profiles or small-sized profiles, bars and rounds. This additional process is called *secondary rolling*, resulting in the final products.

For example, in order to obtain the typical I-shaped profiles, the semi-worked products, at a temperature slightly above 1200°C (or 2192°F), are sent to the rolling train and its initially rectangular cross-section is worked until the desired shape is obtained. Figure 1.4 shows some of the intermediate cross-sections during the rolling process, until the final I-shape product is obtained.

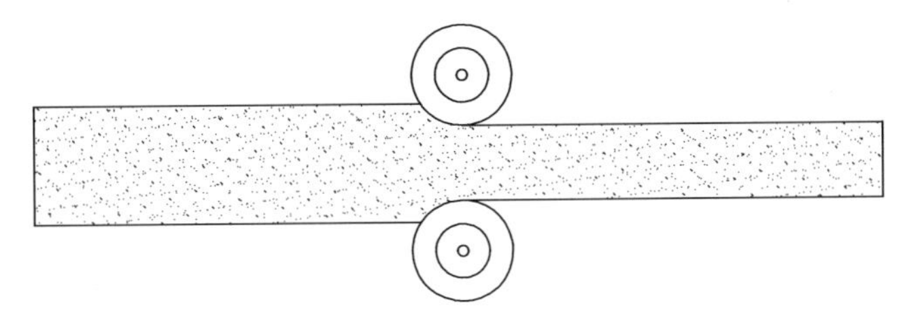

Figure 1.3 Rolling process.

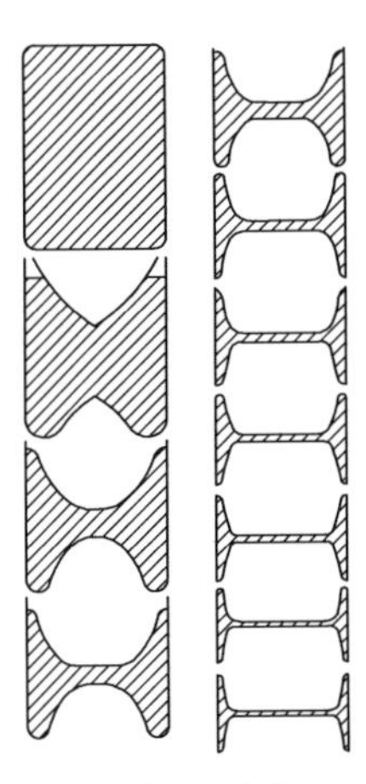

Figure 1.4 Intermediate steps of the rolling process for an I-shape profile.

The rolling process improves the mechanical characteristics of the final product, thanks to the compressive forces applied by the rollers and the simultaneous thinning of the cross-section that favours the elimination of gases and air pockets that might be initially present. At the same time, the considerable deformations imposed by the rolling process contribute to refine the grain structure of the material, with remarkable advantages regarding homogeneity and strength. In such processes, in addition to the amount of deformations, also the rate of deformations is a very important factor in determining the final characteristics of the product.

Cold rolling is performed at the ambient temperature and it is frequently used for non-ferrous materials to obtain higher strengths through hardening at the price of an often non-negligible loss of ductility. When cold-rolling requires excessive strains, the metal can start showing cracks before the desired shape is attained, in which case additional cycles of heat treatments and cold forming are needed (Section 1.3).

The forming processes by *bending and shear* consist of bending thin sheets until the desired cross-section shape is obtained. Typical products obtained by these processes are cold-formed profiles, for which the thickness must be limited to a few millimetres in order to attain the desired deformations. Figure 1.5 shows the intermediate steps to obtain hollow circular cold-formed profiles by means of continuous formation processes.

It can be seen that the coil is pulled and gradually shaped until the desired final product is obtained. Figure 1.6 instead shows the main intermediate steps of the punch-and-die process to obtain some typical profiles currently used in structural applications. With this second working technique, thicker sheets can be shaped into profiles with thicknesses up to 12–15 mm (0.472–0.591 in.), while the limit value of the coil thickness for continuous formation processes is approximately 5 mm (0.197 in.). As an example, Figure 1.7 shows some intermediate steps of the

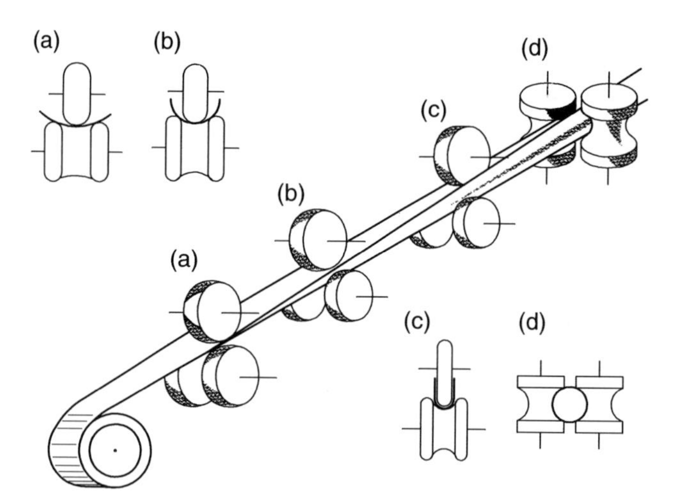

Figure 1.5 Continuous formation of circular hollow cold-formed profiles.

Figure 1.6 Punch-and-die process for cold-formed profiles.

cold-formation process of a stiffened channel profile, with regular perforations, typically used for steel storage pallet racks and shelving structures.

Another important category of steel products obtained with punch-and-die processes is represented by metal decking, currently used for slabs, roofs and cladding.

Figure 1.7 Cold-formation images of a stiffened channel profile.

1.3 Thermal Treatments

Steel products, just like other metal products, can be subject to special *thermal treatments* in order to modify their molecular structure, thus changing their mechanical properties. The basic molecular structures are *cementite, austenite* and *ferrite.* Transition from one structure to another depends on temperature and carbon content. The main thermal treatments commonly used, which are briefly described in the following, are *annealing, normalization, tempering, quenching, pack-hardening* and *quenching and tempering:*

- *annealing* is the thermal cycle that begins with the heating to a temperature close to or slightly above the critical temperature (corresponding to the temperature at which the ferrite-austenite transition is complete); afterwards the temperature is maintained for a predetermined amount of time and then the material is slowly cooled to ambient temperature. Generally, annealing leads to a more homogenous base material, eliminating most defects due to solidifying process. Annealing is applied to either ingots, semi-worked products or final products. Annealing of worked products is useful to increase ductility, which might be reduced by hardening during the mechanical processes of production, or to release some residual stresses related to non-uniform cooling or production processes. In particular, annealing can be used on welded parts that are likely to be mired by large residual stresses due to differential cooling;
- *normalization* consists of heating the steel to a temperature between 900 and 925°C (approximately between 1652 and 1697°F), followed by very slow cooling. Normalization eliminates the effects of any previous thermal treatment;
- *tempering* is a thermal process that, similar to annealing, consists of heating the material slightly above the critical temperature followed by a sudden cooling, aimed at preventing any readjustment of the molecular matrix. The main advantage of the tempering process is represented by an increase of hardness that is, however, typically accompanied by a loss of ductility of the material;
- *quenching* consists of heating the tempered part up to a moderate temperature for an extended amount of time, improving the ductility of the material;
- *pack-hardening* is a process that consists of heating of a part when in contact with solid, liquid or gaseous materials that can release carbon. It is a surface treatment that is employed to form a harder layer of material on the outside surface (up to a depth of several millimetres), in order to improve the wearing resistance;

Quenching and tempering can be applied sequentially, resulting in a remarkable strength improvement of ordinary carbon steels, without appreciably affecting the ductility of the product. High strength bolts used in steel structures are typically quenched and tempered.

1.4 Brief Historical Note

Iron refinement has taken place for millennia in partially buried furnaces, fuelled by bellows resulted in a spongy iron mass, riddled of impurities that could only be eliminated by repeated hammering, resulting in *wrought iron*. That product had modest mechanical properties and could be welded by *forging*; that is, by heating the parts to join to a cherry red colour (750–850°C or 1382–1562°F) and then pressing them together, typically by hammering. Wrought iron products could be superficially hardened by *tempering* them in a bath of cold water or oil and the final product was called *steel*. Note that these terms have different implications nowadays.

In thirteenth century Prussia, thanks to an increase in the height of the interred furnaces and the consequent increase in the amount of air forced in the oven by hydraulically actuated bellows, the maximum attainable temperatures were increased. Consequently, a considerably different material from steel was obtained, namely *cast iron*. Cast iron was a brittle material that, once cooled, could not be wrought. On the other hand, cast iron in its liquid state could be poured into moulds, assuming whatever shape was desired. A further heating in an open oven, resulting in a carbon-impoverished alloy, allowed for *malleable iron* to be obtained.

In the past, the difficulties associated with the refinement of *iron ore* have limited the applications of this material to specific fields that required special performance in terms of strength or hardness. Applications in construction were limited to ties for arches and masonry structures, or connection elements for timber construction. The *industrial revolution* brought a new impulse in metal construction, starting in the last decades of the eighteenth century. The invention of the steam engine allowed hydraulically actuated bellows to be replaced, resulting in a further increase of the air intake and the other significant advantage of locating the furnaces near iron mines, instead of forcing them to be close to rivers. In 1784, in England, Henry Cort introduced a new type of furnace, the *puddling furnace*, in which the process of eliminating excess carbon by oxidation took place thanks to a continuous stirring of the molten material. The product obtained (*puddled iron*) was then hammered to eliminate the impurities. An early rolling process, using creased rollers, further improved the quality of the products, which was worked into plates and square cross-section members. Starting in the second half of the nineteenth century, several other significant improvements were introduced. In 1856, at the Congress of the British Society for the Scientific Progress, Henry Bessemer announced his patented process to rapidly convert cast iron into steel. Bessemer's innovative idea consisted of the insufflation of the air directly into the molten cast iron, so that most of the oxygen in the air could directly combine with the carbon in the molten material, eliminating it in the form of carbon oxide and dioxide in gaseous form.

The first significant applications of cast iron in buildings and bridges date back to the last decades of the eighteenth century. An important example is the cast iron bridge on the Severn River at Ironbridge Gorge, Shropshire, approximately 30 km (18.6 miles) from Birmingham in the UK. It is an arched bridge and it was erected between 1775 and 1779. The structure consisted of five arches, placed side by side, over a span of approximately 30 m (98 ft), each made of two parts representing half of an arch, connected at the key without nails or rivets.

The expansion of the railway industry, with the specific need for stiff and strong structures capable of supporting the large weights of a train without large deformations, provided a further spur to the development of bridge engineering. Between 1844 and 1850 the Britannia Bridge (Pont Britannia) on the Menai River (UK) was built; this bridge represents a remarkable example of a continuously supported structure over five supports, with two 146 m (479 ft) long central spans and two 70 m (230 ft) long side spans. The bridge had a closed tubular cross-section, inside which the train would travel, and it was made of puddled iron connected by nails. Robert Stephenson, William Fairbairn and Eaton Hodgkinson were the main designers, who had to tackle a series of problems that had not been resolved yet at the time of the design. Being a statically indeterminate structure, in order to evaluate the internal forces, B. Clapeyron studied the structure applying the three-moment equation that he had recently developed. For the static behaviour of the cross-section, based on experimental tests on scaled models of the bridge, N. Jourawsky suggested some stiffening details to prevent plate instability. The Britannia Bridge also served as a stimulus to study riveted and nailed connections, wind action and the effects of temperature changes.

With respect to buildings, the more widespread use of metals contributed to the development of framed structures. Around the end of the 1700s, cast iron columns were made with square, hollow circular or a cross-shaped cross-section. The casting process allowed reproduction of the classical shapes of the column or capital, often inspired by the architectural styles of the ancient Greeks or Romans, as can be seen in the catalogues of column manufacturers of the age. The first applications of cast iron to bending elements date back to the last years of the 1700s and deal mostly with floor systems made by thin barrel vaults supported by cast iron beams with an inverted T cross-section. During the first decades of the nineteenth century studies were commissioned to identify the most appropriate shape for these cast iron beams. Hodgkinson, in particular, reached the conclusion that the optimal cross-section was an unsymmetrical I-shape with the compression flange up to six times smaller than the tension flange, due to the difference in tensile and compressive strengths of the material. Following this criterion, spans up to 15 m could be accommodated.

The first significant example of a structure with linear cast iron elements (beams and columns) is a seven-storey industrial building in Manchester (UK), built in 1801. Nearing halfway through the century, the use of cast iron slowed to a stop, to be replaced by the use of steel. Plates and corner pieces made of puddle iron had been already available since 1820 and in 1836 I-shape profiles started to be mass produced.

More recent examples of the potential for performance and freedom of expression allowed by steel are represented by tall buildings and skyscrapers. The prototype of these, the *Home Insurance Building*, was built in 1885 in Chicago (USA) with a 12 storey steel frame with rigid connections and masonry infills providing additional stiffness for lateral forces. In the same city, in 1889, the *Rand–McNally* building was erected, with a nine-storey structural frame entirely made of steel.

Early in the twentieth century, the first skyscrapers were built in Chicago and New York (USA), characterized by unprecedented heights. In New York in 1913, the *Woolworth Building* was built, a 60-storey building reaching a height of 241 m (791 ft); in 1929 the *Chrysler Building* (318 m or 1043 ft) was built and in 1930 the *Empire State Building* (381 m or 1250 ft) was built. Other majestic examples are the steel bridges built around the world: in 1890, near Edinburgh (UK) the *Firth of Forth* Bridge was built, possessing central spans of 521 m (1709 ft), while in 1932 the *George Washington Bridge* was built in New York; a suspension bridge over a span of 1067 m (3501 ft).

Many more references can be found in specialized literature, both with respect to the development of iron working and the history of metal structures.

1.5 The Products

A first distinction among steel products for the construction industry can be made between *linear* and *plane products*. The formers are mono-dimensional elements (i.e. elements in which the length is considerably greater than the cross-sectional dimensions).

Plane products, namely sheet metal, which are obtained from plate by an appropriate working process, have two dimensions that are substantially larger than their thickness. Plane products are used in the construction industry to realize floor systems, roof systems and cladding systems. In particular, these products are most typical:

- *ribbed metal decking for bare steel applications*, furnished with or without insulating material, used for roofing and cladding applications. These products are typically used to span lengths up to 12 m or 39 ft (ribbed decking up to 200 mm/7.87 in. depth are available nowadays). In the case of roofing systems for sheds, awnings and other relatively unimportant buildings, non-insulated ribbed decking is usually employed. The extremely light weight of these systems makes them very sensitive to vibrations. These products are also commercialized with added insulation (Figure 1.8), installed between two outer layers of metal decking (as a sandwich panel). For special applications, innovative products have been manufactured, such as the ribbed arched element shown in Figure 1.9, meant for long-span applications
- *ribbed decking products for concrete decks*: these products are usually available in thicknesses from 0.6 to 1.5 mm (0.029–0.059 in.) and with depths from 55 mm (2.165 in.) to approximately 200 mm (7.87 in.). A typical application of these products is the construction of composite or non-composite floor systems: typically, the ribbed decking is never less than 50 mm (2 in., approximately) deep and the thickness of the concrete above the top of the ribs is never less than 40 mm (1.58 in.) thick. The ribbed decking element functions as a stay-in-place form and may or may not be accounted for as a composite element to provide strength to the floor system (Figure 1.10). If composite action is desired, the ribbed decking may have additional ridges and other protrusions in order to guarantee shear transfer between steel and concrete. When composite action is not required, the ribbed decking can be smooth and it just functions as a stay-in-place form. In either case, welded wire meshes or bi-directional reinforcing bars should be placed at the top fibre of the slab to prevent cracking due to creep and shrinkage or due to concentrated vertical loads on the floor.

The choice of cladding and the detailing of ribbed decking elements for roofing and flooring systems (both bare steel and composite) are usually based on tables provided by the manufacturers. For instance, in manufacturers' catalogues tables are generally provided in which the main

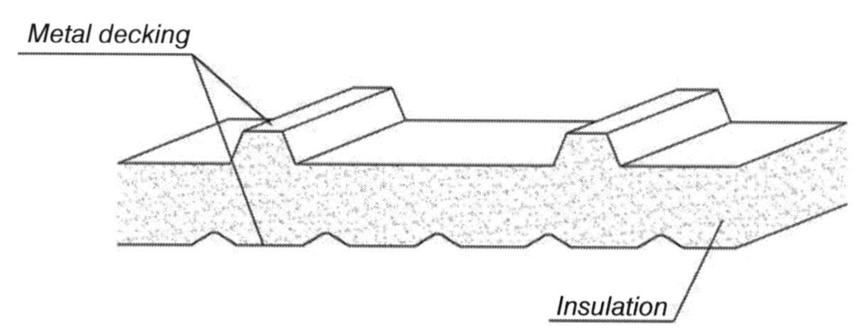

Figure 1.8 Typical insulated element.

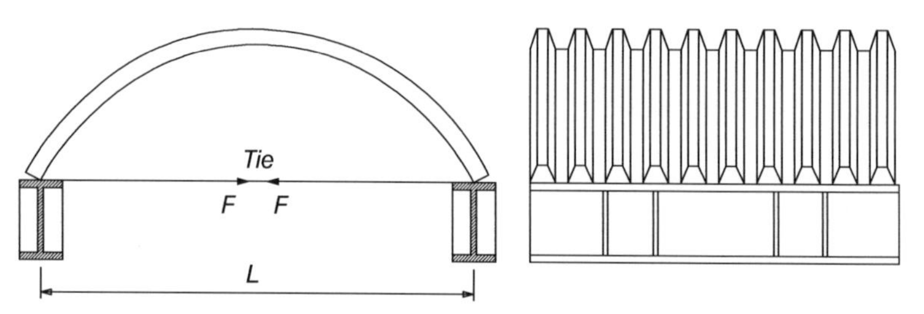

Figure 1.9 Example of a special ribbed decking product.

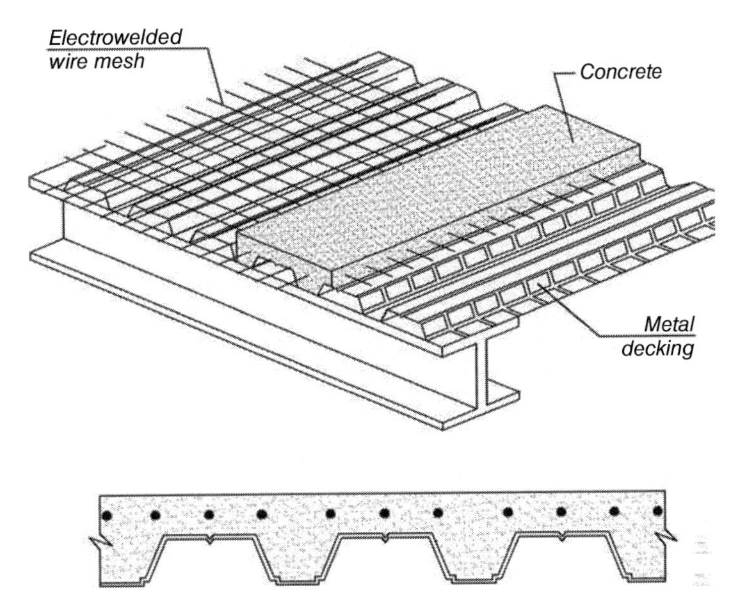

Figure 1.10 Typical steel-concrete composite floor system.

utility data from the commercial and structural points of view are presented: the weight per unit area, the maximum span as a function of dead and live loads and the maximum deflection as a function of the support configuration. Figure 1.11 schematically shows an example of the typical tables developed by manufactures for a bare steel deck: the product is provided with different thicknesses (from 0.6 to 1.5 mm or 0.029 to 0.059 in.): for each thickness, the maximum load is shown as a function of the span.

An aspect that is sometimes overlooked in the design phase is the fastening system of the cladding or roofing panels to the supporting elements, which has to transfer the forces mainly associated with snow, wind and thermal loads. Depending on the configuration of a cladding or a roofing panel with respect to the direction of wind, it can be subject to either a positive or a negative pressure. In the case of cladding, negative (upward) pressures are typically less demanding than positive (downward) pressures. Similarly, negative pressures on roofing systems are typically less controlling than snow or roof live loads. This said, the fastening details between cladding or roofing panels and their supporting elements must be appropriately sized, also taking into account the fact that in the corner regions of a building, or in correspondence to discontinuities such as windows or ceiling openings, local effects might arise causing large values of positive or negative pressures, even when wind speeds are not particularly elevated (Figure 1.12). Concerning thermal variations, it is necessary to make sure that the panels and the fastening systems are capable of sustaining increases or decreases of temperature, mostly due to sun/UV exposure. A rule of thumb that can be followed for maximum ranges of temperature variation, applicable to panels of different colours, in hypothetical summer month and a south-west exposure, is as follows:

- ±18°C (64.4°F) for reflecting surfaces;
- ±30°C (86°F) for light coloured surfaces;
- ±42°C (107.6°F) for dark coloured surfaces.

The fastening systems usually comprise screws with washers to distribute loads more evenly. In some instances, local deformations of thin decks can occur at the fastening locations, causing a potential for leaks.

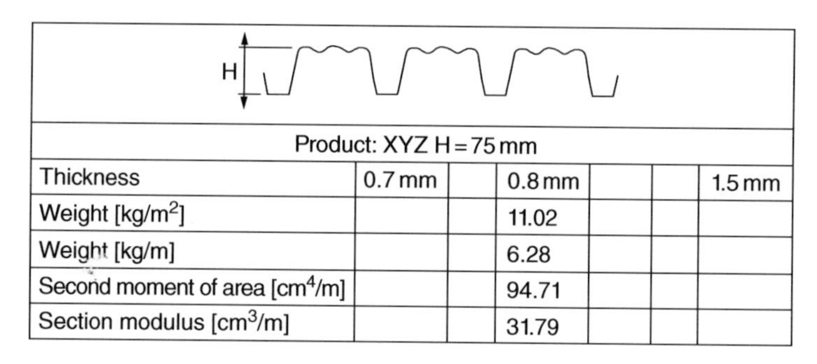

Product: XYZ H = 75 mm						
Thickness	0.7 mm		0.8 mm			1.5 mm
Weight [kg/m^2]			11.02			
Weight [kg/m]			6.28			
Second moment of area [cm^4/m]			94.71			
Section modulus [cm^3/m]			31.79			

	Distance between supports: span length [m]									
Thickness	1.50	1.75	2.00	2.25	2.75	3.00	3.25	3.50	3.75	5
0.6 *mm*			443							
0.7 *mm*			550							
0.8 *mm*			660							
1.0 *mm*			922							
1.2 *mm*			1151							
1.5 *mm*			1147							
	Distance between supports: span length [m]									
Thickness	1.50	1.75	2.00	2.25	2.75	3.00	3.25	3.50	3.75	5
0.6 *mm*			554							
0.7 *mm*			688							
0.8 *mm*			832							
1.0 *mm*			1152							
1.2 *mm*			1438							
1.5 *mm*			1846							

Figure 1.11 Example of a design table for a bare steel ribbed decking product.

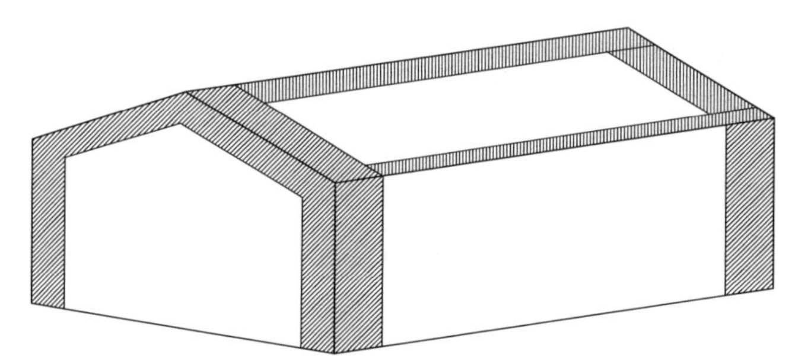

Figure 1.12 Regions that are typically subject to local effects of wind loads.

1.6 Imperfections

The behaviour of steel structures, and thus the load carrying capacity of their elements, depends, sometimes very significantly, on the presence of imperfections. Depending on their nature, imperfections can be classified as follows:

- mechanical or structural imperfections;
- geometric imperfections.

1.6.1 Mechanical Imperfections

The term *mechanical* or *structural imperfections* indicates the presence of *residual stresses* and/or the lack of homogeneity of the mechanical properties of the material across the cross-section of the element (e.g. yielding strength or failure strength varying across the thickness of flanges and web). Residual stresses are a self-equilibrating state of stress that is locked into the element as a consequence of the production processes, mostly due to non-uniform plastic deformations and to non-uniform cooling. If reference is made, for example, to a hot-rolled prismatic member at the end of the rolling process, the temperature is approximately around 600°C (1112°F); the cross-sectional elements with a larger exposed surface and a smaller thermal mass, will cool down faster than other more protected or thicker elements. The cooler regions tend to shrink more than the warmer regions, and this shrinkage is restrained by the connected warmer regions. As a consequence, a stress distribution similar to that shown in Figure 1.13b takes place, with tensile stresses that oppose the shrinkage of the perimeter regions and compressive stresses that equilibrate them in the inner regions. When the warmer regions finally cool down, plastic phenomena contribute to somewhat reduce the residual stresses (Figure 1.13c). Once again, the perimeter regions that have reached the ambient temperature restrain the shrinkage of the inner regions during their cooling process and as a consequence, once cooling has completed, the outside regions are subject to compressive stresses, while the inside regions show tensile stresses (Figure 1.13d).

Figure 1.14 shows the distributions of residual stresses during the cooling phase after the hot-rolling process for a typical I-beam profile and in particular, the phases span from (a), end of the hot-rolling process, to (d), the instant at which the whole profile is at ambient temperature. The magnitude and the distribution of residual stresses depend on the geometric characteristics of the cross-section and, in particular, on the width to thickness ratio of its elements (flanges and webs).

For I-shaped elements, Figure 1.15 shows the distribution of residual stresses (σ_r) as a function of the width/thickness ratio of the cross-sectional elements: terms h and b refer to the height of the profile and to the width of the flange, respectively, while t_w and t_f indicate web and flange thickness, respectively. Stocky profiles; that is, those that have a height/width ratio not greater than 1.2, show tensile residual stresses in the middle of the flanges and compressive residual stresses at the extremes of the flanges, while in the web there can be either tensile or compressive residual stresses, depending on the geometry. For slenderer profiles with $h/b \geq 1.7$, the middle part of the flanges show prevalently tensile residual stresses, while compressive residual stresses can be found in the middle region of the web.

Residual stresses can affect the load carrying capacity of member, especially when they are subject to compressive forces. For larger cross-sections, the maximum values of the residual stresses can easily reach the yielding strength of the material.

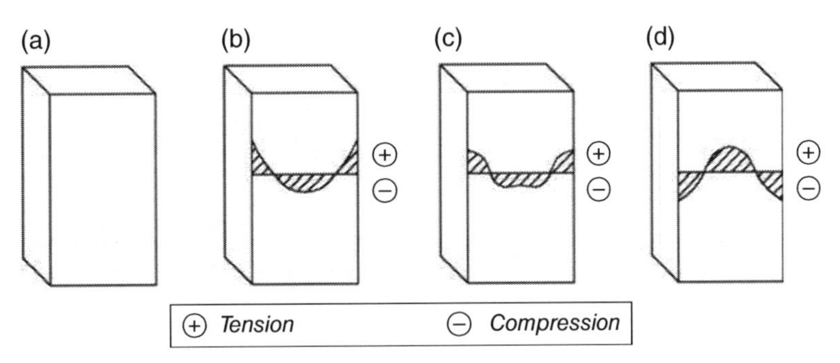

Figure 1.13 Residual stress distribution in a hot-rolled rectangular profile during the cooling phase (temporary from a to d).

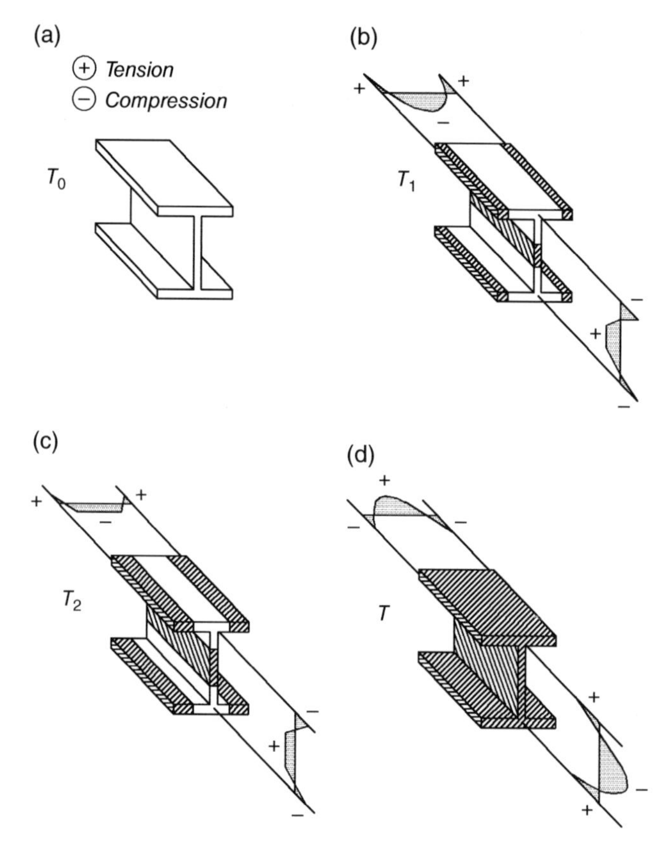

Figure 1.14 Distribution of residual stresses during the cooling phase of an I-shape.

In the case of cold-formed profiles and plates, the raw product is a hot- or cold-rolled sheet. If the rolling process is performed at ambient temperature, the outermost fibres, in contact with the rollers, tend to stretch, while the central fibres remain undeformed. As a consequence, a self-equilibrated residual state of stress arises, such as the one shown in Figure 1.16, due to the differential elongation of the fibres in the cross-section.

In the case of hot-rolling of a plate, the residual stresses develop similarly to those presented for the rectangular (Figure 1.13) and for the I-shaped (Figure 1.14) sections.

In the case of cold-formed profiles or metal decks, an additional source of imperfections is the cold-formation process. The bending processes in fact alter the mechanical properties of the material in the vicinity of the corners. In order to permanently deform the material, the process brings it beyond its yielding point so that the desired shape can be attained. As an example, Figure 1.17 shows the values of the yielding strength (f_y) and of the ultimate strength (f_u) for the virgin material compared to the same values for the cold-formed profile at different locations. It is apparent how the cold-formation process increases both yielding and failure strengths, with a larger impact on the yielding strength.

From the design standpoint, recent provisions on cold-formed profiles, among which part 1–3 of Eurocode 3 (EN-1993-1-3) allows account for a higher yielding strength of the material, due to the cold-formation process, when performing the following design checks:

h/b	Cross-section		σ_r (web)	σ_r (flange)	t_w/h	t_w/b	t_f/h	t_f/b
≤ 1.2	b, h, t_w, t_f	a	T, C, T	C, T, C	0.032 ÷ 0.040	0.032 ÷ 0.040	0.045 ÷ 0.061	0.045 ÷ 0.060
≤ 1.2		b	T	C, T, C	0.075 ÷ 0.100	0.078 ÷ 0.112	0.091 ÷ 0.162	0.093 ÷ 0.182
≤ 1.2		c	C, T, C	C	0.062 ÷ 0.068	0.068 ÷ 0.073	0.104 ÷ 0.114	0.113 ÷ 0.121
>1.2 <1.7		c			0.031 ÷ 0.032	0.042 ÷ 0.048	0.048 ÷ 0.051	0.062 ÷ 0.080
>1.2 <1.7		d	T, C, T	C, T, C	0.030	0.046	0.051	0.077
≥1.7		d			0.018 ÷ 0.028	0.039 ÷ 0.056	0.025 ÷ 0.043	0.063 ÷ 0.085
≥1.7		e	T, C, T	T				

Figure 1.15 Distribution of residual stresses in hot-rolled I-shapes.

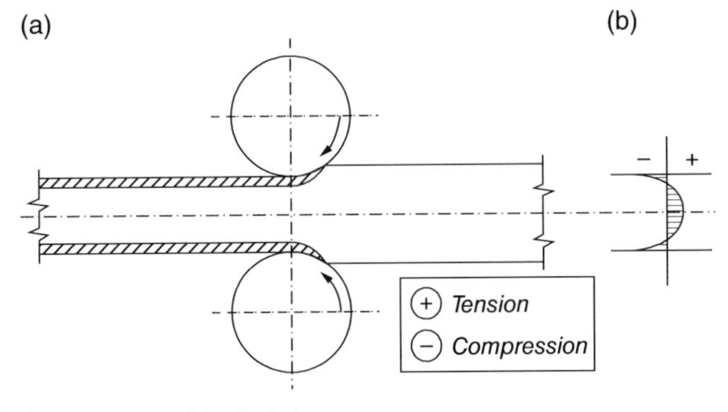

Figure 1.16 Residual stresses in a cold-rolled plate.

- design of tension members;
- design of compression members of class 1, 2 and 3, in accordance with the criteria described in Chapter 4 (Cross-Section Classifications), that is fully engaged cross-sections, in the absence of local buckling;

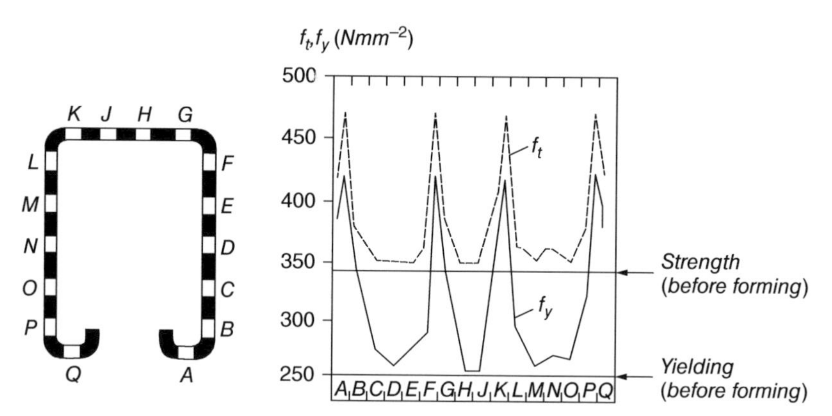

Figure 1.17 Variation of the mechanical properties of the material after cold-formation.

- design of flexural members with compression elements of class 1, 2 and 3 (i.e. with fully engaged compression elements, in the absence of local buckling).

The stub column test (Section 1.7.2) can be used to experimentally evaluate the increase of strength of a cold-formed member; alternatively, the post-forming average yielding strength f_{ya} can be evaluated based on the virgin material's yielding and ultimate strength (f_{yb} and f_u, respectively) as follows:

$$f_{ya} = f_{yb} + \frac{(f_u - f_{yb}) \cdot k \cdot n \cdot t^2}{A_g} \tag{1.4a}$$

$$f_{ya} \le \frac{f_{yb} + f_u}{2} \tag{1.4b}$$

in which coefficient k accounts for the type of process ($k = 5$ in all the cases except for the continuous formation with rollers for which $k = 7$ has to be adopted), A_g is the gross area of the cross-section, n is the number of 90° bends with an inner radius $r \le 5\,t$ (bends at angles different than 90° are taken into account with fractions of n) and t is the thickness of the plate or coil before forming.

The average value of the increased yielding strength f_{yb} cannot be used when calculating the effective cross-section area, or when designing members that, after the cold forming process, have been subject to heat treatments such as annealing, which reduce the residual stresses due to cold forming.

1.6.2 Geometric Imperfections

The term *geometric imperfections* refers to those differences that can be found between the theoretical shape and real size of the members, or of the structural systems as a whole, and the actual members or as-built structure. In particular, geometric imperfections can be subdivided into:

- cross-sectional imperfections;
- member imperfections;
- structural system imperfections.

Cross-sectional imperfections are related to the dimensional variation of the cross-sectional elements with respect to the nominal dimensions and can be ascribed essentially to the production process. Different values of area, moments of inertia and section moduli can influence the performance of the cross-section (e.g. in terms of load-carrying capacity or bending moment resistance). Tolerances are established by standards for the final products, not only in terms of maximum difference between actual and nominal linear dimensions, but also with reference to:

- perpendicularity tolerance between cross-sectional elements;
- tolerances with respect to axes of symmetry;
- straightness tolerance.

Figure 1.18 shows few examples of parameters to be measured for the tolerance checks for an I-shaped section.

Among *member imperfections*, the *longitudinal* (*bow*) imperfection is certainly the most important. It consists essentially of a deviation of the axis of the element from the ideal straight line and is caused by the production process. This out-of-straightness defect can cause load eccentricity, as well as an increased susceptibility to buckling phenomena.

Structural system imperfections can be ascribed to various causes, such as variability in the lengths of framing members, lack of verticality of columns and of horizontality of beams, errors in the location of foundations, errors in the placement of the connections and so on. These imperfections must be carefully accounted for during the global analysis phase. In a very simplified but efficient way, additional fictitious forces (notional loads) can be applied to the structure to reproduce the effects of imperfections. For example, the lack of verticality of columns in sway frames is accounted for by adding horizontal forces to the perfectly vertical columns (Figure 1.19), proportional to the resultant vertical force F_i acting on each floor.

This design simplification can be explained directly with reference to a cantilever column of height h with an out-of-plumb imperfection and subject to a vertical force N at the top. The additional bending moment M due to the lack of verticality, expressed by angle φ (Figure 1.20), can be approximated at the fixed end as:

$$M = N[h \cdot \tan(\varphi)] \tag{1.5}$$

Within the small displacement hypothesis (thus approximating *tan* (φ) with the angle φ itself), the effect of this imperfection can be assimilated to that of a fictitious horizontal force F acting at the top of the column and causing the same bending moment at the base of the column. The magnitude of F is thus given by:

$$F = \frac{M}{h} = N\phi \tag{1.6}$$

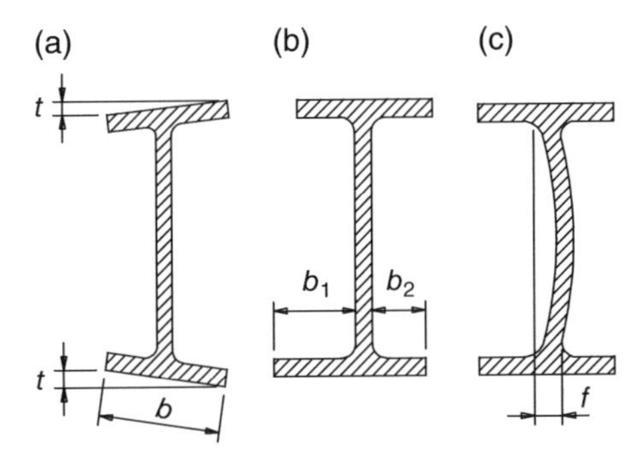

Figure 1.18 Additional tolerance checks for I-shapes: (a) perpendicularity tolerance, (b) symmetry tolerance and (c) straightness tolerance.

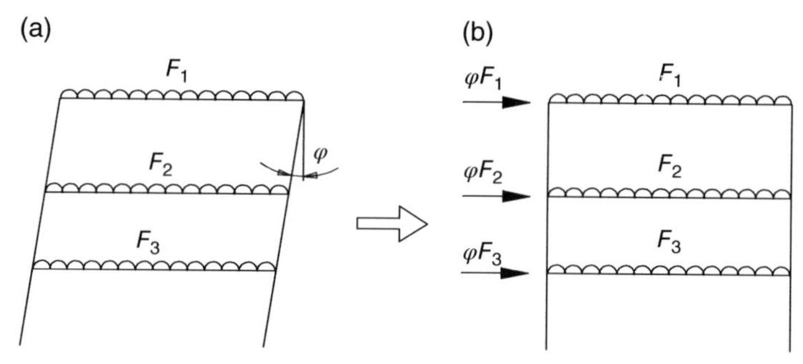

Figure 1.19 Horizontal notional loads equivalent to the imperfections for a sway frame.

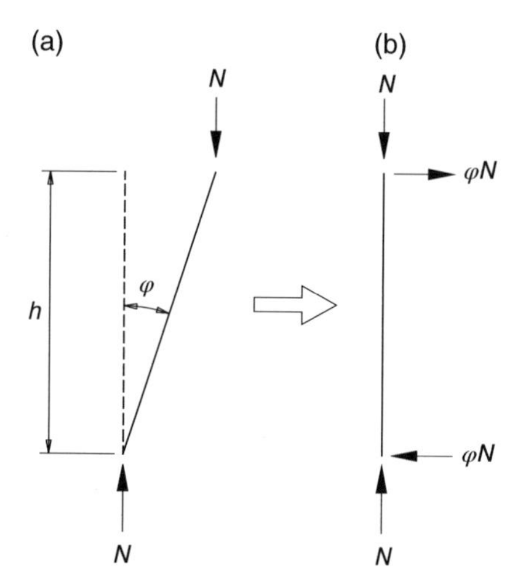

Figure 1.20 Imperfect column (a) and horizontal equivalent force (b).

1.7 Mechanical Tests for the Characterization of the Material

An in-depth knowledge of the mechanical characteristics of steel, as well as of any other structural material, is of paramount importance for design verification checks. Additionally, besides the mandatory tests performed at the factory on base materials and worked products, it is often important to perform laboratory tests on coupons cut from plane and linear *in-situ* products in order to validate the design hypotheses with actual material characteristics.

For each laboratory test there are very specific standardization requirements. Globally *ISO* (*International Organization for Standardization*) and in Europe *CEN* (*European Committee for Standardization*) standardization requirements are provided, whereas in the US, the *ASTM* is the governing body, emanating standards that contain detailed instructions on the geometry of the coupons, on the testing requirements, on the equipment to be used and on the presentation and use of the test results.

Among the most important tests for the characterization of steel there are: chemical analysis, macro- and micro-graphic testing. In particular, chemical analysis is very important to determine the main properties of steel, among which are weldability, ductility and resistance to corrosion, and to determine the percentage of carbon and other desired and undesired alloying elements.

Some alloying elements have no direct impact on the material strength, but play a key role in the determination of other properties, such as weldability and corrosion resistance. As discussed in the introductory section, in addition to carbon and iron, impurities can be present that can have a detrimental effect on the behaviour of the material, such as favouring brittleness. Since it is virtually impossible and uneconomical to completely eliminate such impurities, it is important to verify that their content is within acceptable limits. Due to these considerations, based on the grade of steel considered, the standards specifying material characteristics (EN 10025, ASTM A992, ASTM A36, ASTM A490 are some examples) contain tables defining the maximum percent content of some alloying elements (typically, carbon – C, silica – Si, phosphorous – P, sulfur – S and nitrogen – N) or a range of acceptability for other alloying elements (such as manganese – Mn, chromium – Cr, molybdenum – Mo and copper – Cu).

Chemical analyses can be performed either on the molten material (ladle analysis) or on the final product (product analysis), even after it has been erected, by means of a sample site extraction. It is possible that the limits prescribed for the chemical makeup of the material can be different, based on whether the analysis has been performed on the ladle material or on the final product (in general, the values prescribed for the analysis on molten material are more stringent than the ones on the final product).

The weldability property is directly related to a carbon equivalent value (CEV), based on the results of the analysis on the ladle material, defined as follows:

$$\mathrm{CEV} = \mathrm{C} + \frac{\mathrm{Mn}}{6} + \frac{\mathrm{Cr} + \mathrm{Mo} + \mathrm{V}}{5} + \frac{\mathrm{Ni} + \mathrm{Cu}}{15} \tag{1.7}$$

in which C indicates the percentage content of carbonium, Mn for manganese, Cr for chromium, Cu for copper, Mo for molybdenum and Ni for nickel.

In order to ensure good weldability characteristics, the material should have as low a CEV as possible, with maximum values prescribed by the various standards.

The macrographic test is performed to establish the de-oxidation and the de-carbonation indices of steel, related to weldability. The micrographic test allows analysis of the crystalline structure of steel and its grain size and the ability to relate some mechanical characteristics of the material to its micro-structure as well as to investigate the effects that thermal treatments have on the material.

In the following, a brief description of some of the most important mechanical laboratory tests performed on structural steel is presented.

1.7.1 Tensile Testing

The most important and well-known mechanical test is the *uniaxial tensile test*. This test allows measurement of some important mechanical characteristics of steel (yield strength, ultimate strength, percentage elongation at failure and the complete stress-strain curve, as discussed in Section 1.1). The test consists of the application of a tensile axial force to a sample obtained according to specific standards (EN ISO 6892-1 and ASTM 370-10). The tensile force is applied with an intensity that increases with an established rate, recording the extension Δ over a gauge length L_0 in the middle of the sample (Figure 1.21).

The stress σ is calculated dividing the measured applied force by the nominal cross-sectional area of the coupon (A_{nom}), while the strain ε is calculated by means of change of the gauge length:

$$\varepsilon = \frac{\Delta}{L_0} = \frac{L_d - L_0}{L_0} \tag{1.8}$$

in which L_d is the distance between the gauge marks during loading.

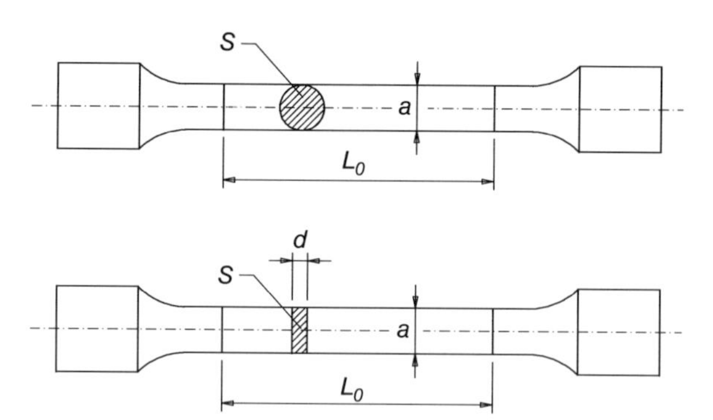

Figure 1.21 Typical sample for rolled products.

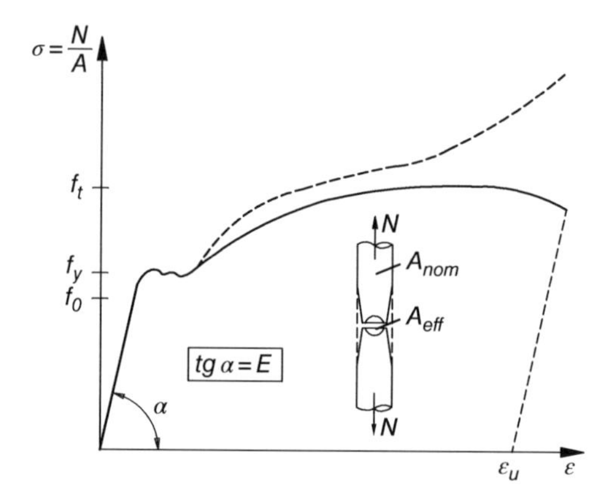

Figure 1.22 Typical stress-strain (σ–ε) relationship for structural steels.

For steel materials with a carbon percentage of up to 0.25%, that is for structural steels, the typical stress-strain relationship is shown in Figure 1.22. The initial branch of the curve is very close to linear elastic.

From the slope of the initial branch of the σ–ε curve, the longitudinal elastic modulus or Young's modulus, can be calculated as $E = \tan(\alpha)$. Once the value of the stress indicated with f_0 in the figure is reached, which can be defined as the limit of proportionality, there is no more direct proportionality between stress and strain, but the material still behaves elastically. Corresponding to a stress f_y, yielding occurs and the stress-strain response is characterized by a slightly undulating response that is substantially horizontal due to the onset plastic deformations (Figure 1.23).

It is worth noting that low-carbon steels usually show two distinct values of the yielding stress: an upper yielding point, R_{eH}, after which the strains increase with a local decrease of the stress, and a lower yielding point, R_{eL}, at which there are no appreciable reductions in the stress associated with an increase in strain. The upper yielding point R_{eH} is significantly affected by the load rate, unlike the lower yielding point, which is substantially independent of the rate and is thus usually taken as the yielding strength to be used for design, that is $f_y = R_{eL}$.

Until the yielding stress is reached, the transverse deformations of the coupon due to Poisson's effect are very small. The effective cross-sectional area of the coupon (A_{eff}) is considered, with a small approximation, to be equal to the nominal cross-sectional area ($A_{eff} = A_{nom}$). For higher levels of the applied force, the transverse deformations are not negligible anymore, but for the sake of practicality the stress is always calculated making reference to the nominal area of the

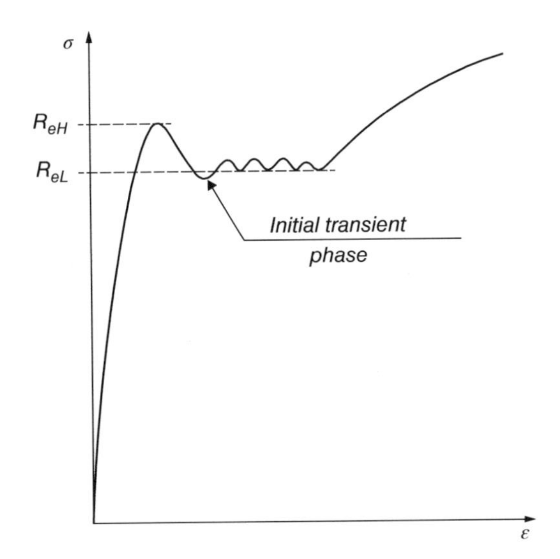

Figure 1.23 Upper and lower yielding points for structural steel.

undeformed cross-section (A_{nom}). As a consequence, the resulting stress-strain diagram results in the solid-line curve in Figure 1.22, which is characterized by a *softening* branch with increasing stresses corresponding to increasing strains, which is the *hardening* branch. This branch ends when the transverse deformations of the coupons stop being uniform along the length of the coupon, and start focusing in a small region towards the middle of the coupon itself. This phenomenon is identified as *necking* (reduction of area) and one of the immediate consequences is that an increase in strain now corresponds to a decrease in stress, until the coupon fails. If the effective cross-sectional area is used (A_{eff}), the resulting stresses would be always increasing until failure, because even if the carried force decreases, so does the cross-sectional area (dashed curve in Figure 1.22), showing hardening all the way up to failure.

The failure strength f_u is based on the maximum value of the applied load during the test, whereas the failure strain ε_u, more commonly measured as the percent elongation at failure, is evaluated according to Eq. (1.8), putting the two parts of the broken coupon back together so that a ultimate length L_u between the gauge points can be measured.

Usually, structural steels are required to have a sufficient elongation at failure so that an adequate ductility can be expected, allowing for large plastic deformations without failure. In the absence of ductility, a considerable amount of design simplifications provided in all specifications could not be used, significantly complicating all design tasks.

The constitutive law, and consequently the material mechanical characteristics, depends on the loading rate and on the temperature at which the tensile test is performed (usually ambient temperature). With an increase in temperature, the performance parameters of steel decrease sensibly, including a reduction of the modulus of elasticity, yielding strength and of the failure strength. Above approximately 200°C (392°F), the yielding phenomenon tends to disappear in favour of a basically monotonic stress-strain curve (Figure 1.24).

1.7.2 Stub Column Test

The stub column test, also known as the global compression test, is performed on stubs cut from steel profiles (Figure 1.25) sufficiently short so that global buckling phenomena will not affect the results. This test, used in the past mainly in the US, is of great interest, because it allows

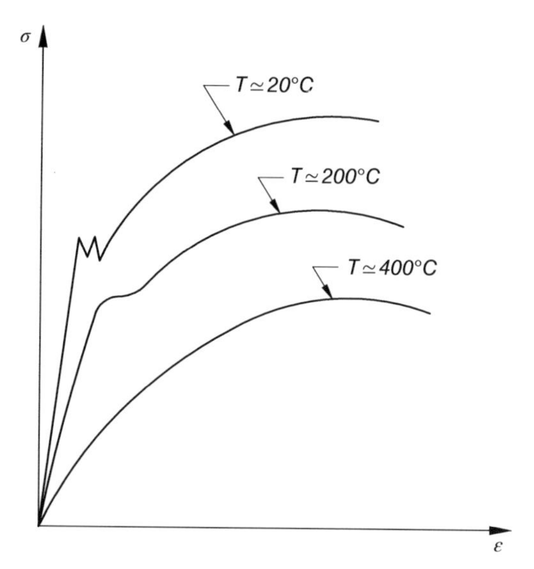

Figure 1.24 Influence of temperature on the constitutive law of steel.

Figure 1.25 Testing of a specimen in a stub column test.

measurement of a stress-strain curve for the whole cross-section of a member, not just for a coupon cut from it.

The stub column test, in fact, provides the mechanical properties of the materials averaging out the structural imperfections of the profile due, for instance, to the presence of residual stresses or to different yielding of failure strengths in various parts of the profile (web, flanges, etc.). Some profiles, in fact, due to the production process, may show a variation of mechanical properties across the thickness and also have a non-uniform distribution of residual stresses. An equivalent yielding strength ($f_{y,eq}$) can be evaluated as a function of the experimental load that causes yielding of the specimen ($P_{y,exp}$) and of the cross-sectional area (A) as follows:

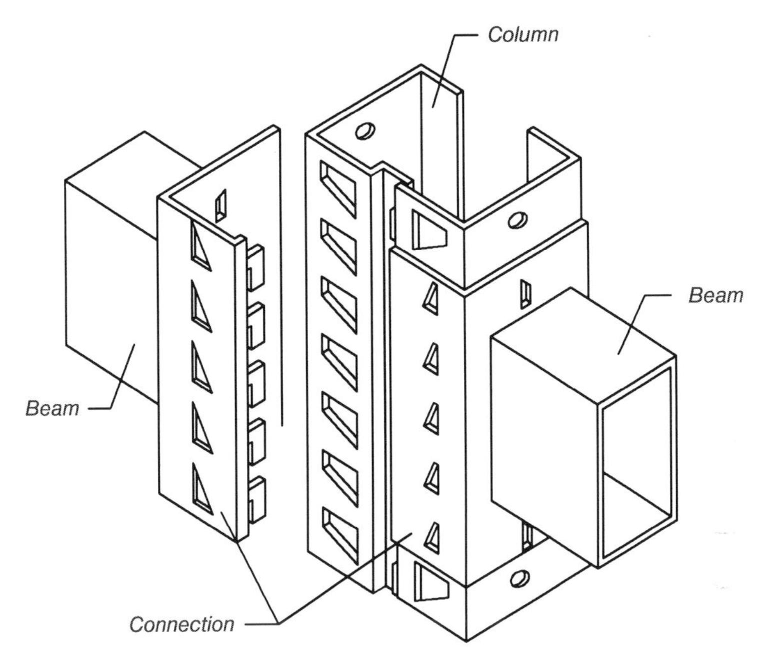

Figure 1.26 Typical components of adjustable storage pallet racks.

$$f_{y,eq} = \frac{P_{y,\text{exp}}}{A} \tag{1.9}$$

The stub column test of stocky elements is very important to determine the performance characteristics, especially when the cross-sectional geometry is particularly complex. As typical examples, industrial storage rack systems can be considered, in which the column, typically a thin-walled cold-formed member, has a regular pattern of holes to facilitate modular connections (Figure 1.26) and thus does not have uniform cross-sectional area over its length.

For such elements, the load carrying capacity is affected by local and distortional buckling phenomena, due to the small thickness of the profiles and to the use of open cross-sections. Often, due to the non-uniform cross-section of these elements, there are no theoretical approaches to evaluate their behaviour. In these circumstances, the experimental ratio of the failure load to the yielding load can be used to equate the element in question with an equivalent uniform cross-section member and then use the theoretical equations available for that case. In the case of profiles with regular perforation systems, based on the experimental axial load capacity (P_{exp}) and on the material yielding strength (f_y), an equivalent cross-sectional area can be determined as:

$$A_{eq} = \frac{P_{\text{exp}}}{f_y} \tag{1.10}$$

1.7.3 Toughness Test

The toughness test measures the amount of energy required to break a specially machined specimen, evaluating the toughness of the material, that is its ability to resist impact and in general to avoid brittle behaviour. The standardized test utilizes a gravity-based pendulum device (Charpy's

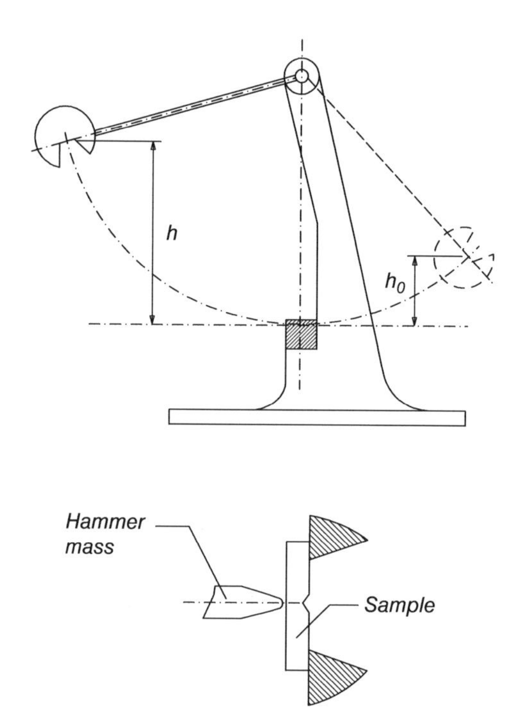

Figure 1.27 Charpy V-notch test.

pendulum) and the specimen is a rectangular bar with a suitable notch having a standardized shape (Figure 1.27). The impact is provided by a hammer suspended above the specimen that is released starting at a relative height h. Upon impacting the specimen, which is restrained by two supports at its ends, the hammer continues its swing climbing on the opposite side to a new relative height h_0 (with $h_0 < h$). The difference between h and h_0 is proportional to the energy absorbed by the specimen, E_p, that is:

$$E_p = G(h - h_0) \tag{1.11}$$

in which G is the weight of the hammer.

Toughness is measured by the ratio between the energy E_p and the area of the notched cross-section of the specimen. The tougher is the metal, the smaller the height h_0.

Toughness values depend on the shape of the specimen and in particular on the details of the notch. Among standardized notch types, it is worth mentioning the types: type *KV*, type K_{cu}, type *Keyhole*, type *Messenger* and type *DVM*. Usually toughness decreases as the mechanical strength increases and it is greatly influenced by the testing temperature, which affects the crack formation and propagation.

A temperature value can be identified, referred to as *transition temperature*, below which toughness is reduced so much to be unacceptable, due to the excessive brittleness of the material. For special applications (structures in extremely cold climates, freezing plants, etc.), metals with a very low transition temperature must be used. Toughness is expressed in energy units, usually Joules, at a specified temperature. Sometimes, the code used to identify toughness (e.g. JR, J0 or J2) follows the identification of the steel type. For structural steel, the minimum toughness required is usually 27 J, as already briefly discussed in Section 1.1. Table 1.5 contains an example of required toughness values for various European designations.

Table 1.5 Codes used for toughness requirement (Charpy V-notch).

	Minimum value of energy		
Test temperature (°C)	27 J	40 J	60 J
20	JR	LR	KR
0	J0	L0	K0
−20	J2	L2	K2
−30	J3	L3	K3
−40	J4	L4	K4
−50	J5	L5	K5
−60	J6	L6	K6

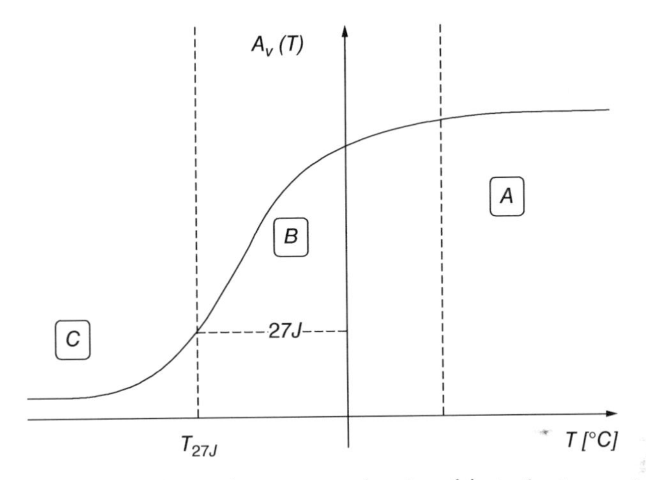

Figure 1.28 Energy associated with the toughness test as a function of the testing temperature.

For welded steel construction, and especially for those structures subject to low temperatures, it is advisable to choose steels with good toughness at low temperatures. Thermo-mechanical rolling typically produces these kinds of steel. It is also worth keeping in mind that good toughness also corresponds to good weldability.

Despite the fact that the ductility of a particular class of steel can be evaluated by means of laboratory tests, the same material in special conditions could show a fragile behaviour associated with a sudden failure at low stresses, even below yielding.

Fragile behaviour depends on several factors. Among these, the temperature at which the element is subject to in-service can cause this type of failure. With reference to the Charpy V-notch test, indicating with $A_V(T)$ the work performed by the hammer as a function of the test temperature (T), a diagram similar to the one in Figure 1.28 can be obtained, characterized by the following three regions:

- region A, corresponding to higher temperatures, with higher toughness values, indicating a material capable of undergoing large plastic deformations;
- region C, corresponding to lower temperatures, with very small toughness values and thus elevated brittleness;
- region B, between regions A and C, is the transition zone and is characterized by a very variable behaviour, with a rapid decrease in toughness as the temperature decreases.

Brittle failure can also be influenced by the rate of increase of stresses, as there is the possibility of localized overstresses that could practically prevent the onset of plastic deformations, causing sudden failures. The width of the three regions in Figure 1.28 is a function of the chemical composition of the steel. In particular, the transition temperature can be lowered by acting on the content of carbon, manganese and nickel, and/or with annealing or quenching and tempering heat treatments.

1.7.4 Bending Test

The bending test is used to evaluate the capacity of the material to withstand large plastic deformations at ambient temperature without cracking. The specimen, usually with a solid rectangular cross-section (but circular or rectangular solid specimens can also be used) is subject to a plastic deformation by means of a continuous bending action without load reversal. In detail, as shown in Figure 1.29, the specimen is placed over two roller bearings with radius R and then a force is applied by means of another roller with diameter D until the ends of the specimen form an angle α with respect to each other.

The values of R and D depend on the size of the specimen. At the end of the test, the specimen's bottom face is examined to ascertain that no cracks have formed.

1.7.5 Hardness Test

Hardness, for metals, represents the resistance that the material opposes to the penetration of another body and thus allows gathering of information on the resistance to scratching, to abrasion, to friction wear and to localized pressure.

The hardness test measures the capacity of the material to absorb energy and can also provide an estimate of the material strength. The test itself consists in the measurement of the indentation left on the specimen surface by a steel sphere that is pressed onto the specimen with a predetermined amount of force for a predetermined amount of time (Figure 1.30).

Depending on the shape of the tip penetrating device, there are various hardness tests that are chosen based on the material to be tested. Among these, the *Brinell Hardness Test*, the *Vickers Hardness Test* and the *Rockwell Hardness Test* are the most important.

The ISO 18265 norm, 'Metallic Materials Conversion of Hardness Values', has been specifically written to provide conversion values among the various types of hardness tests.

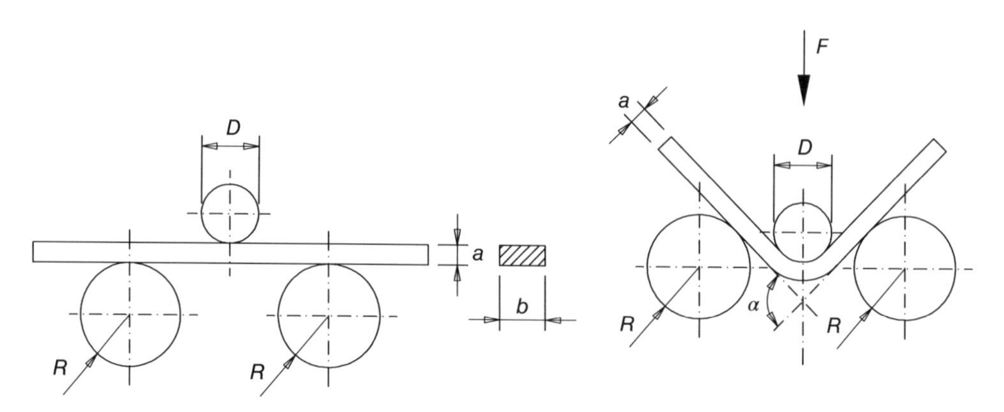

Figure 1.29 Bending test.

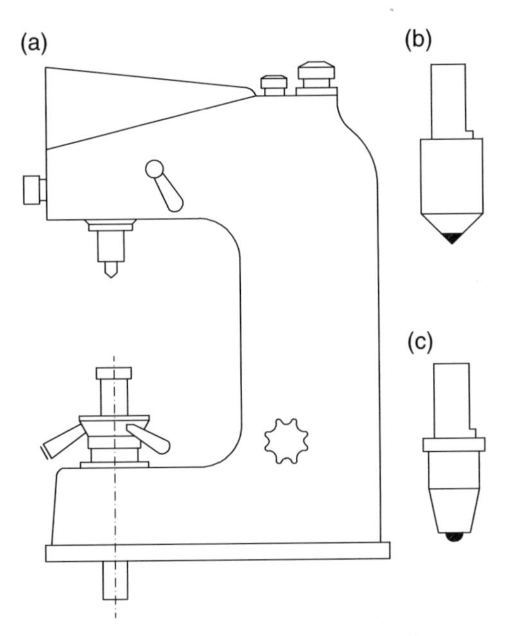

Figure 1.30 Hardness test: (a) durometer, (b) conical tip and (c) spherical tip.

Thanks to the somewhat direct relationship between hardness and strength, hardness testing is sometimes used to evaluate the tensile strength of metal elements in the field when a destructive test is not an option. In the past, several research projects have been conducted to establish a correlation between hardness and tensile strength in some materials. It is worth mentioning that, in 1989, the Technical Report ISO/TR 10108 '*Steel-Conversion of Hardness Values to Tensile Strength Values*' was published, reporting the range of tensile strength values corresponding to experimentally measured hardness.

CHAPTER 2

References for the Design of Steel Structures

2.1 Introduction

A structure has to be designed and executed in such a way that, during its intended life, it will support all load applied, with an appropriate degree of reliability and in an economical way. A very important task of each designer is to size the skeleton frame to be safe for the entire life of the structure. This condition is guaranteed if, from the construction stage until decommissioning due to old age, this condition is satisfied for each component of the structure:

$$\text{Effects of actions} < \text{Resistance} \tag{2.1}$$

With regard to internal forces and moments, on the basis of the structural model, designers must select the loads and individuate the load combinations of interest. It is essential to estimate the loads acting on the structure defining their values appropriately, that is without exaggeration (otherwise, the resulting system could be too heavy and un-economical) but at the same time avoiding load values that are too low, which would lead to unsafe design.

As far as the resistance is concerned, suitable design limits are fixed by codes, which refer to the performance of the cross-section as well as of the parts of whole structural systems. Designers have to evaluate them correctly.

A fundamental requisite of design is that the structure must be safe throughout its use, otherwise all the subjects contributing to the safety of building are responsible, that is designer, project manager, builders, acceptance test engineers and so on.

The concept of building safety is very old. Babylonian king, Hammurabi, about 4000 years ago imposed the law of retaliation to builders, with a penalty proportional to the social class of membership of the parties involved. In this code, which is the first significant example of treaty law, it was prescribed that:

- *If a builder builds a house for someone and completes it, the owner should give him a fee of two shekels in money for each sar of surface (rule 228).*
- *If a builder builds a house for someone, and does not construct it properly, and the house which he built falls and kills its owner, then the builder should be put to death (rule 229).*
- *If it kills the son of the owner the son of that builder should be put to death (rule 230).*
- *If it kills a slave of the owner, then he should pay slave for slave to the owner of the house (rule 231).*

Structural Steel Design to Eurocode 3 and AISC Specifications, First Edition. Claudio Bernuzzi and Benedetto Cordova.

It is important to note the attention given to the concept of durability by this very old code: even then it was assumed that a fundamental requirement of the construction was no damage should occur during its entire life.

About two centuries ago, Napoleon Bonaparte introduced the concept of responsibility extended to the first decade of age of the construction. In addition to the builder, the presence of a technician (e.g. the designer) was required too, sharing all the responsibilities and, eventually prison, in case of collapse or damage within a decade from putting the structure into service.

2.1.1 European Provisions for Steel Design

In 1975, the Commission of the European Community decided on an action programme in the field of construction, based on article 95 of the Treaty, with the objective of eliminating all the technical obstacles to trade and the harmonization of technical specifications. Within this action programme, the Commission took the initiative to establish a set of harmonized technical rules for the design of construction works that, in the first stage, would serve as an alternative to the national rules in force in the Member States and, ultimately, would replace them. For 15 years, the Commission, with the help of a Steering Committee with Representatives of Member States, conducted the development of the Eurocodes programme, which led to the first generation of European codes in the 1980s.

The Structural Eurocode programme comprises the following standards consisting of 10 parts:

EN 1990 – Eurocode 0: Basis of structural design;
EN 1991 – Eurocode 1: Actions on structures;
EN 1992 – Eurocode 2: Design of concrete structures;
EN 1993 – Eurocode 3: Design of steel structures;
EN 1994 – Eurocode 4: Design of composite steel and concrete structures;
EN 1995 – Eurocode 5: Design of timber structures;
EN 1996 – Eurocode 6: Design of masonry structures;
EN 1997 – Eurocode 7: Geotechnical design;
EN 1998 – Eurocode 8: Design of structures for earthquake resistance;
EN 1999 – Eurocode 9: Design of aluminium structures.

The Eurocode standards provide common structural design rules for everyday use for the design of whole structures and component products of both traditional and innovative nature. Unusual forms of construction or design conditions are not specifically covered and additional expert consideration will be required by the designer in such cases.

The National Standards implementing Eurocodes comprise the full text of the Eurocode (including any annexes), as published by CEN, which may be preceded by a National Title Page and National Foreword, and may be followed by a National Annex. This may only contain information on those parameters that are left open in the Eurocode for national choice, known as Nationally Determined Parameters, to be used for the design of buildings and civil engineering works to be constructed in the country concerned, that is:

- values and/or classes where alternatives are given in Eurocodes,
- values to be used where a symbol only is given in Eurocodes,
- country specific data (geographical, climatic, etc.), for example, a snow map,
- the procedure to be used where alternative procedures are given in Eurocodes,
- references to non-contradictory complementary information to assist the user to apply Eurocodes.

There is a need for consistency between the harmonized technical specifications for construction products and the technical rules for works.

The EN 1993 (in the following identified as EC3 or Eurocode 3) is intended to be used with Eurocodes EN 1990 (Basis of Structural Design), EN 1991 (Actions on structures) and EN 1992 to EN 1999, when steel structures or steel components are referred to.

EN 1993-1 is the first of six parts of EN 1993 (Design of Steel Structures), to which this book refers to. It gives generic design rules intended to be used with the other parts EN 1993-2 to EN 1993-6. It also gives supplementary rules applicable only to buildings. EN 1993-1 comprises 12 subparts EN 1993-1-1 to EN 1993-1-12 each addressing specific steel components, limit states or materials. In the following, the list of all the EN 1993 documents is presented:

EN 1993-1: Eurocode 3: Design of steel structures – Part 1, which is composed by:
- EN 1993-1-1: Eurocode 3: Design of steel structures – Part 1-1: General rules and rules for buildings;
- EN 1993-1-2: Eurocode 3: Design of steel structures – Part 1-2: General rules – Structural fire design;
- EN 1993-1-3: Eurocode 3 – Design of steel structures – Part 1-3: General rules – Supplementary rules for cold-formed members and sheeting;
- EN 1993-1-4: Eurocode 3 – Design of steel structures – Part 1-4: General rules – Supplementary rules for stainless steels;
- EN 1993-1-5: Eurocode 3 – Design of steel structures – Part 1-5: Plated structural elements;
- EN 1993-1-6: Eurocode 3 – Design of steel structures – Part 1-6: Strength and Stability of Shell Structures;
- EN 1993-1-7: Eurocode 3 – Design of steel structures – Part 1-7: Plated structures subject to out of plane loading;
- EN 1993-1-8: Eurocode 3: Design of steel structures – Part 1-8: Design of joints;
- EN 1993-1-9: Eurocode 3: Design of steel structures – Part 1-9: Fatigue;
- EN 1993-1-10: Eurocode 3: Design of steel structures – Part 1-10: Material toughness and through-thickness properties;
- EN 1993-1-11: Eurocode 3 – Design of steel structures – Part 1-11: Design of structures with tension components;
- EN 1993-1-12: Eurocode 3 – Design of steel structures – Part 1-12: Additional rules for the extension of EN 1993 up to steel grades S 700;

EN 1993-2: Eurocode 3 – Design of steel structures – Part 2: Steel Bridges;
EN 1993-3-1: Eurocode 3 – Design of steel structures – Part 3-1: Towers, masts and chimneys – Towers and masts;
EN 1993-3-2: Eurocode 3 – Design of steel structures – Part 3-2: Towers, masts and chimneys – Chimneys;
EN 1993-4-1: Eurocode 3 – Design of steel structures – Part 4-1: Silos;
EN 1993-4-2: Eurocode 3 – Design of steel structures – Part 4-2: Tanks;
EN 1993-4-3: Eurocode 3 – Design of steel structures – Part 4-3: Pipelines;
EN 1993-5: Eurocode 3 – Design of steel structures – Part 5: Piling;
EN 1993-6: Eurocode 3 – Design of steel structures – Part 6: Crane supporting structures.

Numerical values for partial factors and other reliability parameters are recommended as basic values that provide an acceptable level of reliability. They have been selected assuming that an appropriate level of workmanship and quality management applies.

2.1.2 United States Provisions for Steel Design

The main specification to apply for the design of steel structures in United States is ANSI/AISC 360-10 'Specification for Structural Steel Buildings' that addresses steel constructions as well as composite constructions: steel acting compositely with reinforced concrete. This specification states design requirements (stability and strength) for steel members and composite constructions, design of connections, fabrication and erection, Quality Control and Quality Assurance.

AISC 360-10 does not address minimum loads to be used: this topic is covered by ASCE/SEI 7-10 'Minimum Design Loads for Buildings and Other Structures' in the absence of an applicable specific local, regional or national building code.

AISC 360-10 does not cover seismic design: for seismic resistant structures the specifications to be applied are: ANSI/AISC 341-10 'Seismic Provisions for Structural Steel Buildings' and ANSI/AISC 358-10 'Prequalified Connections for Seismic Applications'. The first one gives additional rules for design and fabrication of steel structure to be used in seismic areas, the second one gives design methods for designing connections to be used in seismic resistant structures.

Finally, AISC 303-10 'Code of Standard Practice for Steel Buildings and Bridges' addresses design, purchase, fabrication and erection of structural steel.

Very useful tools for the designer are the AISC manuals: mainly the *AISC 325 Steel Construction Manual* and *AISC 327 Seismic Design Manual*, which discuss very interesting design examples to help in design activity.

2.2 Brief Introduction to Random Variables

All the variables involved in the design phase, both for determining resistance and stress distribution in cross-sections and members, are random in nature and not deterministic. As an example, with reference to the strength of materials, the imperfect homogeneity always present in samples for laboratory tests, as well as in structural *in-situ* elements, prevents association of an univocal value to resistance properties (such as, for example the yielding stress or the ultimate strength). Similarly, the set of internal forces and moments on structural members due to acting loads cannot be determined exactly because of the major sources of uncertainty and approximation involved in the parameters used for their definition.

Random variables are characterized by a number that expresses the probability (indicated as *prob* or p_r) of their occurrence. In the following, Y indicates the considered random variable (e.g. the measurement associated to a length, to a force, to weight or to the value of the force acting on a structure) and y represents the generic value assumed; the probability is identified by the term *prob* or p_r. The analytical treatment is based principally on the following two functions:

Relative Probability Density Function (PDF) (Figure 2.1), $f_Y(y)$, defined as:

$$f_Y(y)\mathrm{d}y = prob\{y < Y \leq y + \mathrm{d}y\} = p_r\{y < Y \leq y + \mathrm{d}y\} \tag{2.2}$$

Cumulate Density Function (CDF), $F_Y(y)$, defined as:

$$F_Y(y) = prob\{Y \leq y\} = p_r\{Y \leq y\} \tag{2.3}$$

The PDF describes the relative likelihood for this random variable to take on a given value. The probability of the random variable falling within a particular range of values is given by the integral of this variable density over that range that is given by the area under the density function but above the horizontal axis and between the lowest and greatest values of the range. The PDF is non-negative everywhere, and its integral over the entire space is equal to one.

$$\int_{-\infty}^{\infty} f_Y(y)\mathrm{d}y = 1 \tag{2.4}$$

The cumulative PDF (CDF) represents the probability that the random variable in question takes a value not exceeding y and is linked to the PDF from the integral relationship:

$$F_Y(y) = \int_{-\infty}^{y} f_Y(c)\mathrm{d}c \tag{2.5}$$

To better understand the correspondence between the functions PDF and CDF, it can be considered Figure 2.2. The area under the PDF function in the range between $-\infty$ and y_1, or equally between $-\infty$ and y_2, finds a corresponding value of the abscissa of the CDF, $F_Y(y_1)$ and $F_Y(y_2)$, respectively.

As an example of distribution of random variable, the density probability function of the weight per unit volume of both the concrete and the steel material are presented in Figure 2.3. Note that the curve of the concrete is extended on a portion of the abscissa axis appreciably wider than that

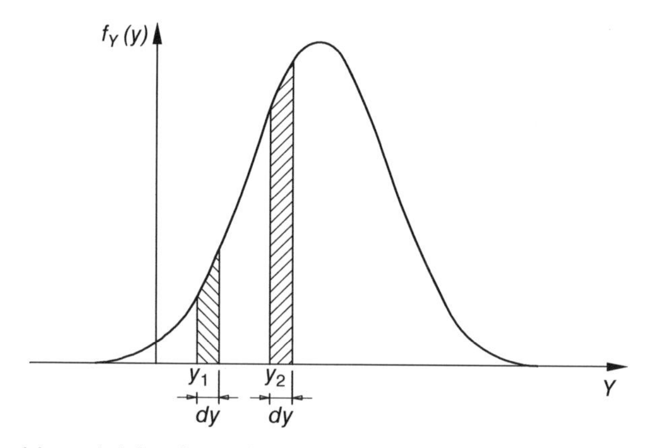

Figure 2.1 Example of the probability density function (PDF).

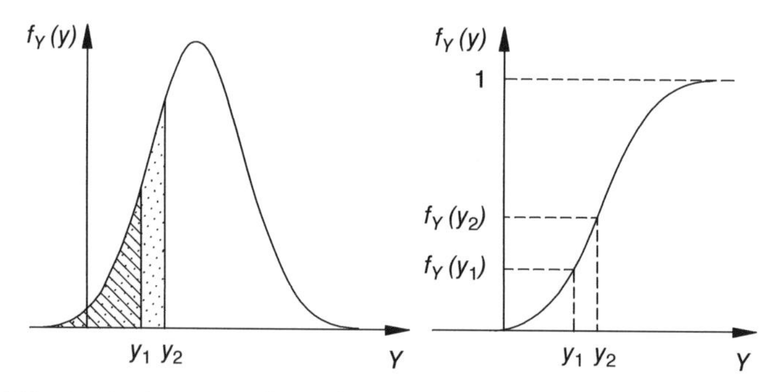

Figure 2.2 Probability density function and cumulate density function.

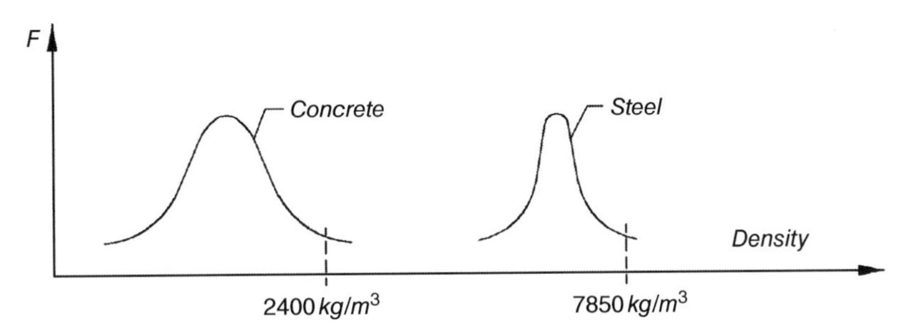

Figure 2.3 Probability density function for the weight per unit volume of the concrete and the steel.

corresponding to the steel, due to the relevant heterogeneity of the first material with respect to the latter.

As already mentioned, all the quantities interested in the check of Eq. (2.1) are random variables, and the data to be used in the design phase should be chosen based on reasonable probability values (or equally, acceptable risk levels) in relation to what they express. Actually, some of them (for example geometrical data or the eccentricity of loads) are, in most cases, taken as deterministic in order to simplify design. As discussed in the following, for the resistance, reference is made to low probabilistic values; that is to values with a high probability of being exceeded (95%). Otherwise, if actions are considered, high values (i.e. values with a reduced probability of being exceeded: 5%) are assumed for design.

In Figure 2.3 the values currently adopted for concrete and steel weight of the structural elements are indicated. (2400 kg/m^3 (149.83 lb/ft^3) to 7850 kg/m^3 (490.06 lb/ft^3) for concrete and steel, respectively). These values, like all the weights per unit of volume of both structural and non-structural elements usually correspond to the value that has a 95% probability of not being exceeded, or the 5% probability of being exceeded, and are defined 95% fractile or 95% characteristic values.

2.3 Measure of the Structural Reliability and Design Approaches

In the last century there has been a significant evolution of the philosophy of the structural reliability and, as a consequence, of the design methods. Nowadays, very sophisticated approaches are available, able to account for the variability of the main parameters governing design. It should be noted that these calculation methods currently in use are characterized by precision awareness. Because of the uncertainties of different type and nature that intervene in the design, the structure is always characterized by a well-defined level of risk. As a consequence, it is not possible to design a structure characterized by a zero level of failure probability: every structure has a probability of failure strictly depending by design, erection phase and maintenance during its use. To better understand this concept, reference can be made to Figure 2.4, where probability of failure and costs are measured by the abscissa and ordinate axis, respectively, of the considered reference systems.

If the initial cost of construction is considered (curve for the erection phase), also including design, a 100% safe structure is associated with infinite costs. The decrease in the cost of construction corresponds to an increase in failure probability. Furthermore, costs associated with repairing phases during the construction life have to be considered, when both moderate (curve b) and severe (curve c) damages could occur. It can be noted that costs associated with these damages increase with increasing failure probability. By adding the initial construction cost with the one of moderate or severe damages, the resulting curves (*d* and *e*, respectively) are characterized

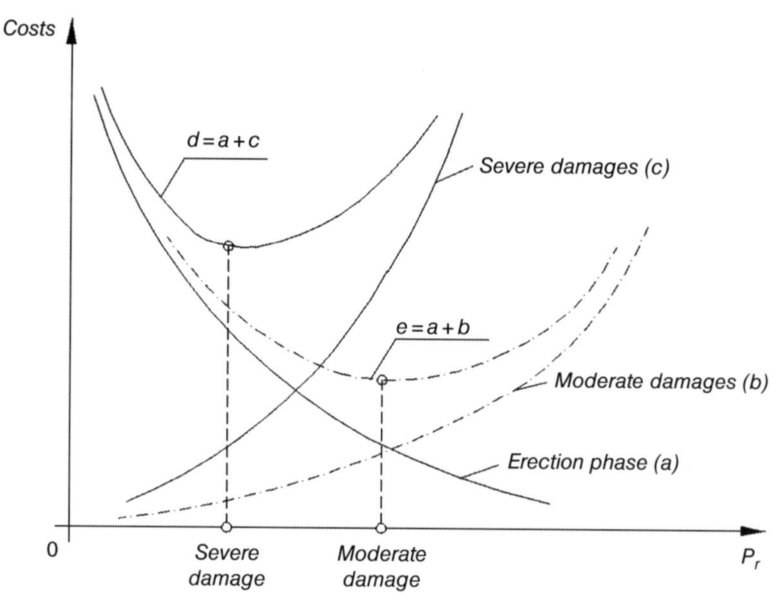

Figure 2.4 Example of the relationship between costs and probability of an unsuccessful outcome.

by a similar trend, with a minimum zone representing a good compromise between cost and overall level of risk (optimal design with the maximum benefits and the minimum costs).

In this brief discussion about reliability measurement and design methods, reference is made to resistance (R) and the effects of actions (E), which are considered generic random variables, neglecting all the functional relationships that contribute to define them.

There are two significant relationships between R and E to measure the structural safety:

- the reliability level, or the safe margin, Z, which is defined by the difference between R and E:

$$Z = R - E \tag{2.6}$$

- the safety factor, γ, which is given by the ratio between R and E

$$\gamma = R/E \tag{2.7}$$

The reliability level of a structure can be directly estimated by the exact methods, distinguished in level IV and level III methods, depending on whether or not the human factor is considered to be a cause of failure. Considering the resistance, R, and the effects of actions, E, as independent random variables, they can be represented in a three-dimensional reference system where the vertical axis expresses the associated PDF of failure (Figure 2.5). Furthermore, level IV methods are reliability methods that compare a structural prospect with a reference prospect. Principles of engineering economic analysis under uncertainty are consequently taken into account by considering cost and benefits of construction, maintenance, repair and consequences of failure. Methods of the III level include numerical integration, approximate analytical methods and simulation methods.

In the plane R–E, the safe zone ($R > E$) and the unsafe zone ($R < E$) are delimited by the bisector plane containing point with $R = E$ and the subtended volume identifies the failure probability (P_r), that is it directly measures the safety of the construction. The use of such methods is extremely complex and expensive in terms of required resources as well as high engineering competencies and is justified only for structures of exceptional importance, such as, for example, nuclear power plants or large-scale infrastructure.

The level II methods are characterized by a degree of complexity lower than those just presented and are based on the assumption that resistance and effects of actions are statistically independent from each other. With reference to the safety margin Z (Eq. (2.6)), the failure probability is represented by the area under the Probability Density Function (PDF) curve up to the point of abscissa $Z = 0$, that is the hatched area in Figure 2.6 and it is expressed by the value of F_Z (0), as:

$$P_r = F_Z(0) = \int_{-\infty}^{0} f_Z(z)\,\mathrm{d}z \tag{2.8}$$

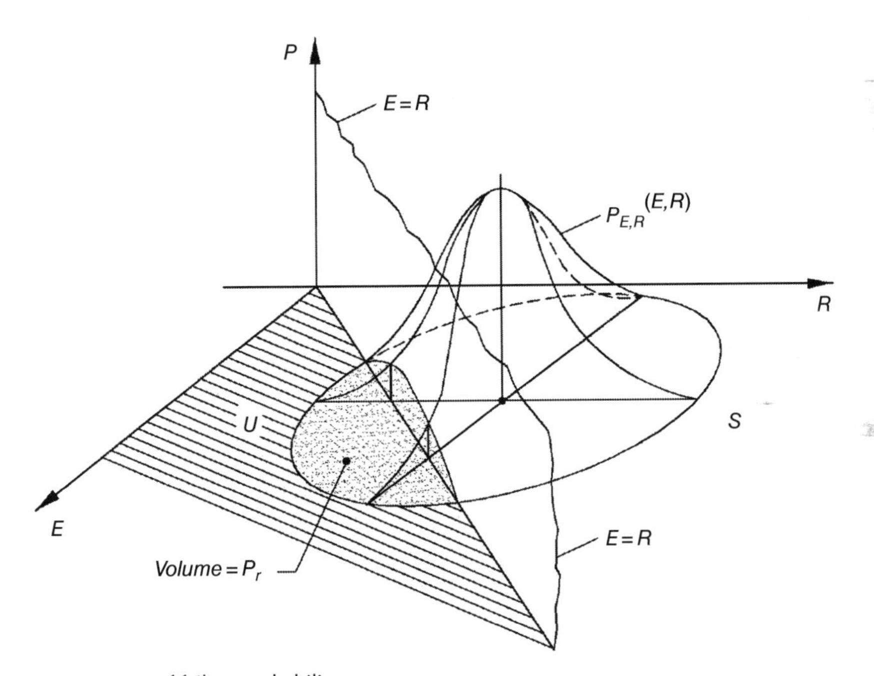

Figure 2.5 Assessment of failure probability.

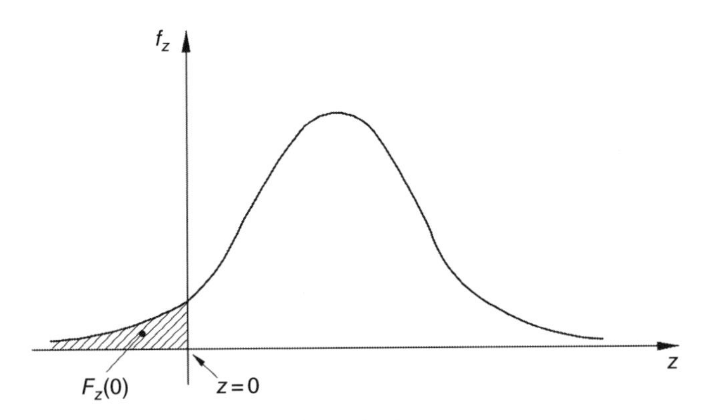

Figure 2.6 Safe domain in accordance with level II methods.

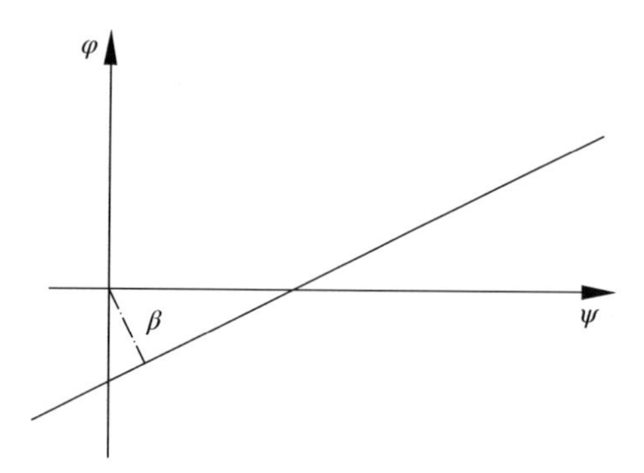

Figure 2.7 Safety index.

Assuming that resistance and the effects of actions both have a Gaussian type distribution, it can be convenient instead to make reference to the safety margin (Z) to a normalized safety margin, u, defined as:

$$u = \frac{z - \eta_z}{\sigma_z} \tag{2.9}$$

where η_z is the average value of the Z and σ_z is its standard deviation.

Failure probability depends on the CDF of variable U through the relationship:

$$P_r = 1 - F_U(\beta) \tag{2.10}$$

where β is the so-called safety index.

Increasing the safety index, the failure probability decreases. The graphical mean of β is presented in Figure 2.7, where the normalized resistance (φ) and the normalized effects of actions (Ψ), are respectively defined as:

$$\varphi = \frac{r - \eta_R}{\sigma_R} \tag{2.11a}$$

$$\psi = \frac{e - \eta_E}{\sigma_E} \tag{2.11b}$$

It can be noted that β is the smallest distance from the straight line (or generally from the hyperplane) forming the boundary between the safe domain and the failure domain, that is the domain defined by the failure event. It should be noted that this definition of the reliability index does not depend on the limit state function but rather the boundary between the safe domain and the failure domain. The point on the failure surface with the smallest distance to origin is commonly denoted the design point or most likely failure point.

Unlike top level methods (level III and IV), the level II approaches assess the value of β and verify that it complies with the limiting design conditions established by codes that are generally defined by using level III methods.

Level I methods are the ones directly used for routine design, which are also identified as extreme value methods. It is assumed that all relevant variables x_k can be modelled by random variables (or stochastic processes). Both resistance (R) and the effects of the actions (E) are assumed to be statically independent and they can be expressed as:

$$R = g_R(x_1, x_2, x_3, \ldots, x_m) \tag{2.12a}$$

$$E = g_E(x_{m+1}, x_{m+2}, \ldots, x_n) \tag{2.12b}$$

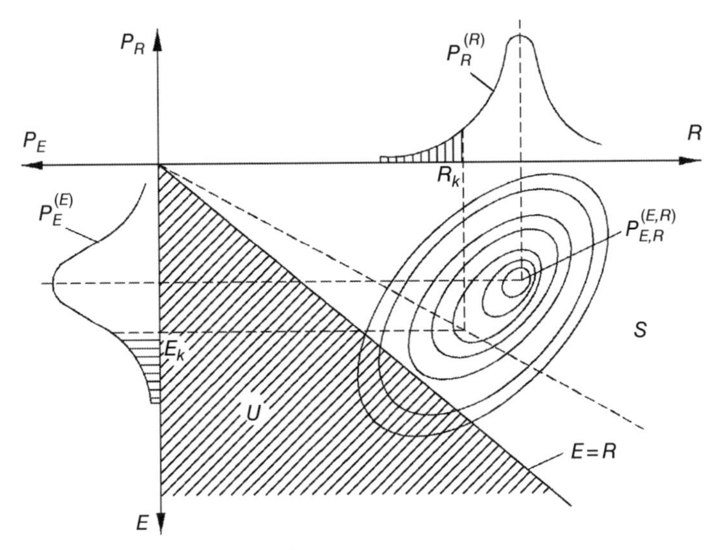

Figure 2.8 Verification in accordance with first level approaches.

These methods refer to suitable values that are lower than the mean ones for the resistance and greater than the mean ones for the effects of actions (Figure 2.8). Inserting these values in the previous equations, the design value R_d and E_d are hence directly defined. The measure of security is satisfied if E does not exceed R and this condition implicitly guarantees that the limits of allowed probability are not exceeded.

The so-called semi-probabilistic limit state method belongs to the family of level I methods for practical design purposes. Its name is due to the fact that the variability of R and E is taken into account in a simplified way by introducing appropriate safety coefficients γ. They can be distinguished into reduction coefficients, γ_m (such as the ones for materials), and amplifying coefficients, γ_f (such as the ones for actions). *Limit state* is defined as a state at which the structure (or one of its key components) cannot longer perform its primary function or no longer meets the conditions for which it was conceived. A distinction is required between ultimate limit states and serviceability limit states.

The limit states that concern the safety of people and/or the safety of the structure are classified as *ultimate limit states.*

The limit states that concern the functioning of the structure or structural members under normal use, the comfort of people and the appearance of the construction works, are classified as *serviceability limit states.* The verification of serviceability limit states should be based on suitable criteria concerning the following aspects:

- deformations that affect appearance, comfort of users or the functioning of the structure (including the functioning of machines or services) or the damage caused to finishes or non-structural members;
- vibrations that cause discomfort to people or that limit the functional effectiveness of the structure;
- damage that is likely to adversely affect the appearance, the durability or the functioning of the structure.

The level 0 methods, which have been widely applied in the past to structural design, do not consider the probabilistic aspects. The most known and used method belonging to this family is the so-called method of allowable stresses, which requires a direct comparison between the values of the maximum stress and the allowable resistance of the material in any critical point of the more stressed cross-section of members. In general, the following criticisms can be moved to the allowable stress design (ASD) method:

- safety factors are large which can create the mistaken belief that designers have always very high safety margins;
- it is impossible to determine the probability of collapse of the structure;
- the verification of the local stress state is based on assumptions excessively simplified because some important phenomena are not considered that influence it (with reference to reinforced concrete structures, e.g. the effects of inelastic deformations and cracking phenomena are ignored);
- the effects induced by forces cannot be clearly distinguished from those produced by distortions;
- it is not possible to organize checks against the various events that designers have to avoid when they do not depend directly from the acting stress level (e.g. corrosion, fire etc.).

2.4 Design Approaches in Accordance with Current Standard Provisions

As the aim of this book is to provide an introduction to the design of the steel structures, no data are reported in the following about the basis of structural design. However, some base indications are required for the determination of the effects on the buildings and for the evaluation of the resistance of the structural components.

2.4.1 European Approach for Steel Design

The design value F_d of an action F can be expressed in general terms as:

$$F_d = \gamma_f \psi F_k \tag{2.13}$$

where γ_f is a partial factor for the action taking into account the possibility of unfavourable deviations from the representative value, ψ is the combination coefficient and F_k is the characteristic value of the action.

The combination ψ coefficients (Table 2.1) take into account the fact that maximum actions cannot occur simultaneously. Representative values of a variable action are:

- $\psi_0 F_k$: rare combination value used for the ultimate limit state verification and irreversible serviceability limit states;
- $\psi_1 F_k$: a frequency value used for the verification of ultimate limit states involving accidental actions and for verifications of reversible serviceability limit states. As an example, for civil and commercial buildings the frequency value is chosen so that the time in which it is exceeded is 0.01 (1%) of the reference period; for road traffic loads on bridges (the frequency value is generally assessed on the basis of a return period of 1 week);
- $\psi_2 F_k$: the quasi-permanent value used for the verification of ultimate limit states involving accidental actions and for the verification of reversible serviceability limit states. Quasi-permanent values are also used for the calculation of long-term effects.

The actions must be combined in accordance with the considered limit states in order to achieve the most unfavourable effects, taking into account the reduced probability of their simultaneous actions with the respective most unfavourable values. The load combinations of the semi-probabilistic limit state method, which coincide with the ones to adopt for the verification at the ultimate limit states, are symbolically expressed by the relationship:

$$F_d = \gamma_{G1} G_1 + \gamma_{G2} G_2 + \gamma_P P_k + \gamma_{Q1} Q_{1k} + \sum_{i=2}^{n} \gamma_{Qi} (\psi_{0i} Q_{ik}) \quad (2.14)$$

where the signs + and Σ mean the simultaneous application of the respective addenda, the coefficients γ represent amplification factors, the subscript k indicates the characteristic value while G identifies the permanent loads (subscript 1 for permanent structural and subscript 2 for non-structural), Q is the variable loads, P is the possible action of prestressing and ψ_{0i} is the coefficient of the combination of actions.

Table 2.2 provides the γ values of the coefficients of the actions to be assumed for the ultimate limit state checks. The symbols in the table have the following meaning:

- γ_{G1} partial factor for the weight of the structure, as well as the weight of the soil;
- γ_{G2} partial coefficient of the weights of non-structural elements;
- γ_{Qi} partial coefficient of the variable actions.
- γ_P prestressing coefficient, taken equal to unity ($\gamma_P = 1.0$).

Table 2.1 Proposed values of the ψ combination coefficients (from Table A1.1 of EN 1990).

Action	ψ_0	ψ_1	ψ_2
Imposed loads in buildings, category (see EN 1991-1-1)	0.7	0.5	0.3
Category A: domestic, residential areas			
Category B: office areas	0.7	0.5	0.3
Category C: congregation areas	0.7	0.7	0.6
Category D: shopping areas	0.7	0.7	0.6
Category E: storage areas	1.0	0.9	0.8
Category F: traffic area, vehicle weight ≤ 30kN	0.7	0.7	0.6
Category G: traffic area, 30 kN < vehicle weight ≤ 160 kN	0.7	0.5	0.3
Category H: roofs	0	0	0
Snow loads on buildings (see EN 1991-1-3)[a]	0.7	0.5	0.2
Finland, Iceland, Norway, Sweden			
Remainder of CEN member states, for sites	0.7	0.5	0.2
Located at altitude H > 1000 m a.s.l.			
Remainder of CEN member states, for sites	0.5	0.2	0.0
Located at altitude H ≤ 1000 m a.s.l.			
Wind loads on buildings (see EN 1991-1-4)	0.6	0.2	0.0
Temperature (non-fire) in buildings (see EN 1991-1-5)	0.6	0.5	0.0

[a] For countries not mentioned, see relevant local conditions.
Note: The ψ values may be set by the National Annex.

Table 2.2 Proposed values of action coefficient γ_F for verification at ultimate limit states.

		γ_F	EQU	A1-STR	A2-GEO
Structural dead load	Favourable	γ_{G1}	0.9	1.0	1.0
	Unfavourable		1.1	1.3	1.0
Non-structural dead load	Favourable	γ_{G2}	0.0	0.0	0.0
	Unfavourable		1.5	1.5	1.3
Variable action	Favourable	γ_{Qi}	0.0	0.0	0.0
	Unfavourable		1.5	1.5	1.3

The following ultimate limit states have to be verified where they are relevant:

- *EQU*: loss of static equilibrium of the structure or any part of it considered as a rigid body, where minor variations in the value or the spatial distribution of actions from a single source are significant and the strengths of construction materials or ground are generally not governing;
- *STR*: internal failure or excessive deformation of the structure or structural members, including footings, piles, basement walls and so on, where the strength of construction materials of the structure governs;
- *GEO*: failure or excessive deformation of the ground where the strengths of soil or rock are significant in providing resistance;
- *FAT*: fatigue failure of the structure or structural members.

With reference to the serviceability limit states, the following load combinations have to be considered:

- *characteristic combination*, which is usually adopted for the irreversible limit states is defined as:

$$F_d = G_1 + G_2 + P_k + Q_{1k} + \sum_{i=2}^{n} (\psi_{0i} Q_{ik}) \tag{2.15a}$$

- *frequent combination*, which is normally used for reversible limit states is defined as:

$$F_d = G_1 + G_2 + P_k + \psi_{1i} Q_{1k} + \sum_{i=2}^{n} (\psi_{2i} Q_{ik}) \tag{2.15b}$$

- *quasi-permanent combination*, which is normally used for long-term effects and the appearance of the structure is defined as:

$$F_d = G_1 + G_2 + P_k + \sum_{i=1}^{n} (\psi_{2i} Q_{ik}) \tag{2.15c}$$

In addition, when relevant, the following action combination has to be considered:

- *combinations of actions for seismic design situations:*

$$E + G_1 + G_2 + P + \sum_{i=1}^{n} \cdot \psi_{2i} \cdot Q_{ik} \tag{2.16a}$$

- *combinations of actions for accidental design situations:*

$$G_1 + G_2 + P + A_d + \sum_{i=1}^{n} \cdot \psi_{2i} \cdot Q_{ik} \tag{2.16b}$$

As to the design resistances, the design values X_d of material properties are defined as:

$$X_d = \frac{X_k}{\gamma_{Mj}} \tag{2.17}$$

where X_k and γ_{Mj} are the characteristic value and the material safety coefficient.

Partial factors γ_{Mj} for buildings may be defined in the National Annex. The following numerical values are recommended for buildings:

- $\gamma_{M0} = 1.00$, to be used for resistance checks;
- $\gamma_{M1} = 1.00$, to be used for stability checks;
- $\gamma_{M2} = 1.25$, to be used for connection design.

In EU countries, different values can be recommended by the National document for the application of Eurocode and, in particular, designers have generally to adopt, in accordance with the indication of several EU countries, $\gamma_{M0} = \gamma_{M1} = 1.05$ and $\gamma_{M2} = 1.35$.

2.4.2 United States Approach for Steel Design

Differently from Eurocodes, AISC 360-10 allows use of the semi-probabilistic limit state method as well as the working stress (allowable stress) design method. The first method is called Load and Resistance Factor Design (LRFD) and the second one ASD. The two methods are specified as alternatives and the ASD method is maintained for those who have been using it in the past (senior engineers), before LRFD method was introduced.

The conceptual verification Eq. (2.1) is expressed in AISC 360-10 in the following way:

(a) for LRFD:

$$R_u \leq \phi R_n \tag{2.18}$$

(b) for ASD:

$$R_u \leq R_n/\Omega \tag{2.19}$$

where R_u is the required strength, R_n is the nominal strength, ϕ is the resistance factor, ϕR_n is the design strength, Ω is the safety factor and R_n/Ω is the allowable strength.

Required strength (RS) has to be less or equal to design strength (LRFD) or allowable strength (ASD) and has to be computed for appropriate loading combinations. RS assumes different values if evaluated in accordance with LRFD or with ASD. Actually, ASCE/SEI 7 document (*Minimum Design Loads for Buildings and Other Structures*) proposes two different sets of loading combinations for the two methods. Loading combinations for the two methods are:

LRFD method	ASD method
(1) $1.4D$	(1) D
(2) $1.2D + 1.6L + 0.5(L_r$ or S or $R)$	(2) $D + L$
(3) $1.2D + 1.6(L_r$ or S or $R) + (L$ or $0.5W)$	(3) $D + (L_r$ or S or $R)$
(4) $1.2D + 1.0W + L + 0.5(L_r$ or S or $R)$	(4) $D + 0.75L + 0.75(L_r$ or S or $R)$
(5) $1.2D + 1.0E + L + 0.2S$	(5) $D + (0.6W$ or $0.7E)$
(6) $0.9D + 1.0W$	(6a) $D + 0.75L + 0.75(0.6W) + 0.75(L_r$ or S or $R)$
	(6b) $D + 0.75L + 0.75(0.7E) + 0.75S$
(7) $0.9D + 1.0E$	(7) $0.6D + 0.6W$
	(8) $0.6D + 0.7E$

where D is the dead load, E is the earthquake load, L is the live load, L_r is the roof live load, R is the rain load, S is the snow load and W is the wind load.

Required strength has to be determined by structural analysis for the appropriate load combinations listed previously. The specification allows for elastic, inelastic or plastic structural analysis. Nominal strength calculation depends on the member stresses (tension, compression, flexure and shear) and will be addressed more in detail in the following chapters. The resistance factor ϕ is less than or equal to 1.0. It reduces the nominal strength taking into account approximations in the theory, variations in mechanical properties and dimension of members.

The safety factor Ω is greater than or equal to 1.0 and is used in the ASD method, and reduces the nominal strength to allowable strength.

Both ϕ and Ω factors vary according to the kind of internal forces acting on the member (tension, compression, flexure, shear), as shown in the following chapters.

CHAPTER 3

Framed Systems and Methods of Analysis

3.1 Introduction

A building is a fairly complex structural system, where linear members (beams, columns and diagonals), floors and roof, diaphragms and cladding interact with each other via different connection types. In steel constructions, especially for residential, office and industrial buildings, the skeleton frame has a three-dimensional configuration, which is generally well-distinguished (Figure 3.1) from secondary (non-structural) components.

With reference to framed systems regular in plane and in elevation (Figure 3.2), that is to very common configurations in steel construction practice, it is possible and convenient to base the design on suitable planar models (Figure 3.3). As a consequence, if the assumption of a rigid floor is satisfied, as generally occurs in routine design, the sizing phase can result simplified and, at the same time, characterized by a satisfactory degree of safety. Hence, it appears of paramount importance to correctly size the structural components, simultaneously guaranteeing full correspondence between the model calculation and the structure.

The steel framed systems can be classified with respect to different criteria, each of them associated with specific aims. By focusing attention on the most commonly used criteria, which are also considered by international standards, the following discriminating elements can be selected for frame classification:

- *the structural typology*: braced and unbraced frames can be identified, on the basis of the presence/absence of a specific structural system (i.e. the bracing system) able of transfer to the foundation all the horizontal actions;
- *the frame stability to lateral loads*: no-sway and sway frames can be identified on the basis of the influence of the second order effects on the structural response;
- *the degree of flexural continuity associated with joints*: simple, rigid and semi-continuous frame models can be selected on the basis of the structural performance of beam-to-column joints as well as base-plate connections.

It is important to observe that these three classification criteria are independent from each other and provide direct information on the design path to follow.

Structural Steel Design to Eurocode 3 and AISC Specifications, First Edition. Claudio Bernuzzi and Benedetto Cordova.

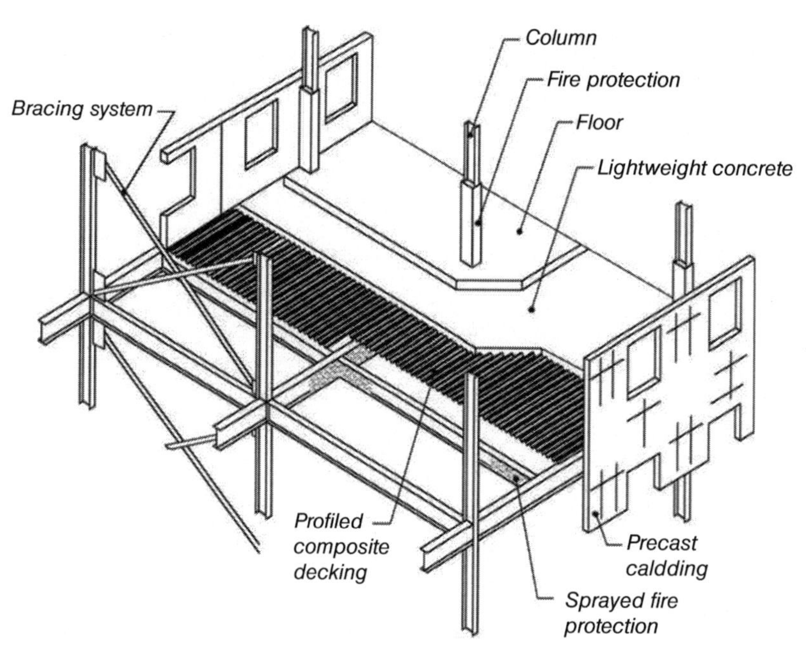

Figure 3.1 Building and framed system.

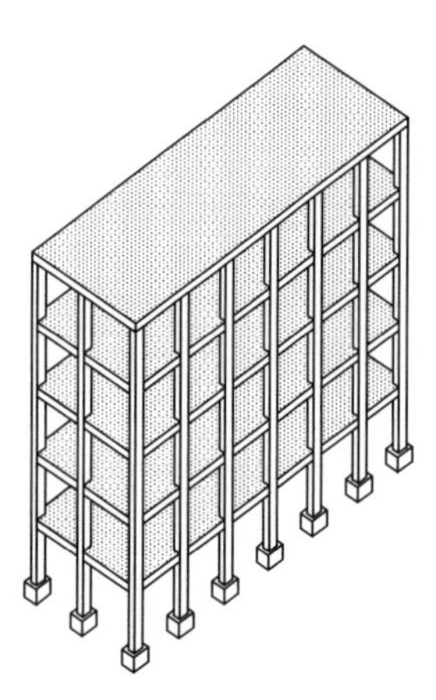

Figure 3.2 Three-dimensional framed system.

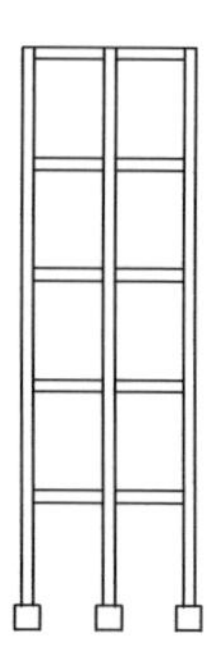

Figure 3.3 Planar frame model.

3.2 Classification Based on Structural Typology

The distinction between braced frames and unbraced frames is due to the presence or absence, respectively, of the structural system (i.e. the bracing system) able to transfer to the foundation all the horizontal actions mainly due to wind, earthquakes and any geometrical imperfection. The bracing systems can be identified from an engineering point of view as the part of the structure able to reduce the transverse displacements of the overall structural system by at least 80%. Similarly, the frame can be considered braced if the stiffness of the bracing system is at least five times greater than the one related to the remaining part of the frame.

The bracing system can be typically realized by means of reinforced concrete elements, such as concrete cores, used typically for containing stairs and/or uplifts (Figure 3.4a), or concrete shear walls (Figure 3.4b), or by using specific steel systems (Figure 3.4c). In absence of bracing systems, the skeleton frame is unbraced (Figure 3.4d) and some of the structural elements, already designed to sustain all the vertical loads, also transfer directly the horizontal loads to the foundation. The bracing system must be designed to withstand:

- all the horizontal actions directly applied to the frame;
- all the horizontal actions directly applied to the bracing system;
- all the effects associated with initial imperfections of both the bracing systems as well as of the remaining parts of the frame. Usually these effects are taken into account by means of the equivalent geometric imperfections or the equivalent additional horizontal actions (notional loads).

If a frame is suitably braced, the design procedure is significantly simplified: with reference to the generic load condition (Figure 3.5), gravity load can be considered to be acting on the sole framed system while the bracing system is subjected to horizontal and vertical load.

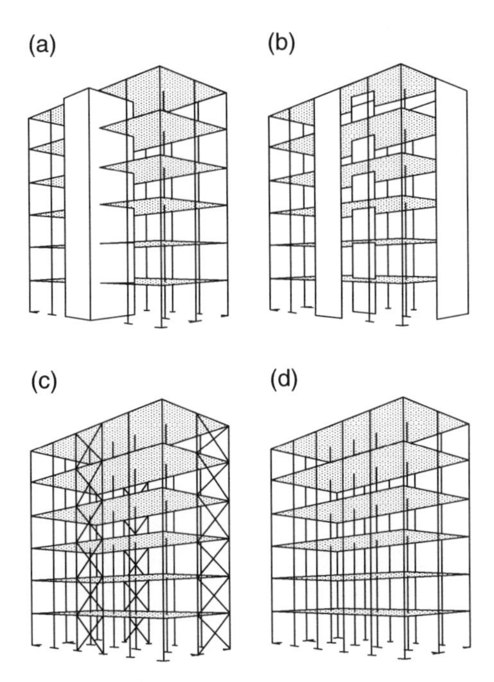

Figure 3.4 Common frame typologies (a–d).

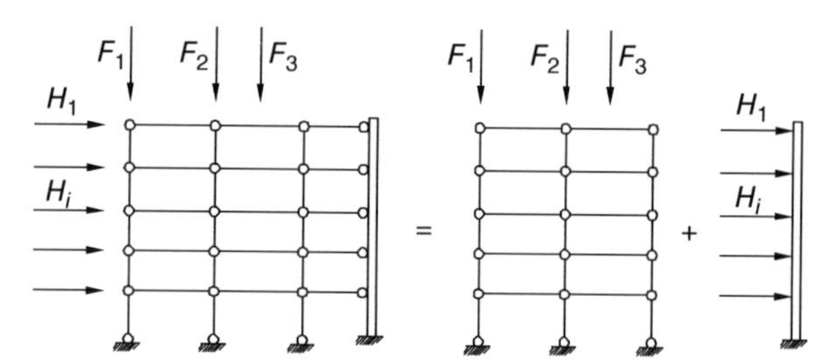

Figure 3.5 Simplified models for the design of braced frames.

3.3 Classification Based on Lateral Deformability

The distinction between sway and no-sway frames is associated with frame lateral stability, that is with the relevance of the second order effects on the structural response in terms of transverse displacements (and consequently, also in terms of additional bending moments and shear actions). From a purely theoretical point of view, any unbraced frame has to be considered, strictly speaking, to be a sway frame, characterized by elements affected by mechanical and geometric imperfections (3.6). For each load condition, transversal displacements are hence expected. Reference is therefore made to the relevance of second order effects and the frame can be classified as:

- *a no-sway frame*: if the lateral displacements are so small to give a negligible increase of internal forces and moments. This situation typically occurs when, in the absence of bracing systems, columns are characterized by great values of the moment of inertia or transverse forces are very small;
- *a sway frame*: if the transverse displacements influence significantly the values of internal forces and moments (especially, shear forces and bending moments). This situation typically occurs when, in the absence of bracing systems, the columns are very slender or very high lateral loads act on the frame.

From an engineering point of view, the second order effects are generally considered negligible if they are less than 10% of those resulting from a first order analysis, which is based on the assumption that internal forces and moments can be determined with reference to the undeformed configuration.

As an example, the first order top transverse displacement (δ) of the cantilever beam in Figure 3.6, loaded by axial (N) and horizontal (F) forces, is given by:

$$\delta = \frac{Fh^3}{3EI} \tag{3.1}$$

On the other hand, the bending moment M at the base of the column (usually by means of equilibrium criteria applied to the non-deformed configuration) assumes the value:

$$M = Fh \tag{3.2a}$$

Furthermore, with reference to the deformed configuration, the bending moment at the fixed end can be expressed as:

$$M = Fh + N\delta = Fh + N\frac{Fh^3}{3EI} \tag{3.2b}$$

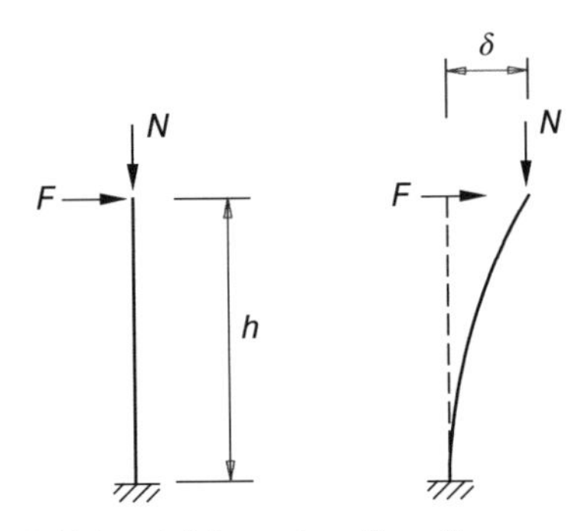

Figure 3.6 Cantilever beam in the initial and deformed configuration.

On the basis of the definition previously introduced, second order effects are not negligible if, with reference to the critical section (i.e. the one restrained at the bottom of the cantilever), the following condition is satisfied:

$$N\frac{Fh^3}{3EI} > 0.1 \cdot (Fh) \tag{3.3}$$

If the frame is braced, then it can be considered to be a no-sway frame, that is the additional internal forces or moments due to lateral displacements can be neglected.

It is important to note that there is no equivalence between the terms *braced frame* and *no-sway frame*, because they refer to two different aspects of the structural behaviour. The former is associated with the structural strength and provides guidance related to the relevant mechanism to transfer of horizontal forces; the latter is related to the transversal deformability.

3.3.1 European Procedure

European provisions related to the methods of analysis are reported in the general part of Eurocode 3 part 1-1 (EN 1993-1-1). A design based on an overall first order analysis is admitted and the frame has to be considered to be a no-sway frame for a given load condition if the value of the elastic critical load multiplier, α_{cr}, fulfils the following conditions, depending on the type of structural analysis:

- *Elastic analysis:*

$$\alpha_{cr} = \frac{F_{cr}}{F_{Ed}} \geq 10 \tag{3.4a}$$

- *Plastic analysis:*

$$\alpha_{cr} = \frac{F_{cr}}{F_{Ed}} \geq 15 \tag{3.4b}$$

where F_{Ed} is the total vertical design load for the considered loading condition and F_{cr} is the elastic critical load associated with the first anti-symmetric deformed buckling shape.

It should be noted that term, F_{cr}, or equivalently the load multiplier, α_{cr}, which govern Eqs. (3.4a) and (3.4b), can be evaluated directly by means of the use of finite element (FE) buckling

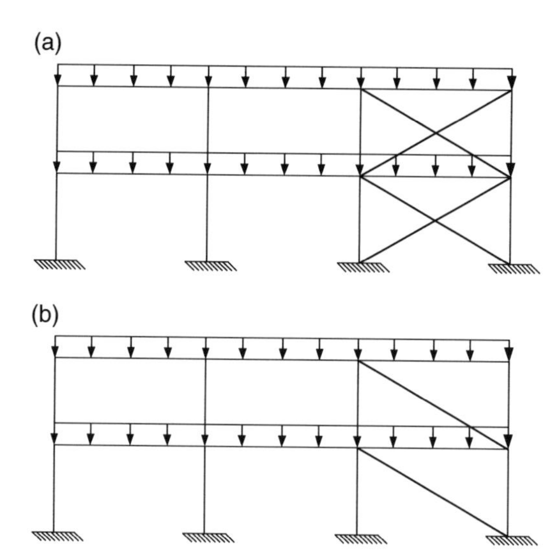

Figure 3.7 Example of planar braced frame with diagonal members resisting to sole tension: (a) the frame and (b) model for the structural analysis.

analysis, which requires the knowledge of the elastic $[K]_E$ and geometric $[K]_G$ stiffness matrices. The latter depends strictly on the value of the axial load in the elements, despite the fact that, in some refined formulations, the dependence from shear forces, bending and torsional moments has also been implemented. It is worth noticing that special care should be paid to the interpretation of the buckling analysis results and, as an example, the frame of Figure 3.7 can be considered, where the X-bracing system is composed by diagonal members only resistant to tension (Figure 3.7a). Most common FE commercial software packages do not consider the fact that, when the compressed diagonals buckle, the structure is, however, safe because of the resistance of the tension diagonals.

Owing to the lack of a FE mono-lateral truss element resisting to tension only (i.e. a rope element) in the library of the most commonly used FE analysis software, a truss element resisting both to tension and compression is usually adopted by designers to model the diagonals (Figure 3.7b). The elastic critical load multipliers associated with deformed shapes similar to the ones in Figure 3.8 could be obtained, which are related to buckling shapes out of any interest for design purposes: when the diagonals in the model buckle, however, the structure is safe due the presence, in each panel, of the other diagonal in tension (not modelled). As a consequence, the sole buckling deformed shape of interest for designers in the one related to the overall buckling mode presented in Figure 3.9.

A simplified approach, which is based on the studies conducted by Horne between 1970 and 1980 and is nowadays proposed by modern steel design codes, allows us to assess the elastic critical load, F_{cr}, of a sway frame (Figure 3.10), regular in plant and elevation, as:

$$F_{cr} \cong \min\left(\frac{h \cdot H}{\delta}\bigg|_i\right) \tag{3.5}$$

where δ is the inter-storey drift evaluated by means of a first order analysis, h is the inter-storey height and H is the resulting horizontal force at the base of the inter-storey (horizontal reaction at the bottom of the storey of both the horizontal loads and fictitious horizontal loads).

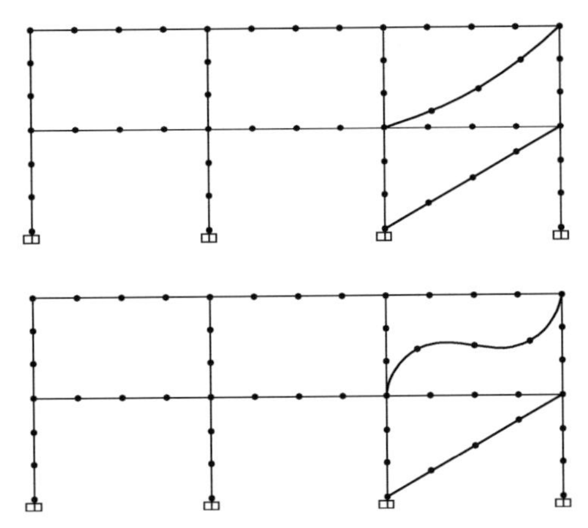

Figure 3.8 Typical buckling modes that are not relevant for design purposes.

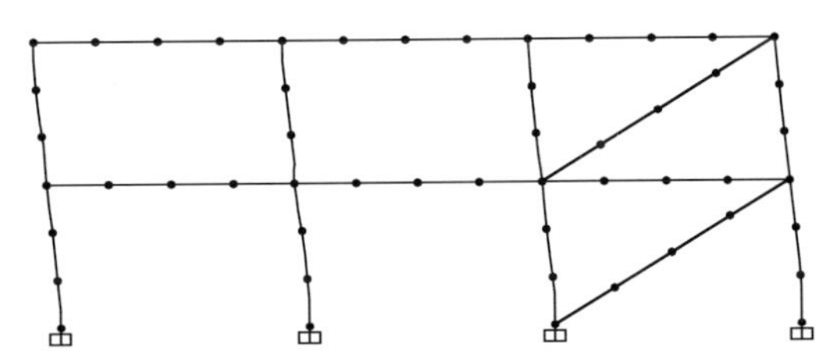

Figure 3.9 Deformed configuration due to lateral frame instability.

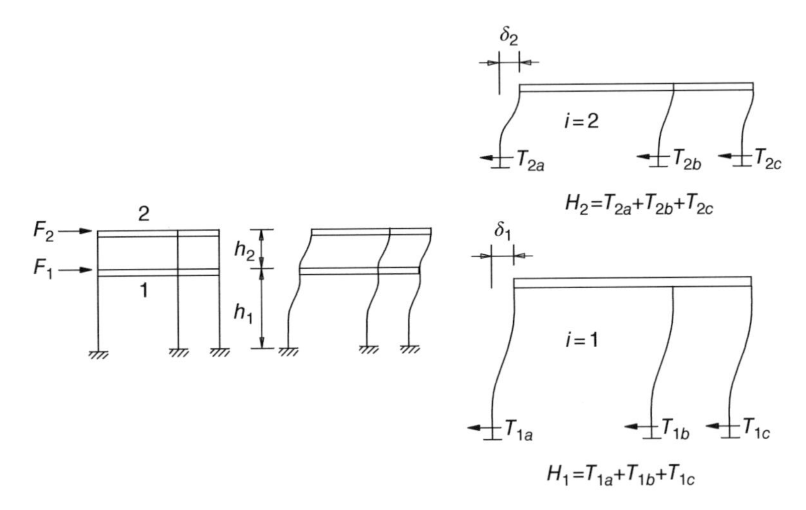

Figure 3.10 Application of Horne's method.

The frame is classified as sway-frame and, as a consequence the deformed shape does not significantly affect the values of internal forces and bending moments as well as the frame reactions, if, for each storey:

$$\text{using elastic analysis}: \frac{F_{Ed}}{F_{cr}} \cong \max\left(\left.\frac{\delta \cdot V}{h \cdot H}\right|_i\right) \le 0.1 \tag{3.6a}$$

$$\text{using plastic analysis}: \frac{F_{Ed}}{F_{cr}} \cong \max\left(\left.\frac{\delta \cdot V}{h \cdot H}\right|_i\right) \le 0.067 \tag{3.6b}$$

3.3.2 AISC Procedure

The AISC Specifications admit an approach based on first order theory when all the following assumptions are satisfied:

(a) the structure supports gravitational loads through vertical elements and/or walls and/or frames;

(b) in all stories the following condition is fulfilled:

$$\alpha \cdot \Delta_{second\text{-}order,\max} / \Delta_{first\text{-}order,\max} \le 1.5 \tag{3.7a}$$

where $\Delta_{second\text{-}order,\max}$ is the maximum second order drift, $\Delta_{first\text{-}order,\max}$ is the maximum first order drift, and $\alpha = 1$ for load and resistance factor design (LFRD) and $\alpha = 1.6$ for allowable stress design (ASD);

(c) the required axial compressive strengths of all members whose flexural stiffness contributes to lateral stability of the structure satisfy the limitation:

$$\frac{\alpha P_r}{P_y} \le 0.5 \tag{3.7b}$$

where P_r is the required axial compression strength, computed using LFRD or ASD load combinations and P_y $(= F_y A_g)$ is the axial yield strength.

A more detailed illustration of AISC methods for overall analysis, based on first order as well as second order theory, is shown in Chapter 12, Section 12.3.

3.4 Classification Based on Beam-to-Column Joint Performance

The degree of flexural continuity associated with the beam-to-column joints influences significantly the response of the whole structural system. In detail, as discussed in Section 15.5, reference is made to the joint response in terms of M–Φ curve, which is intended to be the relationship between the moment M at the beam end and the relative rotation Φ between the beam and column (Figure 3.11a). The following frame typologies can be identified:

simple frame: each joint can be modelled via a perfect hinge allowing a relative rotation between the beam end and the column without any transmission of bending moment (curve a in Figure 3.11b). In this case, due to the presence of hinges in correspondence of each

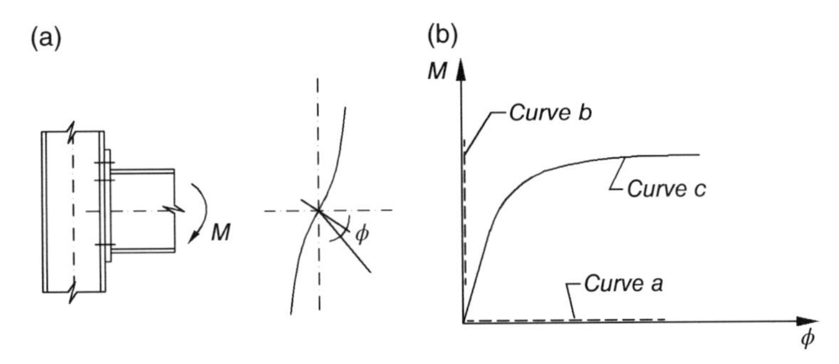

Figure 3.11 (a) Definition of the moment and of the rotation for a joint. (b) Typical moment-rotation relationships.

beam-to-column joint, a specific bracing system is always required to provide lateral stability to the frame, as discussed more in detail in Section **3.7**;

rigid frame or frame with rigid nodes: each joint does not allow any relative rotation between the beam end and the column and bending moments are transferred at the joint locations (curve *b* in Figure 3.11b);

semi-continuous frame; that is, a frame with semi-rigid joints: each joint allows a relative rotation between the beam end and the column transmitting bending moments at the joint location (curve *c* in Figure 3.11b).

From the theoretical point of view, as widely observed one century ago experimentally and reported in detail in technical literature, each type of beam-to-column joint is characterized by a well-defined degree of flexural stiffness and bending resistance. Nowadays, the semi-continuous frame model is included in the most updated standard steel codes, such as the European and the United States specifications. Furthermore, it should be noted that the influence of the actual joint behaviour could sometimes be irrelevant for a safe design and, as a consequence, the ideal models of simple and rigid frames still maintain their validity for a quite wide class of structures. From a practical point of view, the choice of the joint design model depends on degree of flexural continuity provided by joints, which is influenced not only by the joint detailing (see Chapter 15) but also by the whole structure and, in particular, by the characteristics of the beam connected by joints.

3.4.1 Classification According to the European Approach

Eurocode 3 deals with beam-to-column joint classification in its part 1-8 (EN 1993-1-8: Eurocode 3: Design of steel structures – Part 1-8: Design of joints). In particular, the joint classification depends on joint performances in terms of rotational stiffness (S_j). On the basis of the value of the initial joint rotational stiffness ($S_{j,ini}$) the following types of joints can be identified:

- *rigid joints* (region 1 in Figure 3.12) if:

$$S_{j,ini} \geq k_b \frac{EI_b}{L_b} \tag{3.8a}$$

where E is the modulus of elasticity, I_b and L_b are the moment of inertia and the length of the beam, respectively, and term k_b takes into account the type of frame ($k_b = 8$ for braced frames and $k_b = 25$ unbraced frames);

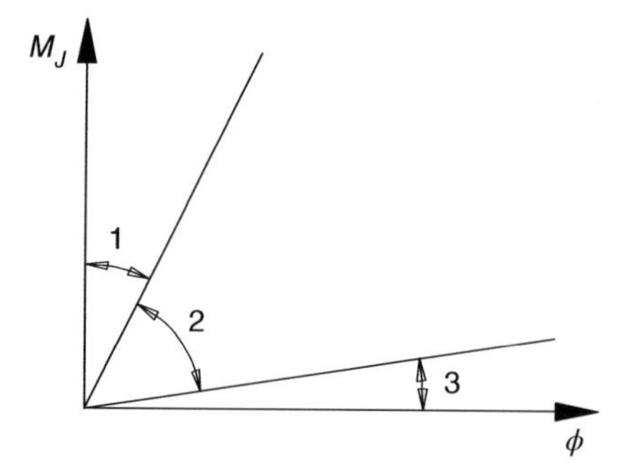

Figure 3.12 Domains for joint classification.

- *semi-rigid* (region 2 in Figure 3.12) if:

$$0.5\frac{EI_b}{L_b} \leq S_{j,ini} \leq k_b\frac{EI_b}{L_b} \tag{3.8b}$$

- *pin or flexible joints* (region 3 in Figure 3.12) if:

$$S_{j,ini} \leq 0.5\frac{EI_b}{L_b} \tag{3.8c}$$

Furthermore, joints are also classified on the basis of their bending resistance ($M_{j,Rd}$), which is compared with that of the connected beam ($M_{pl,Rd}$). The following types of joints can be identified:

- full strength joint, if:

$$M_{j,Rd} \geq M_{pl,Rd} \tag{3.9a}$$

- partial strength joints, if:

$$0.25M_{pl,Rd} \leq M_{j,Rd} \leq M_{pl,Rd} \tag{3.9b}$$

- pin or flexible joints, if:

$$M_{j,Rd} \leq 0.25M_{pl,Rd} \tag{3.9c}$$

Both stiffness and resistance joint classification criteria can be usefully linked. From the joint moment-rotation (M–Φ) relationship, it appears to be convenient to make reference to the $\bar{m} - \bar{\phi}$ relationship where non-dimensional moment ($\bar{m}$) and rotation ($\bar{\phi}$) are, respectively, defined as:

$$\bar{m} = \frac{M}{M_{pl,Rd}} \tag{3.10a}$$

$$\bar{\phi} = \phi\frac{EI_b}{L_b M_{pl,Rd}} \tag{3.10b}$$

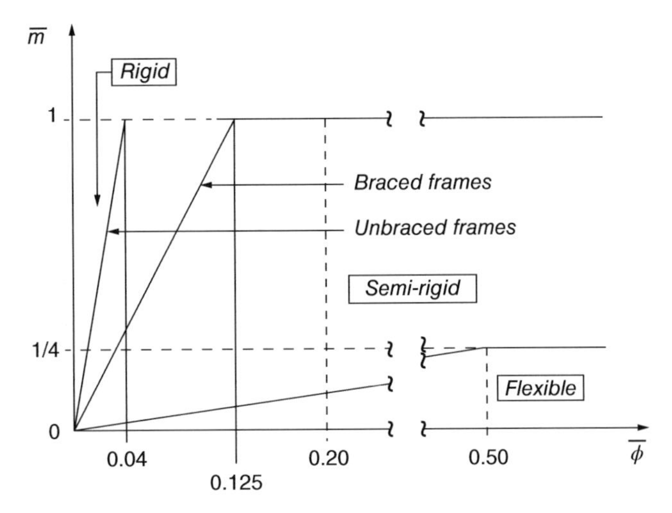

Figure 3.13 European joint classification criteria.

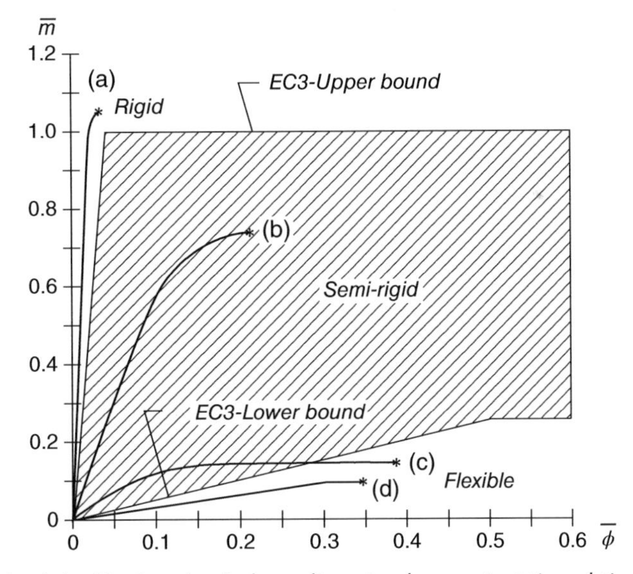

Figure 3.14 Example of classification of typical non-dimensional moment-rotation relationships.

The $\bar{m}-\bar{\phi}$ joint curve has to be compared with the code requirements for stiffness and resistance, which allow for the definition of the three separated regions presented in Figure 3.13 and refer to both cases of braced and unbraced frames.

Examples of joint classification are presented in Figure 3.14 where it is possible to identify:

curve (a): rigid full strength joint;
curve (b): semi-rigid partial strength joint;
curve (c): semi-rigid joint for stiffness and pin joint for resistance;
curve (d): pin joint.

3.4.2 Classification According to the United States Approach

Joint classification is reported in the Chapter B.6 'Design of Connections' of AISC 360-10. As Eurocode 3, AISC Specifications identify three different types of beam-to-column connection:

(a) *Simple connections*, transmitting a negligible bending moment (pin joints);
(b) *Fully restrained (FR) moment connections*, transferring bending moment with a very negligible rotation between the beam and the column at the joint location (rigid joints);
(c) *Partially restrained (PR) moment connections*, transferring bending moment but the rotation between the connected members is non-negligible (semi-rigid joints).

More details can be found in the AISC Commentary where the classification of joints is proposed similarly to Eurocode 3 on the basis of joint response in terms of rotational stiffness (K) and flexural bending resistance (M_n). Furthermore, unlike Eurocode, AISC suggests to compute instead of the initial stiffness the secant stiffness evaluated at service load, because the last one is considered more representative of the actual behaviour of the connection. The secant stiffness (pedix S) is defined as:

$$K_S = \frac{M_S}{\theta_S} \tag{3.11}$$

where θ_S is the rotation experienced when service loads act.

By expressing all the parameters in S.I. units, depending on the value of the secant joint rotational stiffness (K_S) the following type of joints can be identified:

- *fully restrained (rigid) connections* (zone 1 in Figure 3.15) if:

$$K_S \geq 20\frac{EI_b}{L_b} \tag{3.12a}$$

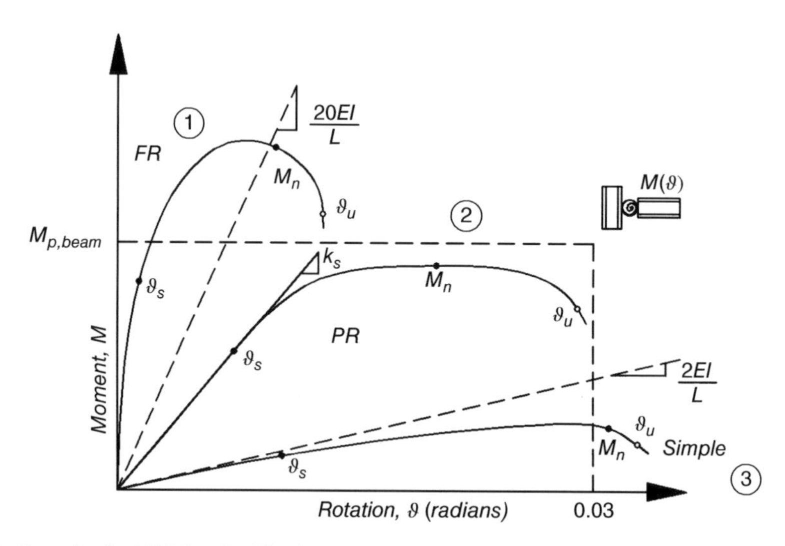

Figure 3.15 Domains for US joint classification.

where E is the modulus of elasticity, I_b and L_b are the moment of inertia and the length of the beam, respectively;

- *partially restrained (semi-rigid) connections* (zone 2 in Figure 3.15) if:

$$2\frac{EI_b}{L_b} \leq K_S \leq 20\frac{EI_b}{L_b} \tag{3.12b}$$

- *simple (pin or flexible) connections* (zone 3 in Figure 3.15) if:

$$K_S \leq 2\frac{EI_b}{L_b} \tag{3.12c}$$

Furthermore, joints are classified also on the basis of their bending resistance (M_n). AISC states that M_n can be theoretically evaluated on the basis of a suitable ultimate limit-state model or directly obtained from experimental test on specimen adequately representative of the beam-to-column joints. If moment-rotation curve does not show a defined peak load, then M_n can be taken as the moment at the rotation of 0.02 rad (20 mrad).

According to strength and ductility (rotation capacity), joint performance is compared with the plastic moment of the beam ($M_{p,beam}$) and the following types of connections are identified:

- full strength connections, if:

$$M_n \geq M_{p,beam} \tag{3.13a}$$

- partial strength connections, if, in correspondence to the value of rotation of 0.02 rad:

$$0.20M_{p,beam} \leq M_n \leq M_{p,beam} \tag{3.13b}$$

- pin or flexible connections, if, in correspondence of the value of rotation of 0.02 rad:

$$M_n \leq 0.20M_{p,beam} \tag{3.13c}$$

With reference to moment connections (full or partial strength), AISC stresses the importance of guaranteeing an adequate level of ductility (rotation capacity) not only for seismic zones but also for connections under static loading, due to the levels of rotation sometimes required by the plastic analysis approach (see Section *3.6.1*) In particular, connection rotation capacity is defined as the rotation exhibited when the strength of the connection is dropped to 80% of M_n, or 0.03 rad (30 mrad) if there is no significant drop of M_n for larger rotations. The rotation limit of 0.03 rad is not only required for special moment frames in seismic provisions but also for moment-resisting frames under static loads.

3.4.3 Joint Modelling

The assumption of the semi-continuous frame model leads to a distribution of internal forces and bending moments, which is intermediate between those associated with the simple and rigid frame models. Figure 3.16 refers to a portal frame under gravity beam load: for the three different

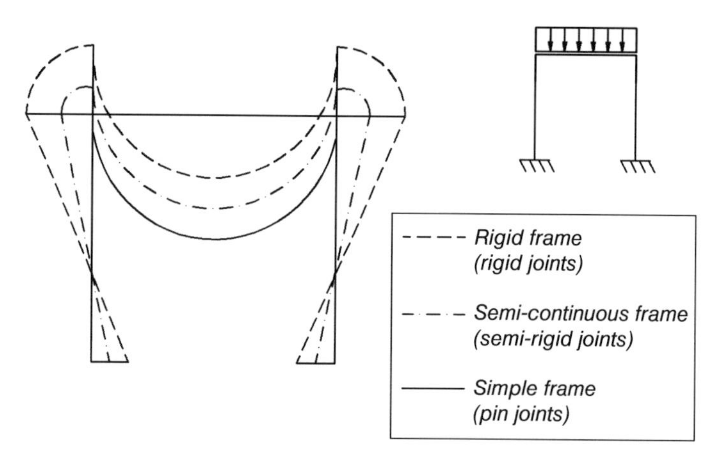

Figure 3.16 Influence of the degree of continuity provided by beam-to-column joints on the distribution of internal moments.

beam-to-column joint types (i.e. hinged, semi-rigid and rigid joint) the distribution of the bending moment diagram along the beam and the columns are represented in the same figure. It can be noted that:

- in the case of the rigid frame model, a severe distribution of bending moments acts on columns, while the bending moment distribution on the beam underestimates the mid-span moment and overestimates the beam end moments. In this case also the overall frame stiffness is significantly over-estimated, leading to neglect of the importance of second order effects;
- in the case of a simple frame, the distribution of bending moments on the beam is the most severe with respect to the distribution associated with the semi-continuous model, leading to overestimate the midspan moment but it implies the design of columns subjected to sole axial load (i.e. bending moments on columns are neglected);
- in the case of a semi-continuous frame, that is considering the actual joint behaviour as previously mentioned, the moment distribution is contained within the range associated with the simple and rigid frame model bending distributions. Furthermore, it is important to remark that this approach allows for the most correct design of all the structural components, hence leading to an optimal use of the material but also to an increase of safety degree of the design. Figure 3.17 proposes some practical solutions to model beam-column semi-rigid joints, which are characterized by different degrees of accuracy and complexity. A simplified but efficient approach, especially when high levels of accuracy are not required in the analysis, consists of the use of an equivalent short beam (Figure 3.17a). Its moment of inertia I_b can be defined assuming the equivalence of the flexural beam behaviour and the rotational stiffness one via the relationship:

$$I_b = \frac{S \cdot L_b}{E} \tag{3.14}$$

where L_b is the length interested by the joints, which is typically considered half of the column width ($h_c/2$), E is the Young's modulus and S is the joint stiffness.

More refined formulations are available in recent finite element analysis packages with a library offering directly the rotational spring element (Figure 3.17b) or allowing for non-linear analysis, hence making it possible to correctly simulate joint behaviour, eventually also including the rigid behaviour of the nodal panel zone (Figure 3.17c).

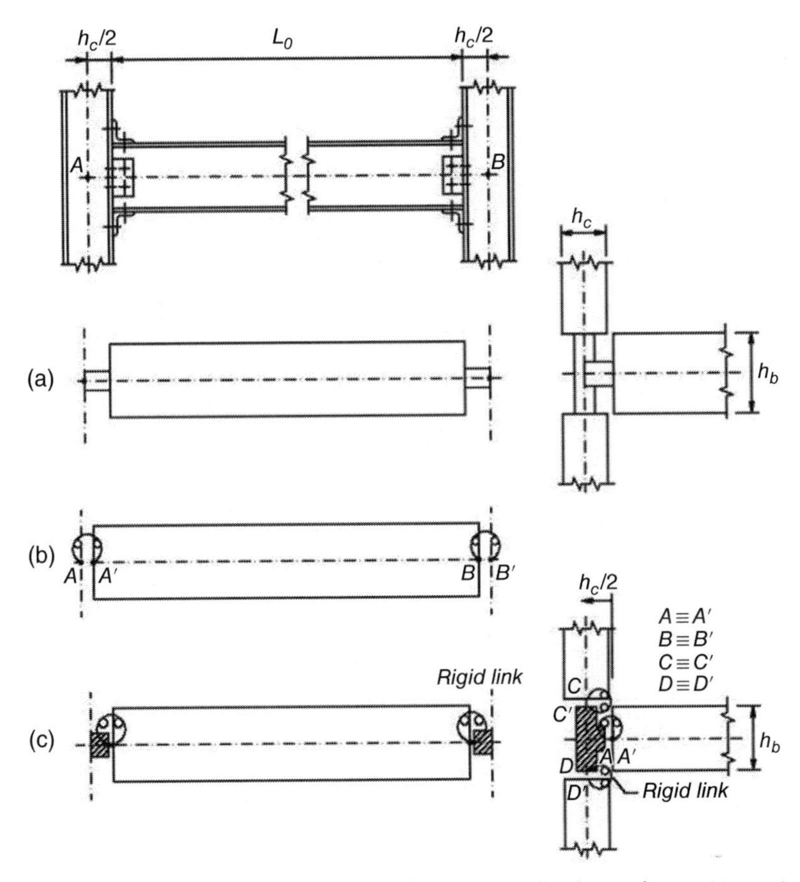

Figure 3.17 Approaches for the modelling of semi-rigid joints: equivalent beam element (a), rotational spring (b) and springs plus rigid bars (c).

3.5 Geometric Imperfections

As already mentioned in Section 1.6.2, *geometric imperfections* can be subdivided into cross-sectional, member and structural system imperfections.

In the following, main requirements of the considered design codes are proposed.

3.5.1 The European Approach

In accordance with the European design procedure, the initial out-of-straightness is considered by means of a system of equivalent horizontal forces. Defining e_0 as the maximum out-of-straightness imperfection with respect to the ideal configuration (Figure 3.18), it is possible to obtain a system of uniformly distributed loads of a magnitude q that can generate a maximum bending moment equal to the moment that would be caused by imperfections in the presence of axial forces N_{Ed} (calculated as $N_{Ed} \cdot e_0$), as follows:

$$q = \frac{8 e_0 N_{Ed}}{L^2} \tag{3.15}$$

Depending on the analysis method used and on the choice of stability curve (see Chapter 6), the EC3 reference values for this imperfection are shown in Table 3.1. It should be noted that the member imperfection values should be reduced by the National Annex, via a factor k, if lateral-torsional buckling is accounted for in second order analysis. EC3 recommends use of $k = 0.5$ in the absence of a more detailed value.

Eurocode 3 allows us to neglect the effects of out-of-straightness imperfections when large lateral forces are applied to the structural system. In particular, for sway frames, the effects of imperfections can be neglected if:

$$N_{Ed} < 0.25 N_{cr} \tag{3.16}$$

in which N_{Ed} is the axial force acting on the element and N_{cr} is the critical elastic buckling load for the member.

Structural system imperfections can be ascribed to various causes, such as variability in lengths of the framing members, lack of verticality of the columns and of horizontality of the beams, errors in the location of the base restraints, errors in the placement of the connections and so on.

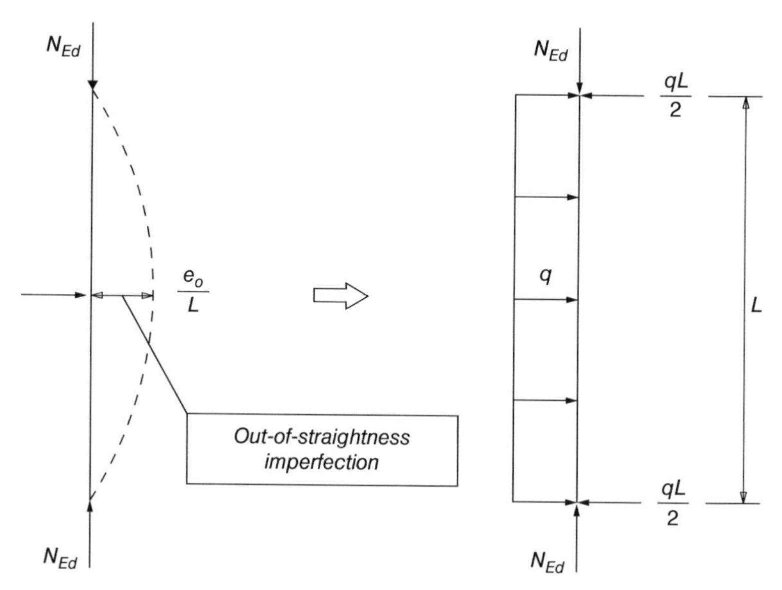

Figure 3.18 Horizontal forces equivalent to out-of-straightness imperfection.

Table 3.1 Maximum values of member imperfections.

Stability curve	e_0/L (global elastic analysis)	e_0/L (global plastic analysis)
a_0	1/350	1/300
a	1/300	1/250
b	1/250	1/200
c	1/200	1/150
d	1/150	1/100

As already mentioned in Section 1.6.2, structural system imperfections must be taken into account carefully during the global analysis phase. This can be done in a simplified way by adding fictitious forces (notional loads) applied to the structure to reproduce suitably the effects of imperfections. For example, the lack of verticality of columns in sway frames is considered by adding horizontal forces to the perfectly vertical columns (Figure 3.19). These horizontal forces are proportional to the resultant vertical force F_i acting on each floor.

This design simplification can be explained considering a cantilever column of height h with an out-of-plumb imperfection and subject to a vertical force N acting at the top. The additional bending moment M that develops due to the lack of verticality (Figure 3.20) can be approximated at the fixed end as:

$$M = N[h \cdot \tan(\varphi)] \tag{3.17}$$

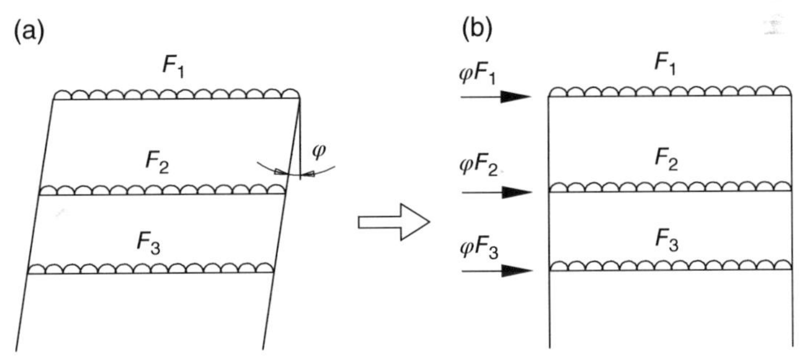

Figure 3.19 Horizontal notional loads equivalent to the imperfections to a sway frame. Imperfect frame (a) and horizontal equivalent forces (b).

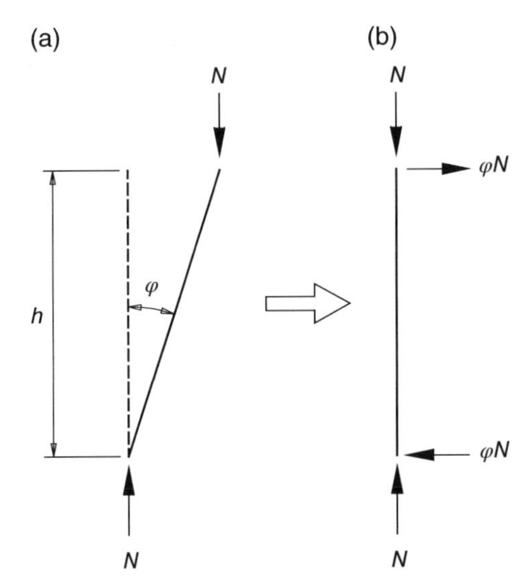

Figure 3.20 Horizontal forces equivalent to the imperfection. Imperfect column (a) and horizontal equivalent forces (b)

Within the small displacement hypothesis (thus approximating $tan(\varphi)$ with the angle φ itself), the effect of the imperfection can be assimilated to that of a fictitious horizontal force F acting at the top of the member and causing the same bending moment at the base of the column. The magnitude of F is thus given by:

$$F = \frac{M}{h} = N\varphi \tag{3.18}$$

The most recent specifications define a global imperfection, in terms of an error of verticality. With reference to EC3, it is expressed by the out-of-plumb angle Φ calculated as:

$$\Phi = \Phi_0 \alpha_h \alpha_m \tag{3.19}$$

Term Φ_0 is assumed equal to 0.005 rad (1/200 rad) and coincides with the value that was traditionally used for steel design.

Coefficients α_h and α_m are reduction factors (not greater than unity) that account for the small probability of all imperfections in the structure adding up unfavourably. Coefficient α_h is defined as a function of the total building height h (in metres), with a limiting range between 0.67 and 1:

$$\alpha_h = \frac{2}{\sqrt{h}} \leq 1 \tag{3.20}$$

Coefficient α_m accounts for the number of bays and it is defined as:

$$\alpha_m = \sqrt{\frac{1}{2}\left(1 + \frac{1}{m}\right)} \tag{3.21}$$

in which m is the number of columns in the frame subject to a design axial force no less than 50% of the average axial force in all columns.

For the evaluation of the effects of the imperfections on the alignment of columns in frames, it is possible to make reference to the calculation scheme in Figure 3.21.

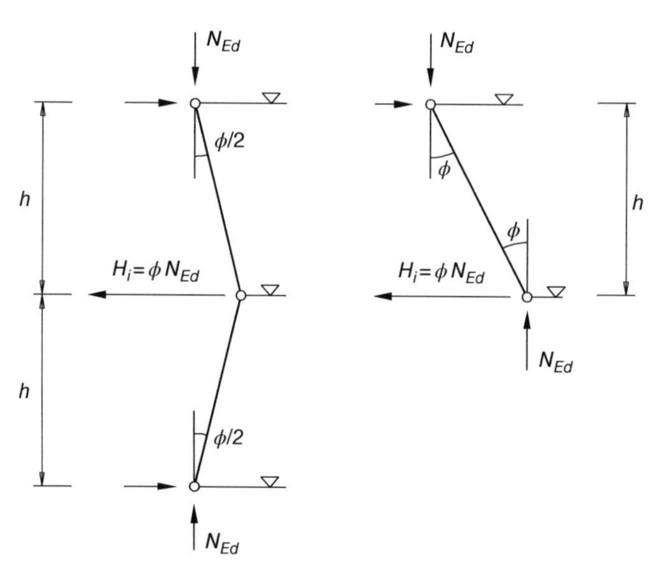

Figure 3.21 Effect of alignment imperfections of columns.

In case of frames subject to large horizontal forces, the effects of the imperfections can be neglected because of the relative magnitude of their effects with respect to the effects of the lateral loads. In particular, if H_{Ed} represents the resultant horizontal forces at the base of all columns in one floor and Q_{Ed} the resultant vertical force acting at the base of all columns in that floor, the imperfections can be neglected if:

$$H_{Ed} \geq 0.15 Q_{Ed} \tag{3.22}$$

3.5.2 The United States Approach

The approach followed by the ANSI/AISC 360-10 Specification, which is used in the United States, is in part quite similar to the one recommended by Eurocode 3. In particular, the following distinction has been made:

- *local imperfections*, such as *out-of-straightness of individual members* are not explicitly taken into account into design analysis, but rather their effects are implicitly included into design equations that involve stability checks;
- *system imperfections*, the most important of which is certainly the out-of-plumb imperfection of columns are explicitly addressed in the Specification.

The US Specification allows the designer to directly include structural system imperfections in the analysis by modelling the structure with the largest admissible shifts (looseness) of the beam-column nodes in the directions that would most affect global stability of the system. As an alternative to the direct modelling of the system imperfections via the use of non-straight members, the Specification also allows the designer to account for them by means of a set of notional loads, applied to the structure, and corresponding to the effects of the imperfections on the stability of the structure. The nominal load approach is valid for any structure, but the Specification provides guidance only for structures that carry gravity forces predominantly through vertical elements, such as columns, walls or frames.

The value of the notional loads, similar to what is prescribed in Eurocode 3, is as follows:

$$N_i = 0.002 Y_i \tag{3.23}$$

in which N_i is the notional load applied at level i and Y_i is the total gravity load applied at level i. The 0.002 coefficient corresponds to an initial out-of-plumbness ratio of a column of 1/500 (associated with an angle of 2 mrad). The Specification allows for designer to use a different value of the ratio if it can be appropriately justified. These notional loads must be applied to the structure in a configuration such that it will have the most influence on the stability of the building. The Specification also allows the designer to apply the notional loads to combinations including gravity loads only, and not lateral loads, for situations in which the ratio of maximum second order drift to maximum first order drift is equal to or less than 1.7. In addition to the consideration of initial imperfections, the Direct Analysis Method outlined in the Specification (see Section 12.3.1) also requires that the stiffness of all elements that contribute to the stability of the structure must be reduced to 80%. Furthermore, an additional reduction factor must be applied to the flexural stiffness of all members contributing to the stability of the structure (e.g. beams and columns in a fully restrained moment-resisting frame). This additional reduction factor, indicated by τ_b, is a function of the ratio of the required axial compressive strength to the yielding compressive strength of the member considered. When the axial force demand is less or equal to 50% of the axial yielding strength of the member, no further reduction is necessary (i.e. $\tau_b = 1$). If the demand is higher, then the stiffness reduction factor is calculated as:

$$\tau_b = 4\frac{P_r}{P_y}\left[1 - \frac{P_r}{P_y}\right] \tag{3.24}$$

As an alternative to using the stiffness reduction factor τ_b, the Specification also allows us to apply an additional notional load equal to $0.001Y_i$ at all levels and for all load combinations.

3.6 The Methods of Analysis

Structural analysis is a very important phase of the design aimed at the evaluation of displacements as well as internal forces and bending and torsional moments associated with the most significant load combinations. As a consequence, appropriate analysis approaches characterized by different degrees of accuracy and complexity have to be selected by designers, depending on the importance of the framed system, as well as by the required level of structural safety.

The most simple methods of analysis are the *first order elastic methods*, that is the ones based on the assumptions of a linear elastic constitutive law of the material, small displacements and infinitesimal deformations: as an example, the slope deflection method is the most commonly used method belonging to this family and the virtual work principle is the most known by engineers. In accordance with these methods, the internal forces and moments on the members of the structure are determined by making reference to the undeformed configuration, that is, only the equilibrium equations are used neglecting the compatibility conditions, or more practically, the effects of deformations. The assumptions adopted by the first order elastic methods significantly simplify the behaviour of the structures, because of the presence of mechanical and geometrical non-linearities. In particular, in several cases it is not possible to ignore:

- *the mechanical non-linearity*, which is due to the actual material constitutive law, already introduced in Section 1.1 with reference to the material constitutive law. The stress-strain curve is in fact typically non-linear, assumed in a simplified way as an elastic-perfect plastic relationship or as an elastic-plastic with strain hardening. Furthermore, beam-to-column and base-plate joint moment-rotation curves are typically non-linear.
- *the geometrical non-linearity*, which is due to the slenderness of the structure or of its components: the second order effects are the consequences of the lateral displacements of the structures that can increase significantly shear forces and bending moments with respect to the one obtained by imposing equilibrium conditions on the undeformed structure.

The layout of the types of analysis which can be currently executed for routine design is presented in Figure 3.22.

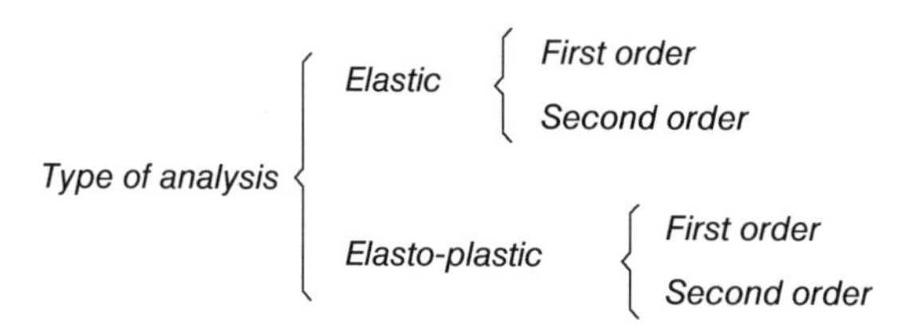

Figure 3.22 Layout of the types of structural analysis.

3.6.1 Plasticity and Instability

The mechanical non-linearity significantly influences the response of both elements and structural systems. With reference to the simply supported beam in Figure 3.23, for which an elastic perfectly plastic constitutive law is assumed for the steel material, the distribution of stress and strain in the generic cross-section can be obtained from the well-known St Venant's theory and it is bi-triangular (distribution 1) in the elastic range. The load can be further increased up to the yielding stress, which corresponds to the achievement of the elastic moment of the cross-section (M_{el}). As the load is increased further, the spread of the plasticity in the cross-section has direct influence on the increase of deformability of the member because of the presence of plastic zones (where the Young's modulus is considered to be zero). The cross-sectional plastification process continues with the effect that the yielded area becomes larger and larger, spreading in towards the centre of the cross-section (distributions 3 and 4, being very limited with reference to the elastic one), up to the achievement of the beam plastic moment (M_{pl}). The corresponding stress distribution is characterized by the plasticity spread across the whole cross-section, which corresponds to the formation of a plastic hinge (distribution 4), that is a hinge which is activated when M_{pl} is reached.

When the plastic hinge at the mid-span cross-section is activated, the structure becomes statically undetermined, having three hinges along a straight line and the load corresponding to the formation of this third hinge being the maximum one associated with the resistance of the beam; that is this load value represents the load carrying capacity of the beam. The benefits in term of increment of the carrying capacity with reference to the classical elastic approach, limiting the load carrying capacity to the load value associated with M_{el}, is directly represented by the ratio M_{pl}/M_{el}.

Furthermore, it should be noted that more relevant benefits in terms of increment of the load carrying capacity could be obtained with reference to statically indeterminate structures. As an example, the fixed end beam in Figure 3.24a can be considered, which is subjected to a uniformly

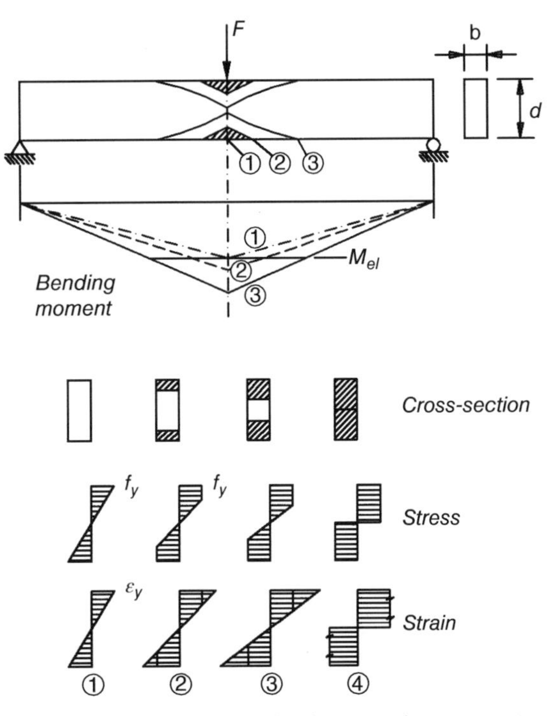

Figure 3.23 Simple supported beam: stress and strain distribution in the cross-section at the mid-span.

distributed load q increasing from zero until collapse is achieved. A perfect rigid plastic behaviour is assumed to represent the moment-curvature (M–χ) beam relationship (Figure 3.24b).

In the elastic range, the bending moment is characterized by a parabolic distribution with the end moment values twice the mid-span moment. By increasing the load, bending moment values increase proportionally till the value of the plastic beam moment, M_{pl}, is reached at the beam ends, which can be associated with the load q_{cp} defined as:

$$q_{cp} = \frac{12M_{pl}}{L^2} \tag{3.25}$$

The bending moment distribution, corresponding to this situation is presented in Figure 3.25. At the beam ends, two plastic hinges have been activated and the beam now can be considered statically determinate.

With reference to the fixed-end beam in the Figure 3.23, it should be noted that when the load activates these first plastic hinges, the structure does not collapse. Due to the presence of these two plastic hinges the deformability is significantly increased with reference to the one associated with the initial built-up one in elastic range. A further increment of the uniform load, Δq, can be sustained until the plastic beam bending resistance of the mid-span cross-section is achieved (Figure 3.26), which corresponds to a load increment Δq_u, defined as:

$$\Delta q_u = \frac{4M_{pl}}{L^2} \tag{3.26a}$$

When this third plastic hinge is activated, three hinges are located along the longitudinal beam axis, which corresponds to a complete collapse mechanism (Figure 3.27). Collapse load, q_u, is hence defined as:

$$q_u = q_{cp} + \Delta q_u = \frac{16M_{pl}}{L^2} \tag{3.26b}$$

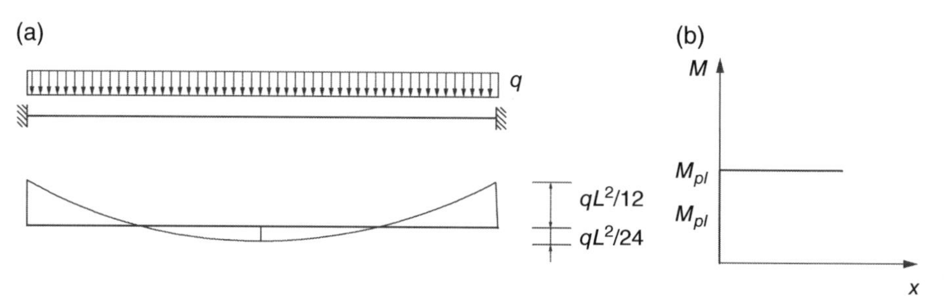

Figure 3.24 Bending moment distribution in a built-in beam (a) and typical moment-curvature relationship of the beam cross-section (b).

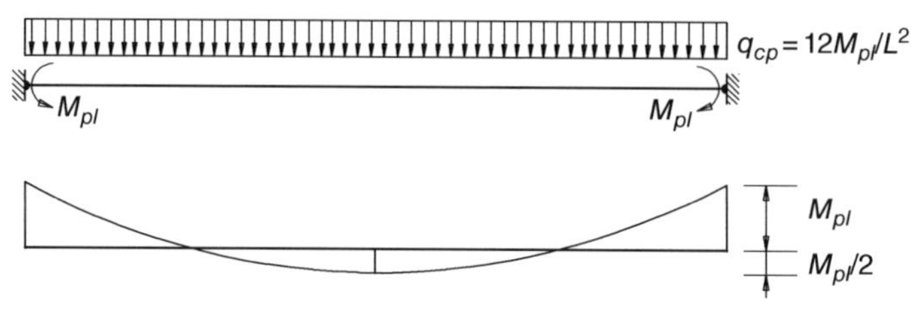

Figure 3.25 Bending moment diagram at the formation of the first two plastic hinges.

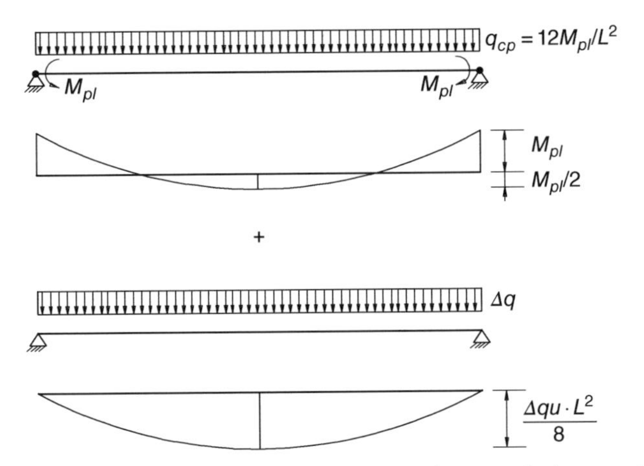

Figure 3.26 Bending moment distribution after the first two plastic hinges at the beam ends.

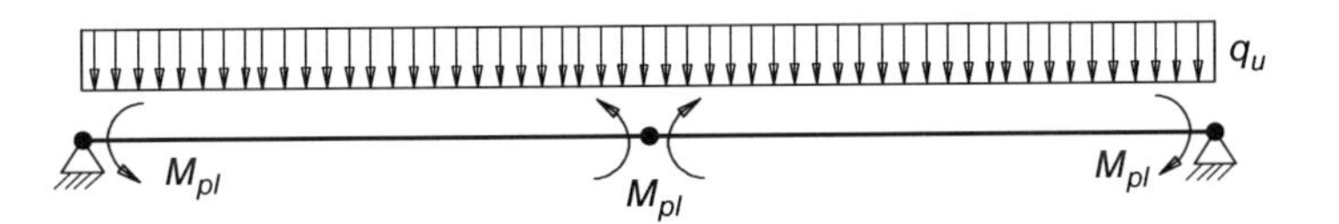

Figure 3.27 Collapse mechanism for the built-in beam.

In case of statically indeterminate structures, like the one in this example, it should be noted that a complete collapse mechanism could be developed only if the member cross-section where the plastic hinge develops has an adequate ductility level, that is, it is able to provide adequate rotational capacity. The first two plastic hinges at the beam ends transform the fixed ends in simple supports on which plastic bending moments act as externally applied loads: as a consequence, the simply supported beam, subjected to a further distributed load Δq_u at its ends (where plastic hinges are located) undergoes rotation θ_{pl}, which can be evaluated, on the basis of the well-known theory of structures, as:

$$\theta_{pl} = \frac{M_{pl}L}{6EI} \tag{3.26c}$$

where E and I are the modulus of elasticity and the moment of inertia of the beam, respectively.

These two examples show that a design approach based on limits associated with the achievement of the yielding stress could result in something very conservative, which corresponds to the achievement of the elastic bending moment (M_{el}) in correspondence to one cross-section of the whole structure. Benefits associated with the spreading of plasticity are not negligible: with reference to simply supported beams, collapse is achieved when the maximum value of the bending moment is equal to the beam plastic moment (M_{pl}) and the associated increment of the load carrying capacity is usually expressed via the shape coefficient, α_{shape}, defined as:

$$\alpha_{shape} = \frac{M_{pl}}{M_{el}} = \frac{W_{pl}}{W_{el}} \tag{3.27}$$

where W_{el} and W_{pl} are the elastic and the plastic section modulus of the cross-section, respectively.

Furthermore, it should be noted that, with reference to the second example, as generally occurs in case of statically indeterminate structures, the increment of the load carrying capacity with

Table 3.2 Value of the shape coefficient (α_{shape}) for the European I beams (IPE) and European wide flange beams (HE) for steel grade S235, S275, S355, S420 and S460.

Standards IPE beams[a]		Wide flange HEA beams					Wide flange HEB beams[a]		Wide flange HEM beams[a]	
			S 235 S 275	S 355	S 420	S 460				
IPE 80	1.15	HEA 100	1.14	1.14	1.14	1.14	HEB 100	1.16	HEM 100	1.24
IPE 100	1.15	HEA 120	1.13	1.13	1.13	1.13	HEB 120	1.15	HEM 120	1.22
IPE 120	1.15	HEA 140	1.12	1.12	1.12	1.12	HEB 140	1.14	HEM 140	1.20
IPE 140	1.14	HEA 160	1.11	1.11	1.11	1.11	HEB 160	1.14	HEM 160	1.19
IPE 160	1.14	HEA 180	1.11	1.11	1.11[b]	1.11[b]	HEB 180	1.13	HEM 180	1.18
IPE 180	1.14	HEA 200	1.10	1.10	1.10[b]	1.10[b]	HEB 200	1.13	HEM 200	1.17
IPE 200	1.14	HEA 220	1.10	1.10	1.10[b]	1.10[b]	HEB 220	1.12	HEM 220	1.17
IPE 220	1.13	HEA 240	1.10	1.10	1.10[b]	1.10[b]	HEB 240	1.12	HEM 240	1.18
IPE 240	1.13	HEA 260	1.10	1.10[b]	1.10[b]	1.10[b]	HEB 260	1.12	HEM 260	1.17
IPE 270	1.13	HEA 280	1.10	1.10[b]	1.10[b]	1.10[b]	HEB 280	1.11	HEM 280	1.16
IPE 300	1.13	HEA 300	1.10	1.10[b]	1.10[b]	1.10[b]	HEB 300	1.11	HEM 300	1.17
IPE 330	1.13	HEA 320	1.10	1.10	1.10[b]	1.10[b]	HEB 320	1.12	HEM 320	1.17
IPE 360	1.13	HEA 340	1.10	1.10	1.10	1.10[b]	HEB 340	1.12	HEM 340	1.16
IPE 400	1.13	HEA 360	1.10	1.10	1.10	1.10	HEB 360	1.12	HEM 360	1.16
IPE 450	1.13	HEA 400	1.11	1.11	1.11	1.11	HEB 400	1.12	HEM 400	1.16
IPE 500	1.14	HEA 450	1.11	1.11	1.11	1.11	HEB 450	1.12	HEM 450	1.15
IPE 550	1.14	HEA 500	1.11	1.11	1.11	1.11	HEB 500	1.12	HEM 500	1.15
IPE 600	1.14	HEA 550	1.11	1.11	1.11	1.11	HEB 550	1.12	HEM 550	1.15
—	—	HEA 600	1.12	1.12	1.12	1.12	HEB 600	1.13	HEM 600	1.15
—	—	HEA 650	1.12	1.12	1.12	1.12	HEB 650	1.13	HEM 650	1.15
—	—	HEA 700	1.13	1.13	1.13	1.13	HEB 700	1.13	HEM 700	1.15
—	—	HEA 800	1.13	1.13	1.13	1.13	HEB 800	1.14	HEM 800	1.15
—	—	HEA 900	1.14	1.14	1.14	1.14	HEB 900	1.15	HEM 900	1.15
—	—	HEA1000	1.15	1.15	1.15	1.15	HEB1000	1.15	HEM1000	1.16
—	—	HEA1100	1.16	1.16	1.16	1.16	HEB1100	1.16	HEM1100	1.17

[a] For all the considered steel grades.
[b] Using this steel grade plastic moment cannot be reached due to local buckling phenomena ($\alpha_{shape} = 1$).

respect to the one associated with a design approach based on the achievement of the elastic bending moment (M_{el}) is significantly greater than term α_{shape}, owing to the moment re-distribution. Table 3.2 shows the values of the shape factor (α_{shape}) for the most common types of hot-rolled profiles for different EU steel grades. It should be noted that some profiles cannot reach the plastic moment because of their local buckling and therefore the value of the shape factor obtained via Eq. (3.27) cannot be used and reference must be made to unity, as better explained in Chapter 4.

In case or regular frames, the typical collapse mechanisms that could occur (Figure 3.28) are:

- beam mechanism;
- panel (lateral) mechanism;
- mixed mechanism.

More details related to the rigid plastic analysis of regular sway frames with semi-rigid beam-to-column and base plate joints are proposed in Chapter 15.

A very important phenomenon affecting member behaviour and, as a consequence, the whole structural performance, is the local buckling that typically affects thin-walled members, that is members with components characterized by a high value ratio between the component width and thickness. This form of instability affects the compressed portions of the cross-section and is characterized by a deformed shape with half-waves of amplitude comparable with the transverse dimensions of the section of the element (Figure 3.29).

The types of analysis previously described are characterized by a different degree of complexity and are able to simulate the frame response more or less accurately. As an example, Figure 3.30 relates to a frame subjected to a gravity load on each floor (P is the resulting value of the

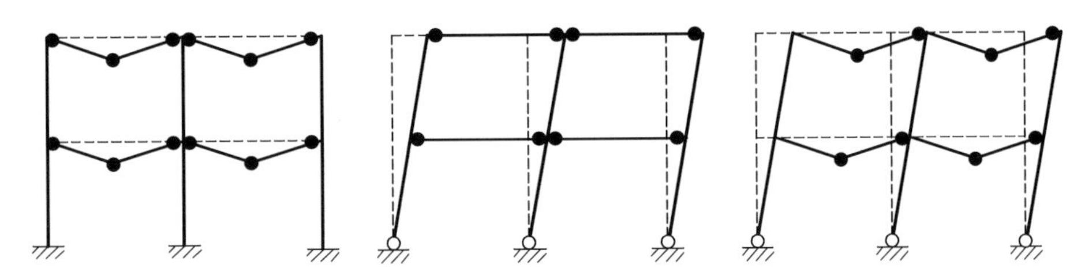

Figure 3.28 Typical collapse mechanisms for steel frames.

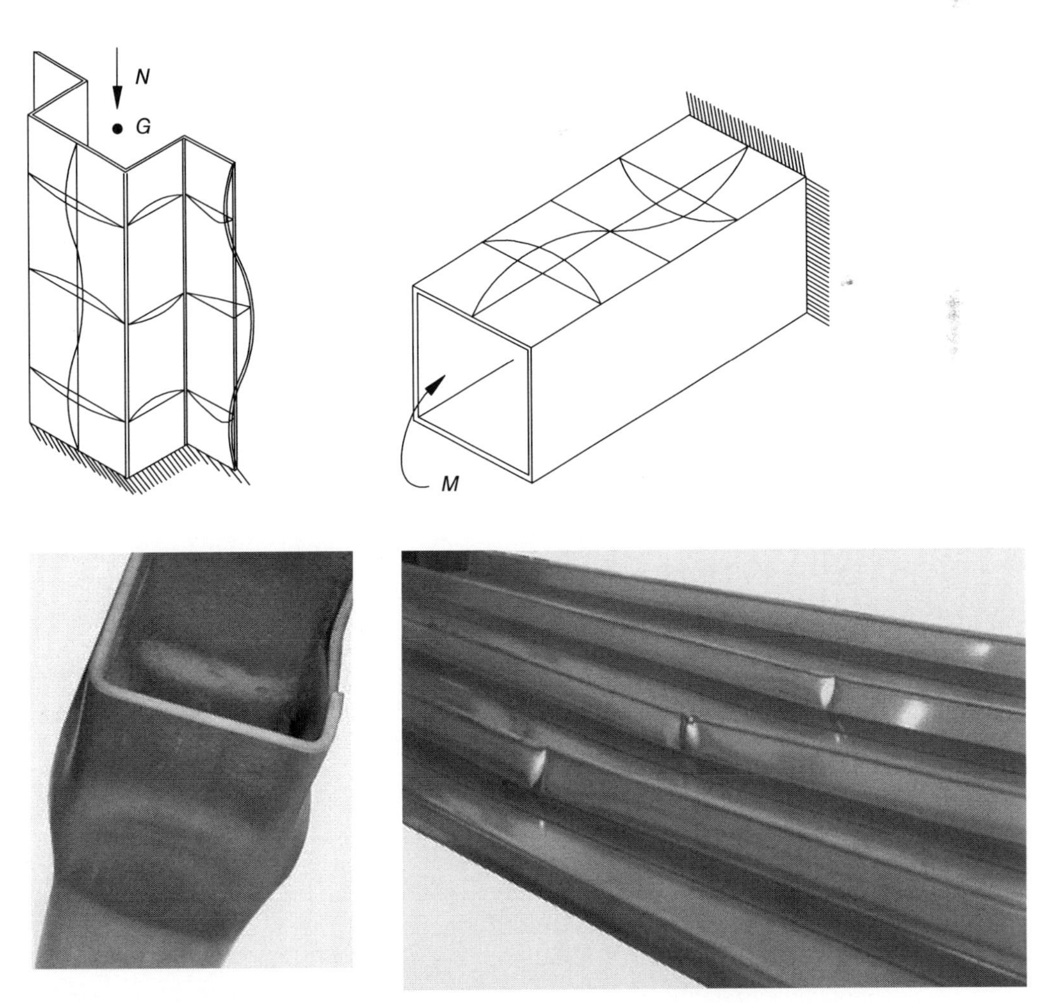

Figure 3.29 Examples of local buckling of thin walled members.

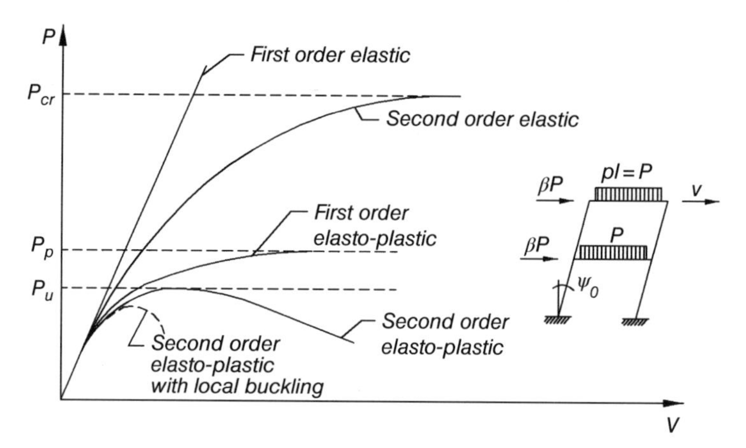

Figure 3.30 Influence of the type of analysis on the response of sway frames.

distributed gravity load on each beam) and a horizontal force, βP, applied on each floor (β is constant and depends on the considered load condition).

The relationships between the applied load P and the lateral top displacement v are presented in the figure, which have been determined via different approaches for the structural analysis (Figure 3.22). Increasing the value of the applied loads from zero, it can be noted that the responses associated with the different types of analysis coincide to each other in their initial portion, that is for the lowest values of P. Remarkable differences can be noted when the yield of the material is achieved and/or the influence of the second order effects becomes non-negligible. In particular, the curve associated with second order elastic analysis tends asymptotically to the elastic critical load for sway mode P_{cr}. By means of first order elasto-plastic analysis the collapse load, P_p, associated with a complete collapse mechanism is generally greater than the corresponding one obtained from an elasto-plastic second order analysis in which the failure occurs by interaction between plasticity and instability.

It should be noted that, for each considered load condition, the deformed shape of the frame is always characterized by transversal displacements and hence a second order elastic analysis is always recommended to approximate the actual frame response. However, errors associated with the assumption of small displacements and infinitesimal deformations, on which elastic first order analysis is based, should be very negligible in many cases. As a consequence, this last type of analysis, characterized by remarkable simplicity, can be used for routine design when the considered load level is significantly far from the critical buckling load, that is, the condition expressed by Eq. (3.4) for the European approach or by Eq. (3.7) for the US approach are fulfilled.

The choice of the method of analysis for steel framed systems depends not only on the structural typology and on the relevance of second order effects on frame response, but also on the type of cross-section of members as well as on the size of each of its components (flange, web, stiffener, etc.). In case of thin-walled members, that is members of its cross-section components with high values of ratio width over thickness, the local instability phenomena might occur in the elastic range, hence preventing the spread of plasticity in the cross-section, that is, the achievement not only of the plastic moment but also of the elastic moment.

3.6.1.1 Remarks on the European Practice

Eurocode 3 proposes a criterion for the classification of cross-sections based on the slenderness ratio (width over thickness ratio) of each compressed component of the cross-section, as well as on other factors, described more in detail in Chapter 4. In particular, with reference to the flexural response in terms of relationships between the moment (M) and the curvature (χ), the following four classes of cross-sections (Figure 3.31) are defined:

- Class 1 cross-sections, which are those able to guarantee a plastic hinge providing adequate rotational capacity for plastic analysis without reduction of the resistance (*plastic* or *ductile sections*);
- Class 2 cross-sections, which are those able to guarantee, as a class 1 cross-section does, plastic moment resistance, but have limited rotational capacity because of local buckling (*compact sections*);
- Class 3 cross-sections, which are those able to sustain yielding stresses only in the more compressed fibres when an elastic stress distribution is considered because of the local buckling phenomena hampering the spread of plasticity along the cross-section (*semi-compact sections*);
- Class 4 cross-sections, which are those subjected to local buckling phenomena before the attainment of yielding stress in one or more parts of the cross-section (*slender sections*).

It should be noted that, in case of compressed member, no distinctions can be observed in the performance of the elements of the first three classes, owing to the stress distribution in axially loaded cross-sections limited to yielding strength.

The possible choices for structural analysis and member verification checks are summarized in Table 3.3. The load carrying capacity of the cross-section has to be evaluated with reference to the

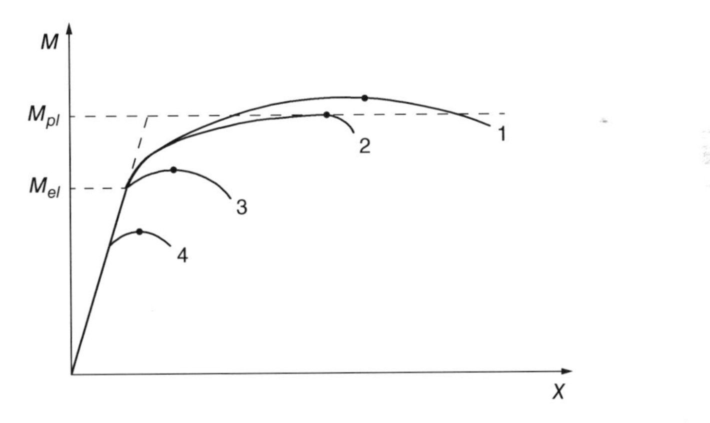

Figure 3.31 Moment-curvature (M–χ) relationships for the different classes of cross-sections considered in accordance with the European approach.

Table 3.3 Methods of analysis and associated approaches for verification checks.

Method of analysis	Approach to evaluate load carrying capacity of cross-section	Cross-section class
(E)	(E)	All[a]
(E)	(P)	Classes 1 and 2
(E)	(EP)	All[a]
(P)	(P)	Class 1
(EP)	(EP)	All[a]

[a] In the case of class 4, reference has to be made to the effective geometric properties (see Chapter 4).

axial load (tension and compression) and to the bending and torsional moment. The following approaches can be adopted:

- *Elastic method* (E): a linear elastic response is assumed till the achievement of the yielding strength. This method can be applied to verify all the cross-section classes; in case of class 4, reference has to be made to the effective geometrical properties;
- *Plastic method* (P): the complete spread of plasticity is assumed in the cross-section, which belongs to classes 1 or 2;
- *Elasto-plastic method* (EP): reference is made to the actual material constitutive law, generally simplified by an elastic-perfectly plastic relationship or with an elastic-plastic with strain hardening relationship.

3.6.1.2 Remarks on the US Practice

AISC cross-section classification criteria are based, as in Eurocode 3, on the *steel grade* and on the *width-to-thickness ratios* distinguished for *stiffened elements* (elements supported along two edges parallel to the direction of the compression force) and *unstiffened elements* (elements supported along only one edge parallel to the direction of the compression force). As discussed in Chapter 4, cross-sections are classified on the basis of type of load acting on the element (i.e. compression and bending).

Members in compression are distinguishable in:

- *slender elements*, which are subject to local buckling, reducing their compression strength;
- *non-slender elements*, never affected by local buckling.

Members in flexure are distinguishable in:

- *compact elements*, which are able to develop a fully plastic stress distribution with an associated rotation capacity of approximately 30 mrad before the onset of local buckling;
- *noncompact elements*, which can develop partial yielding in compression elements before local buckling occurs, but cannot develop a full plastic stress distribution because of local buckling;
- *slender elements*, which have some cross-section components (flanges and/or web) affected by elastic buckling before that yielding is achieved.

Contrary to the European approach, which assumes the same classification criteria for both static and seismic design, it must be remarked that AISC Seismic provisions propose a different classification criteria when profiles are used in seismic areas.

The main analysis method suggested by AISC is the *elastic method* (Chapter C of AISC 360-10): as explained in the following, a linear elastic response is until the achievement of yielding strength. This method can be applied to verify all the cross-section classes. Furthermore, in its Appendix 1, the AISC specifications give more details related to the design by *inelastic analysis*. Any method that uses inelastic analysis is allowed, if some general conditions are fulfilled: among them, second order effects and stiffness reduction due to inelasticity are addressed. These methods may include the use of non-linear finite element analysis; more details about this topic are reported in Chapter 12.

3.6.2 Elastic Analysis with Bending Moment Redistribution

Elastic analysis is based on the assumption that the response is linear, that is the behaviour of the material is in the linear branch of its stress-strain constitutive law, whatever the stress level is. Furthermore, in case of braced frames or non-sway frames, a first order analysis provides results with an adequate accuracy for design purposes. However, if local buckling phenomena do not occur, the resulting design can be quite conservative due to neglecting benefits associated with the spread of plasticity in the cross-sections and in the structural members, as shown in the examples of Figures 3.23 and 3.24. As for the design of concrete reinforced structures, as well as in case of steel frames, an elastic redistribution of the bending moments more accurately approximates the actual bending moment distribution in the post-elastic range.

European provisions (EC3) allow for the plastic redistribution of moments in continuous beams. In particular, at first, an elastic analysis is required: on the basis of the bending moment diagram it can be the case that some peak moments exceed the value of the moment resistance and 15% of the beam plastic resistance is admitted as maximum degree of redistribution. The parts in excess of the vicinity of these peak moments may be redistributed in any member, provided that all these following assumptions are fulfilled:

- all the members in which the moments have been reduced belong to Class 1 or Class 2 (Figure 3.31);
- lateral torsional buckling of the members is prevented;
- the internal forces and bending moments in the structure guarantee the equilibrium under the applied loads.

With reference to continuous beams, usually the redistribution degree is selected to increase the benefits in terms of load carrying capacity associated with plastic design. In particular, if doubly symmetrical beams are used, this approach leads to reduce the difference, after redistribution (subscripts R), between the peak bending negative (hogging) moment, M_R^-, and the peak positive (sagging) one, M_R^+. This method is discussed in the following with reference to a beam having two equal spans in Figure 3.32. If terms M^+ and M^- indicate, the maximum and the minimum values of the elastic bending moments, respectively, the redistribution degree is given by the ratio $\Delta M/M^-$, where ΔM represents the reduction of the bending moment value at the internal support location due to redistribution. An optimal use of the material is guaranteed if $M_R^- = M_R^+$, that is the absolute values of the peak moments after redistribution are equal. As a consequence, on the basis of the equilibrium condition, the reduction ΔM of the negative peak elastic moment M^- can be evaluated in this case as:

$$\Delta M = \frac{|M^-| - |M^+|}{1 + \frac{x}{L}} \tag{3.28}$$

where x indicates the distance between the load application cross-section and the external support and L is the beam length.

After the redistribution, if bending moments are within the elastic limit, the evaluation of the beam deflection can be carried out by traditional elastic analysis methods.

Moment redistribution is allowed also by AISC Specifications (Chapter B, Section B3.7), if the following conditions are fulfilled:

(a) the cross-sections are classified as compact;
(b) the unbraced lengths near the points where plastic hinges occur are limited to the values stated in Section F13.5 of AISC 360-10;
(c) bending moments are not due to loads on cantilevers;
(d) yielding strength, F_y is not greater than 65 ksi (450 MPa);

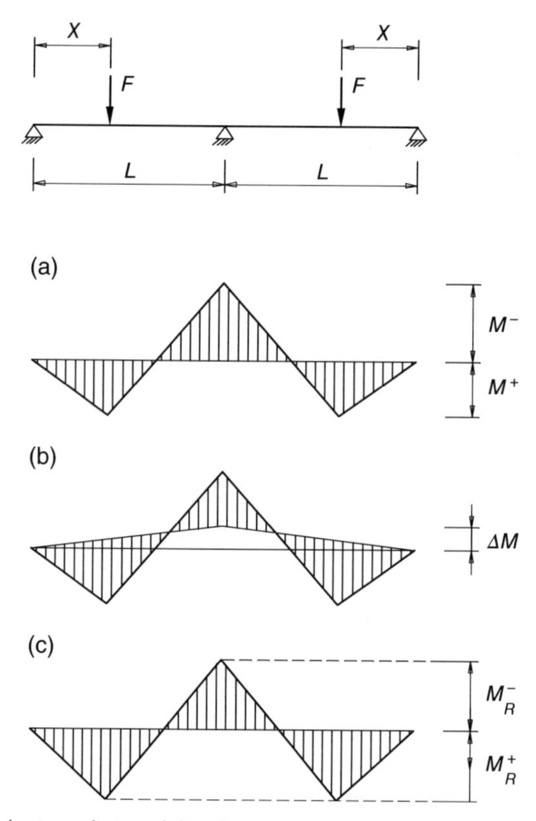

Figure 3.32 Example of elastic analysis with bending moment redistribution.

(e) connections are FR (Fully Restrained);
(f) elastic analysis has been carried out.

If all these conditions are met, then the negative moments can be taken as 9/10 of computed values and the maximum positive moment shall be increased of 1/10 of the average negative moment.

3.6.3 Methods of Analysis Considering Mechanical Non-Linearity

As was introduced previously, the response of steel structures can be remarkably affected by the non-linearity associated with the steel material as well as with components and connections. The frames can be modelled for structural analysis and design using the following approaches:

- elasto-plastic analysis with plastification of the cross-sections and/or joints where plastic hinges are located;
- non-linear plastic analysis considering the partial plastification of members in plastic zones;
- rigid plastic analysis neglecting the elastic behaviour between hinges.

Attention is focused briefly here on the last model, which is the most commonly used and also because of the possibility for it to be employed with calculations by hand, as shown in Chapter 15 for semi-continuous frames. It should be noted that plastic global analysis may be used if the members are able to provide an adequate rotational capacity guaranteed with reference to the required redistributions of bending moments. As a prerequisite for plastic global analysis, the cross-sections of the members where the plastic hinge are located must have adequate rotational capacity, that is, must be able to sustain rotation of no less than that required at the plastic hinge location. Furthermore, this design approach can only be used if at each plastic hinge position the cross-section has efficient lateral and torsional restraints with appropriate resistance to lateral forces and torsion induced by local plastic deformations.

In order to apply the plastic analysis approach in Europe, the steel material must have adequate ductility requirements to undergo plastic rotation without failure: a minimum level of ductility is required, which can be guaranteed if the following conditions are fulfilled:

$$\frac{f_u}{f_y} \geq 1.1 \tag{3.29a}$$

$$\Delta\% \geq 15\% \tag{3.29b}$$

$$\varepsilon_u \geq 15\varepsilon_y \tag{3.29c}$$

where ε represents the strain, f is related to the strength, subscripts y and u are related to yielding and rupture, respectively, and $\Delta\%$ represents the percentage elongation at failure.

It is worth noting that the limits provided by these equations should be more appropriately re-defined in the National Annex and hence the previously mentioned values have to be considered as recommended. Furthermore, Eurocode 3 describes the cross-section requirements for plastic global analysis. In particular, it is explicitly required that at the plastic hinge locations the cross-section of the member containing the plastic hinge must have a rotational capacity that is no less than the one required by calculations. Class 1 is a prerequisite for uniform members to have sufficient rotational capacity at the plastic hinge location. Furthermore, if a transverse force exceeding 10% of the shear resistance of the cross-section is applied to the web at the plastic hinge location, web stiffeners should be provided within a distance along the member of $h/2$ from the plastic hinge position (with h representing the depth of the member at this location).

In case of non-uniform members, that is if the cross-section of the members varies along their length (tapered members), additional criteria have to be fulfilled adjacent to plastic hinge locations. In particular, it is required that:

- the thickness of the web is constant (and not reduced) for a distance each way along the member from the plastic hinge location of at least $2d$ (with d representing the clear depth of the web at the plastic hinge location);
- the compression flange can be classified in Class 1 for a distance each way along the member from the plastic hinge location of not less than the greater between $2d$ and the distance to the adjacent point at which the moment in the member has fallen to 0.8 times the plastic moment resistance at the point concerned.

For plastic design of a frame, regarding cross-section requirements, the capacity of plastic redistribution of moments may be assumed to be sufficient if the previously introduced requirements are satisfied for all members. It should be noted that these requirements can be neglected when methods of plastic global analysis are used, which consider the actual stress-strain constitutive law along the member including (i) the combined effect of local cross-section and (ii) the overall member buckling phenomena.

According to AISC, it is recommended to use steel for members subject to plastic hinging with a specified minimum yield stress not greater than 65 ksi (450 MPa). Furthermore, members where plastic hinges can occur have to be doubly symmetrical belonging to the *compact* class and the laterally unbraced length has to be limited to values reported in Appendix 1 Section 3 of AISC 360-10.

3.6.4 Simplified Analysis Approaches

As introduced previously, the structural analysis of steel frames could require very refined finite element analysis packages capable of taking into account both geometrical and mechanical non-linearity. In the past, refined analysis tools were not available to designers, due to absence or very limited availability of computers and structural analysis software programs. As a consequence, design was carried out using simplified methods to approximate structural response and estimate the set of displacements, internal forces and moments with an adequate degree of accuracy and generally on the safe side. These approaches are, however, still very important because they can be used nowadays for the initial (presizing) phase of the design as well as to check qualitatively the results of more refined finite element structural analysis. In the following, reference is made to the most commonly used simplified approaches for routine design, which are:

- the Merchant–Rankine formula;
- the Equivalent Lateral Force Procedure;
- the Amplified Sway Moment Method.

3.6.4.1 The Merchant-Rankine Formula

In case of sway frames, the ultimate load multiplier, α''_u, for the considered load condition can be evaluated by means of the Merchant–Rankine formula, which allows us to take into account the influence of both plasticity and instability phenomena. In particular, the elastic-plastic second order load multiplier, α''_u, can be directly evaluated via the expression:

$$\frac{1}{\alpha''_u} = \frac{1}{\alpha_{cr}} + \frac{1}{\alpha'_u} \tag{3.30a}$$

where α_{cr} is the critical load multiplier of the frame and α'_u is the elastic-plastic load multiplier associated with a first order rigid or elastic-plastic analysis.

Term α''_u can be re-defined as:

$$\alpha''_u = \frac{\alpha'_u \alpha_{cr}}{\alpha'_u + \alpha_{cr}} \tag{3.30b}$$

The Merchant–Rankine formula should be applied preferably if term α_{cr} ranges between 4 and 10 times α'_u, that is the reference design condition is sufficiently far from the critical one.

3.6.4.2 The Equivalent Lateral Force Procedure

Second order effects, in terms of both displacement and bending moments can be evaluated by using the Equivalent Lateral Force Procedure, which is an example of indirect method for second order analysis via iterative elastic first order analysis. This procedure assumes that no relevant axial deformations occur in the members and the second order effects are due only to the

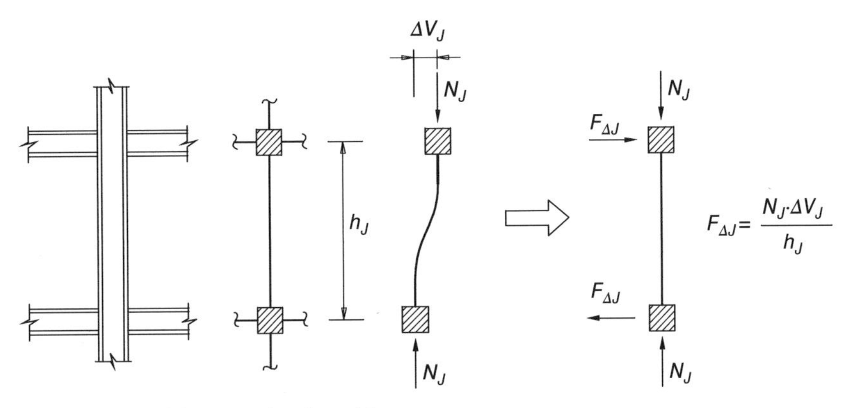

Figure 3.33 Evaluation of the equivalent lateral force.

horizontal displacements. The principle of this procedure is shown in Figure 3.34, which refers to the internal column of a generic inter-storey of a sway frame with a height of h_j. A first order analysis allows us to evaluate the internal axial force N_j and the inter-storey drift Δv_j associated with the deformed shape of the frame, as in the case of the previously introduced cantilever beam (Figure 3.6). With reference to the deformed shape of the column, an additional bending moment can be evaluated as $N_j\ \Delta v_j$, due to the inter-storey drift. This additional bending moment is replaced by an equivalent couple of horizontal forces $F_{\Delta j}$ (i.e. the so-called equivalent lateral force) having intensity equal to $N_j \Delta v_j / h_j$ and acting at the end of the considered column. A new first order elastic analysis is hence required, which is based on the new load condition also including all the terms $F_{\Delta j}$ evaluated for all the columns and new values of the inter-storey drift have to be evaluated; they expected to be greater than the one previously determined. As a consequence of the updated value of the additional bending moment, a new equivalent lateral force has to be added to the initial load condition (Figure 3.33).

The procedure, which can be stopped when the differences between two subsequent steps are very limited, generally requires a few iterations to approximate accurately the effects of the geometrical non-linearity. A very slow convergence (i.e. number of iterations greater than 6 or 7) is due to a load condition too close to the elastic critical one. This means that the application of the method is out of its scope and hence a more refined approach is required to execute second order elastic analysis. The problem of the convergence of this procedure is briefly discussed in the following Example E3.3.

Before starting with the application of the method, the parameter for the convergence check has to be defined: typically, reference is made to the value of the horizontal displacement of the node on the roof, the maximum inter-storey drift or the additional force updating load condition. The flow-chart of the method is presented in Figure 3.34 and the evaluation of the equivalent lateral force is presented in Figure 3.35 with reference to the case of a multi-storey frame. In particular, identifying with terms v and q the transversal absolute displacement and the vertical loads acting on the beams, respectively, and with subscripts *1* and *2* the first and second floor, respectively, the following steps have to be evaluated:

- evaluation of the additional bending moments due to lateral displacements:

$$M_{\Delta 2} = (\Sigma q_2 L_i)(v_2 - v_1) = (\Sigma q_2 L_i)\Delta v_2 \quad (3.31a)$$

$$M_{\Delta 1} = (\Sigma q_1 L_i) v_1 = (\Sigma q_1 L_i)\Delta v_1 \quad (3.31b)$$

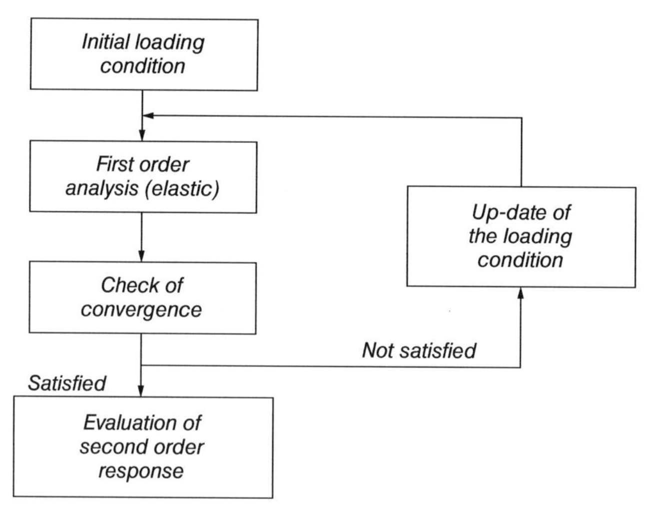

Figure 3.34 Flow-chart of the equivalent lateral force procedure.

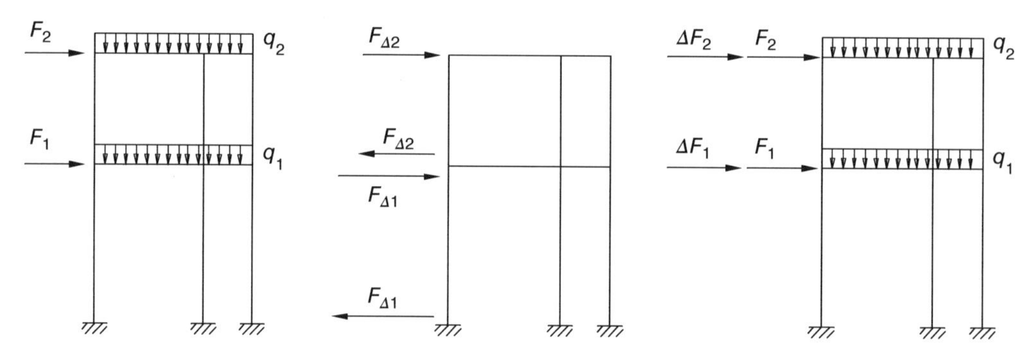

Figure 3.35 Evaluation of the equivalent lateral force in a multi-storey frame.

- evaluation of the equivalent lateral forces for each inter-storey:

$$F_{\Delta 2} = \frac{M_{\Delta 2}}{h_2} \tag{3.32a}$$

$$F_{\Delta 1} = \frac{M_{\Delta 1}}{h_1} \tag{3.32b}$$

- evaluation of the equivalent lateral force to be applied to each storey:

$$\Delta F_2 = F_{\Delta 2} \tag{3.33a}$$

$$\Delta F_1 = F_{\Delta 1} - F_{\Delta 2} \tag{3.33b}$$

3.6.4.3 The Amplified Sway Moment Method

The Amplified Sway Moment Method allows for an indirect allowance for second order effects, which requires a set of first order elastic analysis. Two different procedures can be followed, depending on the frame geometry, which requires structural analysis on different service frames and, in particular, one of the following cases has to be selected:

- symmetrical frame with vertical loads symmetrically distributed;
- other cases.

The approximate evaluation of the second order effects via this method is based on the amplification of the bending moments associated with the lateral displacement of the frame. The amplification factor, β, is defined as:

$$\beta = \frac{1}{1 - \dfrac{V_{Ed}}{V_{cr}}} = \frac{\alpha_{cr}}{\alpha_{cr} - 1} \tag{3.34}$$

where V_{Ed} is the design vertical load for the loading condition of interest and V_{cr} is the associated elastic critical load for sway mode (or, equivalently, α_{cr} represents the elastic critical load multiplier for sway mode associated with the considered load condition, that is $\alpha_{cr} = V_{cr}/V_{Ed}$).

This method can be generally adopted if $V_{Ed}/V_{cr} \leq 0.33$ (in some codes the limit is 0.3) or equivalently if $\alpha_{cr} > 3$.

In case of vertical loads symmetrically distributed on a symmetric frame (Figure 3.36a), the following steps have to be executed:

- first order elastic analysis of the frame loaded only by the vertical load (Figure 3.36b): term M_V represents the associated bending moment distribution;
- first order elastic analysis of the frame loaded by the sole horizontal load (Figure 3.36c): term M_H represents the associated bending moment distribution;
- approximation of the second order bending moment distribution (M^{II}) on the frame as:

$$M^{II} = M_V + \beta \cdot M_H \tag{3.35}$$

In other cases, that is with reference to non-symmetrical frames and/or to vertical loads non-symmetrically distributed (Figure 3.37a), the amplified sway moment method is applied through the following steps:

- first order elastic analysis of the frame loaded by the sole vertical loads with additional fictitious restraints embedding all the horizontal displacements (Figure 3.37b): term R_i represents the horizontal reaction on the generic additional restraint;

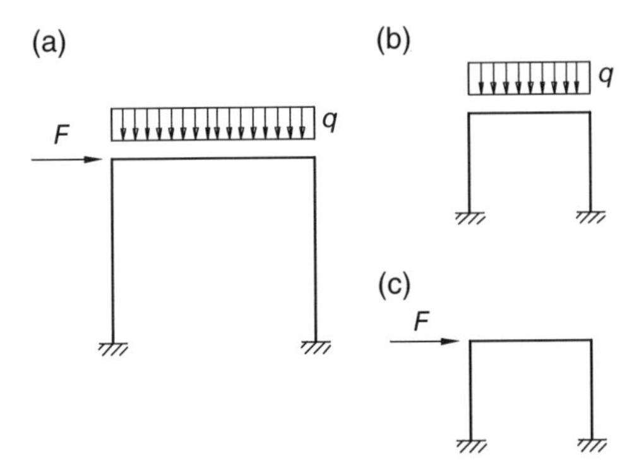

Figure 3.36 Set of frames (a–c) in case of symmetry of both frame and vertical loading condition.

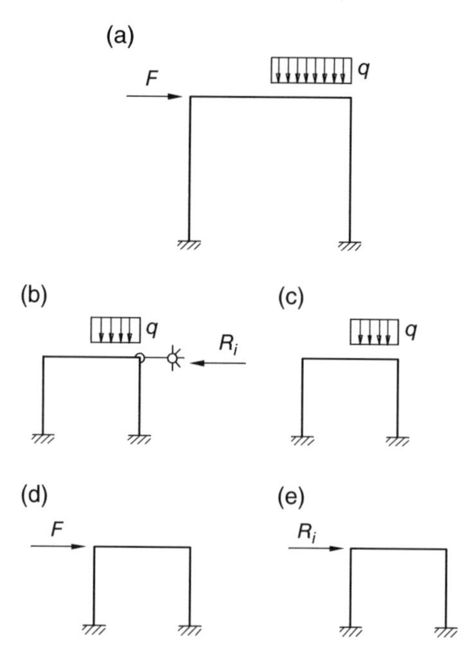

Figure 3.37 Set of frames (a–e) for the application of the amplified sway moment method in case of unsymmetrical structures and/or unsymmetrically distributed vertical loads.

- first order elastic analysis of the frame loaded by the sole vertical loads (Figure 3.37c): term M_V represents the associated bending moment distribution;
- first order elastic analysis of the frame in Figure 3.37d loaded by the horizontal loads: term M_H represents the associated bending moment distribution;
- first order elastic analysis of the frame in Figure 3.37e loaded by forces opposite to the horizontal reactions R_i on the additional restraint: term M_{VF} represents the associated bending moment distribution;
- approximation of the second order bending moment distribution (M^{II}) on the frame is the following:

$$M^{II} = M_V + \beta \cdot M_H + (\beta - 1)M_{VF} \tag{3.36}$$

3.7 Simple Frames

If beam-to-column joints behave like hinges, the structural system is statically undetermined or characterized by an excessive deformability to lateral loads. As a consequence, appropriate bracing systems are definitely required to transfer to the foundation all the horizontal forces acting on the structure (Figure 3.38), which are for routine design cases where generally the forces simulate the effects of wind load, geometrical imperfections and seismic actions.

The most common types of bracing systems adopted in structural steel buildings are (Figure 3.39):

- *X-cross bracing system*, if there is an overlap of the diagonal members at the centre of the braced panel (Figure 3.39a), which are connected at the beam-to-column joint locations. It should be

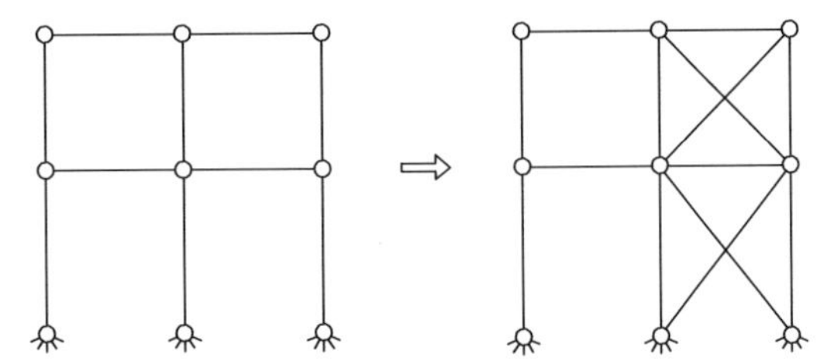

Figure 3.38 Typical simple frame.

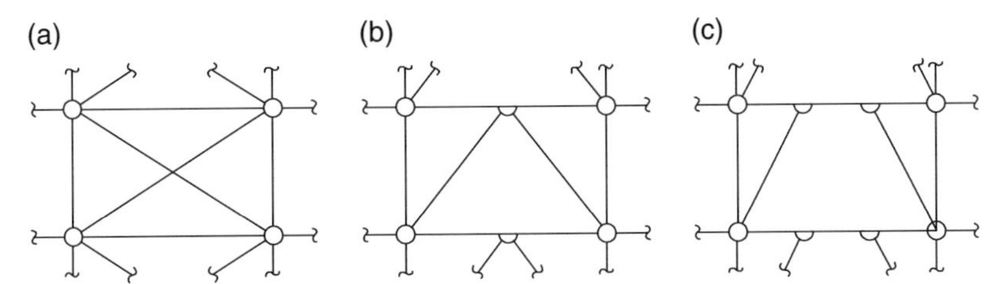

Figure 3.39 Typical bracing panels.

noted that a full use of openings (i.e. doors and windows) should be hampered or limited, by this type of bracing.

- *K-bracing system*, if both diagonal members are connected at the mid-span beam cross-section (Figure 3.39b).
- *eccentric bracing system*, if the diagonal members are connected at different beam cross-sections (Figure 3.39c).

Two different approaches can be adopted for bracing design:

- bracing members are designed considering the fact that diagonals resist both to tension and compression forces. In this case, quite low values of slenderness are required so that the differences in the diagonal responses associated with tension and compression forces are negligible. Cross- or K-bracing systems (Figures 3.39a and 3.40a, respectively) under lateral loads behave as the trussed beams, characterized by diagonal members resisting both tension and compression. In case of eccentric bracing systems, the girder behaves like a beam-column under both hogging and sagging bending moments (Figure 3.41b);
- bracing members are designed considering the sole diagonals under tension, that is the compressed diagonals are not affected by the transfer force mechanisms. This requires very slender diagonal members. Cross-bracing systems also behave as trussed beams in this case while the beam of K (Figure 3.40c) and eccentric (Figure 3.41c) bracing systems are subjected to axial load and bending moments.

It should be noted that hinges in simple frames are usually assumed to be located at the intersection between beam and column longitudinal axes and this assumption significantly simplifies structural analysis. Internal forces and moments can in fact be evaluated with reference to elementary models related to isolated members: typically simply supported beams and a pinned-supported column.

The spatial portal frame in Figure 3.42 can be considered when discussing the correct location of the bracing system, which is a very important aspect for design. In accordance with major code

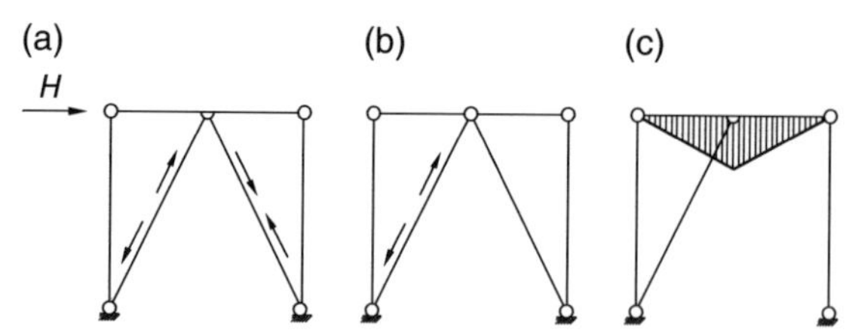

Figure 3.40 Transfer force mechanism in a K-bracing system.

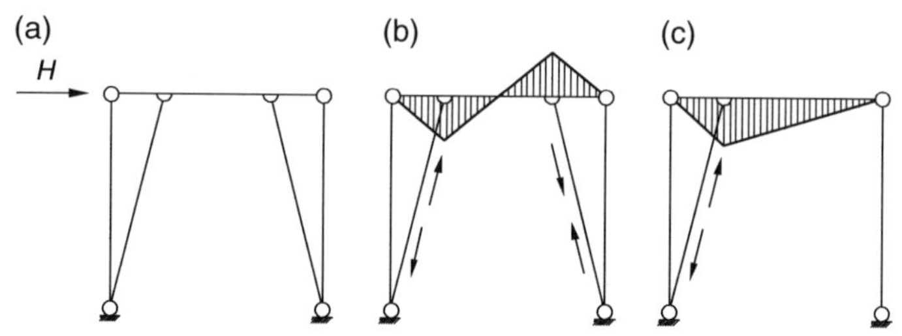

Figure 3.41 Transfer force mechanism in an eccentric bracing system (a–c).

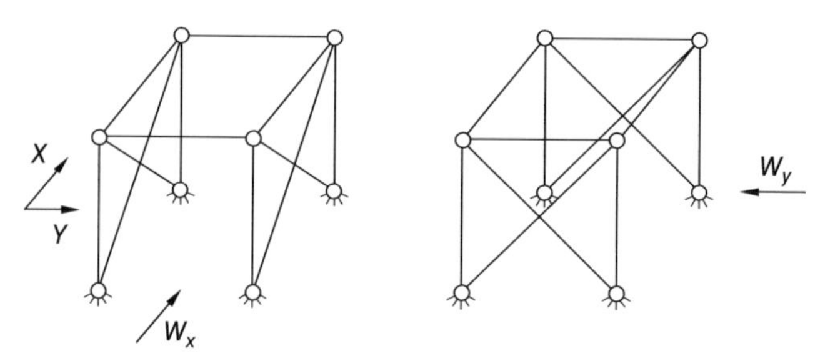

Figure 3.42 Vertical bracing in a three-dimensional portal frame.

provisions, wind effects are generally considered as acting alternately along a principal x- or y-axis and, as a consequence, vertical bracings are required both in transversal and longitudinal directions. The presence of vertical bracings in simple frames does not hamper relative displacement between the top and the bottom of each column: due to the presence of vertical bracing, each frame is only braced efficiently in its vertical plane. However, the connections are spatial hinges and hence the whole spatial framed system is unstable, as is shown in Figure 3.43 where typical deformed shapes associated with the absence of the roof bracings are presented for the portal frame of Figure 3.42 and for the single storey frame in Figure 3.44. To prevent these movements, a horizontal bracing system is hence required on each floor and on the roof. In several cases, the bracing floor should be required only for the erection phase, due to the fact that generally the slab of each floor and the metal decking of the roof should provide sufficient stiffness to transfer horizontal forces to vertical bracing systems. However, it could become generally uneconomical to remove these temporary bracings, which are often conveniently encased in the concrete of the slab or located between the slab and the ceiling.

With regards to the minimum horizontal forces to be transferred to foundations, three vertical bracing systems suitably located are required. Each braced floor, as well as the roof, can in fact be considered as a rigid body in its plane, with 3 degrees of freedom and requiring at least 3 degrees of restraint suitably located. It is usually assumed that each vertical bracing restraints a horizontal

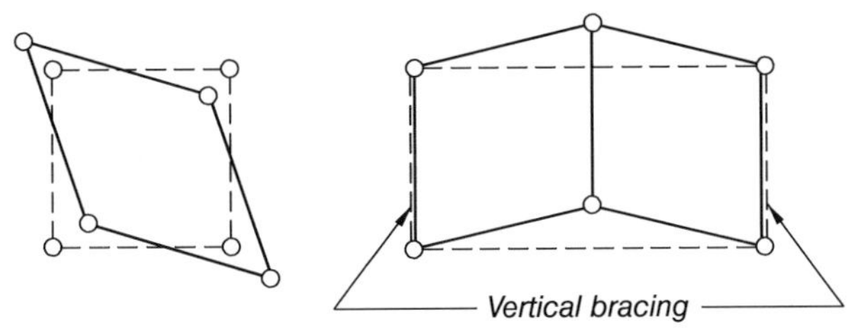

Figure 3.43 Liability due to the absence of horizontal bracings.

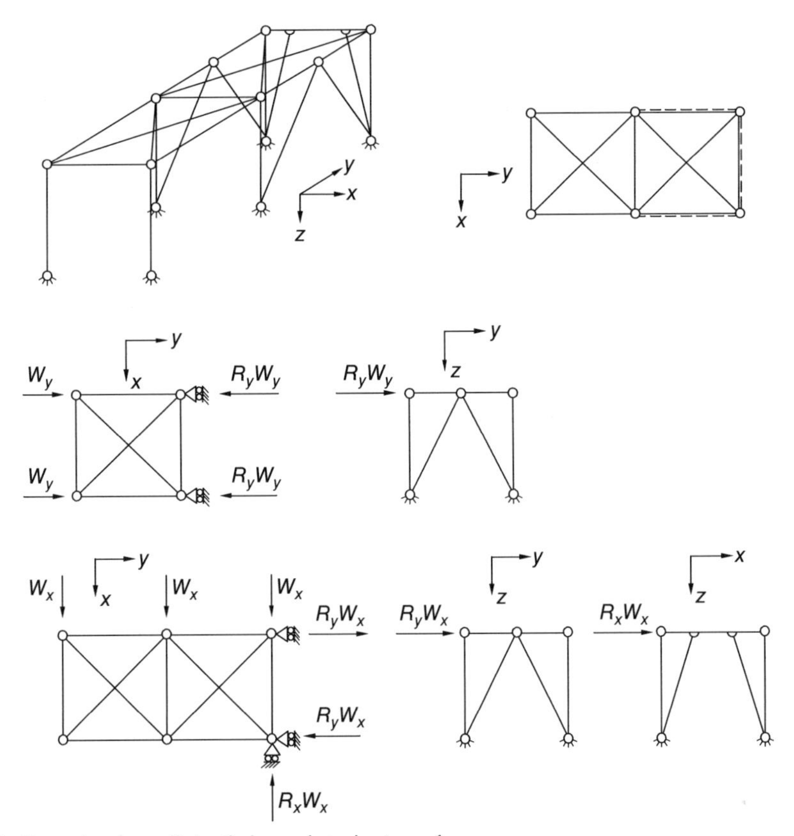

Figure 3.44 Example of an efficiently braced single storey frame.

displacement and hence can be modelled as an elastic support for each floor and the roof, or in a simplified way, as a simple support, neglecting its lateral deformability. As a consequence, at least three restraints, corresponding to three vertical bracing systems have to be appropriately located in order to avoid unstable structural system (Figure 3.44). Furthermore, with reference to the bracing system for each floor, it should be noted that it is not necessary to brace any span delimited by contiguous columns to transfer horizontal load to foundations via vertical bracings. To better explain these concepts, Figure 3.44 can be considered, which shows examples of appropriate bracing systems for single-storey steel frames with one bay in the transversal direction (y) and two bays in the longitudinal direction (x). Wind action, which has been simulated via nodal forces, is indicated with W_x and W_y, depending on the x or y wind direction.

When wind acts along the y direction, the corresponding forces W_y are transferred directly via the vertical longitudinal k-bracings which are symmetrically located and loaded by $R_y W_y$ forces. Otherwise, if wind acts along the x direction, the resulting force W_x is transferred directly via the horizontal bracing to the sole transversal horizontal bracing located along the x direction, which results in loading by an $R_x W_x$ force. In this case, for the overall structural equilibrium, longitudinal vertical bracings are loaded by a couple of forces of intensity equal to $R_y W_x$, which are required to balance the torsion due to the presence of one transversal vertical bracing, eccentrically located in plant with respect to the centroid.

3.7.1 Bracing System Imperfections in Accordance with EU Provisions

As required by major standards, structural analysis of bracing systems providing lateral frame stability has to include the effects of imperfections and generally an equivalent geometric imperfection of the members to be restrained is considered in the form of an initial bow imperfection e_0 defined as:

$$e_0 = \sqrt{\frac{1}{2}\left(1 + \frac{1}{m}\right)} \cdot \frac{L}{500} = \alpha_m \cdot \frac{L}{500} \tag{3.37}$$

where L is the span of the bracing system and m is the number of members to be restrained.

From a practical point of view, instead of modelling a frame with imperfect members, it appears more convenient to make reference to a perfect frame simulating the imperfection via additional notional loads, as previously discussed in Chapter 1.

For convenience, the effects of the initial bow imperfection of the members to be restrained by a bracing system may be replaced by the equivalent force q_s (Figure 3.45) given by:

$$q_s = \frac{8\left(e_0 + \delta_q\right) N_{Ed}}{L^2} \tag{3.38}$$

where δ_q is the in-plane deflection of the bracing system due to the q load plus any external loads calculated from first order analysis and N_{Ed} is the design axial force acting on members.

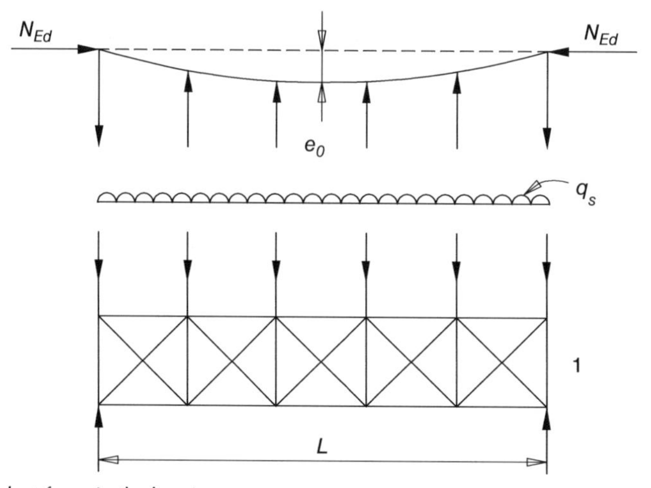

Figure 3.45 Equivalent force in the bracing system.

When the bracing system is required to stabilize the compression flange of a beam of constant height h, the axial force N_{Ed}, representing the action associated with the elements to brace on the bracing systems can be expressed as:

$$N_{Ed} = \frac{M_{Ed}}{h} \tag{3.39}$$

where M_{Ed} is the maximum moment in the beam and h is the distance between flange centroids (or, approximately, the depth of the beam).

If the beam to brace is subjected to external compression, axial force N_{Ed} should include a part of the compression force. Furthermore, at points where members are spliced, bracing system has to be verified to resist to a local force F_d applied to it by each beam or compression member that is spliced at that point, conventionally assumed as equal to:

$$F_d = \frac{\alpha_m \cdot N_{Ed}}{100} \tag{3.40}$$

where N_{Ed} is the axial compressive force (Figure 3.45).

3.7.2 System Imperfections in Accordance with AISC Provisions

The definition of the imperfections of the bracing systems is directly treated in AISC 360-10 Appendix 6 'Stability Bracings for Columns and Beams' that addresses the minimum strength and stiffness that a bracing system must have in order to provide a braced point in a column, beam or beam-column. For bracing systems providing lateral frame stability, the effects of imperfections have to be included in structural analysis by means of an equivalent geometric imperfection of the members to be restrained, in the form of an initial bow imperfection, Δ_0, defined as:

$$\Delta_0 = \frac{L}{500} \tag{3.41a}$$

where L is the span of the bracing system.

Term Δ_0 is independent on the number of compressed members to be restrained and the value $L/500$ is consistent with the maximum frame out-of-plumbness specified in AISC Code of Standard Practice for Steel Buildings and Bridges. In AISC 360-10 Commentary of Appendix 6, a less severe formulation of Eq. (3.41a) is suggested:

$$\Delta_0 = \frac{L}{500\sqrt{n_0}} \tag{3.41b}$$

where n_0 is the number of columns stabilized by the bracing (it corresponds to m of Eq. (3.37)).

In the same Commentary, it is permitted to use a reduced value of Δ_0 defined in Eq. (3.41b) when combining stability forces with the wind or the seismic forces on bracings. Nevertheless, the criteria reported hereafter for defining strength and stiffness of bracing systems, are based on Δ_0 values derived from Eq. (3.41a).

AISC Specifications identify the following two categories of bracing systems:

- *column braces*, which fix locations along column length so that the column unbraced length can be assumed equal to the distance between two adjacent fixed points.
- *beam braces*, which prevent lateral displacement (*lateral bracings*) and/or torsional rotation of the beam (*torsional bracings*). Lateral bracings are usually connected in correspondence with

the beam compression flange, while torsional bracings can be attached at any cross-sectional location. Torsional bracings can either be located at discrete points along the length of the beam, or attached continuously along the length.

For both column and beam bracings, AISC defines:

(a) *relative bracings*, which work efficiently controlling the movement at one end of unbraced length, A, with respect to the other end of unbraced length, B (see Figure 3.46a).
(b) *nodal bracings*, which work efficiently controlling the column or beam movements only at the braced point without any direct interaction with the adjacent braced points (see Figure 3.46b).

For any type of bracing, that is column or beam, nodal or relative bracings, the AISC requires verification of their *strength* and their *stiffness* to give designers minimum values, as specified in the following:

(1) Relative column bracings
(a) *Required strength:*

$$P_{rb} = 0.004 P_r \tag{3.41c}$$

(b) *Required stiffness:*

$$\beta_{br} = k\left(\frac{2P_r}{L_b}\right) \tag{3.41d}$$

(2) Nodal column bracings
(a) *Required strength:*

$$P_{rb} = 0.01 P_r \tag{3.41e}$$

(b) *Required stiffness:*

$$\beta_{br} = k\left(\frac{8P_r}{L_b}\right) \tag{3.41f}$$

where P_r is the required strength in axial compression using LFRD or ASD combinations, L_b is the unbraced length, expressed in inches and $k = 1/\phi = 1/0.75$ (ASD); $k = \Omega = 2.00$ (LFRD).
It is possible to size the brace to provide the lower stiffness determined by using the maximum unbraced length associated with the required strength.

(3) Relative lateral beam bracings
(a) *required strength:*

$$P_{rb} = 0.008 M_r C_d / h_0 \tag{3.41g}$$

(b) *required stiffness:*

$$\beta_{br} = k\left(\frac{4M_r C_d}{L_b h_0}\right) \tag{3.41h}$$

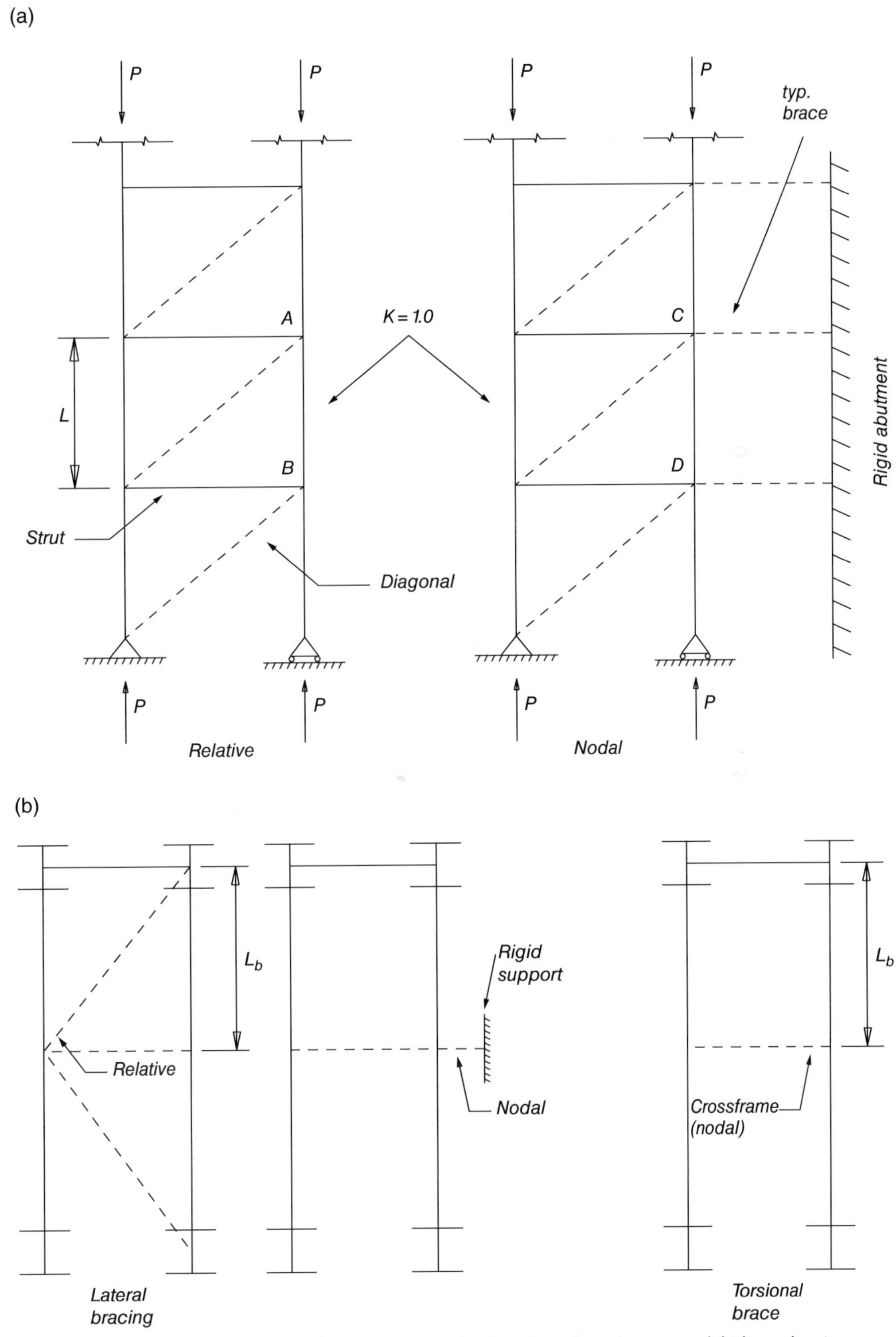

Figure 3.46 Types of bracings according to AISC specifications (a) column bracing and (b) beam bracing.

(4) Nodal lateral beam bracings

(a) *required strength:*

$$P_{rb} = 0.02 M_r C_d / h_0 \tag{3.41i}$$

(b) *required stiffness:*

$$\beta_{br} = k\left(\frac{10 M_r C_d}{L_b h_0}\right) \tag{3.41j}$$

where M_r is the required flexural strength using LFRD or ASD combinations, h_0 is the distance between flange centroids, expressed in inches and $C_d = 1.0$ except for the brace closest to the inflection point in a beam subject to double curvature bending, in which case it when $C_d = 2.0$ and $k = 1/\phi = 1/0.75$ (ASD); $k = \Omega = 2.00$ (LFRD).

(5) Nodal torsional beam bracings

(a) *required strength:*

$$M_{rb} = \frac{0.024 M_r L}{n C_b L_b} \tag{3.41k}$$

where C_b is the modification factor defined in Chapter F of AISC 360-10 (see Chapter 8), L_b is the length of span and n is the number of nodal braced points within the span.

(b) *required stiffness:*

$$\beta_{Tb} = \frac{\beta_T}{\left(1 - \frac{\beta_T}{\beta_{sec}}\right)} \tag{3.41l}$$

where β_T is the overall brace system stiffness and β_{sec} is the web distortional stiffness, including the effect of web transverse stiffeners.

These stiffness values are defined as:

$$\beta_T = k\left(\frac{2.4 L M_r^2}{n E I_y C_b^2}\right) \tag{3.41m}$$

$$\beta_{sec} = \frac{3.3E}{h_0}\left(\frac{1.5 h_0 t_w^3}{12} + \frac{t_{st} b_s^3}{12}\right) \tag{3.41n}$$

where E is the modulus of elasticity of steel, I_y is the out-of-plane moment of inertia, t_w and t_{st} are the thickness of the beam web and of the web stiffener, respectively, $k = 1/\phi = 1/0.75$ (ASD); $k = \Omega = 3.00$ (LFRD) and b_s is the stiffener width for one-sided stiffeners, or twice the individual.

If the torsional bracing is continuous, then the ratio L/n in Eqs. (3.41k) and (3.41m) has to be assumed as equal to unity.

3.7.3 Examples of Braced Frames

With reference to civil and commercial steel buildings, typical examples of braced multi-storey frames are represented in the Figures 3.47–3.49, which also propose the structural schemes to evaluate internal actions and reactions of the horizontal bracing systems. Figures 3.47 and 3.48

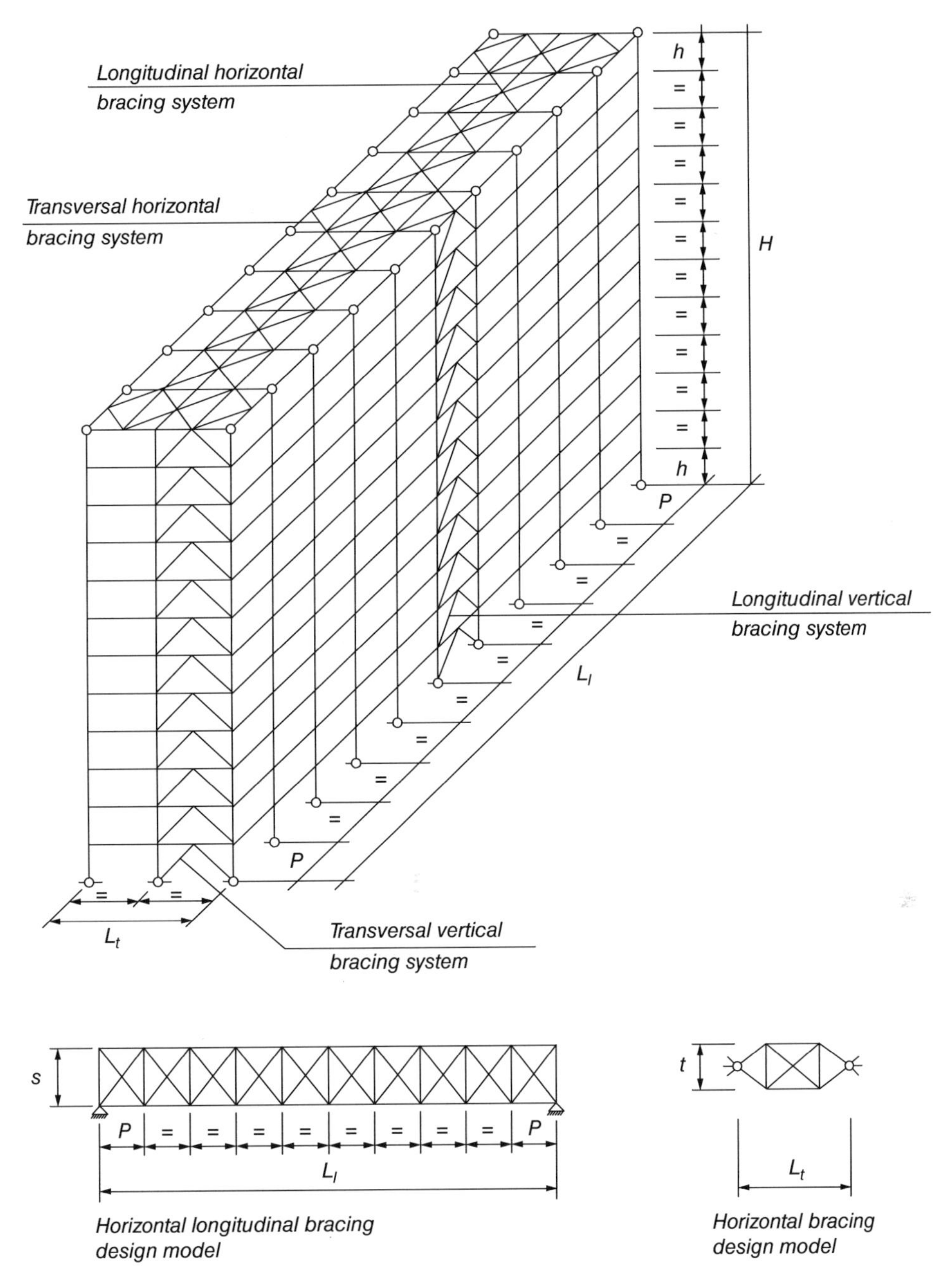

Figure 3.47 Example of a multi-storey braced frame and bracing models for horizontal bracings: two longitudinal and two transversal vertical bracings.

are related to two solutions of skeleton frames braced by steel systems. In the first building, vertical bracings are symmetrically located with reference to the principle axes in plant and the horizontal force transfer mechanism interests the longitudinal or transversal vertical bracing, depending on the considered wind direction. The case proposed in Figure 3.48 is related to the presence of three sole vertical bracings and, as a consequence, the couple of transversal vertical bracings are loaded when wind acts alternatively along both principle directions. Figure 3.49 proposes a multi-storey building braced by a concrete core containing stairs and uplifts. Two solutions are presented

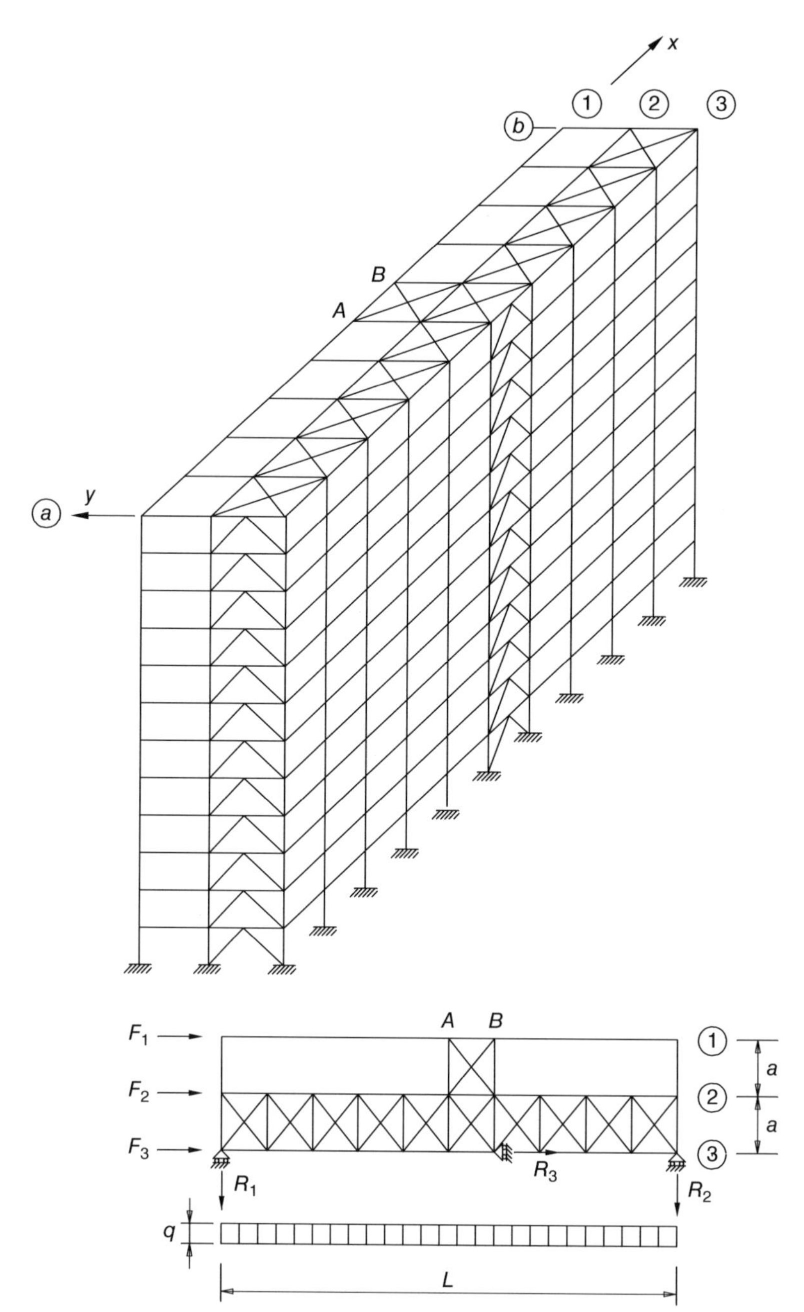

Figure 3.48 Example of a multi-storey braced frame and bracing models for horizontal bracings: one longitudinal and two transversal vertical bracings.

differing for the type of the core cross-section: open or closed (boxed) cross-section. For both cases, pre-sizing (preliminary design) can be developed by considering each wall of the core as an independent cantilever beam embedded at its base.

As for the industrial steelwork, Figure 3.50 represents the most common typologies of single-storey frames. It should be noted that also in the case of rigid frames, the roof bracing system is, however, required (Figure 3.50a) to stabilize triangulated roof beams, that is, to reduce their

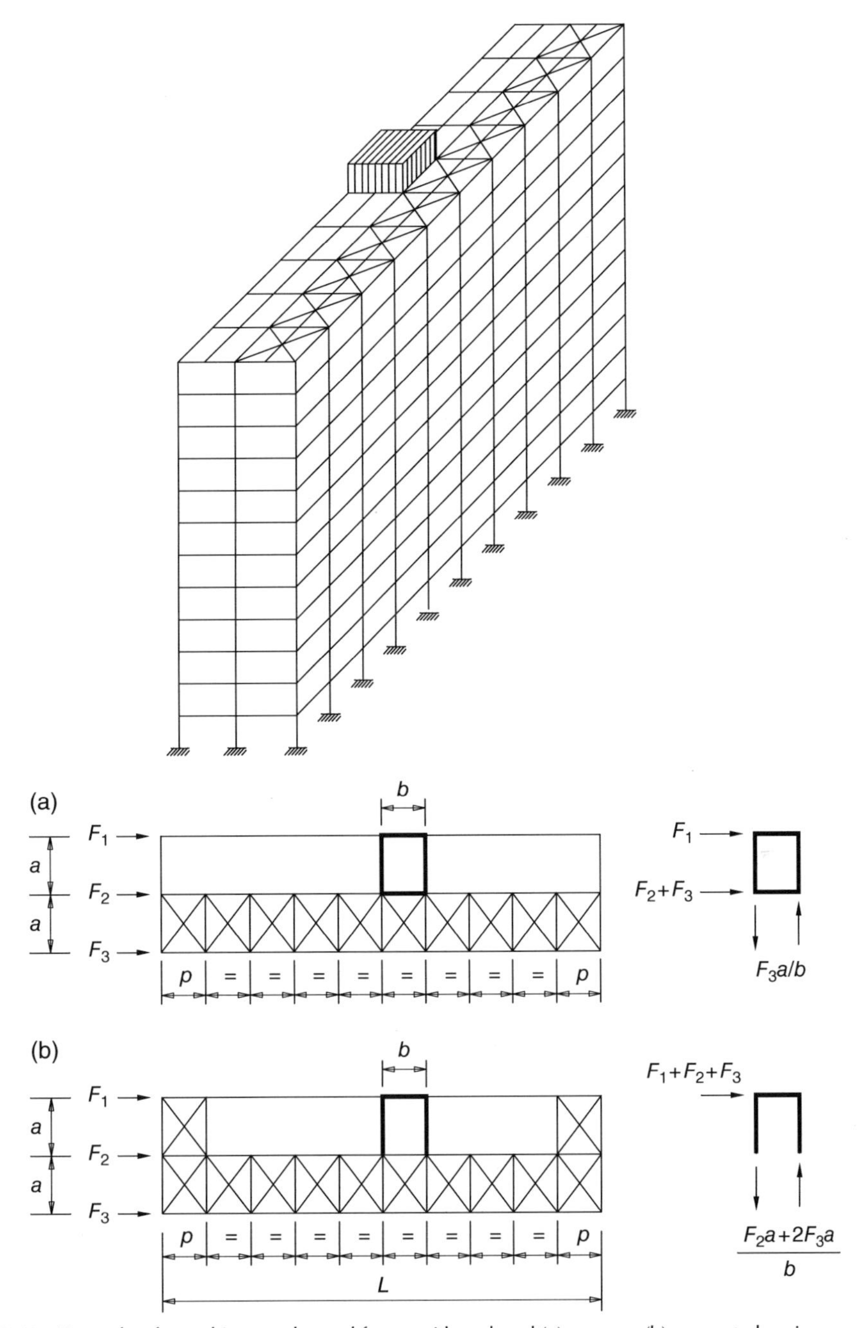

Figure 3.49 Example of a multi-storey braced frame with a closed (a) or open (b) concrete bracing core.

effective out-of-plane length. It is possible to also adopt hybrid structural scheme: a rigid frame in the transversal direction and a simple frame in the longitudinal direction (Figure 3.50b).

The bracing systems on the roof, which are required by the presence of the lattice beams, transfer the horizontal forces to vertical bracings. Figure 3.50c refers to the case of simple frame model to be adopted for the whole single storey building.

It should be noted that the bracing systems are also of fundamental importance in order to provide stability during the erection of the building.

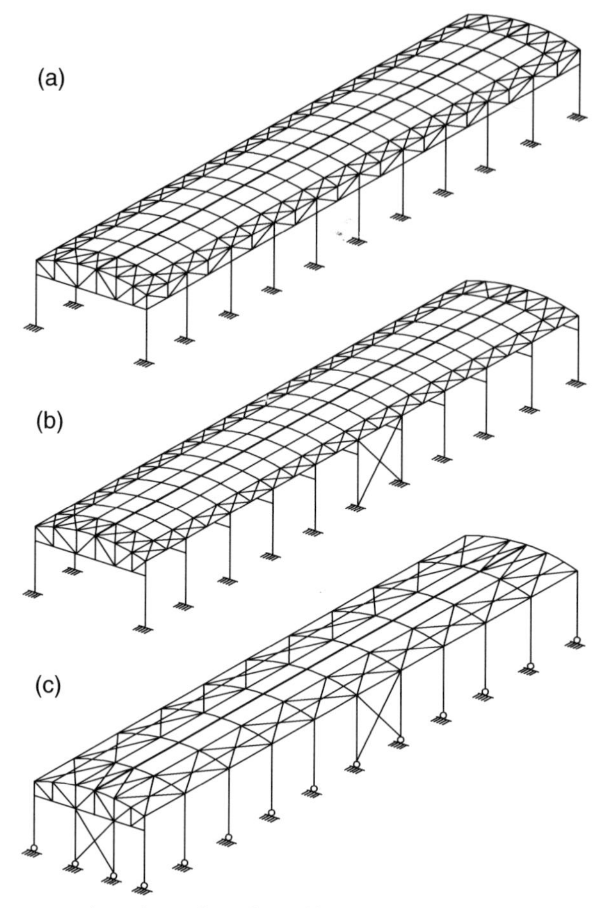

Figure 3.50 Common structural typologies for industrial buildings (a and b).

3.8 Worked Examples

Example E3.1 Individuation of the Bracing System

With reference to the frames in Figure E3.1.1, composed of two rigid portal frames connected via a non-deformable truss element, evaluate whether frame 2 braces frame 1.

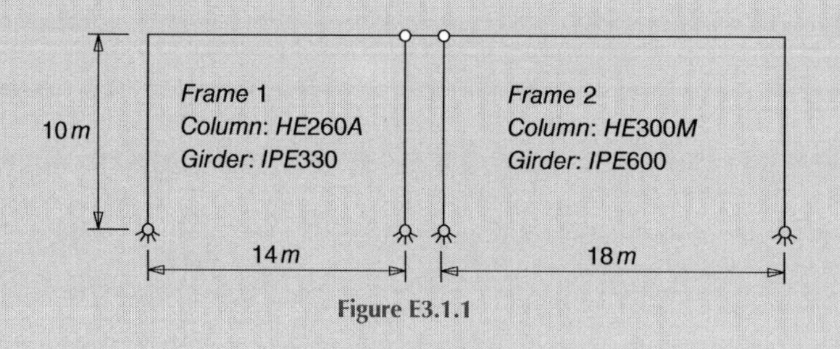

Figure E3.1.1

Procedure
Compare the transversal displacements of each frame (frame 1 and frame 2), generated by horizontal load F applied to the top (Figure E3.1.2).

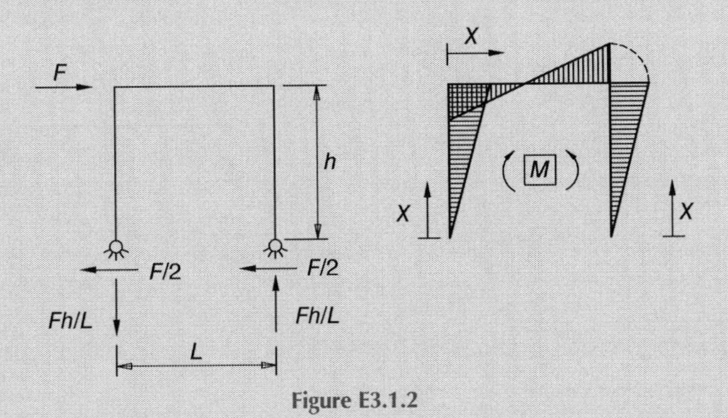

Figure E3.1.2

By neglecting axial member deformability, transversal displacement δ can be evaluated by using the Principle of Virtual Work, as:

$$\delta = \int_{frame} \frac{M(x)M^I(x)}{EI} dx$$

where x is the generic coordinate along the frame, $M(x)$ represents the distribution of the bending moment due to horizontal load F applied to the top frame and $M^I(x)$ is the moment associated with the service load (unitary horizontal load F applied to the top frame).

With reference to Figure E3.1.2 that shows the moment diagram, the expressions of $M(x)$ and $M^I(x)$, distinguished for the components (columns and beams) by the presence of the multiplying factor, F, are:

- for *columns*:

$$M(x) = \frac{F}{2}x \qquad M^I(x) = \frac{1}{2}x$$

- for *beams*:

$$M(x) = \frac{Fh}{2}\left(1 - \frac{2x}{L}\right) \quad M^I(x) = \frac{h}{2}\left(1 - \frac{2x}{L}\right)$$

Alternatively, horizontal displacement could be evaluated approximately, considering beams as perfectly rigid and therefore taking into account the bending deformability of columns only.

Solution
The moment of inertia values of the beam and column cross-sections are:

- *beams*: for IPE 330 $I_b = 11\,770$ cm^4 (282.8 in.^{4}) and for IPE 600 $I_b = 92\,080$ cm^4 (2212 in.^{4});
- *columns*: for HE 260 A $I_c = 10\,450$ cm^4 (251.1 in.^{4}) and for HE 300M $I_c = 59\,200$ cm^4 (1422 in.^{4}).

Frame displacement is computed via the expression:

$$\delta = 2\int_0^h \frac{1}{EI_c}\left(\frac{F}{2}x\right)\cdot\left(\frac{1}{2}x\right)\mathrm{d}x + \int_0^L \frac{1}{EI_b}\frac{Fh}{2}\left(1-\frac{2x}{L}\right)\cdot\frac{h}{2}\left(1-\frac{2x}{L}\right)\mathrm{d}x$$

By solving integrals, it results in:

$$\delta = 2\frac{Fh^3}{12EI_c} + \frac{Fh^2L}{12EI_b}$$

Applying this formula to frame 1, it results $\delta = 0.0123\ F$, hence the frame stiffness is $K_{frame1} = \frac{F}{\delta} = 81.20\frac{\mathrm{kN}}{\mathrm{m}}$ (5.56 kips/ft). For frame 2 it results $\delta = 0.0021\ F$, hence the frame stiffness is $K_{frame2} = \frac{F}{\delta} = 472.51\frac{\mathrm{kN}}{\mathrm{m}}$ (32.38 kips/ft).

The ratio of the lateral stiffness of the two frames is:

$$\frac{K_{frame2}}{K_{frame1}} = \frac{472.51}{81.20} = 5.82$$

So the result is that frame 2 acts efficiently as bracing system for frame 1.

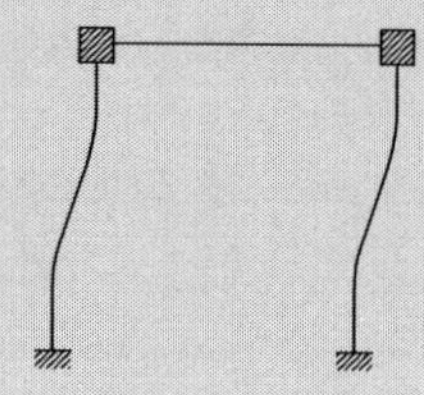

Figure E3.1.3

By neglecting shear deformation of the beam and considering its axial and flexural stiffness as infinite, it is possible to refer to the auxiliary structure in Figure E3.1.3. The approximate stiffness of generic frame K^* is given as:

$$K^* = \frac{F}{\delta} = \frac{12EI_c}{h^3}$$

The ratio between the approximate stiffness of the two frames is:

$$\frac{K^*_{frame2}}{K^*_{frame1}} = \frac{\dfrac{12EI_c^{(2)}}{h^3}}{\dfrac{12EI_c^{(1)}}{h^3}} = \frac{I_c^{(1)}}{I_c^{(2)}} = \frac{59200}{10450} = 5.67$$

This simplified method, neglecting the beam deformation, provides the result that frame 2 acts efficiently as a bracing system for frame 1.

Example E3.2 Selection of the EU Analysis Design Approach

With reference to the cantilever beam in Figure E3.2.1, for which a 100 mm (3.94 in.) width hollow square section (HSS) is used, evaluate the sensibility to transversal displacements (i.e. classify the frame), when:

Case (a): HSS thickness is 5 mm (0.197 in.) and

$$I = 283 \times 10^4 \text{mm}^4 (6.78 \text{ in.}^4);$$

Case (b): HSS thickness is 10 mm (0.394 in.) and

$$I = 474 \times 10^4 \text{mm}^4 (11.39 \text{ in.}^4);$$

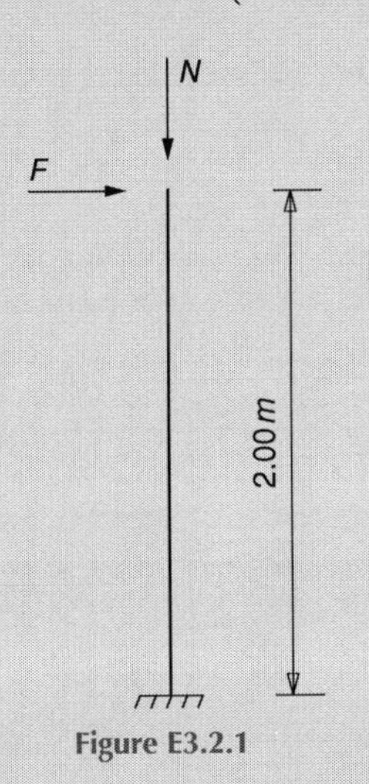

Figure E3.2.1

Free end is loaded by an axial force $N = 50$ kN (11.24 kips) and a transversal force $F = 5$ kN (1.12 kips).

Procedure
According to the Eurocode prescriptions, the elastic buckling load has to be evaluated, for both loading cases, assuming: $L_0 = 2h = 4000$ mm (13.12 ft). Reference is made to Euler's theory and to the following formula:

$$F_{cr} = \frac{\pi^2 EI}{L_0^2}$$

Solution
Main results and indications about the type of analysis are listed in Table E3.2.1.

Table E3.2.1 Indications of analysis type according to European codes.

	Case (a)	Case (b)
F_{Ed}	50 kN (11.24 kips)	50 kN (11.24 kips)
$F_{cr} = \frac{\pi^2 EI}{L_0^2}$	366.59 kN (82.4 kips)	614.01 kN (138.0 kips)
F_{cr}/F_{Ed}	7.33	12.28
Analysis type: elastic	Second order	First order
Analysis type: plastic	Second order	Second order

Using an engineering approach, the bending moment at column base has to be evaluated, taking into account the first order term only, equal to $F \cdot h$, and evaluating second order moment as a function of displacement at column top, δ, with the following formula:

$$N \cdot \delta = N \cdot \left(\frac{Fh^3}{3EI} \right)$$

The calculation results are shown in Table E3.2.2. Also in this case, for the thinner tube a second order elastic analysis is always required.

Table E3.2.2 Indications of the type of analysis using an engineering approach.

	Case a	Case b
$F \cdot h$	10 kNm (7.38 kip-ft)	10 kNm (7.38 kip-ft)
$N \cdot \delta = N \cdot \left(\frac{Fh^3}{3EI} \right)$	1.1218 kNm (0.827 kip-ft)	0.6697 kNm (0.494 kip-ft)
$N \cdot \delta / F \cdot h$	0.1122	0.0700

Example E3.3 Second Order Approximate Analysis via the Equivalent Lateral Force Procedure

With reference to the cantilever beam in Figure E3.3.1, approximate its second order frame response in the following cases:

(a) HE 280 A member: moment of inertia I = 13 670 cm^4 (328.4 in.^{4});
(b) HE 160 M member: moment of inertia I = 5098 cm^4 (122.5 in.^{4}).

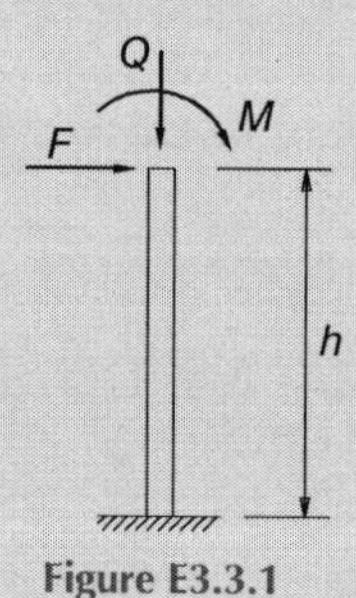

Figure E3.3.1

At the cantilever top Q, F and M act where:

- Q is the axial load equal to 640 kN (143.9 kips)
- F is the lateral force equal to 10 kN (2.25 kips)
- M is the bending moment equal to 40 kNm (29.5 kip-ft).

Procedure
In these applications the influence of the axial deformability of the column on the value of the top lateral displacement is neglected.

In order to appraise the convergence of the approach, top horizontal cantilever displacement is considered defining a tolerance limit equal to 2%.

Case (a): Column realized with a Profile HE 280 A.

The main iterative calculations are herein reported.

- *Iteration* 0: First order elastic analysis of cantilever loaded as stated before.
  - top lateral displacement calculation:

$$\delta_0 = \frac{Fh^3}{3EI} + \frac{Mh^2}{2EI} = 0.0502 \text{ m } (1.98 \text{ in.})$$

  - additional shear force calculation:

$$\Delta F_0 = \frac{Q\delta_0}{h} = 5.35 \text{ kN } (1.20 \text{ kips})$$

  - incremented applied load:

$$F_1 = F + \Delta F_0 = 15.35 \text{ kN}(3.45 \text{ kips})$$

- *Iteration 1*: First order elastic analysis of cantilever consequently to iteration 0.
  - top lateral displacement calculation:

$$\delta_1 = \frac{F_1 h^3}{3EI} + \frac{Mh^2}{2EI} = 0.0636 \text{ m } (2.50 \text{ in.})$$

  - additional shear force calculation:

$$\Delta F_1 = \frac{Q\delta_1}{h} = 6.78 \text{ kN } (1.52 \text{ kips})$$

  - incremented applied load:

$$F_2 = F + \Delta F_1 = 16.78 \text{ kN}(3.77 \text{ kips})$$

  - convergence check:

$$\frac{\delta_1}{\delta_0} = 1.27\,(>2\%)$$

- *Iteration 2*: First order elastic analysis of cantilever loaded consequently to iteration 1.
  - top lateral displacement calculation:

$$\delta_2 = \frac{F_2 h^3}{3EI} + \frac{Mh^2}{2EI} = 0.0672 \text{ m } (2.65 \text{ in.})$$

- additional shear force calculation:

$$\Delta F_2 = \frac{Q\delta_2}{h} = 7.17 \text{ kN (1.61 kips)}$$

- incremented applied load:

$$F_3 = F + \Delta F_2 = 17.17 \text{ kN(3.86 kips)}$$

- convergence check:

$$\frac{\delta_2}{\delta_1} = 1.06 \;\; (>2\%)$$

- *Iteration 3*: First order elastic analysis of cantilever loaded consequently to iteration 2.
  - top lateral displacement calculation:

$$\delta_3 = \frac{F_3 h^3}{3EI} + \frac{Mh^2}{2EI} = 0.0681 \text{ m (2.68 in.)}$$

  - additional shear force calculation:

$$\Delta F_3 = \frac{Q\delta_3}{h} = 7.27 \text{ kN (1.63 kips)}$$

  - incremented applied load:

$$F_4 = F + \Delta F_3 = 17.27 \text{ kN(3.88 kips)}$$

  - convergence check:

$$\frac{\delta_3}{\delta_2} = 1.01 \leq 2\%$$

Second order response of the structure has been evaluated, estimating a top lateral displacement equal to 68.1 mm (2.68 in.), that is greater by about 36% than the one found by first order analysis. Comparing first and second order bending moments at the column base, a difference of about 44% can be observed, that is first order analysis underestimates considerably the bending moment at the column base due to lateral deflection.

Case (b): Column realized with a profile HE 160 M.

Iterative calculations are herein reported, considering in this case a more flexible column.

- *Iteration 0*: First order elastic analysis of cantilever loaded as stated before.
  - top lateral displacement calculation:

$$\delta_0 = \frac{Fh^3}{3EI} + \frac{Mh^2}{2EI} = 0.1345 \text{ m (5.30 in.)}$$

– additional shear force calculation:

$$\Delta F_0 = \frac{Q\delta_0}{h} = 14.35 \text{ kN } (3.23 \text{ kips})$$

– incremented applied load:

$$F_1 = F + \Delta F_0 = 24.35 \text{ kN} (5.47 \text{ kips})$$

- *Iteration 1*: First order elastic analysis of cantilever loaded consequently to iteration 0.
  – top lateral displacement calculation:

$$\delta_1 = \frac{F_1 h^3}{3EI} + \frac{Mh^2}{2EI} = 0.2310 \text{ m } (9.09 \text{ in.})$$

  – additional shear force calculation:

$$\Delta F_1 = \frac{Q\delta_1}{h} = 24.64 \text{ kN } (5.54 \text{ kips})$$

  – incremented applied load:

$$F_2 = F + \Delta F_1 = 34.64 \text{ kN} (7.79 \text{ kips})$$

  – convergence check:

$$\frac{\delta_1}{\delta_0} = 1.72 \ \ (>2\%)$$

- *Iteration 2*: First order elastic analysis of cantilever loaded consequently to iteration 1.
  – top lateral displacement calculation:

$$\delta_2 = \frac{F_2 h^3}{3EI} + \frac{Mh^2}{2EI} = 0.3002 \text{ m } (11.82 \text{ in.})$$

  – additional shear force calculation:

$$\Delta F_2 = \frac{Q\delta_2}{h} = 32.02 \text{ kN } (7.20 \text{ kips})$$

  – incremented applied load:

$$F_3 = F + \Delta F_2 = 42.02 \text{ kN} (9.45 \text{ kips})$$

  – convergence check:

$$\frac{\delta_2}{\delta_1} = 1.30 \ \ (>2\%)$$

- *Iteration 3*: First order elastic analysis of cantilever loaded consequently to iteration 2.
  - top lateral displacement calculation:

$$\delta_3 = \frac{F_3 h^3}{3EI} + \frac{Mh^2}{2EI} = 0.3499 \text{ m } (13.78 \text{ in.})$$

  - additional shear force calculation:

$$\Delta F_3 = \frac{Q\delta_3}{h} = 37.32 \text{ kN } (8.39 \text{ kips})$$

  - incremented applied load:

$$F_4 = F + \Delta F_3 = 47.32 \text{ kN} (10.64 \text{ kips})$$

  - convergence check:

$$\frac{\delta_3}{\delta_2} = 1.17 > 2\%$$

- *Iteration 4*: First order elastic analysis of cantilever loaded consequently to iteration 3.
  - top lateral displacement calculation:

$$\delta_4 = \frac{F_4 h^3}{3EI} + \frac{Mh^2}{2EI} = 0.3855 \text{ m } (15.18 \text{ in.})$$

  - additional shear force calculation:

$$\Delta F_4 = \frac{Q\delta_4}{h} = 41.12 \text{ kN } (9.24 \text{ kips})$$

  - incremented applied load:

$$F_5 = F + \Delta F_4 = 51.12 \text{ kN} (11.49 \text{ kips})$$

  - convergence check:

$$\frac{\delta_4}{\delta_3} = 1.10 \ (> 2\%)$$

It must be underlined that, by reducing the flexural stiffness of the column by about two-thirds, convergence becomes slower, as can be detected by iterations 7 and 8, listed next.

- *Iteration 7*: First order elastic analysis of cantilever loaded consequently to iteration 6 (iterations 5 and 6 have been omitted for the sake of simplicity).
  - top lateral displacement calculation:

$$\delta_7 = \frac{F_7 h^3}{3EI} + \frac{Mh^2}{2EI} = 0.4425 \text{ m } (17.42 \text{ in.})$$

– additional shear force calculation:

$$\Delta F_7 = \frac{Q\delta_7}{h} = 47.20 \text{ kN } (10.61 \text{ kips})$$

– incremented applied load:

$$F_8 = F + \Delta F_7 = 57.20 \text{ kN}(12.86 \text{ kips})$$

– convergence check:

$$\frac{\delta_7}{\delta_6} = 1.03 \ (> 2\%)$$

- *Iteration* 8: First order elastic analysis of cantilever loaded consequently to iteration 7.
  – top lateral displacement calculation:

$$\delta_8 = \frac{F_8 h^3}{3EI} + \frac{Mh^2}{2EI} = 0.4520 \text{ m } (17.80 \text{ in.})$$

  – additional shear force calculation:

$$\Delta F_8 = \frac{Q\delta_8}{h} = 48.21 \text{ kN } (10.84 \text{ kips})$$

  – incremented applied load:

$$F_9 = F + \Delta F_8 = 58.21 \text{ kN}(13.09 \text{ kips})$$

  – convergence check:

$$\frac{\delta_8}{\delta_7} = 1.02$$

Remarks

It should be noted that in case (b) with a more flexible column, the number of iterations required for satisfying convergence criteria is greater than the one of case (a) with a stiffer column. To explain this difference, the curve plotted in Figure E3.3.2 can be considered, the lateral displacement is plotted versus the ratio between the load multiplier (α) over the critical one (α_{cr}).

In case (a), with a stiffer member, a second order approximate analysis is performed for a structure with an elastic flexural buckling load equal to:

$$P_{cr} = \frac{\pi^2 EI}{L_0^2} = \frac{\pi^2 \cdot 210000 \cdot (13670 \cdot 10^4)}{(2 \cdot 6000)^2} = 1967.5 \text{ kN } (442.3 \text{ kips})$$

Axial load acting on the structure (640 kN, 143.9 kips) is considerably lower than buckling load (640/1967.5 = 0.325).

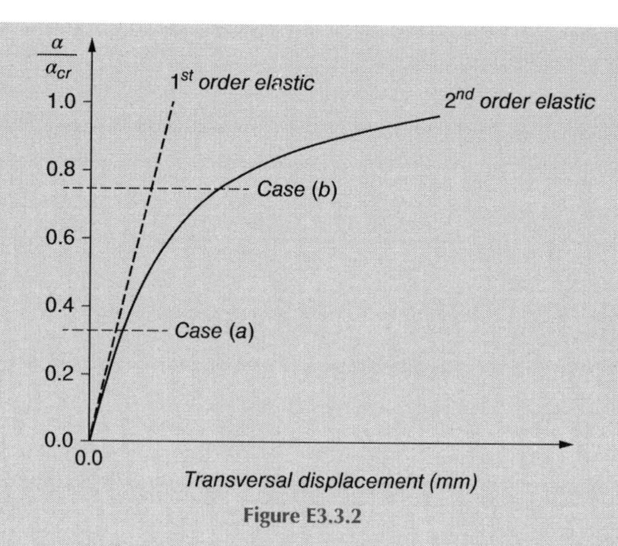

Figure E3.3.2

In case (b), with a less stiff column, a second order approximate analysis is performed for a structure with an elastic flexural buckling load equal to:

$$P_{cr}=\frac{\pi^2 EI}{L_0^2}=\frac{\pi^2\cdot 210000\cdot(5098\cdot 10^4)}{(2\cdot 6000)^2}=733.8\ \text{kN}\ (165.0\ \text{kips})$$

Axial load acting on the structure (640 kN, 143.9 kips) is significantly close to the buckling load (640/733.8 = 0.872).

In Figure E3.3.2 both the cases are summarized in a non-dimensional form for that which concerns the axial load, referring to the actual multiplier over the critical one versus the lateral displacement. The dotted horizontal lines indicate directly the load level at which second order elastic analysis has been conducted for the (a) and (b) cases. It can be noted that, in case (b) in which the applied load is very close to the buckling load, second order displacement is remarkably far greater than for the first order one.

CHAPTER 4

Cross-Section Classification

4.1 Introduction

As previously discussed in Section 1.1, the *steel* material is characterized by a symmetrical mono-axial stress-strain (σ–ε) constitutive law, which can be determined by monotonic tension tests on samples taken from the base material before the working process or from the products in correspondence of appropriate locations. The response of steel members can, however, be significantly different in tension or compression, owing to the relevant influence of the buckling phenomena. The instability of compressed steel members as well as of all the members realized with other materials can be distinguished in:

- *overall buckling* or Euler buckling, which affects the element throughout its length (or a relevant portion of it). More details can be found in Chapter 6 for columns, interested by flexural, torsional and flexural-torsional buckling modes, in Chapter 7 for beams, interested by lateral-torsional buckling modes and in Chapter 9 for beam-columns subjected to a complex interaction between axial and flexural instability;
- *local buckling*, already introduced in Section 3.6.1, which affects the compressed plates forming the cross-section, characterized by relatively short wavelength buckling.

Furthermore, there is a third type of instability, the so-called *distortional buckling*, which has been extensively investigated in recent decades. As the term directly suggests, this buckling mode takes place as a consequence of the distortion of cross-sections (Figure 4.1): with reference to thin-walled members, that is the members mainly interested by this phenomenon, distortional buckling is characterized by relative displacements of the fold-line of the cross-section and the associated wave-length is generally in the range delimited by one of local buckling and one of global buckling.

It should be noted that local and distortional buckling, which can be considered 'sectional modes', can interact with each other and the design of cold formed steel members is very complex. For the European design, the reference is EC3-1-3 (*Supplementary rules for cold-formed members and sheetings*) and for the US the Code provisions governing design of these members are the American Iron and Steel Institute (AISI) 'North American specification for the design of cold-formed of steel members'.

As already mentioned in Chapter 3, the classification of a cross-section is necessary in order to select the appropriate analysis method as well as the suitable approaches to the member

Structural Steel Design to Eurocode 3 and AISC Specifications, First Edition. Claudio Bernuzzi and Benedetto Cordova.

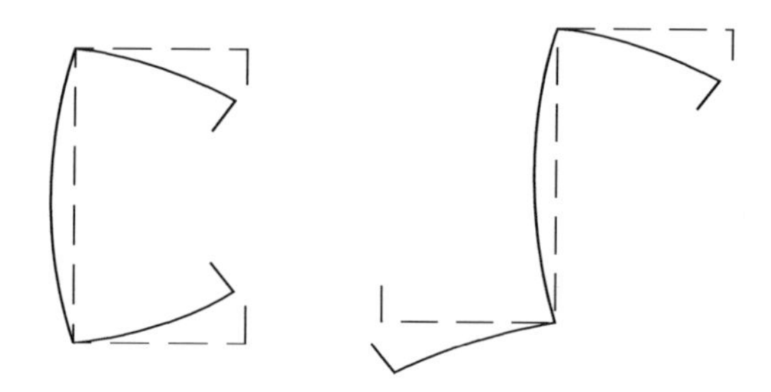

Figure 4.1 Typical deformed cross-section for distortional buckling.

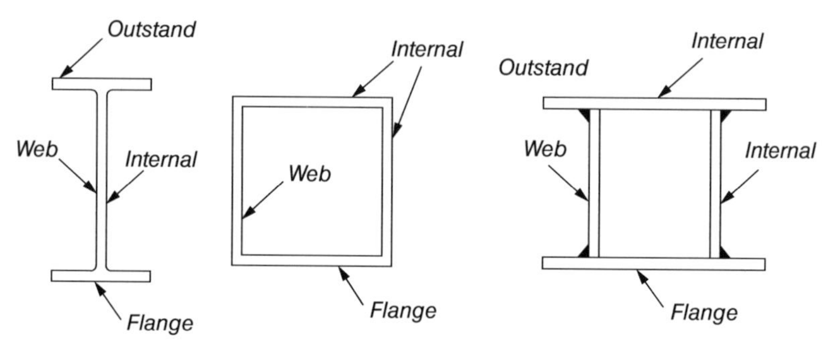

Figure 4.2 Internal or outstand elements.

verification checks. Furthermore, with reference to earthquakes and, consequently, seismic loads, the design strategy is based on the so-called *capacity design*, and the classification of profiles is very important too, owing to the role played by the post-elastic ductile response.

Generally speaking, any cross-section is composed of different plate elements, such as flanges and webs, which fall into in two categories (Figure 4.2):

- *internal* or *stiffened elements*, simply supported along two edges parallel to the direction of the compressive stress (longitudinal axis of the element);
- *outstand* (*external*) or *unstiffened elements*, simply supported along one edge and free on the other edge parallel to the direction of the compressive stress.

The cross-section classification depends mainly on the width-to-thickness ratio of each plate, either totally or partially in compression.

4.2 Classification in Accordance with European Standards

The cross-section classification has already been introduced in Chapter 3 in terms of performances guaranteed by the four classes and the associated moment-curvature relationships have been presented in Figure 3.31. The requirements for the classification criteria are proposed in the general part of EC3 (i.e. EN 1993-1-1): the limiting proportions for compression elements of class 1–3 are presented in the following Tables 4.1–4.3. When any of the compression elements of a cross-section does not fulfil the limits given in these tables, the section is classified as *slender* (class 4) and local buckling must be adequately taken into account into design by defining effective cross-sections.

As can be noted from the tables, the limiting value of the width-to-thickness ratio (b/t) of the generic plate element under compression depends on the steel grade via a suitable reduction material factor $\varepsilon = \sqrt{235/f_y}$, where f_y is the yield strength of the considered steel (expressed in

Table 4.1 Maximum width-to-thickness ratios for compression elements, from EN 1993-1-1: Table 5.2 (sheet 1 of 3).

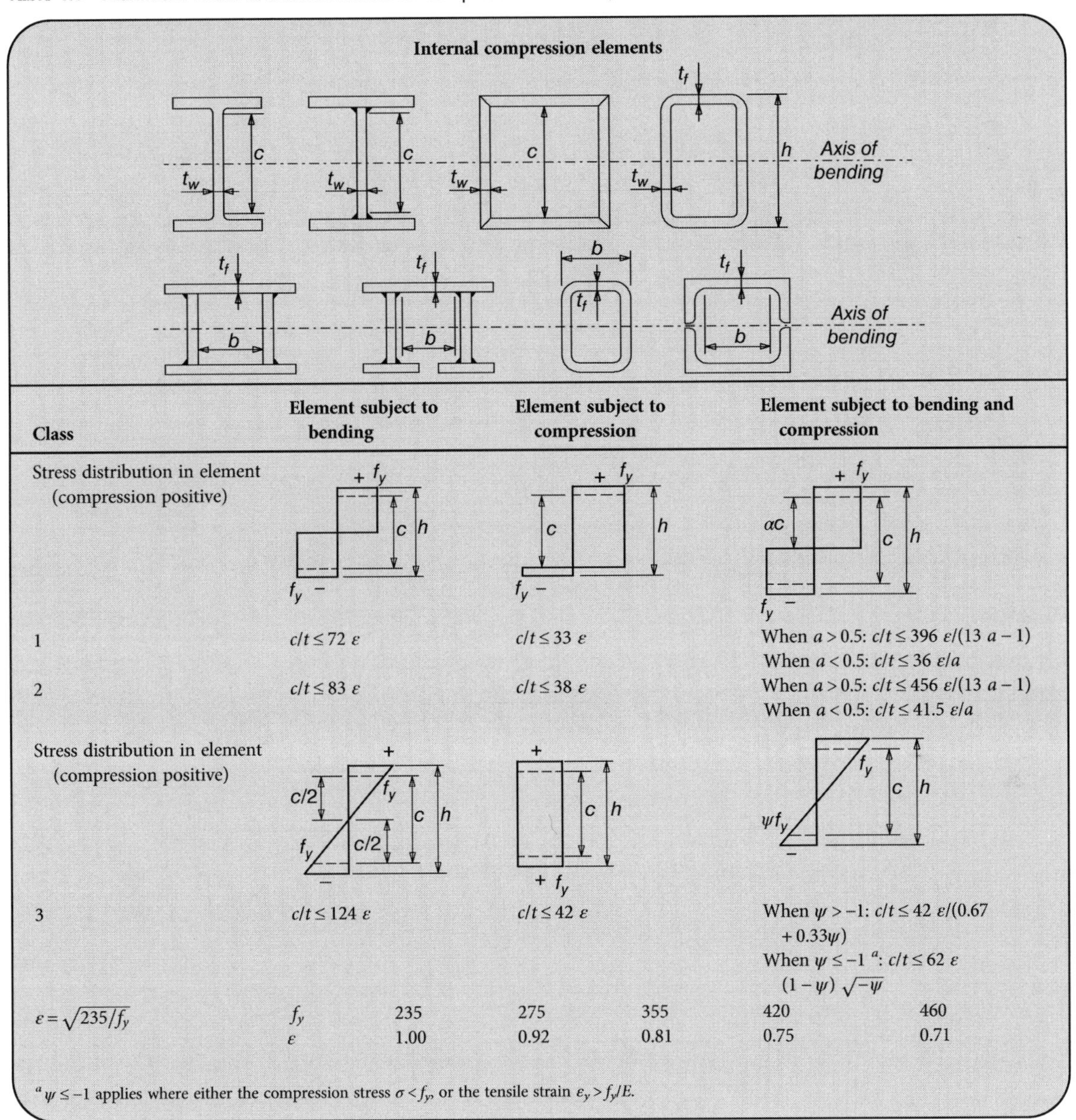

Internal compression elements

Class	Element subject to bending	Element subject to compression	Element subject to bending and compression
Stress distribution in element (compression positive)			
1	$c/t \le 72\,\varepsilon$	$c/t \le 33\,\varepsilon$	When $a > 0.5$: $c/t \le 396\,\varepsilon/(13\,a - 1)$ When $a < 0.5$: $c/t \le 36\,\varepsilon/a$
2	$c/t \le 83\,\varepsilon$	$c/t \le 38\,\varepsilon$	When $a > 0.5$: $c/t \le 456\,\varepsilon/(13\,a - 1)$ When $a < 0.5$: $c/t \le 41.5\,\varepsilon/a$
Stress distribution in element (compression positive)			
3	$c/t \le 124\,\varepsilon$	$c/t \le 42\,\varepsilon$	When $\psi > -1$: $c/t \le 42\,\varepsilon/(0.67 + 0.33\psi)$ When $\psi \le -1$ [a]: $c/t \le 62\,\varepsilon\,(1-\psi)\sqrt{-\psi}$

$\varepsilon = \sqrt{235/f_y}$					
f_y	235	275	355	420	460
ε	1.00	0.92	0.81	0.75	0.71

[a] $\psi \le -1$ applies where either the compression stress $\sigma < f_y$, or the tensile strain $\varepsilon_y > f_y/E$.

MPa). In more general cases, the compression elements forming a cross-section under compression could belong to different classes and the cross-section has to be classified on the basis of the most unfavourable (highest) class of its compression elements. Table 4.1 proposes the classification criteria for internal compression elements, Tables 4.2 and 4.3 are related to the classification of outstand flanges and of angles and tubular circular cross-sections, respectively.

If a cross-section has a class 3 web and class 1 or 2 flanges, it should be classified as a class 3 cross-section, which from the design point of view can achieve the elastic moment without any spreading of plasticity along the cross-section and member. In this case, an alternative is admitted: the web can be treated as an equivalent class 2 web containing a hole in its compressed parts

Table 4.2 Maximum width-to-thickness ratios for compression elements from EN 1993-1-1: Table 5.2 (sheet 2 of 3).

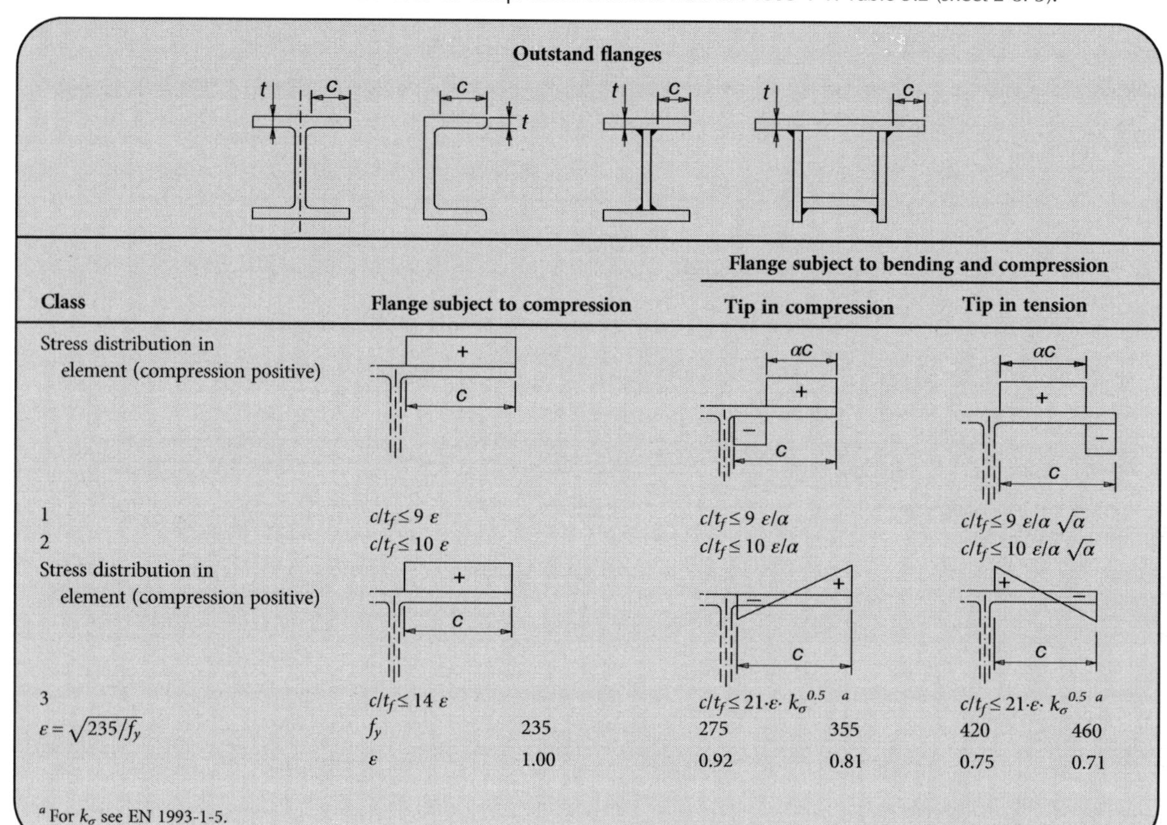

Outstand flanges

Class	Flange subject to compression	Flange subject to bending and compression: Tip in compression	Flange subject to bending and compression: Tip in tension
Stress distribution in element (compression positive)			
1	$c/t_f \le 9\,\varepsilon$	$c/t_f \le 9\,\varepsilon/\alpha$	$c/t_f \le 9\,\varepsilon/\alpha\sqrt{\alpha}$
2	$c/t_f \le 10\,\varepsilon$	$c/t_f \le 10\,\varepsilon/\alpha$	$c/t_f \le 10\,\varepsilon/\alpha\sqrt{\alpha}$
Stress distribution in element (compression positive)			
3	$c/t_f \le 14\,\varepsilon$	$c/t_f \le 21 \cdot \varepsilon \cdot k_\sigma^{0.5}$ [a]	$c/t_f \le 21 \cdot \varepsilon \cdot k_\sigma^{0.5}$ [a]

$\varepsilon = \sqrt{235/f_y}$					
f_y	235	275	355	420	460
ε	1.00	0.92	0.81	0.75	0.71

[a] For k_σ see EN 1993-1-5.

(Figure 4.3). As a consequence, reference can be made to a class 2 profile and the effective parts (i.e. the ones contributing to the cross-section resistance) of this web have a length equal to $20 \cdot \varepsilon \cdot t_w$, where ε is the reduction material factor previously defined. The main advantage of this approach is the possibility of using the plastic verification criteria for cross-section checks.

4.2.1 Classification for Compression or Bending Moment

Classification of cross-sections for axial load or for the bending moment only depends on the geometrical and mechanical parameters, as shown in the worked examples proposed in Section 4.4.

4.2.2 Classification for Compression and Bending Moment

The presence of both the compression and bending moment on the generic plate (or on the cross-section member) generates a stress distribution between that related to pure compression and that associated with the presence of the sole bending moment. In many cases, a plate component belongs to the same class under compression and under flexure: as a consequence, when they act simultaneously, the cross-section class is directly determined, that is it is the same as for components subjected to an axial load and bending moment. Otherwise, in case of classes that are different for compression and bending, the class for compression and bending must be evaluated by considering the values of the design force and moment (N_{Ed} and M_{Ed}, respectively). In the following, reference is made for compression combined with bending for an internal web and this

Table 4.3 Maximum width-to-thickness ratios for compression elements, from EN 1993-1-1: Table 5.2 (sheet 3 of 3).

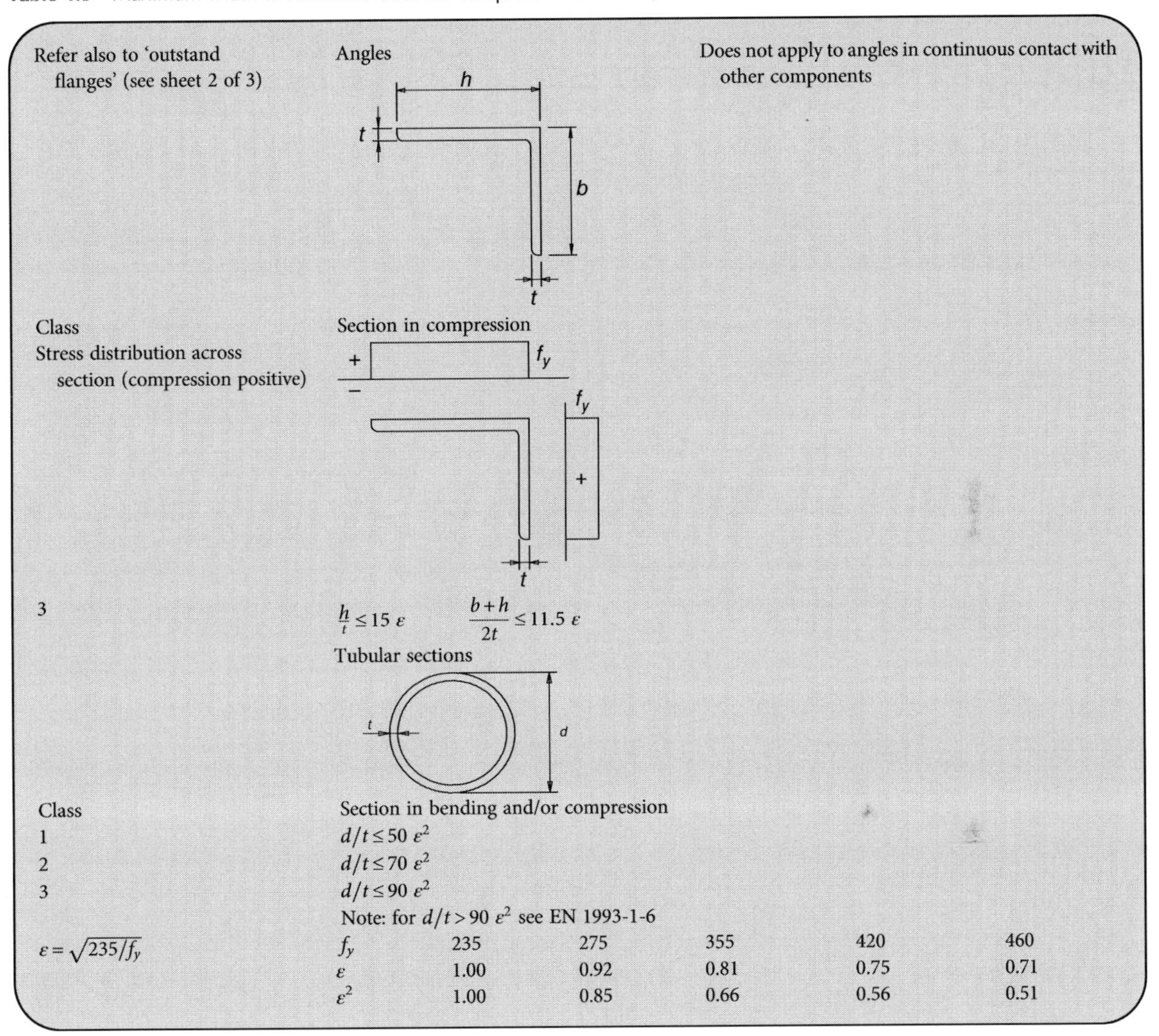

Refer also to 'outstand flanges' (see sheet 2 of 3)	Angles		Does not apply to angles in continuous contact with other components		
Class	Section in compression				
Stress distribution across section (compression positive)					
3	$\frac{h}{t} \leq 15\ \varepsilon$	$\frac{b+h}{2t} \leq 11.5\ \varepsilon$			
	Tubular sections				
Class	Section in bending and/or compression				
1	$d/t \leq 50\ \varepsilon^2$				
2	$d/t \leq 70\ \varepsilon^2$				
3	$d/t \leq 90\ \varepsilon^2$				
	Note: for $d/t > 90\ \varepsilon^2$ see EN 1993-1-6				
$\varepsilon = \sqrt{235/f_y}$	f_y 235	275	355	420	460
	ε 1.00	0.92	0.81	0.75	0.71
	ε^2 1.00	0.85	0.66	0.56	0.51

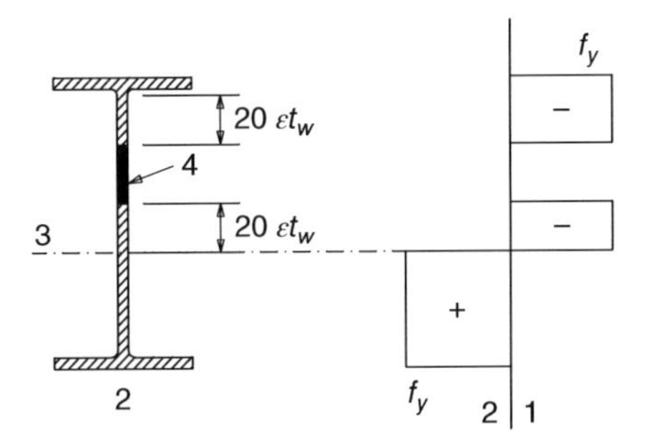

Figure 4.3 Effective class 2 web method: compression (1), tension (2), plastic neutral axis (3) and neglected area (4) of the web.

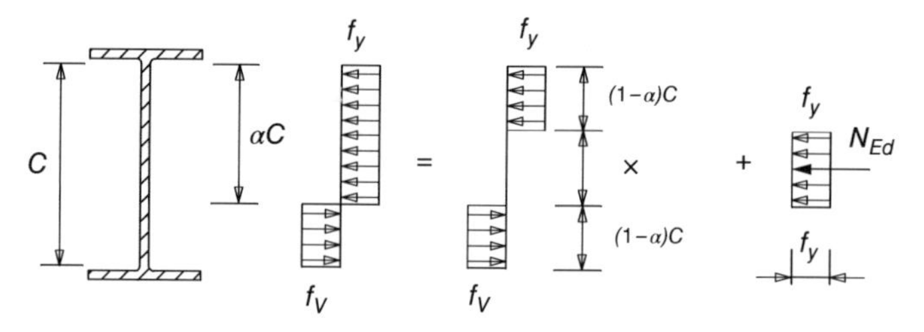

Figure 4.4 Superimposition of the stress distribution diagram associated with a plastic web under compression and bending.

case is typical for doubly symmetrical profiles under eccentric axial load or mono-symmetric profiles under pure bending.

4.2.2.1 Bending and Compression about a Strong Axis

If internal elements (web) are considered, reference has to be made to Table 4.1. Normal stress distribution on the web depends on the value of the design axial load by means of parameter α for profiles able to resist in the plastic range (classes 1 and 2). Otherwise, in case of elastic normal stress distribution, reference has to be made to parameter ψ (classes 3 and 4). With reference to the case of a neutral axis located in the web, α ranges between 0.5 (bending) and 1 (compression) and ψ ranges between -1 (bending) and 1 (compression).

With reference to the normal stress plastic distribution, it should be convenient to use the superposition principle separately considering the stress distribution associated with the axial load (N_{Ed}) and the one associated with the bending moment (M_{Ed}) (see Figure 4.4). It can be noted that N_{Ed} acts on the central part of the web on a zone of extension x, as it results from equilibrium condition:

$$N_{Ed} = (x t_w) f_y \tag{4.1a}$$

where t_w is the thickness of the web and f_y is the yield strength.

The design bending moment ($M_{Ed(N)}$) associated with the stress distribution in Figure 4.4, is given by:

$$M_{Ed(N)} = M_{pl} - \frac{(t_w \cdot x^2) f_y}{4} \tag{4.1b}$$

where M_{pl} is the plastic flexural resistance of the member.

The depth (x) of the web under pure axial load can be expressed as:

$$x = (2\alpha - 1)c \tag{4.2a}$$

where factor α identifies the contribution of the web subjected to compression force, expressed as $\alpha \cdot c$ where c is the web depth.

The axial design load, N_{Ed}, and the correspondent bending design moment, $M_{Ed(N)}$, can be expressed, respectively, as:

$$N_{Ed} = [(2\alpha - 1)c] t_w f_y \tag{4.2b}$$

$$M_{Ed(N)} = M_{pl} - \frac{t_w [(2\alpha - 1)\, c]^2 f_y}{4} \tag{4.2c}$$

Factor α depends strictly on the value of the acting axial load as:

$$\alpha = \frac{1}{2}\left(1 + \frac{1}{c} \cdot \frac{N_{Ed}}{t_w f_y}\right) \tag{4.3}$$

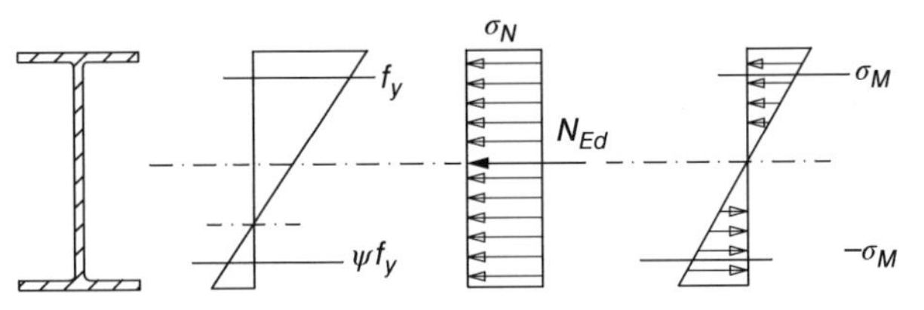

Figure 4.5 Superimposition of the stress distribution diagram associated with an elastic web under compression and bending.

In the case of class 3 and class 4 profiles, the axial force is transferred by the whole components (web and flanges) of the cross-section, which has an elastic response; that is the typical triangular distribution of stresses and strains governed by St Venant's theory. As a consequence, with reference to Figure 4.5, normal stresses σ, depending on the area A and on the section modulus W_{el} of the cross-section, can be divided into σ_N and σ_M, associated with axial load and bending moments, respectively. By using the superposition principle, top and bottom stresses are expressed as:

$$\psi f_y = \sigma_N - \sigma_M \tag{4.4}$$

$$f_y = \sigma_N + \sigma_M \tag{4.5}$$

Design axial load N_{Ed} acting on the member can be used to evaluate σ_N as:

$$\sigma_N = \frac{N_{Ed}}{A} \tag{4.6a}$$

By using the similar approach presented for the plastic cases, the bending moment associated with the stress distribution in Figure 4.5, which is related to an elastic web under compression and bending, $M_{Ed(N)}$, can be obtained as:

$$M_{Ed(N)} = \left(f_y - \sigma_N\right) W_{el} \tag{4.6b}$$

where W_{el} is the elastic section modulus.

From the previous equations, we can obtain:

$$f_y + \psi f_y = 2\frac{N_{Ed}}{A} \tag{4.7}$$

Term ψ can be directly associated with the design axial load (N_{Ed}) as:

$$\psi = 2\frac{N_{Ed}}{A\,f_y} - 1 \tag{4.8}$$

It should be noted that for a direct use of this approach it is possible to define, for each set of standard profiles the value of the axial load N_{Ed} associated with the boundary from class 1 to class 2. In particular, in case of portion of the web with the maximum compression stress greater that the tension one ($\alpha > 0.5$) in absolute values, the classification boundary that is reported in Table 4.1 is:

$$\frac{c}{t} = \frac{396 \cdot \varepsilon}{13 \cdot \alpha - 1} \tag{4.9}$$

where ε has been already introduced to account for the yielding strength $f_{y,}$ being defined as:

$$\varepsilon = \sqrt{\frac{235}{f_y[MPa]}} \tag{4.10}$$

At the boundary between 1 and 2 classes, term α is:

$$\alpha = \frac{396 \cdot \varepsilon + (c/t_w)}{13 \cdot (c/t_w)} \tag{4.11}$$

By substituting α in Eq. (4.2b,c), the value of the limit design axial load (N_{Ed}^{1-2}) and bending moment (M_{Ed}^{1-2}) are defined as:

$$N_{Ed}^{1-2} = \left[2 \cdot \frac{396 \cdot \varepsilon + (c/t_w)}{13 \cdot (c/t_w)} - 1\right] \cdot (c \cdot t_w) \cdot f_y \tag{4.12a}$$

$$M_{Ed}^{1-2} = M_{pl} - \frac{\left\{\left[2 \cdot \frac{396 \cdot \varepsilon + (c/t_w)}{13 \cdot (c/t_w)} - 1\right] \cdot c\right\}^2 \cdot t_w}{4} \cdot f_y \tag{4.12b}$$

Similarly, N_{Ed}^{2-3} and M_{Ed}^{2-3}, which are related the transition between classes 2 and 3, are expressed by:

$$N_{Ed}^{2-3} = \left[2 \cdot \frac{456 \cdot \varepsilon + (c/t_w)}{13 \cdot (c/t_w)} - 1\right] \cdot (c \cdot t_w) \cdot f_y \tag{4.13a}$$

$$M_{Ed}^{2-3} = M_{pl} - \frac{\left\{\left[2 \cdot \frac{456 \cdot \varepsilon + (c/t_w)}{13 \cdot (c/t_w)} - 1\right] \cdot c\right\}^2 \cdot t_w}{4} \cdot f_y \tag{4.13b}$$

The same approach can be adopted with reference to an elastic stress distribution (classes 3 and 4) by taking into account the relation between the axial load and the parameter ψ characterizing the stress distribution of the web under bending and compression. Transition values of axial force and bending moment in correspondence of the classification boundary between the classes 3 and 4 are given by:

$$N_{Ed}^{3-4} = \frac{1}{2} \cdot \left[\frac{42 \cdot \varepsilon - 0.67 \cdot (c/t_w)}{0.33 \cdot (c/t_w)} + 1\right] \cdot A \cdot f_y \tag{4.14a}$$

$$M_{Ed}^{3-4} = \left\{f_y - \frac{f_y}{2} \cdot \left[\frac{42 \cdot \varepsilon - 0.67 \cdot (c/t_w)}{0.33 \cdot (c/t_w)} + 1\right]\right\} \cdot W_{el} \tag{4.14b}$$

The values of axial load and bending moment at the transition between classes 1 and 2 (Eq. (4.12a,b)), classes 2 and 3 (Eq. (4.13a,b)) and classes 3 and 4 (Eq. (4.14a,b)), together with the value of the moment resistance (M_{Lim}) and of the axial load resistance (N_{Lim}) define a M-N domain depending on the cross-section geometry and steel grade. Figure 4.6 refers to the more general situation of the profile in class 1 for bending and in class 4 for compression. Designers have to classify the member on the basis of the design values of the axial force and bending moment; that is on the basis of the position of the generic point P (N_{Ed}, M_{Ed}) in this domain. No univocally defined criteria are codified in EC3 to identify the cross-section class on the basis of N_{Ed} and M_{Ed}. Few alternatives can be adopted by designers. As an example, reference should be made to Figure 4.6 related to the more generic case of member in class 1 under flexure and class 4 under compression: if N and M increase proportionally (path A) the profile results in class 3, otherwise, if the axial force is constant and only the bending moment increases (path B), the profile is classified as class 2.

As a general remark associated with the classification procedure for elements subjected to compression and bending, it has to be noted that the same member could belong to different classes,

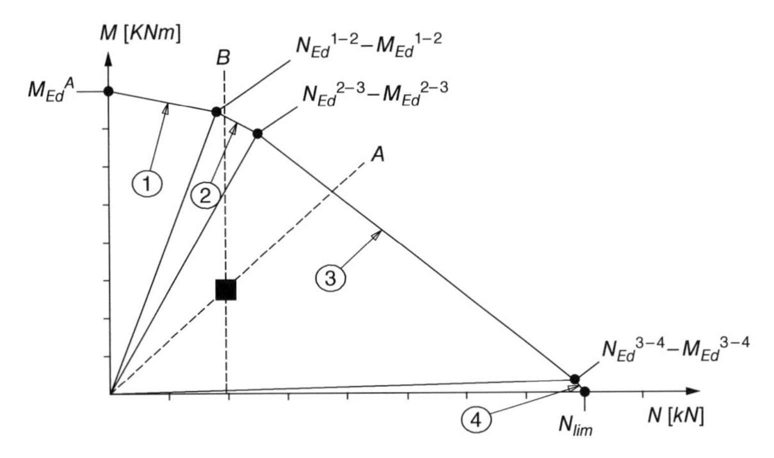

Figure 4.6 Example of classification moment (M)-axial load (N) classification domain.

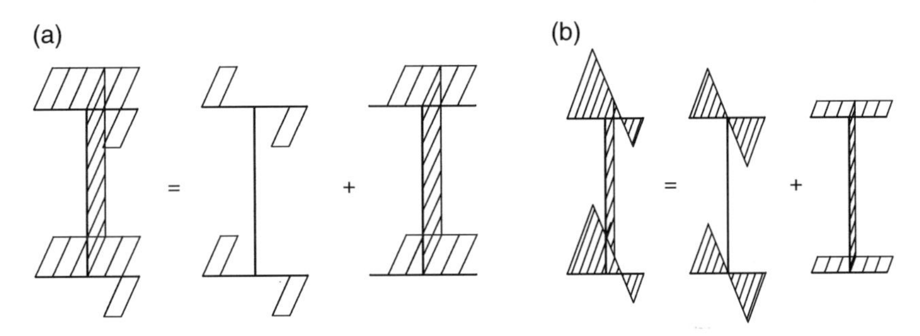

Figure 4.7 Stress distributions due to axial load (a) and bending (b) about a minor axis.

owing to the variability of the values of the acting bending moments along its longitudinal axis, as typically occurs when a beam column in a rigid or semi-continuous frame is considered.

4.2.2.2 Bending and Compression about a Weak Axis

All the concepts previously discussed for bending about the strong axis can be extended to the case of bending about the weak axis. It should be noted that, in case of bending about the weak axis of I- and H-shaped profiles, the classification criterion only has to be applied to the flanges. The web is always in class 1, due to the presence of the neutral axis located at the midline of the thickness of the web. The approach already presented for the definition of the domains M-N can hence be applied and, by using the superposition principle, it can be convenient to separate the state of stress due to axial force from the one associated with the bending moment (Figure 4.7).

4.2.3 Effective Geometrical Properties for Class 4 Sections

As already mentioned, for class 4 cross-sections it is assumed that parts of the area under compression due to local instability phenomena do not have any resistance (lost area): typically, the compressed portions of the cross-sections, which have to be neglected for the resistance checks, are the parts close to the free end of an outstand flange or the central part of an internal compressed element. As an example, reference can be made to Figure 4.8 related to typical cases of cross-section properties reduced for local buckling phenomena when a member is subjected to compression (a) or flexure (b). In the first case, it should be noted that the effective cross-section

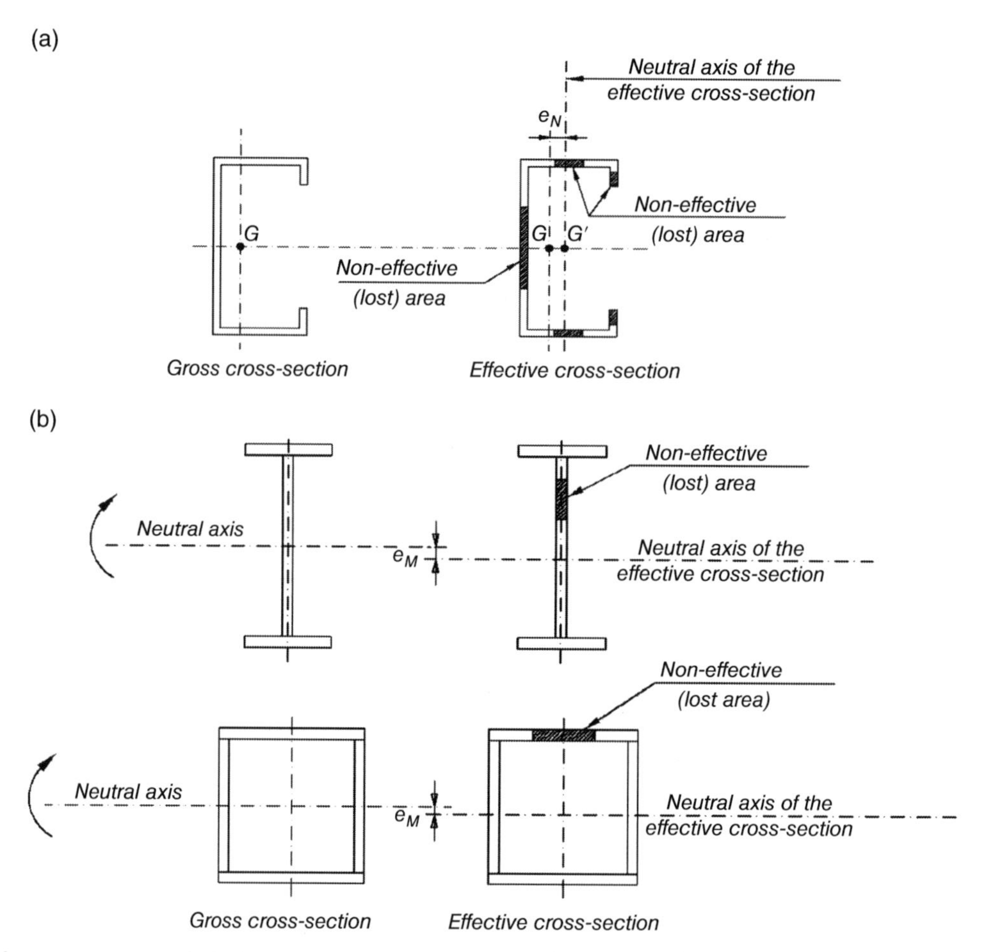

Figure 4.8 Gross and effective cross-sections in the case of axial load (a) and flexure (b).

is subjected to an eccentric axial load due to the shift of the centroid from the gross to the effective cross-section; that is the cross-section is subjected to an additional bending moment. From the design point of view, it is necessary to evaluate the effective cross-section (i.e. gross section minus all the lost parts) in accordance with the procedures specified in EN 1993-1-5 (Design of steel structures – Part 1–5: Plated structural elements). In particular, the references are Tables 4.1 and 4.2 and are reproduced here in Figures 4.9 and 4.10. In case of a class 4 circular hollow cross-section, reference has to be made to EN 1993-1-6 (Design of steel structures – Part 1–6: Strength and Stability of Shell Structures).

The effective area of a compressed plate $A_{c,\text{eff}}$ can be obtained from the gross area, A_c, as:

$$A_{c,eff} = \rho A_c \tag{4.15}$$

Reduction factor ρ is defined as:

- Internal compression elements (webs):

$$\rho = 1.0 \text{ if } \overline{\lambda_p} \le 0.673 \tag{4.16a}$$

$$\rho = \left[\bar{\lambda}_p - 0.055(3+\psi)\right] / \bar{\lambda}_p^{\,2} \le 1 \text{ if } \overline{\lambda_p} > 0.673 \text{ and } (3+\psi) \ge 0 \tag{4.16b}$$

Stress distribution (compression positive)	Effective^p width b_{eff}
σ_1, σ_2, b_{e1}, $\bar{b}$, b_{e2}	$\psi = 1$: $b_{eff} = \rho\, \bar{b}$; $b_{e1} = 0.5\, b_{eff}$ $\quad b_{e2} = 0.5\, b_{eff}$
σ_1, σ_2, b_{e1}, $\bar{b}$, b_{e2}	$1 > \psi \geq 0$: $b_{eff} = \rho\, \bar{b}$; $b_{e1} = \frac{2}{5-\psi} b_{eff}$ $\quad b_{e2} = b_{eff} - b_{e1}$
b_c, b_t, σ_1, σ_2, b_{e1}, b_{e2}, $\bar{b}$	$\psi < 0$: $b_{eff} = \rho\, b_c = \rho \bar{b}/(1-\psi)$; $b_{e1} = 0.4\, b_{eff}$ $\quad b_{e2} = 0.6\, b_{eff}$

$\psi = \sigma_1/\sigma_1$	1	$1 > \psi > 0$	0	$0 > \psi > -1$	−1	$-1 > \psi > -3$
Buckling factor k_σ	4.0	$8.2 / (1.05 + \psi)$	7.81	$7.81 - 6.29\psi + 9.78\psi^2$	23.9	$5.98\,(1-\psi)^2$

Figure 4.9 Rules for the evaluation of the effective width of internal compression elements (a–c: from Table 4.1 of EN 1993-1-5).

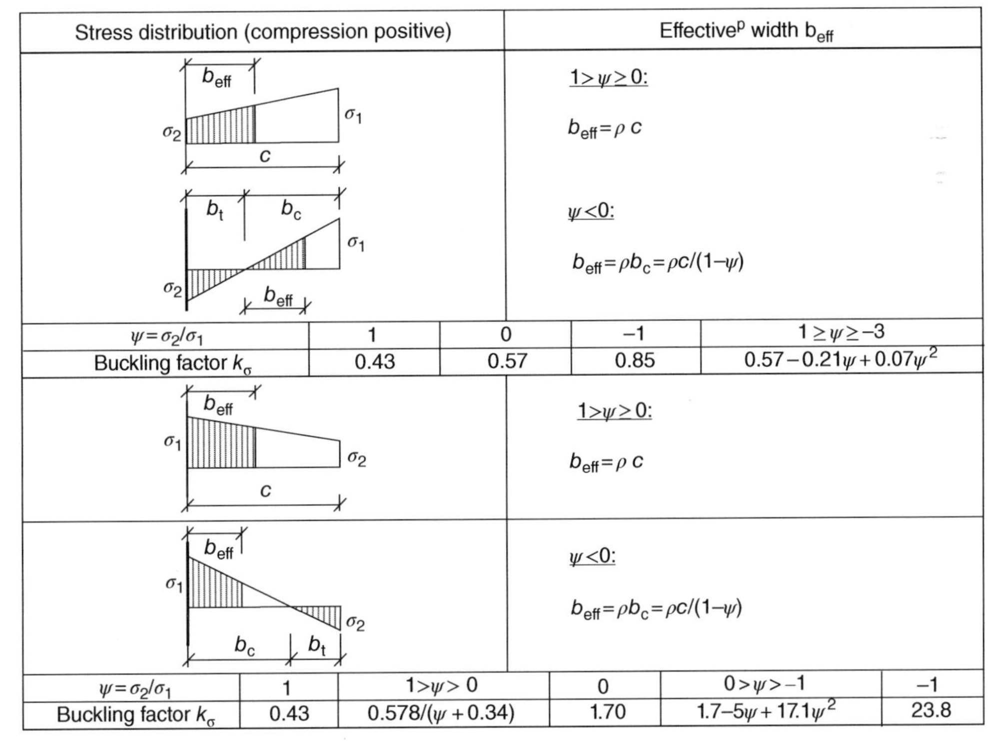

Stress distribution (compression positive)	Effective^p width b_{eff}
b_{eff}, σ_1, σ_2, c	$1 > \psi \geq 0$: $b_{eff} = \rho\, c$
b_t, b_c, σ_1, σ_2, b_{eff}	$\psi < 0$: $b_{eff} = \rho b_c = \rho c/(1-\psi)$

$\psi = \sigma_2/\sigma_1$	1	0	−1	$1 \geq \psi \geq -3$
Buckling factor k_σ	0.43	0.57	0.85	$0.57 - 0.21\psi + 0.07\psi^2$

Stress distribution (compression positive)	Effective^p width b_{eff}
b_{eff}, σ_1, σ_2, c	$1 > \psi \geq 0$: $b_{eff} = \rho\, c$
b_{eff}, σ_1, σ_2, b_c, b_t	$\psi < 0$: $b_{eff} = \rho b_c = \rho c/(1-\psi)$

$\psi = \sigma_2/\sigma_1$	1	$1 > \psi > 0$	0	$0 > \psi > -1$	−1
Buckling factor k_σ	0.43	$0.578/(\psi + 0.34)$	1.70	$1.7 - 5\psi + 17.1\psi^2$	23.8

Figure 4.10 Rules for the evaluation of the effective width of outstanding compression elements (a–d: from Table 4.2 of EN 1993-1-5).

- Outstand compression elements (flanges):

$$\rho = 1.0 \text{ if } \overline{\lambda_p} \le 0.748 \quad (4.17a)$$

$$\rho = \left(\overline{\lambda_p} - 0.188\right)/\overline{\lambda_p}^2 \le 1 \text{ if } \overline{\lambda_p} > 0.748 \quad (4.17b)$$

where:

$$\overline{\lambda_p} = \sqrt{\frac{f_y}{\sigma_{cr}}} = \frac{\bar{b}/t}{28.4\varepsilon\sqrt{k_\sigma}}; \quad \varepsilon = \sqrt{\frac{235}{f_y\left[\mathrm{N/mm^2}\right]}}$$

The width $\bar{b}$ has to be evaluated in accordance with Tables 4.1–4.3, where the reference is made to term c instead of $\bar{b}$.

Term ψ represents the ratio between the values of the plate end stresses while k_σ is the *buckling factor*, which can be evaluated from Figures 4.9 and 4.10 on the basis of the distribution of the normal stresses.

4.3 Classification in Accordance with US Standards

As already mentioned, AISC 360-10 addresses classification of cross-sections in Chapter B, Section B4; the code deals with members subjected to axial load and members subjected to bending in a different way:

- members subjected to axial load are distinguished as *non-slender* or *slender*;
- members subjected to flexure are distinguished as *compact, non-compact* or *slender*.

The classification for members subjected to axial load and bending is absent in the US approach.

Classifications criteria are listed in Table B4.1.a of AISC specifications (reproduced in Table 4.4a) for compressed members and in Table B4.1b (reproduced in Table 4.4b) for members in bending.

Classification criteria are based, as in the EC3 code, on *steel grade* and on *width-to-thickness ratios* for *stiffened elements* (elements supported along two edges parallel to the direction of the compression force, typically webs of I- or C-shaped sections) and *unstiffened elements* (elements supported along only one edge parallel to the direction of the compression force, typically flanges of I- or C-shaped sections).

AISC code defines:

(a) for members subjected to axial load:
λ_r, that is width-to-thickness ratio that defines non-slender/slender limit;

(b) for members subject to flexure:
λ_p, that is width-to-thickness ratio that defines compact/non-compact limit;
λ_r, that is width-to-thickness ratio that defines non-compact/slender limit.

It should be noted that:

- US flange width is one-half of full flange width, while in EC3 it is the outstanding part of the flange (one-half of full flange width less one-half of web thickness less the fillet or corner radius);
- US web width of rolled sections, as in EC3 code, is the clear distance between flanges less the fillet or corner radius at both flanges;

Table 4.4a Width-to-thickness ratios for members subject to axial compression (from Table B4.1a of AISC 360-10).

Case	Description of element	Width to thickness ratio	Limiting width-to-thickness ratio λ_r (non-slender/ slender)	Examples
		Unstiffened elements		
1	Flanges of rolled I-shaped sections, plates projecting from rolled I-shaped sections; outstanding legs of pairs of angles connected with continuous contact, flanges of channels and flanges of tees	b/t	$0.56\sqrt{\frac{E}{F_y}}$	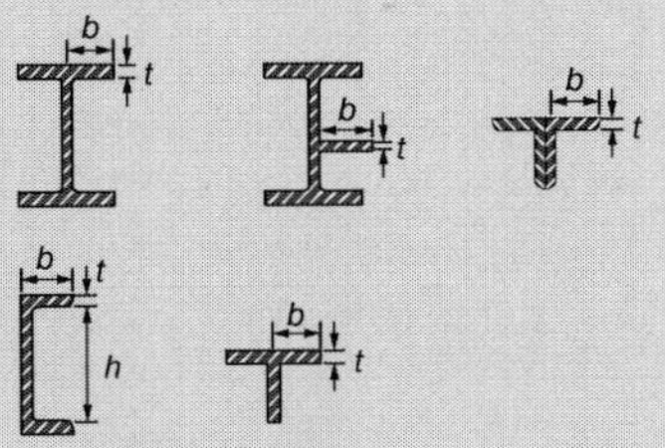
2	Flanges of built-up I-shaped sections and plates or angle legs projecting from built-up I-shaped sections	b/t	$0.64\sqrt{\frac{k_c E}{F_y}}$	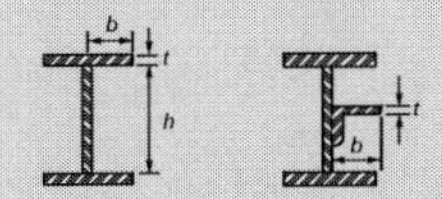
3	Legs of single angles, legs of double angles with separators and all other unstiffened elements	b/t	$0.45\sqrt{\frac{E}{F_y}}$	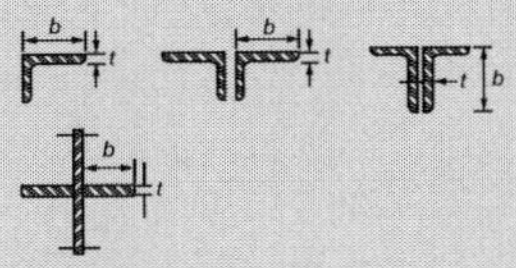
4	Stems of tees	d/t	$0.75\sqrt{\frac{E}{F_y}}$	
		Stiffened elements		
5	Webs of doubly-symmetric I-shaped sections and channels	h/t_w	$1.49\sqrt{\frac{E}{F_y}}$	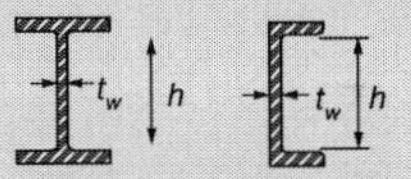
6	Walls of rectangular HSS and boxes of uniform thickness	b/t	$1.40\sqrt{\frac{E}{F_y}}$	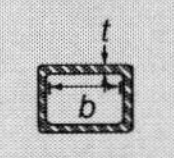
7	Flange cover plates and diaphragm plates between lines of fasteners or welds	b/t	$1.40\sqrt{\frac{E}{F_y}}$	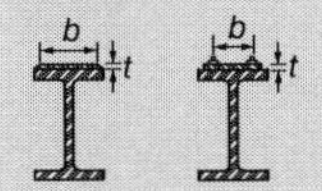
8	All other stiffened elements	b/t	$1.49\sqrt{\frac{E}{F_y}}$	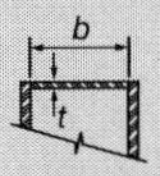
9	Round HSS	D/t	$0.11\frac{E}{F_y}$	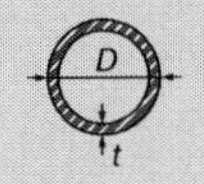

HSS, hollow square section.

- US web width of built-up sections is the clear distance between flanges, while in EC3 it is defined as for hot-rolled sections.

Classification of flanges of built-up members depends not only on width-to-thickness ratio of the flange itself but also on that of the web, by means of parameter k_c:

$$k_c = \frac{4}{\sqrt{h/t_w}} \tag{4.18}$$

Table 4.4b Width-to-thickness ratios for members subject to flexure (from Table B4.1b of AISC 360-10).

Case	Description of element	Width-to-thickness ratio	Limiting width-to-thickness ratio: λ_p (compact/noncompact)	Limiting width-to-thickness ratio: λ_r (noncompact/slender)	Examples
10	Flanges of rolled I-shaped sections, channels and tees	b/t	$0.38\sqrt{\frac{E}{F_y}}$	$1.00\sqrt{\frac{E}{F_y}}$	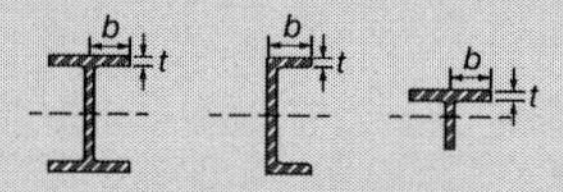
11	Flanges of doubly and singly symmetric I-shaped built-up sections	b/t	$0.38\sqrt{\frac{E}{F_y}}$	$0.95\sqrt{\frac{k_c E}{F_L}}$	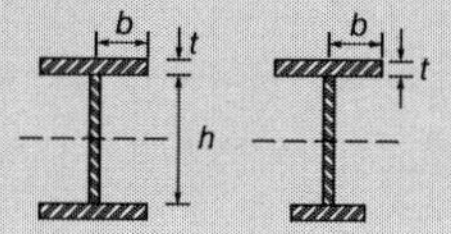
12	Legs of single angles	b/t	$0.54\sqrt{\frac{E}{F_y}}$	$0.91\sqrt{\frac{E}{F_y}}$	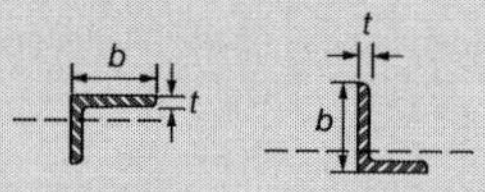
13	Flanges of all I-shaped sections and channels in flexure about the weak axis	b/t	$0.38\sqrt{\frac{E}{F_y}}$	$1.00\sqrt{\frac{E}{F_y}}$	
14	Stems of tees	d/t	$0.84\sqrt{\frac{E}{F_y}}$	$1.03\sqrt{\frac{E}{F_y}}$	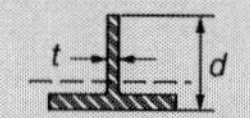
15	Webs of doubly symmetric I-shaped sections and channels	h/t_w	$3.76\sqrt{\frac{E}{F_y}}$	$5.70\sqrt{\frac{E}{F_y}}$	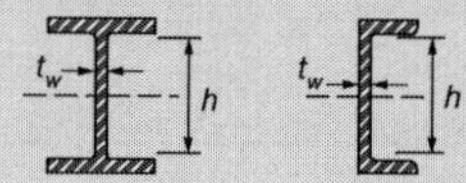
16	Webs of singly symmetric I-shaped sections	h_c/t_w	$\frac{\frac{h_c}{h_p}\sqrt{\frac{E}{F_y}}}{\left(0.54\frac{M_p}{M_y}-0.09\right)^2} \le \lambda_r$	$5.70\sqrt{\frac{E}{F_y}}$	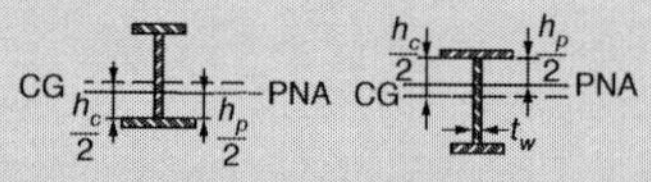
17	Flanges of rectangular HSS and boxes of uniform thickness	b/t	$1.12\sqrt{\frac{E}{F_y}}$	$1.40\sqrt{\frac{E}{F_y}}$	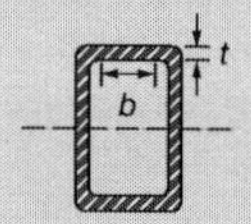
18	Flange cover plates and diaphragm plates between lines of fasteners or welds	b/t	$1.12\sqrt{\frac{E}{F_y}}$	$1.40\sqrt{\frac{E}{F_y}}$	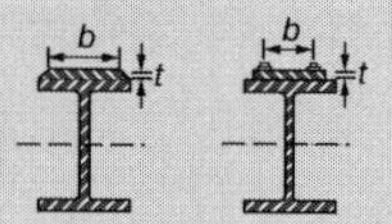
19	Webs of rectangular HSS and boxes	h/t	$2.42\sqrt{\frac{E}{F_y}}$	$5.70\sqrt{\frac{E}{F_y}}$	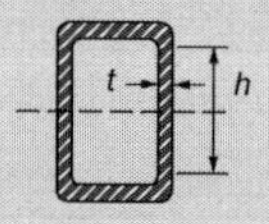
20	Round HSS	D/t	$0.07\frac{E}{F_y}$	$0.31\frac{E}{F_y}$	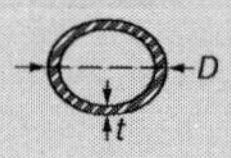

where h and t_w are width and thickness of web panel, respectively.

This parameter accounts for the stiffening effect of web on flange: more slender webs give a lower degree of stiffening on flanges. This effect is not considered in EC3 classification criteria.

4.4 Worked Examples

Example E4.1 Classification of a Member for Compression

Determine the class of IPE 550 profile S 275 steel grade under axial compression load.

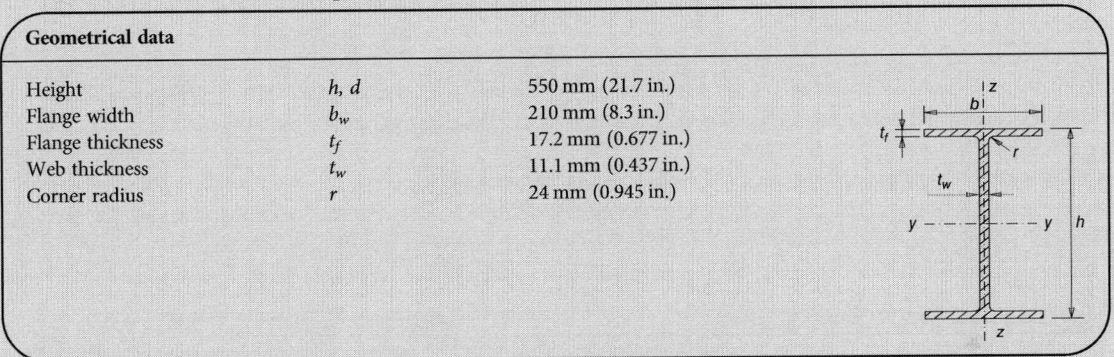

Geometrical data		
Height	h, d	550 mm (21.7 in.)
Flange width	b_w	210 mm (8.3 in.)
Flange thickness	t_f	17.2 mm (0.677 in.)
Web thickness	t_w	11.1 mm (0.437 in.)
Corner radius	r	24 mm (0.945 in.)

Material data:

S 275 steel grade $f_y = 275$ MPa ($F_y = 39.9$ ksi)

EC3 procedure

For S 275 steel grade

$$\varepsilon = \sqrt{\frac{235}{f_y}} = \sqrt{\frac{235}{275}} = 0.924$$

Flange

$$\frac{c}{t} = \frac{b_w - t_w - (2 \cdot r)}{2 \cdot t_f} = \frac{210 - 11.1 - (2 \times 24)}{2 \times 17.2}$$

$$= \frac{150.9}{34.4} = 4.39 \leq 9\varepsilon = 9 \times 0.924 = 8.32$$

Flange is class 1

Web

$$\frac{c}{t} = \frac{h - (2 \cdot t_f) - (2 \cdot r)}{t_w}$$

$$= \frac{550 - (2 \times 17.2) - (2 \times 24)}{11.1}$$

$$= \frac{467.6}{11.1} = 42.2 > 42\varepsilon = 42 \times 0.924 = 38.83$$

Web is class 4

S 275 steel IPE 550 section, subject to axial load, has class 1 flanges and class 4 web, therefore it is classified as *class 4*

AISC procedure

Flange

$$\frac{b}{t} = \frac{b_w}{2 \cdot t_f} = \frac{8.3}{2 \times 0.677} = 6.13 <$$

$$= 0.56\sqrt{\frac{E}{F_y}} = 0.56 \times \sqrt{\frac{29000}{39.9}} = 15.1$$

Flange is non-slender

Web

$$\frac{h}{t_w} = \frac{d - (2 \cdot t_f) - (2 \cdot r)}{t_w}$$

$$= \frac{21.7 - (2 \times 0.677) - (2 \times 0.945)}{0.437}$$

$$= \frac{18.46}{0.437} = 42.2 > 1.49\sqrt{\frac{E}{F_y}}$$

$$= 1.49 \times \sqrt{\frac{29000}{39.9}} = 40.2$$

Web is slender

S 275 steel IPE 550 section, subject to axial load, has non-slender flanges and slender web, therefore, it is classified as slender

Example E4.2 Classification of a Member for Flexure about the Major Axis

Determine class of a HEA 280 profile in S 420 steel grade bent along its major axis.

Geometrical data

Height	h, d	270 mm (10.63 in.)
Flange width	b_w	280 mm (11.02 in.)
Flange thickness	t_f	13 mm (0.512 in.)
Web thickness	t_w	8 mm (0.315 in.)
Corner radius	r	24 mm (0.945 in.)

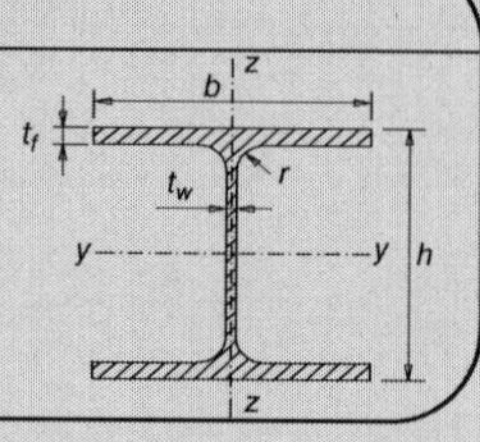

Material data:

S 420 steel grade $f_y = 420$ MPa ($F_y = 60.9$ ksi)

EC3 procedure

For S 420 steel grade

$$\varepsilon = \sqrt{\frac{235}{f_y}} = \sqrt{\frac{235}{420}} = 0.748$$

Flange

$$\frac{c}{t} = \frac{b - t_w - (2 \cdot r)}{2 \cdot t_f} = \frac{280 - 8 - (2 \times 24)}{2 \times 13} = \frac{224}{26} = 8.62$$

Limiting value class 1/class 2

$$\frac{c}{t} = 9 \cdot \varepsilon = 6.732$$

Limiting value class 2/class 3

$$\frac{c}{t} = 10 \cdot \varepsilon = 7.480$$

Limiting value class 3/class 4

$$\frac{c}{t} = 14 \cdot \varepsilon = 10.472$$

Flange is class 3

Web

$$\frac{c}{t} = \frac{h - (2 \cdot t_f) - (2 \cdot r)}{t_w} = \frac{270 - (2 \times 13) - (2 \times 24)}{8} = \frac{196}{8} = 24.5 \le 72 \cdot \varepsilon = 72 \times 0.748 = 53.86$$

Web is class 1

S 420 steel HEA 280 section, subjected to flexure, has class 3 flanges and class 1 web, therefore, it is classified as *class 3*

AISC procedure

Flange

Limiting width-to-thickness ratio compact/non-compact

$$\lambda_p = 0.38\sqrt{\frac{E}{F_y}} = 0.38 \times \sqrt{\frac{29000}{60.9}} = 8.29$$

Limiting width-to-thickness ratio non-compact/slender

$$\lambda_r = 1.0\sqrt{\frac{E}{F_y}} = 1.0 \times \sqrt{\frac{29000}{60.9}} = 21.8$$

$$\frac{b}{t} = \frac{b_w}{2 \cdot t_f} = \frac{11.02}{2 \times 0.512} = 10.76$$

Flange is non-compact

Web: limiting width-to-thickness ratio compact/non-compact

$$\lambda_p = 3.76\sqrt{\frac{E}{F_y}} = 3.76 \times \sqrt{\frac{29000}{60.9}} = 82.0$$

Limiting width-to-thickness ratio non-compact/slender

$$\lambda_r = 5.70\sqrt{\frac{E}{F_y}} = 5.70 \times \sqrt{\frac{29000}{60.9}} = 124.4$$

$$\frac{h}{t_w} = \frac{d - (2 \cdot t_f) - (2 \cdot r)}{t_w} = \frac{10.63 - (2 \times 0.512) - (2 \times 0.945)}{0.315} = \frac{7.716}{0.315} = 24.5$$

Web is compact

S 420 steel HEA 280 section, subjected to flexure, has class non-compact flanges and compact web, therefore, it is classified as *non-compact*

Example E4.3 Classification of a Member under Axial Load and Flexure about the Major Axis

Consider an S 275 steel IPE 600 profile subjected to axial load and bending moments about its strong axis and determine the classification domain.

Geometrical data

Height	h, d	600 mm (23.62 in.)
Flange width	b_w	220 mm (8.66 in.)
Flange thickness	t_f	19 mm (0.748 in.)
Web thickness	t_w	12 mm (0.472 in.)
Corner radius	r	24 mm (0.945 in.)
Area	A	156 cm^2 (24.18 in.^{2})
Elastic modulus	W_e, S	3070 cm^3 (187.3 in.^{3})
Plastic modulus	W_{pl}, Z	3512 cm^3 (214.3 in.^{3})

Material data:

S 275 steel grade $f_y = 275$ MPa ($F_y = 39.9$ ksi)

EC3 procedure

Determine boundary axial load values as explained in Example E4.1

For S 275 steel grade

$$\varepsilon = \sqrt{\frac{235}{f_y}} = \sqrt{\frac{235}{275}} = 0.924$$

Flange

$$\frac{c}{t} = \frac{b - t_w - (2 \cdot r)}{2 \cdot t_f} = \frac{220 - 12 - (2 \times 24)}{2 \times 19}$$

$$= \frac{160}{38} = 4.21 \le 9 \cdot \varepsilon = 9 \times 0.924 = 8.32$$

Flange is class 1

Web in compression

$$\frac{c}{t} = \frac{h - (2 \cdot t_f) - (2 \cdot r)}{t_w}$$

$$= \frac{600 - (2 \times 19) - (2 \times 24)}{12} = \frac{514}{12}$$

$$= 42.83 > 42 \cdot \varepsilon = 42 \times 0.924 = 38.83$$

Web in compression is class 4

Web in flexure

$$\frac{c}{t} = \frac{h - (2 \cdot t_f) - (2 \cdot r)}{t_w} = \frac{600 - (2 \times 19) - (2 \times 24)}{12}$$

$$= \frac{514}{12} = 42.83 < 72 \cdot \varepsilon = 72 \times 0.924 = 66.56$$

AISC procedure

For members subjected to flexure and axial load AISC prescribes to compute classification in case of axial load only and in case of bending moment only

(1) Axial load

Flange

$$\frac{b}{t} = \frac{b_w}{2 \cdot t_f} = \frac{8.66}{2 \times 0.748} = 5.79 <$$

$$= 0.56 \sqrt{\frac{E}{F_y}} = 0.56 \times \sqrt{\frac{29000}{39.9}} = 15.1$$

Flange is non-slender

Web

$$\frac{h}{t_w} = \frac{d - (2 \cdot t_f) - (2 \cdot r)}{t_w} = \frac{23.62 - (2 \times 0.748) - (2 \times 0.945)}{0.472}$$

$$= \frac{20.23}{0.472} = 42.9 > 1.49 \sqrt{\frac{E}{F_y}} = 1.49 \times \sqrt{\frac{29000}{39.9}} = 40.2$$

Web is slender

S 275 steel IPE 600 section, subjected to axial load, has non-slender flanges and slender web, therefore it is classified as slender

(Continued)

(Continued)

EC3 procedure	AISC procedure
Web in flexure is class 1	(2) Flexure
If member is subjected to axial load and bending, class for the section changes depends on the entity of axial load	Flange Limiting width-to-thickness ratio compact/non-compact $\lambda_p = 0.38\sqrt{\frac{E}{F_y}} = 0.38 \times \sqrt{\frac{29\,000}{39.9}} = 10.24$
The parameter α for transition from class 1 to class 2 is therefore determined $\alpha = \frac{396 \cdot \varepsilon + (c/t_w)}{13 \cdot (c/t_w)} = \frac{396 \times 0.924 + 42.83}{13 \times 42.83} = 0.7341$	Limiting width-to-thickness ratio non-compact/slender $\lambda_r = 1.0\sqrt{\frac{E}{F_y}} = 1.0 \times \sqrt{\frac{29\,000}{39.9}} = 26.96$ $\frac{b}{t} = \frac{b_w}{2 \cdot t_f} = \frac{8.66}{2 \times 0.748} = 5.79$ Flange is compact
In case of pure flexure, limit moment M_{Lim} coincides with plastic moment, and is determined as follows $M_{\text{Lim}} = M_{pl} = W_{pl} \cdot f_y = 3512 \times 275 \cdot 10^{-3}$ $= 965.8\,\text{kNm}\ (712.3\,\text{kip-ft})$	Web Limiting width-to-thickness ratio compact/non-compact $\lambda_p = 3.76\sqrt{\frac{E}{F_y}} = 3.76 \times \sqrt{\frac{29\,000}{39.9}} = 101.4$
Compute with Eqs. (4.12a) and (4.12b) limit values for axial load and bending moment at transition from class 1 to class 2 $N_{Ed}^{1-2} = [2 \cdot \alpha - 1] \cdot (c \cdot t_w) \cdot f_y$ $= [2 \times 0.7341 - 1] \times (514 \times 12) \times 275 \cdot 10^{-3}$ $= 794.2\,\text{kN}\ (174.5\,\text{kips})$ $M_{Ed}^{1-2} = M_{pl} - \frac{\left\{\left[2 \cdot \frac{396 \cdot \varepsilon + (c/t_w)}{13 \cdot (c/t_w)} - 1\right] \cdot c\right\}^2 t_w}{4} f_y$ $= 965.8 - \frac{\left\{\left[2 \times \frac{396 \times 0.924 + 42.83}{13 \times 42.83} - 1\right] \times 514\right\}^2 \times 12}{4}$ $\times 275 \cdot 10^{-6} = 965.8 - 47.8 = 918.0\,\text{kNm}\ (178.5\,\text{kip-ft})$	Limiting width-to-thickness ratio non-compact/slender $\lambda_r = 5.70\sqrt{\frac{E}{F_y}} = 5.70 \times \sqrt{\frac{29\,000}{39.9}} = 153.7$ $\frac{h}{t_w} = \frac{d - (2 \cdot t_f) - (2 \cdot r)}{t_w} = \frac{23.62 - (2 \times 0.748) - (2 \times 0.945)}{0.472}$ $= \frac{20.23}{0.472} = 42.9$ Web is compact S 275 steel IPE 600 section, subjected to flexure, has class compact flanges and compact web, therefore, it is classified as compact
In the same way, limit values for axial load and bending moment at transition from class 2 to class 3 shall be $N_{Ed}^{2-3} = \left[2 \cdot \frac{456 \cdot \varepsilon + (c/t)}{13 \cdot (c/t)} - 1\right] \cdot (d \cdot t_w) \cdot f_y$ $= \left[2 \cdot \frac{456 \times 0.924 + 42.83}{13 \cdot 42.83} - 1\right] \times (514 \times 12) \times 275 \cdot 10^{-3}$ $= [2 \times 0.8337 - 1] \cdot (514 \times 12) \times 275 \cdot 10^{-3}$ $= 1131.9\,\text{kN}\ (254.5\,\text{kips})$	Being the section non-slender for axial load only, the nominal compressive strength P_n shall be $P_n = F_{cr} A_g = \left[0.658^{\frac{QF_y}{F_e}}\right] QF_y A_g$

(Continued)

<table>
<tr><th>EC3 procedure</th><th>AISC procedure</th></tr>
<tr><td>

$$M_{Ed}^{2-3}=M_{pl}-\frac{\left\{\left[2\cdot\frac{456\cdot\varepsilon+(c/t_w)}{13\cdot(c/t_w)}-1\right]\cdot c\right\}^2\cdot t_w}{4}\cdot f_y$$

$$=965.8-\frac{\left\{\left[2\times\frac{456\times0.924+42.83}{13\times42.83}-1\right]\times514\right\}^2\times12}{4}$$

$$\times275\cdot10^{-6}=965.8-97.1=868.7\,\text{kNm}\,(640.7\,\text{kips-ft})$$

</td><td>

For low slenderness KL/r value, $F_e \gg QF_y$, therefore, this expression can be approximated as

$P_n=F_{cr}A_g=QF_yA_g$

$Q=Q_s\cdot Q_a$

$Q_s=1$ (flanges are non-slender)

</td></tr>
<tr><td>

Using the same method also for elastic behaviour, taking into account the relationship between axial load and parameter ψ, related to elastic stress distribution in web, limit values for axial load and bending moment at transition from class 3 to class 4 are computed according to Eqs. (4.14a) and (4.14b)

</td><td>

Compute Q_a because the web is slender

$$\frac{h}{t_w}=42.9>1.49\sqrt{\frac{E}{f}}=1.49\times\sqrt{\frac{29000}{39.9}}=40.2$$

</td></tr>
<tr><td>

$$N_{Ed}^{3-4}=\frac{1}{2}\cdot\left[\frac{42\cdot\varepsilon-0.67\cdot(c/t)}{0.33\cdot(c/t)}+1\right]\cdot A\cdot f_y$$

$$=\frac{1}{2}\cdot\left[\frac{42\times0.924-0.67\times(42.83)}{0.33\times42.83}+1\right]\times156\times275\cdot10^{-1}$$

$$=\frac{1}{2}\cdot(0.7154+1)\times156\times275\cdot10^{-1}$$

$$=3682\,\text{kN}\,(827.7\,\text{kips})$$

</td><td>

($f=F_{cr}=39.9$ ksi)

</td></tr>
<tr><td>

$$M_{Ed}^{3-4}=\left\{f_y-\frac{1}{2}\cdot\left[\frac{42\cdot\varepsilon-0.67\cdot(c/t_w)}{0.33\cdot(c/t_w)}+1\right]\cdot f_y\right\}\cdot W_{el}$$

$$=\left\{275-\frac{1}{2}\times\left[\frac{42\times0.924-0.67\times42.83}{0.33\times42.83}+1\right]\times275\right\}$$

$$\times3069\cdot10^{-6}$$

$$=(275-235.87)\times3069\cdot10^{-6}=120.1\,\text{kNm}\,(88.6\,\text{kip-ft})$$

</td><td>

Then:

$$b_e=1.92t\sqrt{\frac{E}{f}}\left[1-\frac{0.34}{(b/t)}\sqrt{\frac{E}{f}}\right]$$

$$=1.92\times0.472\times\sqrt{\frac{29000}{39.9}}\times\left[1-\frac{0.34}{42.8}\times\sqrt{\frac{29000}{39.9}}\right]$$

$$=19.2\,\text{in.}<h=20.23\,\text{in.}\,(488\,\text{mm}<514\,\text{mm})$$

</td></tr>
<tr><td>

The compressive strength for axial load only N_{pl} computed using gross area is

$N_{pl}=A\cdot f_y=156\times275\cdot10^{-1}=4290\,\text{kN}\,(964.4\,\text{kips})$

</td><td>

Effective area A_e

$A_e=A_g-(h-b_e)t_w=24.18-(20.23-19.2)\times0.472$

$=23.69\,\text{in.}^2(152.8\,\text{cm}^2)$

</td></tr>
<tr><td>

When N_{pl} is larger than limit axial load at transition from class 3 to class 4, there are no limitations for compressive strength

</td><td>

$Q_a=A_e/A_g=23.69/24.18=0.98$

</td></tr>
<tr><td>

The section is class 4 due to high web slenderness, therefore, compressive strength shall be computed using the effective area (for a definition, see Section **4.3**.3, for the calculation, see Example E6.1)

$A_{eff}=149.5\,\text{cm}^2(23.17\,\text{in.}^2)$

</td><td>

$Q=Q_s\cdot Q_a=1\times0.98=0.98$

</td></tr>
<tr><td>

Then member compressive strength for axial load only N_{Lim} shall be:

$N_{Lim}=A_{eff}\cdot f_y=149.5\times275\cdot10^{-1}=4112\,\text{kN}\,(924.4\,\text{kips})$

</td><td>

$P_n=QF_yA_g=0.98\times39.9\times24.18=945.5$ kips (4206 kN)

</td></tr>
</table>

Example E4.4 Classification of a Member for Flexure about the Minor Axis

Determine the class of an S 460 steel HEA 280 profile in flexure around its major axis.

Geometrical data

Height	h, d	270 mm (10.63 in.)
Flange width	b_w	280 mm (11.02 in.)
Flange thickness	t_f	13 mm (0.512 in.)
Web thickness	t_w	8 mm (0.315 in.)
Corner radius	r	24 mm (0.945 in.)

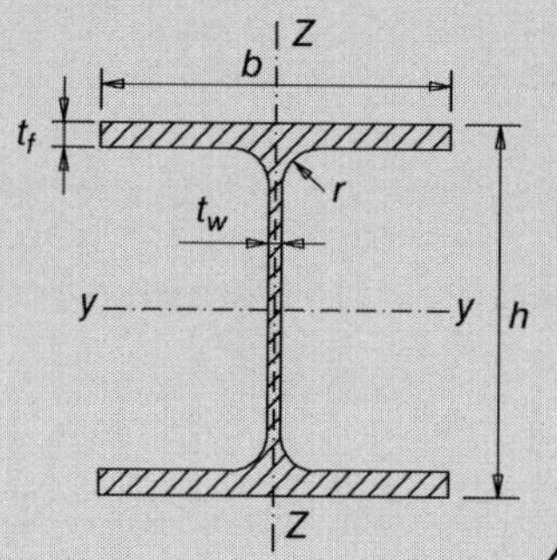

Material data:

S 460 steel grade $f_y = 460$ MPa ($F_y = 66.7$ ksi)

EC3 procedure

For S 460 steel

$$\varepsilon = \sqrt{\frac{235}{f_y}} = \sqrt{\frac{235}{460}} = 0.715$$

Flange

$$\frac{c}{t} = \frac{b - t_w - (2 \cdot r)}{2 \cdot t_f} = \frac{280 - 8 - (2 \times 24)}{2 \times 13} = \frac{224}{26} = 8.62$$

Limit width-to-thickness ratio from class 1 to class 2

$$\frac{c}{t} = 9 \cdot \varepsilon = 6.432$$

Limit width-to-thickness ratio from class 2 to class 3

$$\frac{c}{t} = 10 \cdot \varepsilon = 7.146$$

Limit width-to-thickness ratio from class 3 to class 4

$$\frac{c}{t} = 14 \cdot \varepsilon = 10.01$$

Flange is class 3

Web

Web buckling is not a limit state for flexure around a minor axis, so web classification is not applicable in this case

S 460 steel HEA 280 section, subjected to flexure around minor axis, has class 3 flanges and class 2 web, therefore, it is classified as *class 3*

AISC procedure

Flange

$$\frac{b}{t} = \frac{b_w}{2 \cdot t_f} = \frac{11.02}{2 \times 0.512} = 10.76$$

Limiting width-to-thickness ratio compact/non-compact

$$\lambda_p = 0.38\sqrt{\frac{E}{F_y}} = 0.38 \times \sqrt{\frac{29000}{66.7}} = 7.92$$

Limiting width-to-thickness ratio non-compact/slender

$$\lambda_r = 1.0\sqrt{\frac{E}{F_y}} = 1.0 \times \sqrt{\frac{29000}{66.7}} = 20.9$$

$$\frac{b}{t} = \frac{b_w}{2 \cdot t_f} = \frac{11.02}{2 \times 0.512} = 10.76$$

Flange is non-compact

Web

Web buckling is not a limit state for flexure around a minor axis, so web classification is not applicable in this case

S 460 steel HEA 280 section, subjected to flexure around minor axis, has class 3 flanges, therefore, it is classified as *class 3*

Example E4.5 Comparison between US and EC3 Classification Approaches

As shown in the chapter, similarities and differences can be detected in the EU and US classification criteria. The example herein proposed presents a direct comparison related to the application of both codes to the same cross-section.

Consider an I-shaped doubly symmetrical built-up welded section, bent about the major axis as in Figure E4.5.1, composed of a 600 × 10 mm web plate and 20 mm-thick flanges. Increase the flange width b_f and, for each value, compute the section classification, the nominal flexural strength M_n according to AISC specifications, and the characteristic resistance bending moment $M_{y,Rk}$ according to EC3.

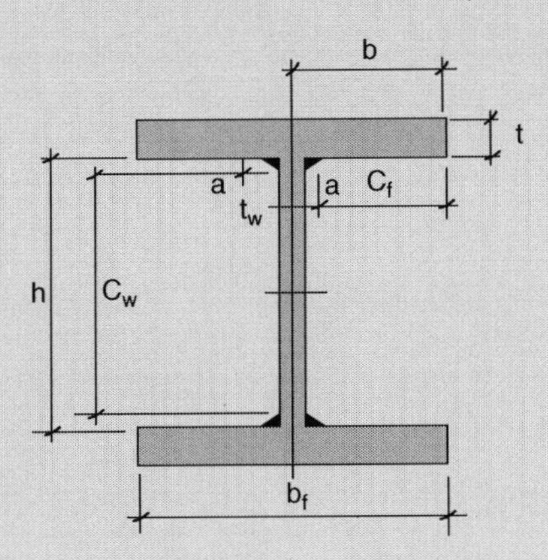

Figure E4.5.1 Geometrical parameters.

Note: M_n and $M_{y,Rk}$ are the names by which in AISC 360-10 and EC3, respectively, the product of relevant value for section modulus (plastic, elastic or effective) and relevant value for minimum steel stress (yielding or critical) is indicated. Design values for flexural strength are obtained, in both codes, by multiplying such values with proper safety factors: ϕ_b and Ω_b in AISC, γ_{M0} in EC3, and they represent flexural strength of a beam section if lateral torsional buckling of the beam is prevented (see Chapter 7).

Geometrical parameters:

i = 600 mm (23.6 in.)
t_w = 10 mm (0.39 in.)
b_f = variable
t = 20 mm (0.79 in.)
a = 7 mm (0.28 in.)
A = 30 000 mm^2 (46.5 in.^{2})
I_y = 248 720 cm^4 (5976 in.^{4})

Material properties:

$$\text{Steel grade: ASTM A992}\quad F_y = 50\text{ ksi}(345\text{ MPa})\quad F_u = 65\text{ ksi}(448\text{ MPa})$$

(1) Compute web class:
 (a) According to EC3:

$$c_w = h - 2a = 600 - 2 \times 7 = 586\text{ mm}\,(23.1\text{ in.})$$

$$\frac{c_w}{t_w} = \frac{586}{10} = 58.6 < 72\sqrt{\frac{235}{f_y}} = 72 \times \sqrt{\frac{235}{345}} = 59.4 \rightarrow \text{class 1}$$

(b) According to AISC 360-10:

$$\frac{h}{t_w} = \frac{600}{10} = 60 < 3.76\sqrt{\frac{E}{F_y}} = 3.76 \times \sqrt{\frac{29000}{50}} = 90.6 \rightarrow \text{Compact}$$

(2) Compute values of b_f that define boundaries between classes

(a) According to EC3:

- boundary between class 1 and class 2:

$$c_f = 9\sqrt{\frac{235}{f_y}} \cdot t = 9 \times \sqrt{\frac{235}{345}} \times 20 = 148 \text{ mm}$$

$$b_f = 2c_f + 2a + t_w = 2 \times 148 + 2 \times 7 + 10 = 321 \text{ mm (12.6 in.)}$$

- boundary between class 2 and class 3:

$$c_f = 10\sqrt{\frac{235}{f_y}} \cdot t = 10 \times \sqrt{\frac{235}{345}} \times 20 = 165 \text{ mm}$$

$$b_f = 2c_f + 2a + t_w = 2 \times 165 + 2 \times 7 + 10 = 354 \text{ mm (13.9 in.)}$$

- boundary between class 3 and class 4:

$$c_f = 14\sqrt{\frac{235}{f_y}} \cdot t = 14 \times \sqrt{\frac{235}{345}} \times 20 = 231 \text{ mm}$$

$$b_f = 2c_f + 2a + t_w = 2 \times 231 + 2 \times 7 + 10 = 486 \text{ mm (19.1 in.)}$$

(b) According to AISC 360-10:

- boundary between compact and non-compact:

$$b_f = 2b = 2 \cdot 0.38\sqrt{\frac{E}{F_y}} \cdot t = 2 \times 0.38 \times \sqrt{\frac{29000}{50}} \times 20 = 366 \text{ mm (14.4 in.)}$$

- boundary between non-compact and slender:

$$k_c = \frac{4}{\sqrt{h/t_w}} = \frac{4}{\sqrt{600/10}} = 0.52$$

$$F_L = 0.7F_y = 0.7 \times 50 = 35 \text{ ksi (241 MPa)}$$

$$b_f = 2b = 2 \cdot 0.95\sqrt{\frac{k_c E}{F_L}} \cdot t = 2 \times 0.95 \times \sqrt{\frac{0.52 \times 29000}{35}} \times 20 = 788 \text{ mm (31.0 in.)}$$

(3) Compute $M_{y,Rk}$ and M_n for the previously computed values of b_f:

(a) Compute $M_{y,Rk}$ according to EC3. In Tables E4.5.1a (with S.I. units) and E4.5.1b (with US units) values for $M_{y,Rk}$ are listed. They have been computed for b_f values previously chosen. In addition, b_f = 600 mm (23.6 in.) and 700 mm (27.6 in.) have been added. Values in the last three columns are the applicable values for $M_{y,Rk}$. Applicable formulas are actually:

$$\text{For class 1 and 2}: M_{y,Rk} = W_{pl,y} \cdot f_y \tag{E4.5.1}$$

$$\text{For class 3}: M_{y,Rk} = W_{el,y} \cdot f_y \tag{E4.5.2}$$

$$\text{For class 4}: M_{y,Rk} = W_{eff,y} \cdot f_y \tag{E4.5.3}$$

Table E4.5.1a Values of $M_{y,Rk}$ for different b_f values – EC3 (S.I. units).

					$M_{y,Rk}$ (kNm)		
b_f (mm)	Class	$W_{el,y}$ (cm^3)	$W_{eff,y}$ (cm^3)	$W_{pl,y}$ (cm^3)	$W_{el,y} \cdot f_y$	$W_{eff,y} \cdot f_y$	$W_{pl,y} \cdot f_y$
321	1/2	4420	4420	4880	1525	1525	1684
354	2/3	4816	4816	5290	1662	1662	1825
366	3	4961	4961	5438	1712	1712	1876
486	3/4	6403	6403	6926	2209	2209	2389
600	4	7773	6838	8340	2682	2359	2877
700	4	8974	7119	9580	3096	2456	3305
788	4	10 032	7311	10 731	3461	2522	3702

Table E4.5.1b Values of $M_{y,Rk}$ for different b_f values – EC3 (US units).

					$M_{y,Rk}$ (kip-ft)		
b_f (in.)	Class	$W_{el,y}$ (in.^{3})	$W_{eff,y}$ (in.^{3})	$W_{pl,y}$ (in.^{3})	$W_{el,y} \cdot f_y$	$W_{eff,y} \cdot f_y$	$W_{pl,y} \cdot f_y$
12.6	1/2	269.7	269.7	297.8	1125	1125	1242
13.9	2/3	293.9	293.9	322.8	1226	1226	1347
14.4	3	302.7	302.7	331.8	1263	1263	1385
19.1	3/4	390.7	390.7	422.7	1630	1630	1763
23.6	4	474.3	417.3	508.9	1979	1741	2123
27.6	4	547.6	434.4	584.6	2285	1813	2439
31.0	4	612.2	446.1	654.8	2554	1861	2732

Note that for b_f = 354 mm (13.9 in.) two values for $M_{y,Rk}$ are valid: 1662 kNm (1226 kip-ft) and 1825 kNm (1347 kip-ft), because this value for b_f is the boundary between a class 2 flange and a class 3 flange, and therefore between a class 2 section and a class 3 section, so that $M_{y,Rk}$ can be computed with Eq. (E4.5.1) in the first case and Eq. (E4.5.2) in the second case, and there is no continuity between the two formulas.

Calculations of $W_{el,y}$ and $W_{pl,y}$ are straightforward.

Calculation of $W_{eff,y}$ shall be performed according to Section 4.2.3. As an example, compute $W_{eff,y}$ for b_f = 600 mm (23.6 in.). Refer to Figure E4.5.2 for symbols.

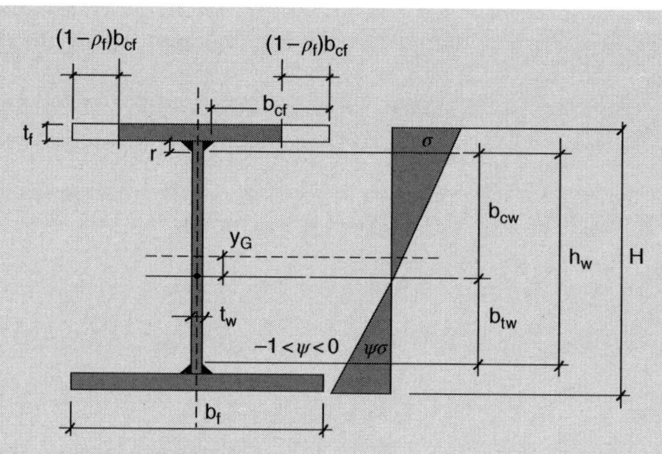

Figure E4.5.2 Geometrical parameters for calculation of $W_{eff,y}$.

$$\psi = 1 \text{(constant stress along the flange)}$$

$$\rightarrow k_\sigma = 0.57 - 0.21\psi + 0.07\psi^2 = 0.57 - 0.21 \times 1 + 0.07 \times 1^2 = 0.43$$

$$b_{cf} = 0.5(b_f - t_w - 2r) = 0.5 \times (600 - 10 - 2 \times 7) = 288\,\text{mm}$$

$$\bar{\lambda}_{pf} = \sqrt{\frac{f_y}{\sigma_{cr}}} = \frac{b_{cf}/t_f}{28.4\varepsilon\sqrt{k_\sigma}} = \frac{288/20}{28.4 \times \sqrt{235/345} \times \sqrt{0.43}} = 0.937 > 0.673$$

$$\rho_f = (\bar{\lambda}_{pf} - 0.188)/\bar{\lambda}_{pf}^2 = (0.937 - 0.188)/0.937^2 = 0.853$$

Effective area:

$$A_{eff,x} = A - 2(1-\rho_f)b_{cf}t_f = 30000 - 2 \times (1 - 0.853) \times 288 \times 20 = 28307\,\text{mm}^2 = 283.1\,\text{cm}^2$$

Compute shifting y_G of the centroid of effective section with respect to the gross one:

$$y_G = \frac{\left[2(1-\rho_f)b_{cf}t_f\right]\left(\frac{H}{2} - \frac{t_f}{2}\right)}{A - 2(1-\rho_f)b_{cf}t_f} = \frac{[2 \times (1-0.853) \times 288 \times 20]\left(\frac{640}{2} - \frac{20}{2}\right)}{30000 - 2 \times (1-0.853) \times 288 \times 20}$$

$$= 18.52\,\text{mm} = 1.852\,\text{cm}$$

Effective moment of inertia:

$$I_{eff,y} = I_y - 2\frac{1}{12}(1-\rho_f)b_{cf}t_f^3 - 2(1-\rho_f)b_{cf}t_f\left(\frac{H}{2} - \frac{t_f}{2}\right)^2 - A_{eff}y_G^2$$

$$= 248720 - 2\frac{1}{12}(1-0.853) \times 288 \times 20^3 \cdot 10^{-4} +$$

$$-2 \times (1-0.853) \times 288 \times 20 \times \left(\frac{640}{2} - \frac{20}{2}\right)^2 \cdot 10^{-4} +$$

$$-283.1 \times 1.852^2 = 231490\,\text{cm}^4\,(5562\,\text{in.}^4)$$

Effective modulus of section:

$$W_{eff,y} = \frac{I_{eff,y}}{\frac{H}{2} + y_G} = \frac{231\,490}{\frac{640 \cdot 10^{-1}}{2} + 1.852} = 6838\ \text{cm}^3\ (417.3\ \text{in.}^3)$$

(b) Compute M_n according to AISC 360-10. For the chosen b_f values, section belongs to compact ($b_f \le 366$ mm) and non-compact ($366\ \text{mm} \le b_f \le 788$ mm) classes.

- if the section is compact, M_n is computed with Eq. (7.61):

$$M_n = M_p = F_y Z$$

- if the section is non-compact, M_n is computed with Eq. (7.69):

$$M_n = M_p - (M_p - 0.7F_y S_x)\left(\frac{\lambda - \lambda_{pf}}{\lambda_{rf} - \lambda_{pf}}\right)$$

- if the section is slender, M_n is computed with Eq. (7.70):

$$M_n = \frac{0.9 E k_c S_x}{\lambda^2}$$

where:

$k_c = 0.52$ (see before);

$\lambda = b_f / 2t_f$ is the width-to-thickness ratio (variable);

$$\lambda_{pf} = 0.38\sqrt{E/F_y} = 0.38 \times \sqrt{29000/50} = 9.2$$

Term λ_{pf} is the limiting width-to-thickness ratio for a compact flange;

$$\lambda_{rf} = 0.95\sqrt{k_c E/F_L} = 0.95 \times \sqrt{0.52 \times 29000/35} = 19.7$$

Term λ_{rf} is the limiting width-to-thickness ratio for a non-compact flange.
Results are reported in Tables E4.5.2a (S.I. units) and E4.5.2b (US units).

Table E4.5.2a Values of M_n for different b_f values – AISC (S.I. units).

b_f (mm)	Class	S (cm^3)	Z (cm^3)	$\lambda = \frac{b_f}{2t_f}$	M_p (kNm)	M_n (kNm)
321	C	4420	4880	8.0	1684	1684
354	C	4816	5290	8.9	1825	1825
366	C/NC	4961	5438	9.2	1876	1876
486	NC	6403	6926	12.2	2389	2150
600	NC	7773	8340	15.0	2877	2324
700	NC	8974	9580	17.5	3305	2406
788	NC/S	10 032	10 731	19.7	3702	2425

Table E4.5.2b Values of M_n for different b_f values – AISC (US units).

b_f (in.)	Class	S (in.^{3})	Z (in.^{3})	$\lambda=\frac{b_f}{2t_f}$	M_p (kip-ft)	M_n (kip-ft)
12.6	C	269.7	297.8	8.0	1242	1242
13.9	C	293.9	322.8	8.9	1347	1347
14.4	C/NC	302.7	331.8	9.2	1385	1385
19.1	NC	390.7	422.7	12.2	1763	1587
23.6	NC	474.3	508.9	15.0	2123	1715
27.6	NC	547.6	584.6	17.5	2439	1776
31.0	NC/S	612.2	654.8	19.7	2732	1790

It can be noted that for $b_f = 366$ mm (14.4 in.) section turns from compact to non-compact but values of M_n, computed with formulas for compact and non-compact sections are the same: no discontinuity in formulas.

For $b_f = 788$ mm (31.0 in.) cross-section changes from non-compact to slender. Computing M_n with formula for slender sections results in:

$$M_n = \frac{0.9Ek_cS_x}{\lambda^2} = \frac{0.9 \times 29000 \times 0.52 \times 612.2}{19.7^2}/12 = 1790\,\text{kip-ft}$$

It should be noted that there is no discontinuity in formulas passing from non-compact to slender sections.

Finally, in Figure E4.5.3 $M - b_f$ curves are reported, where M represents M_n (AISC) and $M_{y,Rk}$ (EC3). It should be noted that:

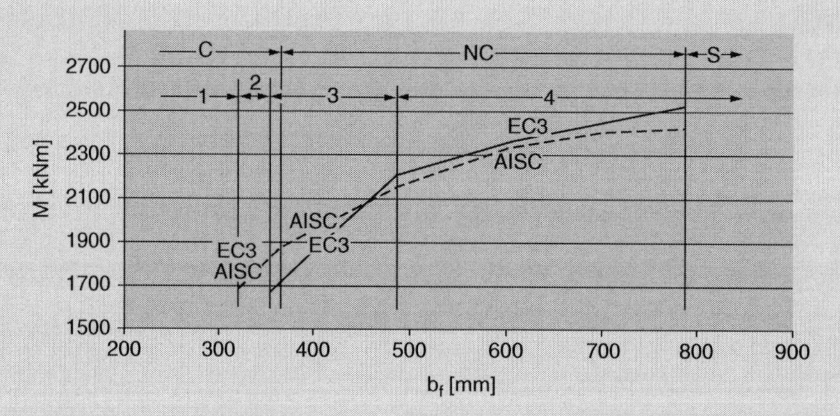

Figure E4.5.3 Comparison between M_n (AISC) and $M_{y,Rk}$ (EC3) computed for various values of b_f.

(a) all the EC3 curves represent a discontinuity in $M_{y,Rk}$ values passing cross-section from class 2 to class 3, due to **hard (sharp)** change of formula, that passes from plastic to elastic modulus without transition;

(b) there are no discontinuities in AISC formulas passing from compact to non-compact and to slender sections;

(c) AISC compact class corresponds to EC3 classes 1 and 2: nominal flexural strength values obtained using the two codes are the same in this range;

(d) non-compact class covers approximately EC3 class 3 but extends on initial part of class 4 according to EC3;

(e) term λ_{rf}, width-to-thickness ratio computed for flanges of welded sections separates non-compact from slender sections, depending on parameter k_c that takes into account the influence of flange and web local buckling. As a consequence, M_n curves computed for non-compact welded flanges depend also on web width-to-thickness ratio: a less slender web (with a higher width-to-thickness ratio) would lead to higher M_n values.

CHAPTER 5
Tension Members

5.1 Introduction

Usually members in tension are made with hot-rolled profiles, typically angles or channels: in other cases, cold-formed profiles can be conveniently used. Load carrying capacity of tension members is essentially governed by:

- distribution of the residual stresses due to the manufacturing process;
- connection details of the element ends.

The load carrying capacity at the connection location depends on the effective area (Figure 5.1). When the force transfer mechanism is analysed in correspondence with the cross-section centroid, the effective (or net) area corresponds to the gross area appropriately reduced for the presence of holes. In case of staggered holes, the effective area has to be assumed to be the minimum between the effective one evaluated with reference to a straight section and the one associated with a suitable multi-linear piece line passing through the holes.

The effective cross-sectional area has to be evaluated according to standards provisions. The design of members under tensile force can be based on the selection of a member with a cross-section greater than the minimum area A_{min}, which can be evaluated on the basis of tensile design load N, such as:

$$A_{min} = \frac{N}{f_d} \tag{5.1}$$

where f_d is the design tension limit strength.

5.2 Design According to the European Approach

Members in tension subjected to the design axial force N_{Ed} must satisfy the following condition at every section, in accordance with European provisions:

$$N_{Ed} \le N_{t,Rd} \tag{5.2}$$

Structural Steel Design to Eurocode 3 and AISC Specifications, First Edition. Claudio Bernuzzi and Benedetto Cordova.

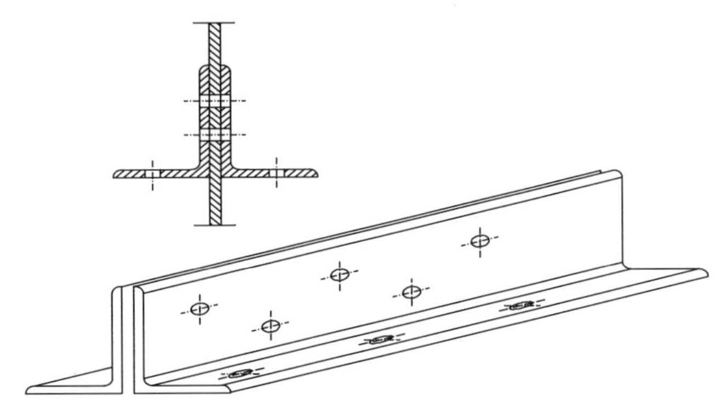

Figure 5.1 Connection detail for a member in tension.

Design tension resistance, $N_{t,Rd}$ of the cross-section has to be assumed to be the minimum between the plastic resistance of the gross cross-section, $N_{pl,Rd}$, and the ultimate resistance of the net cross-section in correspondence of the connection, $N_{u,Rd}$, which are, respectively, defined as:

$$N_{pl,Rd} = \frac{A \cdot f_y}{\gamma_{M0}} \tag{5.3a}$$

$$N_{u,Rd} = 0.9 \cdot \frac{A_{\text{net}} \cdot f_u}{\gamma_{M2}} \tag{5.3b}$$

where A and A_{net} represent the gross area and the net area in correspondence of the holes, respectively, and f_y and f_u are the yield and ultimate strength, respectively, with γ_{M0} and γ_{M2} representing the material partial safety factors.

It should be noted that term $N_{pl,Rd}$ is associated with ductile failure due to the attainment of the yield strength, while $N_{u,Rd}$, is related to a brittle failure in the connection section (governed by the attainment of the ultimate strength). In case of seismic loads, the well-established capacity design approach requires a ductile behaviour of member under tension (i.e. $N_{u,Rd} > N_{pl,Rd}$), which could be guaranteed if:

$$A_{\text{net}} \geq \frac{f_y}{f_u} \cdot \frac{\gamma_{M2}}{\gamma_{M0}} \cdot \frac{A}{0.9} \tag{5.4}$$

With reference to single or coupled angles connected via one leg, the effective area to be considered to evaluate the tensile load carrying capacity, assuming the force transfer mechanism is associated with only one leg.

When a single angle is used, reference has to be made to the criterion reported in EN 1993-1-8: a single angle in tension connected by a single row of bolts in one leg may be treated as concentrically loaded over an effective net section for which the design ultimate resistance, $N_{u,Rd}$, has to be determined as:

- with one bolt (Figure 5.2a):

$$N_{u,Rd} = \frac{2.0 \cdot (e_2 - 0.5 d_0) \cdot t \cdot f_u}{\gamma_{M2}} \tag{5.5a}$$

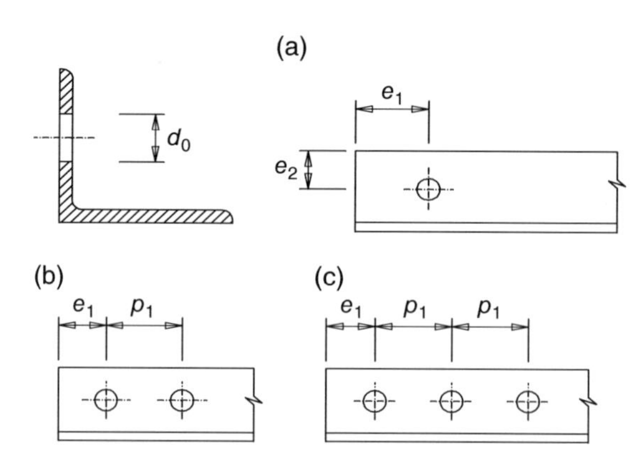

Figure 5.2 Single angle connected by one leg via (a) one bolt, (b) two bolts and (c) three bolts.

Table 5.1 Reduction factors β for angles connected via a single leg.

Pitch p_1	$\leq 2.5\ d_0$	$\geq 5\ d_0$
Two bolts	$\beta_2 = 0.4$	$\beta_2 = 0.7$
Three bolts or more	$\beta_3 = 0.5$	$\beta_3 = 0.7$

- with two bolts (Figure 5.2b):

$$N_{u,Rd} = \frac{\beta_2 \cdot A_{\text{net}} \cdot f_u}{\gamma_{M2}} \tag{5.5b}$$

- with three or more bolts (Figure 5.2c):

$$N_{u,Rd} = \frac{\beta_3 \cdot A_{\text{net}} \cdot f_u}{\gamma_{M2}} \tag{5.5c}$$

where e_2 is the distance from the axis of the hole to the outer edge of the element in the direction orthogonal to the force, d_0 is the diameter of the hole, terms β_2 and β_3 are reduction factors depending on the pitch p_1 as given in Table 5.1.

For an intermediate value of p_1 the value of β may be determined by linear interpolation and γ_{M2} is the safety coefficient and A_{net} is the effective area.

It should be noted that for an unequal-leg angle connected by its smaller leg, the resisting area A_{net} should be taken as equal to the net section area of an equivalent equal-leg angle of leg size equal to that of the smaller leg.

In case of staggered holes for fasteners (Figure 5.3), collapse could occur along a multi-linear path and the total area to be deduced for the evaluation of the net area (A_{net}) has to be considered the greater between:

- the maximum sum of the sectional area of the holes (A_f) in any cross-section perpendicular to the member axis;

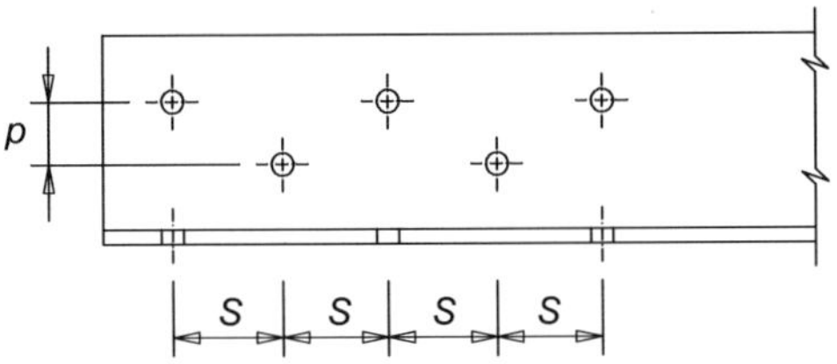

Figure 5.3 Typical connection in tension with staggered holes.

- the sum of the sectional areas of all holes in any diagonal or multi-linear line extending progressively across the member or part of the member less $s^2t/(4p)$ for each gauge space in the chain of holes, which can be expressed as:

$$t \cdot n \cdot d_0 - \sum \frac{s^2 \cdot t}{4 \cdot p} \tag{5.6}$$

where t is the thickness, n is the number of holes along the considered line, d_0 is the hole diameter and terms p and s have to be assumed in accordance with Figure 5.3.

Term p indicates the staggered pitch, which corresponds to the spacing of the centres of continuous holes in the chain measured parallel to the axis member. Term s measures the spacing of the centre of the same two holes measured perpendicular to the member axis.

5.3 Design According to the US Approach

LRDF approach

Tension member design in accordance with the US provisions for *load and resistance factor design* (LRFD) satisfies the requirements of AISC Specification when the *design tensile strength* $\phi_t P_n$ of each structural component equals or exceeds the *required tensile strength* P_u determined on the basis of the LRFD load combinations, that is:

$$P_u \le \phi_t P_n \tag{5.7}$$

where ϕ_t is the *tensile resistance factor and* P_n represents the *nominal tensile strength*

ASD approach

Design according to the provisions for *allowable strength design* (ASD) satisfies the requirements of AISC Specification when the *allowable tensile strength* P_n/Ω_t of each structural component equals or exceeds the *required tensile strength* P_a determined on the basis of the ASD load combinations, that is:

$$P_a \le P_n/\Omega_t \tag{5.8}$$

where Ω_t is the *tensile safety factor* and P_n represents the *nominal tensile strength*

P_n has to be determined as the minimum value obtained according to the limit states of *tensile yielding* and *tensile rupture*:

$$P_n = \min\{P_{n,y}; P_{n,u}\} \tag{5.9}$$

(a) For tensile yielding in the member gross section (ductile failure), the resistance is defined as:

$$P_{n,y} = F_y A_g \tag{5.10}$$

where F_y is the *specified minimum yield stress* and A_g is the *gross area* of the member. In this case $\Omega_t = 1.67$ and $\phi_t = 0.90$.

(b) For tensile rupture in the member net section (brittle failure), the resistance is defined as:

$$P_{n,u} = F_u A_e \tag{5.11}$$

where F_u is the *specified minimum tensile strength* and A_e is the *effective net area* of the member. In this case $\Omega_t = 2.00$ and $\phi_t = 0.75$.

As a comparison with Eurocode approach, the product between the minimum tensile strength and the effective net area is reduced by $0.9/\gamma_{M2} = 0.9/1.25 = 0.72$ (see Eq. (5.3b)) that is a bit more severe than $\phi_t = 0.75$.

The effective net area A_e has to be determined as follows:

(a) For tension members where the tension load is transmitted directly to each of the cross-sectional elements by fasteners:

$$A_e = A_n \tag{5.12}$$

where A_n is the net area of the member, computed as indicated in Section 5.2 but considering any bolt hole 1/16 in. (2 mm) greater than the nominal dimension of the hole.

(b) For tension members where the tension load is transmitted to some but not all of the cross-sectional elements by fasteners or welds:

$$A_e = A_n U \tag{5.13}$$

where U is the *shear lag factor*, determined as shown in Table 5.2.

For welded members also A_n should be determined as shown in Table 5.2.

In case of bracing members in seismic zones, for Special Concentrically Braced Frames (SCBF), the effective area must not be less than the gross area (see AISC 341-10, F2.5b(3)).

Table 5.2 Shear leg factors for connections to tension members (from Table D3.1 of AISC 360-10).

	Description of element		Shear lag factor, U	Example
1	All tension members where the tension load is transmitted directly to each of the cross-sectional elements by fasteners or welds (except as in Cases 4, 5 and 6)		$U=1$	—
2	All tension members, except plates and HSS, where the tension load is transmitted to some but not all of the cross-sectional elements by fasteners or longitudinal welds or by longitudinal welds in combination with transverse welds. (Alternatively, for W, M, S and HP, Case 7 may be used. For angles, Case 8 may be used.)		$U=1-\bar{x}/l$	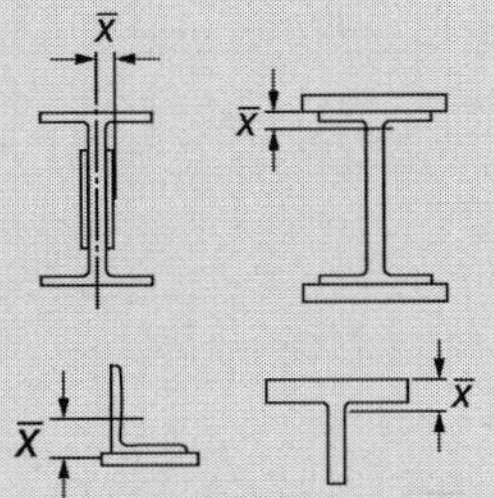 $\bar{x}$ $\bar{x}$ $\bar{x}$ $\bar{x}$
3	All tension members where the tension load is transmitted only by transverse welds to some but not all of the cross-sectional elements		$U=1$ and $A_n=$ area of the directly connected elements	—
4	Plates where the tension load is transmitted by longitudinal welds only		$U=1$ if $l\geq 2w$ $U=0.87$ if $2w>l\geq 1.5w$ $U=0.75$ if $1.5w>l\geq w$	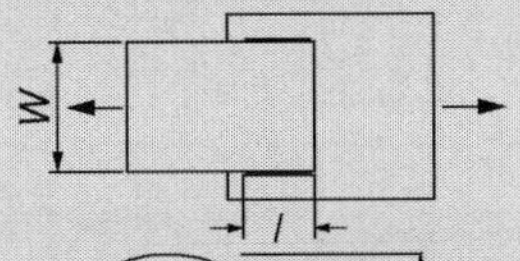 W l
5	Round HSS with a single concentric gusset plate		$U=1$ if $l\geq 1.3D$ $U=1-\bar{x}/l$ if $D\leq l<1.3D$; $\bar{x}=D/\pi$	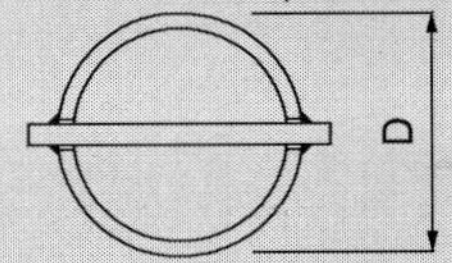 D
6	Rectangular HSS	With a single concentric gusset plate	$U=1-\bar{x}/l$ if $l\geq H$ $\bar{x}=\dfrac{B^2+2BH}{4(B+H)}$	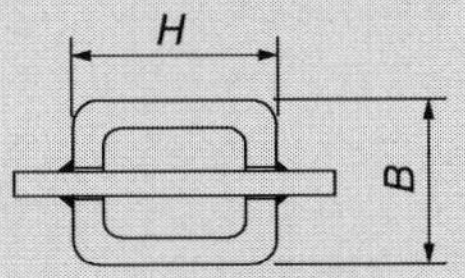 H B
		With two side gusset plates	$U=1-\bar{x}/l$ if $l\geq H$ $\bar{x}=\dfrac{B^2}{4(B+H)}$	—
7	W, M, S or HP Shapes or Tees cut from these shapes. (If U is calculated per Case 2, the larger value is permitted to be used.)	Flange connected with three or more fasteners per line in the direction of loading	$U=0.90$ if $b_f\geq 2/3d$ $U=0.85$ if $b_f<2/3d$	—
		Web connected with four or more fasteners per line in the direction of loading	$U=0.70$	—

(Continued)

Table 5.2 (*Continued*)

	Description of element		Shear lag factor, U	Example
8	Single and double angles (If U is calculated per Case 2, the larger value is permitted to be used.)	With four or more fasteners per line in the direction of loading	$U = 0.80$	—
		With three fasteners per line in the direction of loading (with fewer than three fasteners per line in the direction of loading, use Case 2)	$U = 0.60$	—

l = length of connection, in. (mm); w = plate width, in. (mm); $\bar{x}$ = eccentricity of connection, in. (mm); B = overall width of rectangular HSS (hollow structural steel) member, measured 90° to the plane of the connection, in. (mm); and H = overall height of rectangular HSS member, measured in the plane of the connection, in. (mm).

5.4 Worked Examples

Example E5.1 Angle in Tension According to EC3

Verify, according to the EC3 Code, the strength of a single equal leg angle L 120 × 10 mm (4.72 × 0.394 in.) in tension connected on one side via one line of two M16 (0.63 in. diameter) bolts in standard holes (Figure E5.1.1, dimensions in millimetres). Bolts connect only one side of the angle to a gusset plate. The angle is subjected to a design axial load N_{Ed} of 350 kN (75.7 kips).

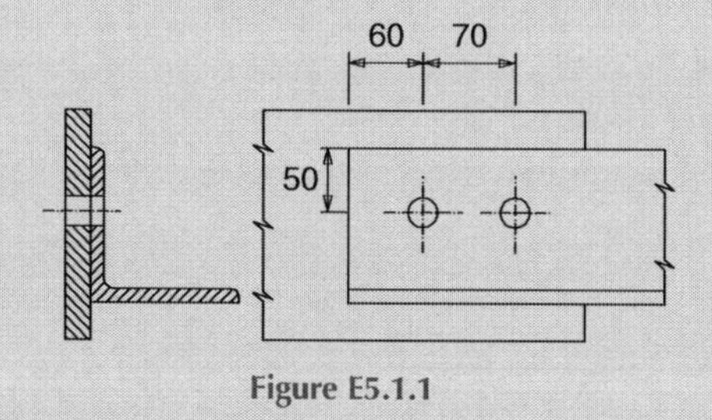

Figure E5.1.1

Material:

$$\text{S235 – EN 10025 – 2} \quad f_y = 235\ \text{MPa}(34\ \text{ksi}) \quad f_u = 355\ \text{MPa}(51.5\ \text{ksi})$$

Geometric properties L 120 × 10 mm (4.72 × 0.394 in.):

$$A_g = 2318\ \text{mm}^2 (3.59\ \text{in.}^2)$$

Bolt diameter : $d = 16$ mm (0.63 in.)
Standard hole : $d_0 = 17$ mm(0.67 in.)
Holes distance : $p_1 = 70$ mm(2.76 in)

Calculate the available tensile yield strength (Eq. (5.3a)):

$$N_{pl,Rd} = \frac{A \cdot f_y}{\gamma_{M0}} = \frac{2318 \times 235}{1.00} \cdot 10^{-3} = 544.7\,\text{kN}\,(122.5\,\text{kips})$$

Calculate the available tensile rupture strength (Eq. (5.3b)):

Being pitch p_1 (= 70 mm) ranging between $2.5 \cdot d_0$ (= 42.5 mm) and $5 \cdot d_0$ (= 85 mm), term β_2, is evaluated from Table 5.1 by linear interpolation and the value of 0.594 is assumed:

$$N_{u,Rd} = \frac{\beta_2 \cdot A_{\text{net}} \cdot f_u}{\gamma_{M2}} = \frac{0.594 \times [2318 - (10 \times 17)] \times 360}{1.25} \cdot 10^{-3} = 367.5\,\text{kN}\,(82.6\,\text{kips})$$

$$\text{Check}: N_{Ed} = 350\,\text{kN} \le N_{t,Rd} = 367.5\,\text{kN}\ \ (75.7 < 82.6\,\text{kips})\text{OK}$$

Example E5.2 Joint of a Tension Chord of a Trussed Beam According to EC3

Verify, in accordance with EC3, the splice connection in Figure E5.2.1 (dimensions in millimetres), which connects the end of two members of the chord of a trussed beam and transfers a design axial tension load N_{Ed} of 2250 kN (506 kips).

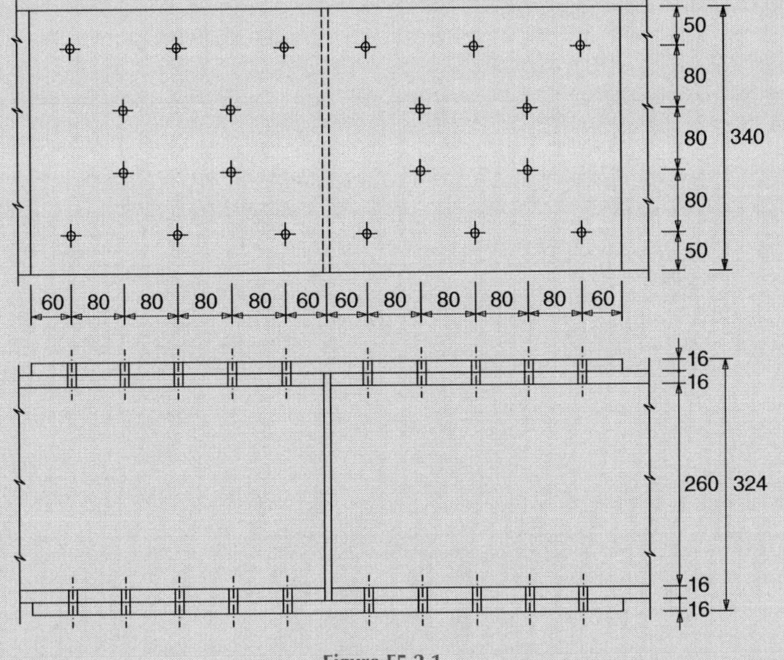

Figure E5.2.1

The flanges of the beam are composed by 340 × 16 mm (13.4 × 0.63 in.) plates and a plate 260 × 12 mm (10.2 × 0.472 in.) forms the beam web. Single cover plates 340 × 16 mm (13.4 × 0.63 in.) are bolted to the beam flange in normal holes (d_0 = 26 mm = 1.02 in.).

Material:

$$\text{S235} - \text{EN 10025} - 2\ \ f_y = 235\ \text{MPa}(34\ \text{ksi})\ \ f_u = 355\ \text{MPa}(51.5\ \text{ksi})$$

Area of the chord:

$$A_b = 2 \cdot b_f \cdot t_f + b_w \cdot t_w = 2 \times 340 \times 16 + 260 \times 12 = 14000\text{mm}^2\left(21.7\,\text{in.}^2\right)$$

Area of the cover plates:

$$A_g = 2 \times 340 \times 16 = 10880\,\text{mm}^2\left(15.88\ \text{in.}^2\right)$$

$$\text{Bolt diameter}: \quad d = 24\,\text{mm}(0.945\ \text{in.})$$
$$\text{Standard hole}: \quad d_0 = 26\,\text{mm}(1.02\ \text{in.})$$

Verification for plastic collapse. Calculate the tensile yield strength of the cover plates (being their area lower than the one of the H-shaped profile):

$$N_{pl,Rd} = \frac{A_f \cdot f_y}{\gamma_{M0}} = \frac{10880 \times 235}{1.00} \cdot 10^{-3} = 2556.8\,\text{kN}\ (575\ \text{kips})$$

$$\text{Check}: N_{Ed} = 2250\,\text{kN} \leq N_{pl,Rd} = 2556.8\ \text{kN}\ (506 < 575\ \text{kips})\ \text{OK}$$

Verification for brittle collapse. Due to the presence of staggered holes (Figure E5.2.2), the total area to deduce to the gross resisting area has to be considered the minimum between (Eq. (5.6)):

$$2 \times (26 \times 16) = 832\ \text{mm}^2\ (1.29\,\text{in.}^2)$$

$$\text{and}\ t \cdot n \cdot d_0 - \sum \frac{s^2 \cdot t}{4 \cdot p} = \left[4 \times (26 \times 16) - \left(2 \times \frac{80^2 \times 16}{4 \times 80}\right)\right] = 1024\ \text{mm}^2\ (1.587\,\text{in.}^2)$$

$$A_{\text{net}} = 10880 - 2 \times 1024 = 8832\ \text{mm}^2\ \left(13.69\,\text{in.}^2\right)$$

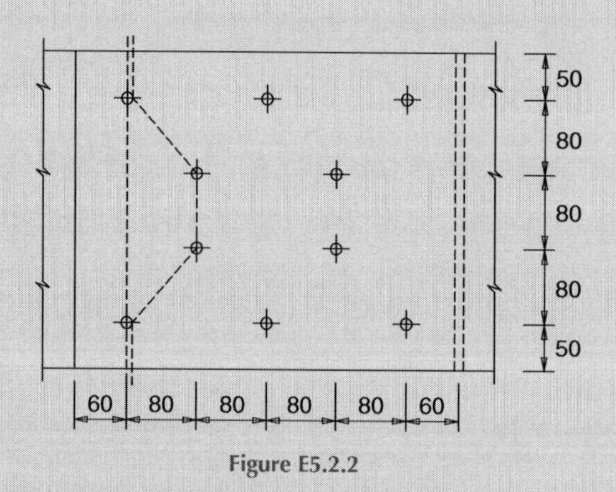

Figure E5.2.2

It should be noted that coefficient 2 is due to the presence of two cover plates.

$$N_{u,Rd} = 0.9 \cdot \frac{A_{\text{net}} \cdot f_u}{\gamma_{M2}} = 0.9 \times \frac{8832 \times 355}{1.25} \cdot 10^{-3} = 2257.5\ \text{kN}\ \ (507.5\,\text{kips})$$

$$\text{Check}: N_{Ed} = 2250\ \text{kN} \leq N_{u,Rd} = 2257.5\ \text{kN}\ \ (506 < 507.5\,\text{kips})\ \text{OK}$$

Example E5.3 Single Angle Tension Member, Connected on One Side by Bolts, According to AISC 360-10

Verify, according to AISC Code, both ASD and LRFD, the strength of a L5 × 5 × ⅜ (L127 × 127 × 9.5), ASTM A36, with one line of two ⅝ in. diameter bolts in standard holes.

Bolts connect only one side of the angle to a gusset. The angle is subjected to a dead load of 15 kips (66.7 kN) and a live load of 30 kips (133.4 kN) in tension (Figure E5.3.1).

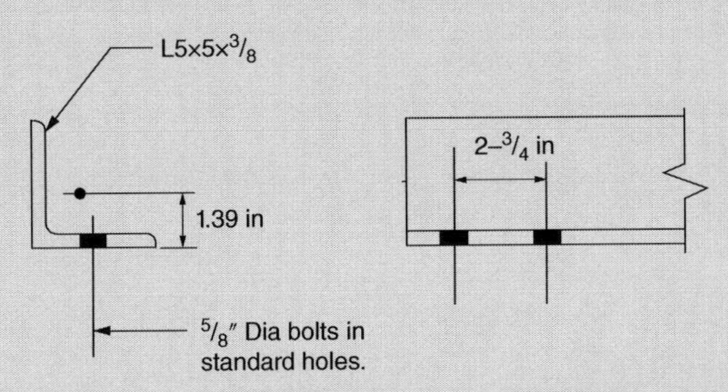

Figure E5.3.1

Material:

$$\text{ASTM A36}\quad F_y = 36\ \text{ksi}(248\ \text{MPa})\quad F_u = 58\ \text{ksi}(400\ \text{MPa})$$

Geometric properties L5 × 5 × ⅜ (L127 × 127 × 9.5):

$$A_g = 3.61\ \text{in.}^2\ (2329\ \text{mm}^2)\ \bar{y} = \bar{x} = 1.39\ \text{in.}\ (35.3\ \text{mm})$$

$$\text{Bolt diameter}: d = {}^5/_8\ \text{in.}(15.9\ \text{mm})$$
$$\text{Standard hole}: d_h = 11/16\ \text{in.}(17.5\ \text{mm})$$
$$\text{Holes distance}: l = 2 - {}^3\!/\!_4\ \text{in.}(70\ \text{mm})$$

Calculate the required tensile strength:

$$\text{LFRD}:\ P_u = 1.2 \times 15 + 1.6 \times 30 = 66\ \text{kips}(294\ \text{kN})$$
$$\text{ASD}:\ P_a = 15 + 30 = 45\ \text{kips}(200\ \text{kN})$$

Calculate the available tensile yield strength:

$$P_n = F_y A_g = 36 \times 3.61 = 130\ \text{kips}\ (578\ \text{kN})$$

$$\text{LFRD}: \phi_t P_n = 0.90 \times 130 = 117\ \text{kips}\ (520\ \text{kN})$$

$$\text{ASD}: P_n/\Omega_t = 130/1.67 = 77.8\ \text{kips}\ (346\ \text{kN})$$

Calculate the available tensile rupture strength:
Calculate U from Table D3.1 of AISC 360-10 Case 2 (see Table 5.2):

$$U = 1 - \frac{\bar{x}}{l} = 1 - \frac{1.39}{2^3/_4} = 0.50$$

Calculate net area A_n:

$$A_n = A_g - (d_h + {}^1/_{16})t = 3.61 - ({}^{11}/_{16} + {}^1/_{16}) \times {}^3/_8 = 3.33\,\text{in.}^2\left(2148\,\text{mm}^2\right)$$

Standard hole diameter shall be taken as 1/16 in. (2 mm) greater then nominal dimension of the hole (see AISC 360-10, B4.3b).

Calculate effective net area A_e:

$$A_e = A_n U = 3.33 \times 0.50 = 1.66\,\text{in.}^2\left(1074\,\text{mm}^2\right)$$

$$P_n = F_u A_e = 58 \times 1.66 = 95.5\,\text{kips}\,(429\,\text{kN})$$

$$\text{LFRD}: \phi_t P_n = 0.75 \times 96.5 = 72.4\,\text{kips}\,(322\,\text{kN})$$

$$\text{ASD}: P_n/\Omega_t = 96.5/2.00 = 48.3\,\text{kips}\,(215\,\text{kN})$$

The L5 × 5 × ⅜ (L127 × 127 × 9.5) tensile strength is governed by the tensile rupture limit state.

$$\text{LFRD}: \phi_t P_n = 72.4\,\text{kips} > P_u = 66\,\text{kips OK}$$

$$\text{ASD}: P_n/\Omega_t = 48.3\,\text{kips} > P_a = 45\,\text{kips OK}$$

Example E5.4 Development of Example E5.2 in Accordance with AISC 360-10

Compute the available tensile strength of profile and flange cover plates for required tensile strength (LRFD) of 506 kips (2250 kN). Geometrical details are represented in Figure E5.4.1 where dimensions are reported both in millimetres and in inches.

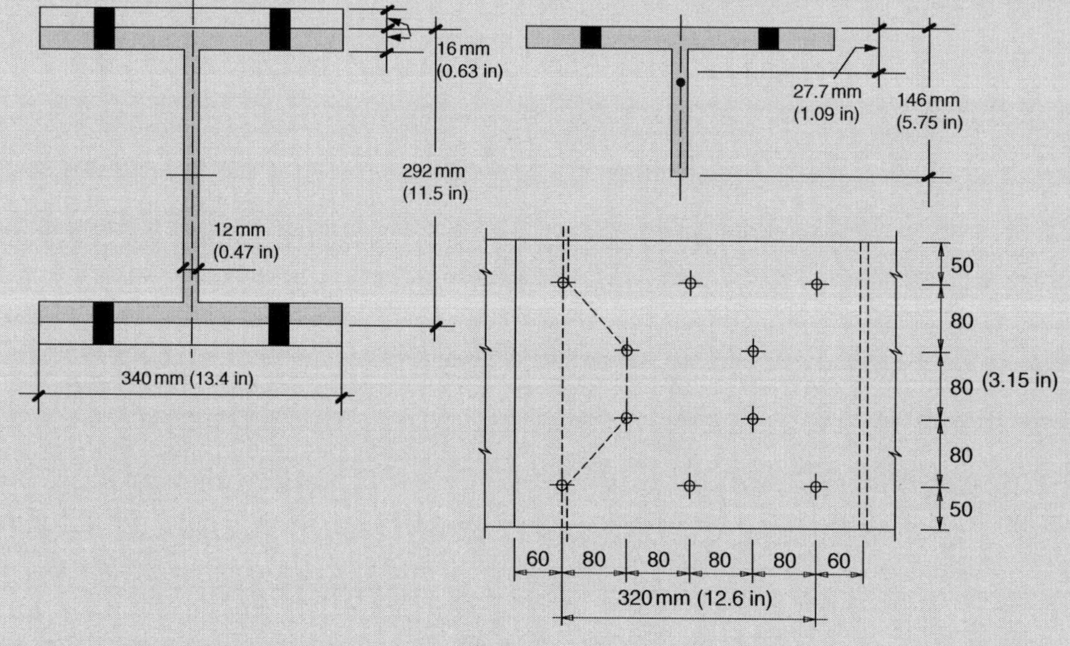

Figure E5.4.1

Material:

$$\text{S235}\quad F_y = 34.1\ \text{ksi}(235\ \text{MPa})\quad F_u = 52.2\ \text{ksi}(360\ \text{MPa})$$

Geometric properties of profile:

$$A_g = 21.7\ \text{in.}^2\ (14\,000\ \text{mm}^2)\ \bar{x} = 5.75\ \text{in.}\ (146\ \text{mm})\ (\text{referred to the tee-section})$$

Geometric properties of plates:

$$A_g = 2 \times 13.4 \times 0.63 = 15.88\ \text{in.}^2 \left(10890\ \text{mm}^2\right)$$

$$\begin{aligned}&\text{Bolt diameter}: d = 0.945\ \text{in.}(24\ \text{mm})\\&\text{Standard hole}: d_h = 1.02\ \text{in.}(26\ \text{mm})\\&\text{Holes distance}: l = 4 \times 3.15 = 12.60\ \text{in.}(4 \times 80 = 320\ \text{mm})(\text{see figure E5.4.1})\end{aligned}$$

Calculate the available tensile yield strength of the profile:

$$P_{n,profile} = F_y A_g = 34.1 \times 21.7 = 740\ \text{kips}\ (3292\ \text{kN})$$

$$\text{LFRD}: \phi_t P_{n,profile} = 0.90 \times 740 = 666\ \text{kips}\ (2963\ \text{kN})$$

$$\text{ASD}: P_{n,profile}/\Omega_t = 740/1.67 = 443\ \text{kips}\ (1971\ \text{kN})$$

Calculate the available tensile yield strength of flange cover plates:

$$P_{n,plates} = F_y A_g = 34.1 \times 16.88 = 576\ \text{kips}\ (2562\ \text{kN})$$

$$\text{LFRD}: \phi_t P_{n,plates} = 0.90 \times 576 = 518\ \text{kips}\ (2304\ \text{kN})$$

$$\text{ASD}: P_{n,plates}/\Omega_t = 576/1.67 = 345\ \text{kips}\ (1535\ \text{kN})$$

The available tensile strength is the minimum value between those of profile and flange cover plates, which must be greater or equal to the required tensile strength:

$$\phi_t P_n = \min\left\{\phi_t P_{n,profile}; \phi_t P_{n,plates}\right\} = \min\{666\ \text{kips};\ 518\ \text{kips}\} = 518 > 506\ \text{kips OK}$$

Verify now the available tensile strength at the connection.
Calculate the available tensile rupture strength of the profile:
Calculate U as the larger of the values from Table D3.1 of AISC 360-10 cases 2 and 7 in Table 5.2:
Case 2 – check as two T-shapes:

$$U = 1 - \frac{\bar{x}}{l} = 1 - \frac{1.09}{12.6} = 0.913$$

Case 7

$$b_f = 13.4\ \text{in.};\ \ d = 11.5\ \text{in.};\ \ b_f \geq (2/3)d \rightarrow U = 0.90$$

Use U = 0.913

Calculate the net area A_n:

$$A_n = A_g - 2\left[t \cdot n \cdot \left(d_h + {}^1/_{16}\right) - \sum \frac{s^2 t}{4p}\right] =$$

$$= 21.7 - 2 \times \left[0.63 \times 4 \times \left(1.02 + {}^1/_{16}\right) - \frac{3.15^2 \times 0.63}{4 \times 3.15}\right]$$

$$= 18.22\,\text{in.}^2 (11750\,\text{mm}^2)$$

The calculations are similar to those of Example E5.2 except that the standard hole diameter should be taken as 1/16 in. (2 mm) greater than nominal dimension of the hole (see AISC 360-10, B4.3b).

Calculate the effective net area A_e:

$$A_e = A_n U = 18.22 \times 0.913 = 15.63\,\text{in.}^2 (10730\,\text{mm}^2)$$

$$P_n = F_u A_e = 52.2 \times 16.63 = 865\,\text{kips}\,(3848\,\text{kN})$$

$$\text{LFRD}: \phi_t P_n = 0.75 \times 865 = 649\,\text{kips}\,(2887\,\text{kN})$$

$$\text{ASD}: P_n/\Omega_t = 865/2.00 = 423\,\text{kips}\,(1926\,\text{kN})$$

The profile tensile strength is 649 kips (2887 kN) and it is governed by the tensile rupture limit state.

Calculate the available tensile rupture strength of flange cover plates:

$$U = 1$$

Calculate the net area A_n:

$$A_n = A_g - 2\left[t \cdot n \cdot \left(d_h + {}^1/_{16}\right) - \sum \frac{s^2 t}{4p}\right] =$$

$$= 16.88 - 2 \times \left[0.63 \times 4 \times \left(1.02 + {}^1/_{16}\right) - \frac{3.15^2 \times 0.63}{4 \times 3.15}\right] = 13.41\,\text{in.}^2 (8652\,\text{mm}^2)$$

Calculate the effective net area A_e:

$$A_e = A_n U = 13.41 \times 1 = 13.41\,\text{in.}^2 \left(8652\,\text{mm}^2\right)$$

$$P_n = F_u A_e = 52.2 \times 13.41 = 700\,\text{kips}\,(3114\,\text{kN})$$

$$\text{LFRD}: \phi_t P_n = 0.75 \times 700 = 525\,\text{kips}\,(2335\,\text{kN})$$

$$\text{ASD}: P_n/\Omega_t = 700/2.00 = 350\,\text{kips}\,(1557\,\text{kN})$$

The flange cover plates tensile strength is 518 kips (2304 kN) and it is governed by the tensile yielding limit state.

The available rupture strength is the minimum value between the rupture strength of profile and flange cover plates, and it must be greater or equal to the required tensile strength:

$$\phi_t P_n = \min\left\{\phi_t P_{n,profile}; \phi_t P_{n,plates}\right\} = \min\{649\,\text{kips};\ 525\,\text{kips}\} = 525 > 506\,\text{kips OK}$$

CHAPTER 6

Members in Compression

6.1 Introduction

A member is considered to be compressed when subjected to an axial force applied at its centroid or if it is loaded by an eccentric axial force with a very small eccentricity. In accordance with the current design practice, eccentricity is considered to be sufficiently small when it is less than 1/1000 of the member length.

6.2 Strength Design

Pure compression on steel members is, in general, associated with instability phenomena due to their inherent slenderness. As a consequence, strength design must often be accompanied by stability design.

6.2.1 Design According to the European Approach

Strength design for a compression member subjected to a centric axial force N_{Ed} at a given cross-section is performed by comparing the demand to the axial resistance capacity $N_{c,Rd}$, that is:

$$N_{Ed} \leq N_{c,Rd} \tag{6.1}$$

The design compressive strength, $N_{c,Rd}$, is defined as a function of the cross-sectional class, identified as:

- cross-sections of class 1, 2 or 3:

$$N_{c,Rd} = \frac{A \cdot f_y}{\gamma_{M0}} \tag{6.2a}$$

- cross-sections of class 4:

$$N_{c,Rd} = \frac{A_{\text{eff}} \cdot f_y}{\gamma_{M0}} \tag{6.2b}$$

Structural Steel Design to Eurocode 3 and AISC Specifications, First Edition. Claudio Bernuzzi and Benedetto Cordova.

in which A and A_{eff} are the gross cross-section area and the effective cross-section area, respectively, f_y is the yielding strength of the material and γ_{M0} is the partial safety factor.

Local instability phenomena only penalize the axial force-carrying capacity for cross-sections belonging to class 4 because failure occurs at a stress level considerably smaller than the yielding stress. When the cross-section of the compression member is characterized by a single axis of symmetry, an additional flexural action ΔM_{Ed} may arise, due to the eccentricity between the gross cross-section centroid (on which the axial force is nominally applied) and the centroid of the resisting cross-section (Figure 4.8).

6.2.2 Design According to the US Approach

The AISC Specification does not distinguish between strength and stability design, but rather follows a unified approach accounting for global and resistance stability effects in the calculation of the design strength. As such, the AISC approach is described within the following section devoted to stability checks.

6.3 Stability Design

For a compression member in absence of imperfections and assuming a linear-elastic constitutive law (*Euler column*), a value of the axial force can be found to trigger element instability, called *elastic critical load*, N_{cr}. This phenomenon can take place flexurally, torsionally or with a combination of a flexural and a torsional behaviour: the Figure 6.1 shows the configuration of a generic cross-section in the undeformed and in the deformed position, respectively, for each of these instability phenomena.

Compression members having typical I- or H-shaped cross-section with two axes of symmetry are generally interested by flexural buckling, owing to the fact that the *torsional buckling*, generally occurs when the column has a very limited length, out of interest for many routine design applications. Cruciform sections, T-sections, angles and, in general, all cross-sectional shapes in which all the elements converge into a single point, are generally sensitive to torsional buckling phenomena. Furthermore, cross-sections with one axis of symmetry are prone to flexural-torsional buckling in many cases instead of the torsional one, owing to the fact that both cross-sectional centroid and shear centre lie on the axis of symmetry but are often not coincident.

If flexural buckling takes place before any other instability phenomena, the associated critical load $N_{cr,F}$ is defined on the basis of equilibrium stability criteria as follows:

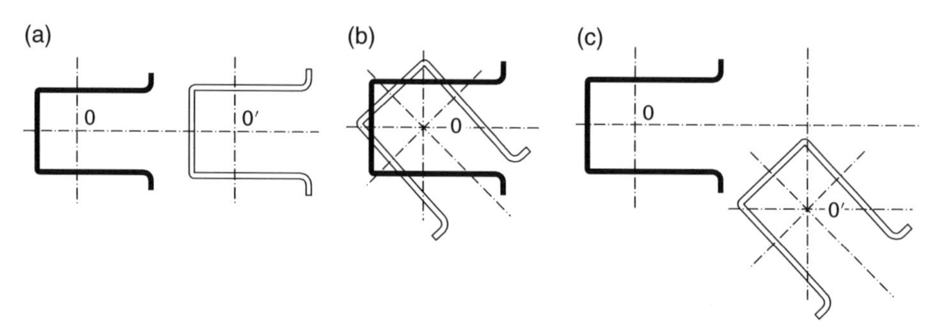

Figure 6.1 Typical global instability configurations: (a) flexural, (b) torsional and (c) flexural-torsional.

$$N_{cr,F} = \min\left\{\frac{\pi^2 EI_y}{L_{0,y}^2}, \frac{\pi^2 EI_z}{L_{0,z}^2}\right\} \tag{6.3}$$

in which E is Young's modulus of the material, I is the moment of inertia, L_0 is the effective length of the member (equal to the actual length of the member for a pinned-pinned column) and subscripts y and z indicate the principal axes of the cross-section.

From the design standpoint, it is sometimes convenient to refer to the critical stress, σ_{cr}, instead of the critical axial load. The critical stress, based on the flexural buckling modes, is defined as follows:

$$\sigma_{cr} = \frac{N_{cr,F}}{A} = \min\left\{\frac{\pi^2 E\rho_y^2}{L_{0,y}^2}, \frac{\pi^2 E\rho_z^2}{L_{0,z}^2}\right\} = \min\left\{\frac{\pi^2 E}{\lambda_y^2}, \frac{\pi^2 E}{\lambda_z^2}\right\} \tag{6.4}$$

in which A is the gross cross-sectional area of the column, ρ is the radius of gyration $\left(\rho = \sqrt{I/A}\right)$ and λ is the slenderness of the compression member ($\lambda = L_0/\rho$).

The slenderness to be used, λ, is the larger of those calculated in the y and z directions, that is $\lambda = \max(\lambda_y, \lambda_z)$. For example, it is possible to make reference to the compression member in Figure 6.2, restrained in different ways in the x–y and x–z planes, where the x-axis is the one along the length of the member. The effective length in the x–y plane has to be taken as $L/4$ (i.e. $L_{0z} = 2.25$ m, 6.38 ft), whereas in the x–z plane it is $L/2$ ($L_{0y} = 4.5$ m, 14.76 ft).

The theory of Euler column has no practical applications for structural design, due to the hypothetical linear-elastic material and to the absence of geometrical imperfections. The behaviour of compression members is always influenced by:

- the non-linear constitutive law for a material that is limited in strength, characterized by a post-elastic branch associating large strains to small increments of stress (for practical purposes, such material can be approximated by an elastic-perfectly plastic law or by an elastic-plastic law with hardening, as discussed in Chapter 1);

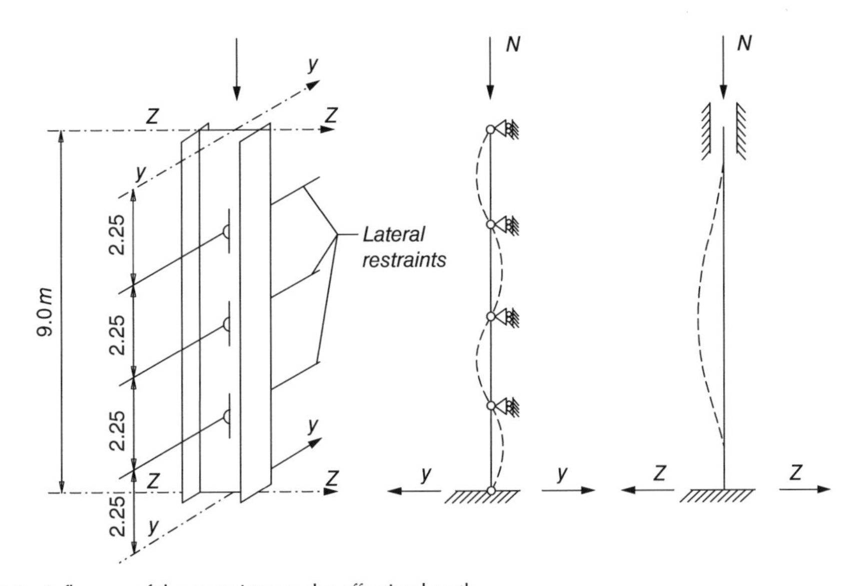

Figure 6.2 Influence of the restraints on the effective length.

- the mechanical and geometrical imperfections, mostly due to the production process and arising during the fabrication and erection phases.

In compression members with a gross cross-section area A in the absence of imperfections but with a strength limited to the material yielding, f_y, the critical load cannot exceed the yielding force (squash load) for the cross-section ($f_y{\cdot}A$). The stability curve associated with this case is shown in Figure 6.3, in terms of the stress (σ) versus slenderness (λ) relationship. The intersection between the curve associated with Eq. (6.4) and the horizontal line corresponding to the yielding stress, f_y, identifies point P, the abscissa of which (λ_p) is called *proportionality slenderness*, and is defined as follows:

$$\lambda_p = \pi\sqrt{\frac{E}{f_y}} \tag{6.5}$$

As an example, Table 6.1 summarizes the values of the proportionality slenderness for the steel grades commonly used in EC3.

The value of the proportionality slenderness for a perfect compression member (a situation quite far from reality) can be immediately associated with the failure mode of the member:

- when $\lambda < \lambda_p$, failure is due to full plasticization of the cross-section (squashing failure);
- when $\lambda > \lambda_p$, failure is due to buckling phenomena;
- when $\lambda = \lambda_p$, failure is due to simultaneous squashing and buckling of the member.

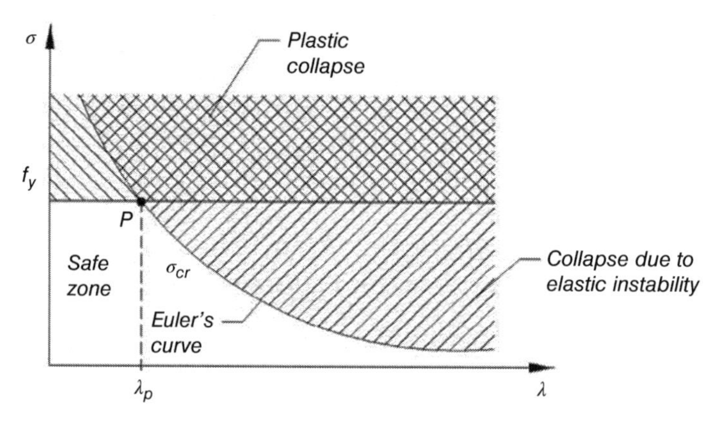

Figure 6.3 Capacity domain in terms of stress (σ) and slenderness (λ) relationship for a compression member.

Table 6.1 Values of the proportionality slenderness.

Steel grade	Elements with thickness ≤ 40 mm (≤1.58 in.)	Elements with thickness > 40 mm (≤1.58 in.)
S 235	$\lambda_p = 93.91$	$\lambda_p = 98.18$
S 275	$\lambda_p = 86.81$	$\lambda_p = 90.15$
S 355	$\lambda_p = 76.41$	$\lambda_p = 78.66$
S 420	$\lambda_p = 70.25$	$\lambda_p = 72.90$
S 460	$\lambda_p = 67.12$	$\lambda_p = 69.43$

In industrial compression members, mechanical and geometrical imperfections are always present, significantly affecting their axial load carrying capacity. In particular, initial imperfections are usually approximated via a sinusoidal deflected shape of the member (bow imperfection): when the applied axial force N increases, the lateral deflection δ increases as well and, as a consequence, the flexural effects due to load eccentricity increase too. The member is really subjected to a combination of axial force and bending moment. As an example, for a simply supported-pinned column the mid-length section is subjected to the largest flexural moment (equal to $N\,\delta$). With an increase in the axial force, the maximum stress in the outer fibres of the cross-section also increases, up to yielding. The response of a real compression member in terms of a force-lateral displacement relationship (Figure 6.4a), initially coincides with the one of an ideal member (perfectly elastic constitutive law) with an initial imperfection. This is due to the fact that the material is initially within its elastic range. The dashed curve in Figure 6.4a tends asymptotically to the elastic critical load value (N_{cr}), and the relationship between the lateral displacement (δ) and the axial force N can be approximated in terms of the initial imperfection of the member (δ_0), as:

$$\delta = \delta_0 \cdot \frac{1}{1 - \dfrac{N}{N_{cr}}} \tag{6.6}$$

The mid-length cross-section is subjected to combined axial and bending stresses, and the maximum stress can be estimated as shown in Figure 6.4b as:

$$\sigma = \frac{N}{A} + \frac{N \cdot \delta}{W} = \frac{N}{A} + \frac{N \cdot \delta_0}{W} \cdot \frac{1}{1 - \dfrac{N}{N_{cr}}} \tag{6.7}$$

Once the yielding stress is locally reached, a non-negligible stiffness reduction occurs related to the lower (ideally null) value of the material elastic modulus in the fibres that exceed the yielding strain. Consequently, the behaviour of an industrial compression member deviates from that of an ideally elastic member, showing an increasingly larger flexural deformability. The load value N_u, smaller than the elastic critical load N_{cr}, corresponds to the attainment of the member load carrying capacity. Further increments of the transverse displacement δ require that equilibrium can only be attained upon a decrement of the applied load.

When comparing the stress-slenderness curves for the ideal member with that for the industrial member, it can be observed that, by incorporating geometrical and mechanical imperfections via term δ_0 and by considering a strength limit for the material, the load-carrying capacity can be

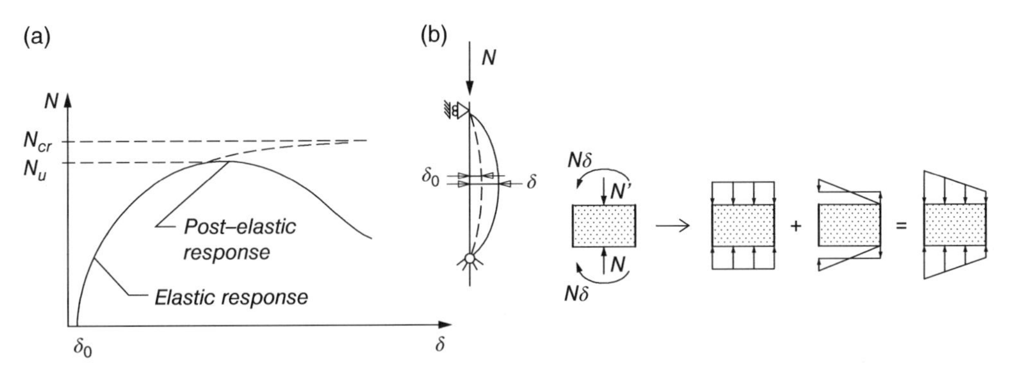

Figure 6.4 Load-transverse displacement relationship for a real compression member with initial imperfection δ_0 (a) and stress state in the mid-length section (b).

greatly reduced with respect to the ideal behaviour (Figure 6.5). The slenderness level at which the load carrying capacity becomes smaller than the squash load reduces from λ_p to approximately $0.2\lambda_p$.

From the design standpoint, a compression member is checked against instability by enforcing a maximum value of stress (capped at the yielding stress) defined as a function of the following parameters:

- *element slenderness*: from a purely theoretical standpoint, in the case of non-sway frames, the effective length of a member of length L varies generally between $0.5L$ and L (Figure 6.6a), whereas columns of sway frames are characterized by effective lengths varying between L and infinity (Figure 6.6b);
- *cross-sectional shape*: depending on the production process and on the shape of the cross-section, residual stresses that develop during production process can affect buckling behaviour and this effect is taken into account via the definition of suitable imperfections (as an example, bow imperfection δ_0);
- *steel grade*: residual stresses can represent a non-negligible fraction of the yielding strength of the material, thus reducing the load-carrying capacity when compared to an ideally stress-free cross-section.

As an alternative to the design check in terms of stress typically associated with allowable stress design approach, it is possible to directly compare the design axial demand with the design load-carrying capacity of the member according to the semi-probabilistic limit-state approach recommended by more recent design standards.

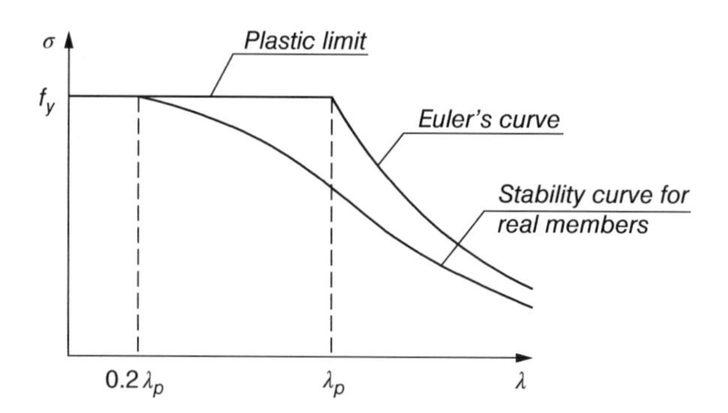

Figure 6.5 Stability curve for a compression member with or without imperfections.

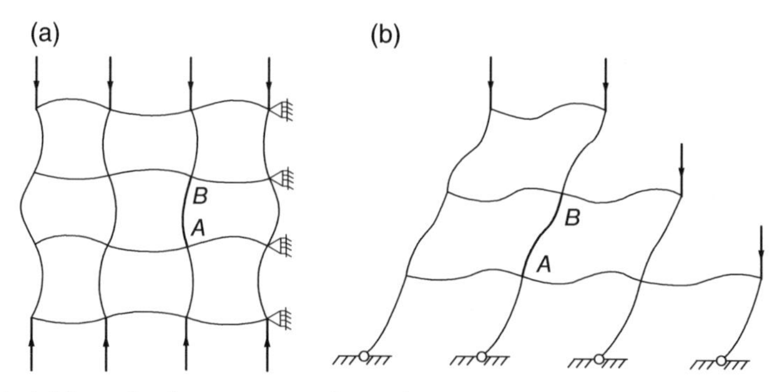

Figure 6.6 Typical deformation for non-sway and sway frames.

Design against instability of compression members requires an accurate calculation of the effective length L_0, which can be defined as the distance between two contiguous inflection points of the buckled shape or, equivalently, the effective length factor k (with $L_0 = k \cdot L$). In the case of isolated columns, as well as in the case of truss members, the effective length can be determined on the basis of simple considerations on the restraints at the member ends and thus the buckled configuration of the member. The most common situations related to isolated members are summarized in Figure 6.7, which can be associated with a practical evaluation of the effective length only when the load is applied at one end of the member.

Modifying the load distribution but maintaining the same resultant applied load, the critical load varies. This corresponds to a change in the slenderness of the considered member or, in other words, to an effective length value different from the one associated with the load applied at one end. As an example, Figure 6.8 shows some numerical results related to a simply-supported pinned member in terms of effective length L_0. Reference is made to the cases of a single load applied at the member end, of two loads applied at equally spaced locations or, more generally, of loads applied at n equally spaced cross-sections with a force of intensity N/n, maintaining the same end restraints for all cases and the same base reaction force N.

It can be noted that increasing the number of the loaded cross-sections, the value of the effective length decreases thus exemplifying also the influence of the load conditions on the stability response of a compression member.

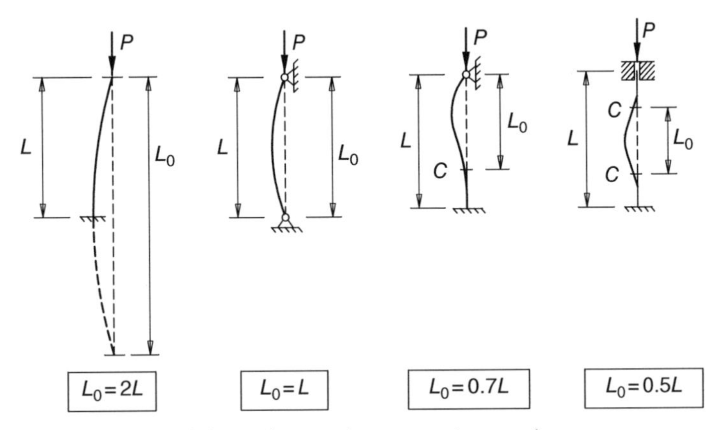

Figure 6.7 Typical cases of buckled shapes for a single compression member.

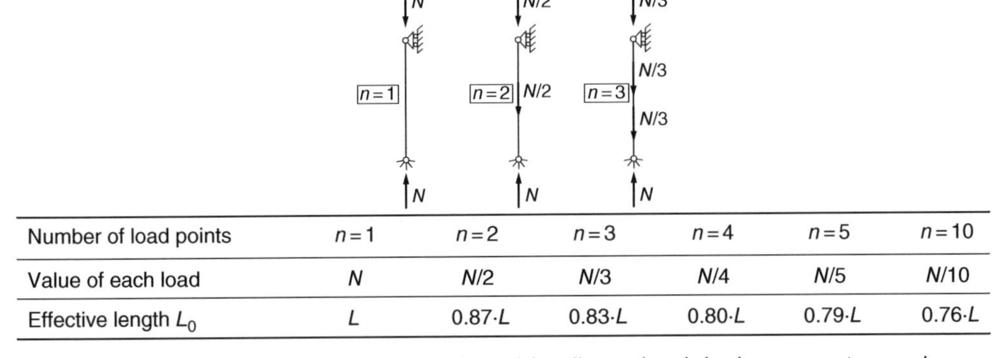

Number of load points	$n=1$	$n=2$	$n=3$	$n=4$	$n=5$	$n=10$
Value of each load	N	$N/2$	$N/3$	$N/4$	$N/5$	$N/10$
Effective length L_0	L	$0.87 \cdot L$	$0.83 \cdot L$	$0.80 \cdot L$	$0.79 \cdot L$	$0.76 \cdot L$

Figure 6.8 Load conditions and corresponding values of the effective length for the compression members considered.

As already mentioned, there are situations in which torsional or flexural-torsional instability take place before the achievement of the flexural instability, depending on the cross-sectional shape of the member considered.

In particular, when considering a generic cross-section with at least one axis of symmetry, the elastic torsional critical load $N_{cr,T}$ can be evaluated as:

$$N_{cr,T} = \frac{1}{\rho_0^2}\left[GI_t + \frac{\pi^2 EI_w}{L_T^2}\right] \tag{6.8a}$$

where E and G are the elastic and shear modulus of the material, respectively, I_t is the torsional coefficient, I_w is the warping coefficient, I_c is the polar moment of inertia with respect to the shear centre and L_T is the effective length of the compression member for torsional buckling.

$$\text{with:}\ \rho_0^2 = \rho_y^2 + \rho_z^2 + y_0^2 + z_0^2 \tag{6.8b}$$

where y_0 and z_0 represent the distance between the shear centre and the cross-section centroid along the y–y and z–z axes, respectively, and ρ is the radius of inertia.

More details on the calculation of the torsional and warping coefficients, as well as a discussion on the shear centre of a cross-section, are provided in Chapter 8 devoted to torsion. For cross-sections with at least one axis of symmetry (y–y axis), the elastic flexural-torsional critical load can be theoretically estimated as:

$$N_{cr,TF} = \frac{N_{cr,y}}{2\beta}\left[1 + \frac{N_{cr,T}}{N_{cr,y}} - \sqrt{\left(1 - \frac{N_{cr,T}}{N_{cr,y}}\right)^2 - 4\left(\frac{y_0}{\rho_0}\right)^2 \frac{N_{cr,T}}{N_{cr,y}}}\right] \tag{6.9a}$$

in which coefficient β is defined as:

$$\beta = 1 - \left(\frac{y_0}{\rho_0}\right)^2 \tag{6.9b}$$

Usually, in design practice only flexural buckling is considered, even though all buckling modes should be investigated, obtaining stability curves such as those shown in Figure 6.9 that refers to an equal-leg angle shape.

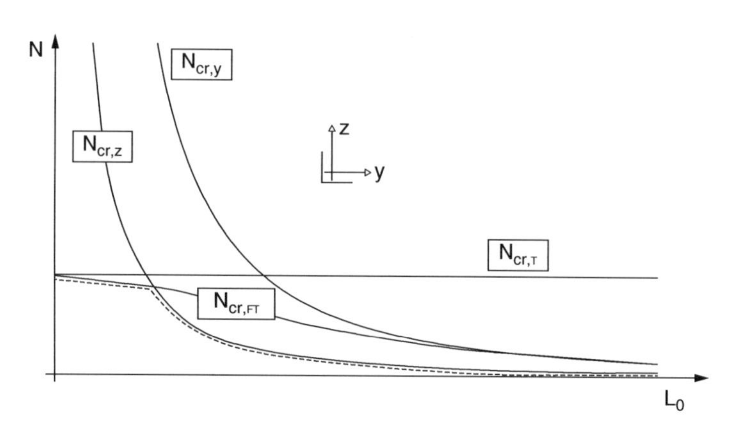

Figure 6.9 Stability curves for an equal-leg angle considering flexural, torsional and flexural-torsional buckling.

In the common case of hot-rolled I-shaped sections, flexural buckling is generally by far the dominant phenomenon. When considering bi-symmetrical cross-sections, only relatively heavy cross-sections, with very low slenderness, can be affected by torsional buckling before the flexural buckling occurs.

6.3.1 Effect of Shear on the Critical Load

The well-established Euler approach neglects the shear deformations for the evaluation of the critical load. In some cases (e.g. short beams or built-up latticed columns), it is important to account for the contribution of shear, which can affect significantly the member response.

The shear deformability of a infinitesimal element of a beam subjected at its ends to a shear force $V(x)$ can be written as a function of the longitudinal coordinate x by means of the shear strain $\gamma(x)$, defined as (Figure 6.10):

$$v'_T = \gamma(x) = \chi_T \frac{T(x)}{GA} \tag{6.10}$$

where χ_T is the shear factor of the cross section (see Chapter 7), A is the cross-sectional area and G is the shear modulus of the material.

The variation of γ (x) along the longitudinal axis of the beam generates an additional curvature that can be expressed as:

$$v''_T(x) = \gamma'(x) = \chi_T \frac{T'(x)}{GA} \tag{6.11}$$

where v_T (x) is the contribution to the transverse deflection due to shear.

If the member under consideration is prismatic (i.e. if its cross-section does not change along the length), by approximating the total curvature with the second derivative of the lateral deflection, it results in:

$$v''(x) = v''_F(x) + v''_T(x) = -\frac{M(x)}{EI} + \chi_T \frac{T'(x)}{GA} \tag{6.12}$$

Within the small displacement hypothesis, we can obtain (Figure 6.11):

$$V(x) = N(x) \cdot v'(x) \tag{6.13a}$$

where $N(x)$ is the internal axial force.

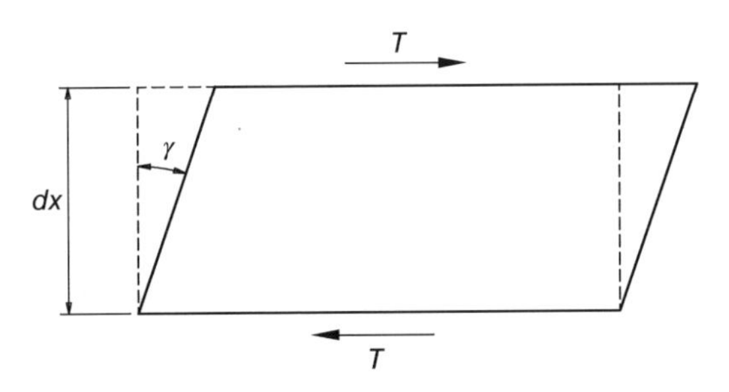

Figure 6.10 Shear deformation of an infinitesimal element of a beam.

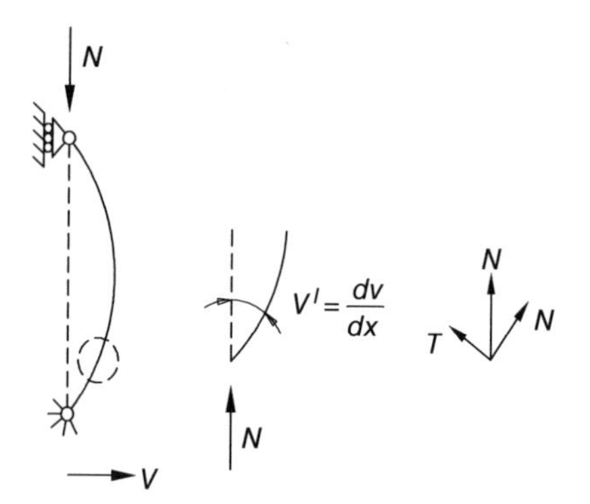

Figure 6.11 Second order effects on a simply supported compression member.

If the axial force is constant (i.e. $N(x) = N = P$), taking the derivative of Eq. (6.13a) this results in:

$$V'(x) = N \cdot v''(x) \tag{6.13b}$$

The elastic curvature equation thus becomes:

$$v''(x) = v''_F(x) + v''_T(x) = -\frac{M(x)}{EI} + \chi_T \frac{N \cdot v''(x)}{GA} \tag{6.14a}$$

Now, by expressing the internal bending moment with reference to the deformed configuration, that is by imposing that $M(x) = N \cdot v(x)$, it results in:

$$v''(x) + \frac{N}{E \cdot I\left(1 - \chi_T \frac{N}{G \cdot A}\right)} v(x) = 0 \tag{6.14b}$$

By defining $\alpha^2 = \frac{N}{E \cdot I\left(1 - \chi_T \frac{N}{G \cdot A}\right)}$, the differential Eq. (6.14b) can be expressed as:

$$v''(x) + \alpha^2 v(x) = 0 \tag{6.15}$$

It is worth mentioning that Eq. (6.15) is formally identical to that used for the determination of the elastic critical load in the presence of purely flexural deformations, with the difference represented by the meaning of term α^2. The solution to Eq. (6.15) is of the form:

$$v(x) = A \cdot \cos(\alpha \cdot x) + B \cdot \text{sen}(\alpha \cdot x) \tag{6.16}$$

in which A and B are the constants of integration that can be calculated on the basis of the boundary conditions, that is based on the restraints at the ends of the column.

For a simply supported compression member (with an effective length equal to L) when $x = 0$ it must be $v(0) = 0$, resulting in:

$$B = 0 \tag{6.17a}$$

Similarly, by imposing a zero transverse displacement at $x = L$ ($v(L) = 0$):

$$A \cdot \sin(\alpha \cdot L) = 0 \tag{6.17b}$$

With reference to Eq. (6.17b), besides the trivial solution (i.e. $A = 0$), it is possible to find a non-trivial solution when $\alpha \cdot L = \pi$. The axial force value that satisfies this condition is the elastic critical load obtained by taking into account the shear deformability of the compression member, indicated in the following with $N_{cr,id}$. Solving for α results in:

$$\alpha^2 = \frac{\pi^2}{L^2} \tag{6.18}$$

Equating, Eq. (6.18) with the definition of α^2 is modified in:

$$N_{cr,id} = \frac{\pi^2 \cdot E \cdot I}{L^2}\left[\frac{1}{1 + \frac{\chi_T}{G \cdot A}\pi^2 \frac{E \cdot I}{L^2}}\right] \tag{6.19}$$

Using the definition of Euler critical load (N_{cr}) obtained neglecting the shear contribution, the expression in Eq. (6.19) becomes:

$$N_{cr,id} = N_{cr}\left[\frac{1}{1 + \frac{\chi_T}{G \cdot A}N_{cr}}\right] = \frac{N_{cr}}{1 + \frac{N_{cr}}{S_v}} = \frac{1}{\frac{1}{N_{cr}} + \frac{1}{S_v}} \tag{6.20}$$

in which the term S_v represents the shear stiffness of the member considered, the contribution that is usually negligible for solid cross-section hot-rolled standard profiles.

When considering built-up compression members, the approach is based on the concept of equivalent slenderness (or interchangeably of equivalent effective length, L_{eq}, which can also be expressed in terms of an effective length amplification factor $k_{\beta,eq}$). The more accurate evaluation of the critical load can be expressed in alternative modes, such as:

$$N_{cr,id} = \frac{\pi^2 EI}{L^2}\left[\frac{1}{1 + \frac{\chi_T}{GA}\pi^2 \frac{EI}{L^2}}\right] = \frac{\pi^2 EI}{\left[1 + \pi^2 \chi_T \left(\frac{\rho}{L}\right)^2 \frac{E}{G}\right] L^2} = \frac{\pi^2 EI}{\left(k_{\beta,eq} L\right)^2} = \frac{\pi^2 EI}{\left(L_{eq}\right)^2} \tag{6.21}$$

where ρ is the radius of gyration of the built-up cross section.

It can be noted that in Eq. (6.21) the effective length amplification factor ($k_{\beta,eq}$) should be defined as:

$$k_{\beta,eq} = \sqrt{1 + \pi^2 \cdot \chi_T \cdot \left(\frac{\rho}{L}\right)^2 \cdot \frac{E}{G}} \tag{6.22}$$

The value of $N_{cr,id}$ can thus be calculated following the same approach used for N_{cr} and the effective length (and, consequently, the slenderness) is correspondingly modified. When geometry, restraints and load conditions are kept the same, the critical load $N_{cr,id}$ is always smaller than the Euler critical load, N_{cr}, being the slenderness value obtained when accounting for the contribution of shear deformations always larger than that obtained with flexural contributions alone.

The same approach of shifting the focus from ideal to real compression members that was discussed for the case of negligible shear deformations can be followed. Namely, reference can be made to stability curves in the specifications that account for the presence of imperfections, residual stresses and potential overages with respect to the yielding strength.

6.3.2 Design According to the European Approach

The stability check for a compression member subjected to a design axial force N_{Ed} is satisfied when the demand is smaller than its design capacity $N_{b,Rd}$, that is if:

$$N_{Ed} \leq N_{b,Rd} \tag{6.23}$$

The design capacity against instability of a compression member is calculated as a function of the cross-sectional class, as follows:

- cross-sections of class 1, 2 or 3:

$$N_{b,Rd} = \chi \cdot A \frac{f_y}{\gamma_{M1}} \tag{6.24a}$$

- cross-sections of class 4:

$$N_{b,Rd} = \chi \cdot A_{eff} \frac{f_y}{\gamma_{M1}} \tag{6.24b}$$

where A is the gross cross-sectional area, A_{eff} is the effective cross-sectional area (accounting for local buckling phenomena), f_y is the yielding strength of the material, χ is a reduction factor and γ_{M1} is the partial safety factor.

More specifically, coefficient χ is the reduction factor for the appropriate buckling mode calculated as follows:

$$\chi = \frac{1}{\varphi + \sqrt{\varphi^2 - \bar{\lambda}^2}} \quad \text{with } \chi \leq 1 \tag{6.25a}$$

in which the coefficient φ is defined as:

$$\varphi = 0.5 \cdot \left[1 + \alpha(\bar{\lambda} - 0.2) + \bar{\lambda}^2\right] \tag{6.25b}$$

where α is the imperfection coefficient (defined in Table 6.2), which is a function of the stability curve chosen according to Table 6.3a for hot-rolled and built-up sections and in Table 6.3b for cold-formed sections.

Factor χ can also be obtained from Table 6.4 by interpolation as a function of the appropriate stability curve based on Table 6.3a or 6.3b, and of the relative slenderness $\bar{\lambda}$, defined as follows:

Table 6.2 Values of α for the various stability curves.

Stability curve	a_0	a	b	c	d
Imperfection coefficient α	0.13	0.21	0.34	0.49	0.76

Table 6.3a Guide for the selection of the appropriate stability curve for hot-rolled and welded sections.

Shape of cross-section	Limits	Axis of instability	Stability curve S 235, S 275, S 355, S 420	Stability curve S 460
Hot-rolled I sections	$h/b > 1.2$, $t_f \leq 40$ mm (1.58 in.)	y-y	a	a_0
		z-z	b	a_0
	40 mm $\leq t_f \leq$ 100 mm (1.58 in. $\leq t_f \leq$ 3.94 in.)	y-y	b	a
		z-z	c	a
	$h/b \leq 1.2$, $t_f \leq$ 100 mm (3.94 in.)	y-y	b	a
		z-z	c	a
	$t_f >$ 100 mm (3.94 in.)	y-y	d	c
		z-z	d	c
Welded I-sections	$t_f \leq$ 40 mm (1.58 in.)	y-y	b	b
		z-z	c	c
	$t_f >$ 40 mm (1.58 in.)	y-y	c	c
		z-z	d	d
Box sections	Hot-rolled	All	a	a_0
	Cold-formed	All	c	c
Welded box sections 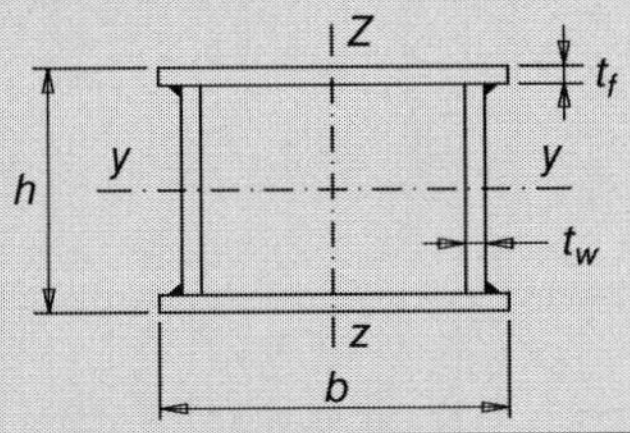	All (unless specified below)	All	b	b
	Thick welds $a > 0.5\ t_f$, $b/t_f < 30$, $h/t_w < 30$	All	c	c
Channels, tees and solid sections 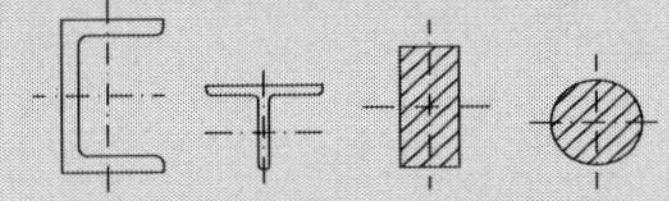		All	c	c
Angles 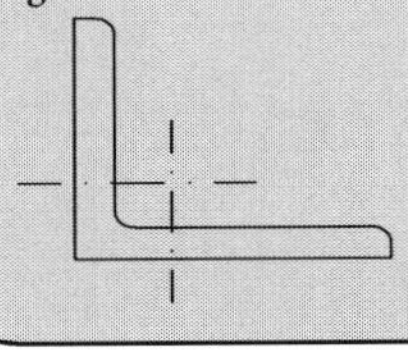		All	b	b

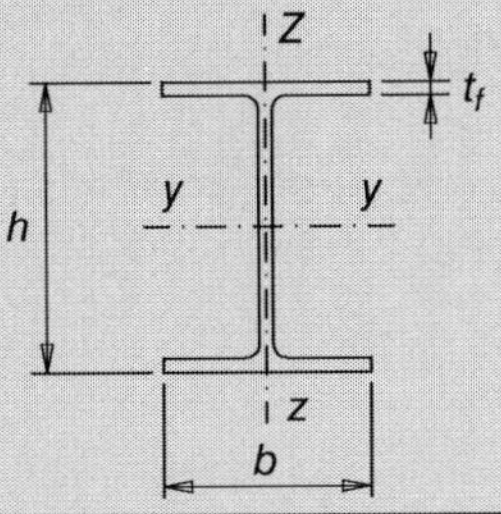

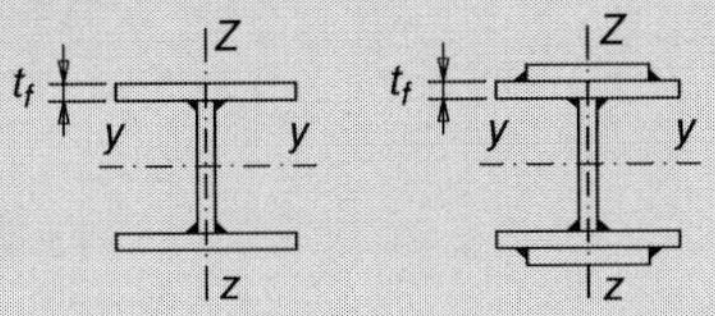

Table 6.3b Guide for the selection of the appropriate stability curve for cold-formed sections.

Shape of cross-section		Axis of instability	Stability curve
	If using f_{yb} (see Section 1.6.1)	All	*b*
	If using f_{ya} (this should be used only if the gross cross-sectional area is the effective area: $A_{eff} = A_g$)	All	*c*
z, y–y, z		*y-y*	*a*
		z-z	*b*
		All	*b*
		All	*c*

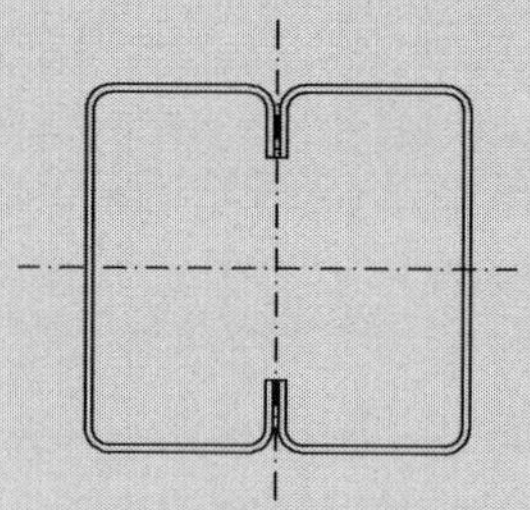

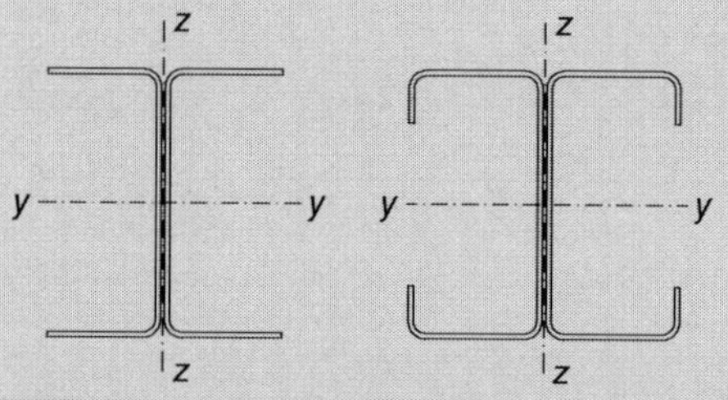

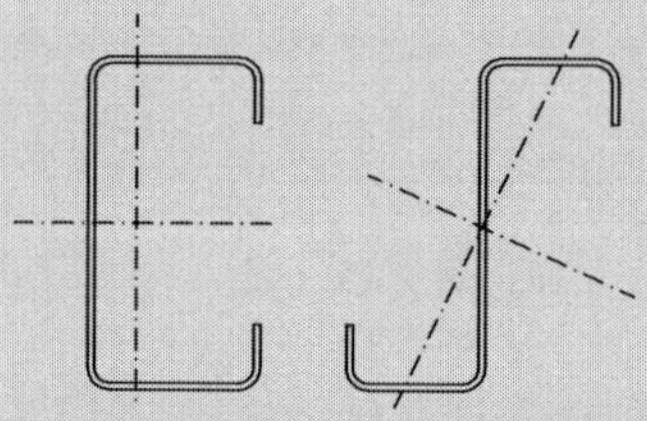

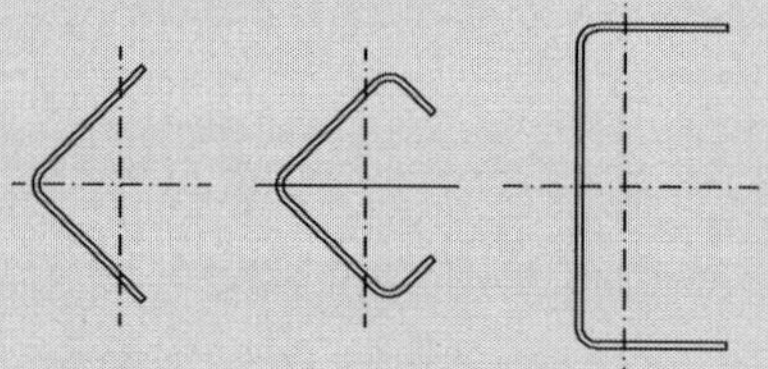

- cross-sections of class 1, 2 or 3:

$$\bar{\lambda} = \sqrt{\frac{A \cdot f_y}{N_{cr}}} \tag{6.26a}$$

- cross-sections of class 4:

$$\bar{\lambda} = \sqrt{\frac{A_{eff} \cdot f_y}{N_{cr}}} \tag{6.26b}$$

in which N_{cr} is the elastic critical load for the appropriate buckling mode (flexural, torsional or flexural-torsional).

Table 6.4 Values of coefficient χ for design checks according to EC3.

$\bar{\lambda}$	Coefficient χ				
	a_0	a	B	c	d
0.0	1.0000	1.0000	1.0000	1.0000	1.0000
0.1	1.0000	1.0000	1.0000	1.0000	1.0000
0.2	1.0000	1.0000	1.0000	1.0000	1.0000
0.3	0.9859	0.9775	0.9641	0.9491	0.9235
0.4	0.9701	0.9528	0.9261	0.8973	0.8504
0.5	0.9513	0.9243	0.8842	0.8430	0.7793
0.6	0.9276	0.8900	0.8371	0.7854	0.7100
0.7	0.8961	0.8477	0.7837	0.7247	0.6431
0.8	0.8533	0.7957	0.7245	0.6622	0.5797
0.9	0.7961	0.7339	0.6612	0.5998	0.5208
1.0	0.7253	0.6656	0.5970	0.5399	0.4671
1.1	0.6482	0.5960	0.5352	0.4842	0.4189
1.2	0.5732	0.5300	0.4781	0.4338	0.3762
1.3	0.5053	0.4703	0.4269	0.3888	0.3385
1.4	0.4461	0.4179	0.3817	0.3492	0.3055
1.5	0.3953	0.3724	0.3422	0.3145	0.2766
1.6	0.3520	0.3332	0.3079	0.2842	0.2512
1.7	0.3150	0.2994	0.2781	0.2577	0.2289
1.8	0.2833	0.2702	0.2521	0.2345	0.2093
1.9	0.2559	0.2449	0.2294	0.2141	0.1920
2.0	0.2323	0.2229	0.2095	0.1962	0.1766
2.1	0.2117	0.2036	0.1920	0.1803	0.1630
2.2	0.1937	0.1867	0.1765	0.1662	0.1508
2.3	0.1779	0.1717	0.1628	0.1537	0.1399
2.4	0.1639	0.1585	0.1506	0.1425	0.1302
2.5	0.1515	0.1467	0.1397	0.1325	0.1214
2.6	0.1404	0.1362	0.1299	0.1234	0.1134
2.7	0.1305	0.1267	0.1211	0.1153	0.1062
2.8	0.1216	0.1182	0.1132	0.1079	0.0997
2.9	0.1136	0.1105	0.1060	0.1012	0.0937
3.0	0.1063	0.1036	0.0994	0.0951	0.0882

As discussed in the previous section, for I-shapes with at least one axis of symmetry, torsional instability can usually be neglected, while for cross-sections having all elements converging into one point (cruciform sections, angles, tees) torsional buckling often governs the design.

When flexural buckling governs the design, the value of relative slenderness $\bar{\lambda}$ can be calculated as follows:

- cross-sections of class 1, 2 or 3:

$$\bar{\lambda} = \sqrt{\frac{A \cdot f_y}{N_{cr}}} = \frac{L_{cr}}{i} \cdot \frac{1}{\lambda_l} \tag{6.27a}$$

- cross-sections of class 4:

$$\bar{\lambda} = \sqrt{\frac{A_{\text{eff}} \cdot f_y}{N_{cr}}} = \frac{L_{cr}}{i} \cdot \frac{\sqrt{\frac{A_{\text{eff}}}{A}}}{\lambda_l} \tag{6.27b}$$

in which λ_1 represents the proportionality slenderness (already defined in Eq. (6.5) and indicated with λ_p), L_{cr} is the effective length of the member under consideration (previously identified as L_0), A and A_{eff} are the gross cross-section and effective area, respectively, and i is the radius of gyration of the cross-section (already identified as ρ).

When the governing buckling mode is torsional or flexural-torsional, the relative slenderness is defined as follows:

- cross-sections of class 1, 2 or 3:

$$\bar{\lambda}_T = \sqrt{\frac{A \cdot f_y}{N_{cr}}} \tag{6.28a}$$

- cross-sections of class 4:

$$\bar{\lambda}_T = \sqrt{\frac{A_{eff} \cdot f_y}{N_{cr}}} \tag{6.28b}$$

in which N_{cr} is defined as:

$$N_{cr} = N_{cr,TF} \text{ with } N_{cr} < N_{cr,T} \tag{6.28c}$$

where $N_{cr,TF}$ in Eq. (6.9a) and $N_{cr,T}$ in Eq. (6.8a) represent the elastic critical load for flexural-torsional and torsional buckling, respectively.

6.3.3 Design According to the US Approach

LRFD approach	**ASD approach**
Design according to the provisions for *load and resistance factor design* (LRFD) satisfies the requirements of AISC Specification when the *design compressive strength* $\phi_c P_n$ of each structural component equals or exceeds the *required compressive strength* P_u determined on the basis of the LRFD load combinations	Design according to the provisions for *allowable strength design* (ASD) satisfies the requirements of AISC Specification when the *allowable compressive strength* P_n/Ω_c of each structural component equals or exceeds the *required compressive strength* P_a determined on the basis of the ASD load combinations
Design has to be performed in accordance with the following equation:	Design has to be performed in accordance with the following equation:
$P_u \le \phi_c P_n$ (6.29)	$P_a \le P_n/\Omega_c$ (6.30)
where ϕ_c is the *compressive resistance factor* ($\phi_c = 0.90$)	where Ω_c is the *compressive safety factor* ($\Omega_c = 1.67$)

The nominal compressive strength P_n is determined as:

$$P_n = F_{cr} A_g \tag{6.31}$$

The *critical stress* F_{cr} is referred to the limit state of *flexural buckling* as well as for *torsional* and *flexural-torsional buckling*. AISC Specifications give different expressions for F_{cr}, which depends on the cross-section type and are different for *non-slender* and *slender element* sections. AISC specifications provides specific rules for the following cases:

(a) *Generic doubly symmetrical members;*
(b) *Particular generic doubly symmetrical members;*
(c) *Singly symmetrical members;*
(d) *Unsymmetrical members;*
(e) *Single angles with $b/t \leq 20$;*
(f) *Single angles with $b/t > 20$;*
(g) *T-shaped compression members.*

(a) *Generic doubly-symmetric members*:
When generic doubly-symmetric members are considered, such as typical hot-rolled wide flange columns, except those listed in (b), only flexural buckling has to be taken into account. The *critical stress* F_{cr} is determined as follows:

$$\text{(a1)When } \frac{KL}{r} \leq 4.71\sqrt{\frac{E}{(QF_y)}} \left(\text{or } \frac{(QF_y)}{F_e} \leq 2.25\right): \quad F_{cr} = \left[0.658^{\frac{(QF_y)}{F_e}}\right](QF_y) \tag{6.32}$$

$$\text{(a2)When } \frac{KL}{r} > 4.71\sqrt{\frac{E}{(QF_y)}} \left(\text{or } \frac{(QF_y)}{F_e} > 2.25\right): \quad F_{cr} = 0.877F_e \tag{6.33}$$

where K is the effective length buckling factor (see Section 6.4), ρ is the ratius of gyration and F_e is the *elastic buckling stress* determined according to the following equation:

$$F_e = \frac{\pi^2 E}{\left(\frac{KL}{r}\right)^2} \tag{6.34}$$

It should be noted that Q is the *net reduction factor* accounting for all slender compression elements. Its value is equal to 1.0 for non-slender elements, so it must be considered for slender elements only, and in this case it has to be determined according to section E6 of AISC 360-10. It is worth mentioning that the design of thin-walled member is beyond the scope of the present volume.

(b) *Particular generic doubly-symmetric members*:
Flexural and torsional buckling has to be considered in case of particular generic doubly-symmetric members, such as cruciform members, built-up members or generic doubly-symmetric members with torsional unbraced length exceeding lateral unbraced length:

The *critical stress* F_{cr} is determined with Eqs. (6.32) and (6.33) but F_e is the minimum of the values computed with Eq. (6.34) and with the following expression accounting for the torsional buckling:

$$F_e = \left[\frac{\pi^2 E C_w}{(K_z L)^2} + GJ\right]\frac{1}{I_x + I_y} \tag{6.35}$$

where G is the shear modulus of elasticity of steel, K_z is the *effective length factor* for torsional buckling, I_x and I_y are the moments of inertia about the principal axes, J is the torsional constant and C_w is the warping constant.

(c) *Singly symmetrical members*:
For singly symmetrical members where y is the axis of symmetry, flexural, torsional and flexural-torsional buckling all have to be considered. The *critical stress* F_{cr} is determined as in (a) (Eqs. (6.32) and (6.33)) and F_e is the minimum of the values computed with Eq. (6.34) for flexural buckling and Eq. (6.36a) that takes into account the torsional buckling:

$$F_e = \left(\frac{F_{ey}+F_{ez}}{2H}\right)\left[1-\sqrt{1-\frac{4F_{ey}F_{ez}H}{\left(F_{ey}+F_{ez}\right)^2}}\right] \quad (6.36a)$$

with:

$$H = 1-\frac{x_0^2+y_0^2}{\bar{r}_0^2}$$

where:

$$F_{ey} = \pi^2 E \bigg/ \left(\frac{K_y L}{r_y}\right)^2 \quad (6.36b)$$

$$F_{ez} = \left[\frac{\pi^2 E C_w}{(K_z L)^2}+GJ\right]\frac{1}{A_g \cdot \bar{r}_0^2} \quad (6.36c)$$

where K_y represents the effective length factors for flexural buckling about the y-axis, K_z is the effective length factor for torsional buckling, r_y is the radius of gyration about the y-axis and $\bar{r}_0$ is the polar radius of gyration about the shear centre, (already proposed in Eq. (6.8b) as ρ_0) defined as:

$$\bar{r}_0^2 = x_0^2+y_0^2+\frac{I_x+I_y}{A_g} \quad (6.36d)$$

where A_g is the gross cross-sectional area of the member and x_o, y_o are the coordinates of the shear centre with respect to the centroid.

It is worth noticing that the torsional (Eq. (6.35)) and flexural-torsional (Eq. (6.36a)) stresses can be obtained also from the Eqs. (6.8a) and (6.9a), respectively.

(d) *Unsymmetrical members*:
For unsymmetrical members, flexural, torsional and flexural-torsional buckling has to be considered. The *critical stress* F_{cr} is determined as in (a), according to Eqs. (6.32) or (6.33), and F_e is the minimum of the values computed with Eq. (6.34) for flexural buckling and the lowest root of the cubic equation:

$$\left(F_e-F_{ex}\right)\left(F_e-F_{ey}\right)\left(F_e-F_{ez}\right)-F_e^2\left(F_e-F_{ey}\right)\left(\frac{x_0}{\bar{r}_0}\right)^2-F_e^2\left(F_e-F_{ex}\right)\left(\frac{y_0}{\bar{r}_0}\right)^2=0 \quad (6.37a)$$

where (other symbols are defined before):

$$F_{ex} = \pi^2 E \bigg/ \left(\frac{K_x L}{r_x}\right)^2 \quad (6.37b)$$

(e) *Single angles with $b/t \leq 20$:*

The limitation $b/t \leq 20$ applies to all currently produced hot-rolled angles. If angles have only one leg fixed to a gusset plate by welding or by a bolted connection with at least two bolts and there are no intermediate transverse loads, they can be treated as axially loaded members, neglecting the effect of load eccentricity and considering flexural buckling only, but *adjusting the member slenderness.* So F_{cr} has to be determined as in cases a1 and a2, and F_e is evaluated in accordance with Eq. (6.34) using for KL/r the following values:

For equal-leg angles or unequal-leg angles connected through the longer leg that are individual members or are web members of planar trusses with adjacent web members attached to the same side of the gusset plate or chord:

when $\frac{L}{r_x} \leq 80$:

$$\frac{KL}{r} = 72 + 0.75\frac{L}{r_x} \tag{6.38}$$

when $\frac{L}{r_x} > 80$:

$$\frac{KL}{r} = 32 + 1.25\frac{L}{r_x} \tag{6.39}$$

For equal-leg angles or unequal-leg angles connected through the longer leg that are web members of box or space trusses with adjacent web members attached to the same side of the gusset plate or chord:

when $\frac{L}{r_x} \leq 75$:

$$\frac{KL}{r} = 60 + 0.80\frac{L}{r_x} \tag{6.40}$$

when $\frac{L}{r_x} > 75$:

$$\frac{KL}{r} = 45 + \frac{L}{r_x} \tag{6.41}$$

where r_x is the radius of gyration about the geometric axis parallel to the connected leg.

(f) *Single angles with $b/t > 20$:*

The limitation $b/t > 20$ applies to fabricated angles. Like any asymmetric member, F_{cr} has to be determined as in (a) and F_e is the minimum of the values computed with Eq. (6.34) for flexural buckling and the lowest root of the cubic Eq. (6.37a).

(g) *T-shaped compression members:*

For T-shaped compression members flexural and flexural-torsional buckling has to be considered. Flexural buckling shall be verified computing F_{cr} according to Eqs. (6.32) and (6.33) computing F_e according to Eq. (6.34). For flexural-torsional buckling F_{cr} has to be determined according to the following equation:

$$F_{cr} = \left(\frac{F_{cry} + F_{crz}}{2H}\right)\left[1 - \sqrt{1 - \frac{4F_{cry}F_{crz}H}{\left(F_{cry} + F_{crz}\right)^2}}\right] \tag{6.42a}$$

where F_{cry} is taken as F_{cr} of Eqs. (6.32) and (6.33) for flexural buckling about the y-axis of symmetry, $KL/r = K_yL/r_y$ and:

$$F_{crz} = \frac{GJ}{A_g \bar{r}_0^2} \tag{6.42b}$$

6.4 Effective Length of Members in Frames

A key aspect related to the design of a compression member of length L is the evaluation of the effective length, L_0, or equivalently of the K length factor, of members in frames (with $L_0 = K{\cdot}L$). In few design cases, reference can be made to isolated members, already briefly discussed in Section 6.3. In general, the authors' option is to use a buckling analysis for determining the minimum α_{cr} value associated with the buckling mode of interest. The critical load N_{cr} can be easily determined via equation $\alpha_{cr} = \frac{N_{cr}}{N_{Ed}}$, with N_{Ed} equal to the design axial load. Finally, from N_{cr}, critical length L_{cr} for flexural buckling can be computed as:

$$L_{cr} = L_0 = \pi\sqrt{\frac{E{\cdot}I}{N_{cr}}} = \pi\sqrt{\frac{E{\cdot}I}{\alpha_{cr}{\cdot}N_{Ed}}} \tag{6.43}$$

Furthermore, reference should be made to simplified approaches currently used in accordance with both the EU and US steel design practice, based on the use of alignment charts. These have been determined with reference to assumptions of idealized conditions which seldom exist in real structures. These assumptions are as follows:

(1) behaviour of the steel material is purely elastic;
(2) all members have constant cross section;
(3) all joints are rigid;
(4) for columns in frames with inhibited sidesway, rotations at opposite ends of the restraining beams are equal in magnitude and opposite in direction, producing single curvature bending;
(5) for columns in frames with uninhibited sidesway, rotations at opposite ends of the restraining beams are equal in magnitude and direction, producing reverse curvature bending;
(6) the stiffness parameter defined as $L\sqrt{P/EI}$ of all the columns is equal;
(7) joint restraint is distributed to the column above and below the joint in proportion to EI/L for the two columns;
(8) all columns buckle simultaneously;
(9) no significant axial compression force exists in the girders.

6.4.1 Design According to the EU Approach

As regards to European design, reference should be made to the old ENV edition of EC3 (ENV 1993-1-1:2004), relative to columns belonging to non-sway or sway frames. To this purpose Figure 6.12 for a non-sway frame, and to Figure 6.13 for a sway frame can be considered. Both figures provide the ratio L_{cr}/L as a function of distribution factors η_1 and η_2, related to the two ends of the column, and defined as follows:

$$\eta_1 = \frac{K_c}{K_c + K_{11} + K_{12}}; \quad \eta_2 = \frac{K_c}{K_c + K_{21} + K_{22}} \tag{6.44}$$

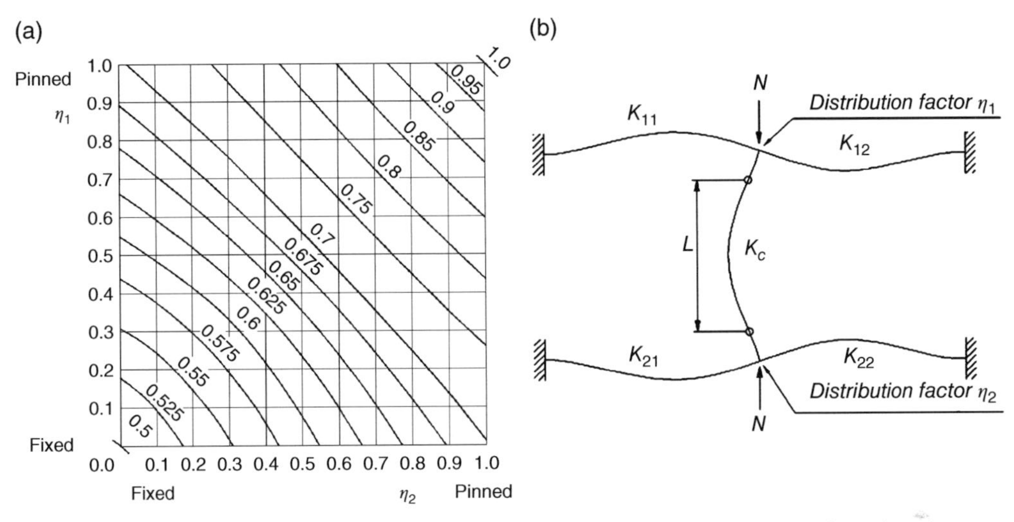

Figure 6.12 Buckling length ratio L_{cr}/L for a column in a non-sway mode (a) and distribution factor for column in a non-sway mode (b).

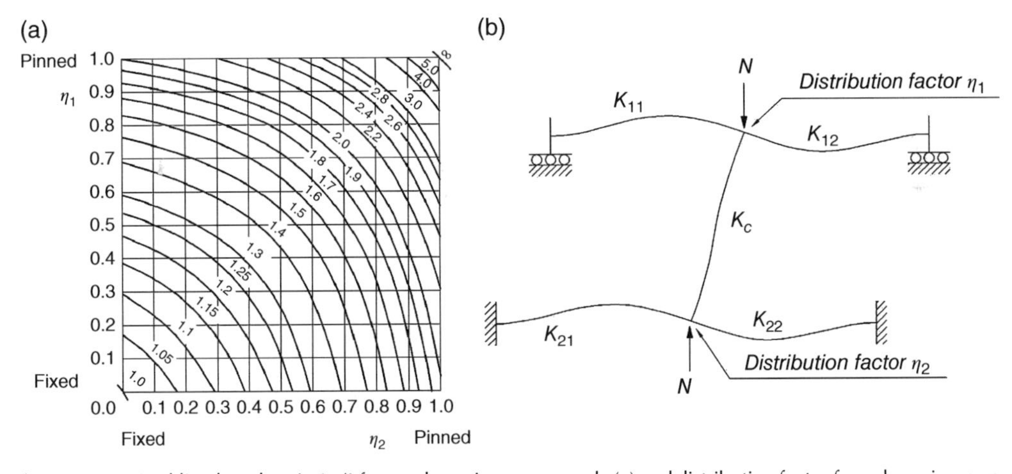

Figure 6.13 Buckling length ratio L_{cr}/L for a column in a sway mode (a) and distribution factor for column in a sway mode (b).

where: $K_c = I_c/h_c$ is the column stiffness coefficient, $K_{ij} = k_{ij}\left(I_{ij}/L_{ij}\right)$ is the effective beam stiffness coefficient and coefficient k_{ij} depends on the beam restraints (Tables 6.5 and 6.6).

Term I expresses the moment of inertia, h and L are the lengths of the column and of the beam, respectively, and subscripts are represented in Figures 6.12 and 6.13.

Formulas (6.44) are valid for columns of single-story buildings. For columns of multi-story buildings, if ratio N/N_{cr} is almost constant among all columns, an alternative approach can be used, which is based on the evaluation of the end column stiffness, defined with reference to

Table 6.5 Effective stiffness coefficient K_{ij} for a beam in a frame without concrete floor slabs.

Conditions of rotational restraint at far end of beam	K_{ij}
Fixed	$1.0 \cdot I_{ij}/L_{ij}$
Pinned	$0.75 \cdot I_{ij}/L_{ij}$
Rotation as at near end (double curvature)	$1.5 \cdot I_{ij}/L_{ij}$
Rotation equal and opposite to that at near end (single curvature)	$0.5 \cdot I_{ij}/L_{ij}$
General case. Rotation θ_a at near end, θ_b at far end	$\left(1+0.5\frac{\theta_b}{\theta_a}\right) \cdot I_{ij}/L_{ij}$

Table 6.6 Reduced beam stiffness coefficients K_{ij} due to axial compression.

Conditions of rotational restraint at far end of beam	K_{ij}
Fixed	$1.0\frac{I_{ij}}{L_{ij}}\left(1-0.4\frac{N}{N_E}\right)$
Pinned	$0.75\frac{I_{ij}}{L_{ij}}\left(1-1.0\frac{N}{N_E}\right)$
Rotation as at near end (double curvature)	$1.5\frac{I_{ij}}{L_{ij}}\left(1-0.2\frac{N}{N_E}\right)$
Rotation equal and opposite to that at near end (single curvature)	$0.5\frac{I_{ij}}{L_{ij}}\left(1-1.0\frac{N}{N_E}\right)$
$N_E = \pi^2 EI_{ij}/L_{ij}^2$ (If N represents tension, it must be considered = 0)	

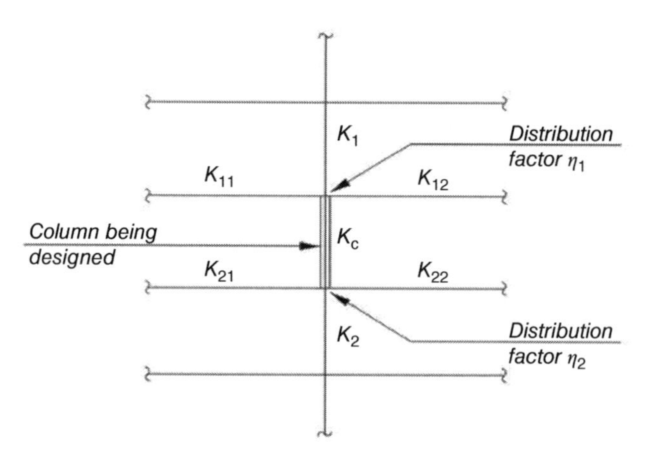

Figure 6.14 Distribution factor for continuous columns.

the restraining effects of the upper and the lower column stiffness (K_1 and K_2, respectively) as (see Figure 6.14):

$$\eta_1 = \frac{K_c + K_1}{K_c + K_1 + K_{11} + K_{12}}; \quad \eta_2 = \frac{K_c + K_2}{K_c + K_2 + K_{21} + K_{22}} \tag{6.45}$$

Alternatively, for direct evaluation via Figure 6.12 or 6.13, the following more conservative approach can be used:

(a) non-sway frames:

$$K = L_{cr}/L = 0.5 + 0.14(\eta_1 + \eta_2) + 0.055(\eta_1 + \eta_2)^2 \tag{6.46}$$

Table 6.7 Effective stiffness coefficient K_{ij} for a beam in a building frame with concrete floor slabs.

Loading conditions for the beam	K_{ij}	
	Non-sway mode	Sway mode
Beam directly supporting concrete floor slabs	$1.0{\cdot}I_{ij}/L_{ij}$	$1.0{\cdot}I_{ij}/L_{ij}$
Other beams with direct loads	$0.75{\cdot}I_{ij}/L_{ij}$	$1.0{\cdot}I_{ij}/L_{ij}$
Beams with end moments only	$0.5{\cdot}I_{ij}/L_{ij}$	$1.5{\cdot}I_{ij}/L_{ij}$

(b) sway frames:

$$K = L_{cr}/L = \sqrt{\frac{1-0.2(\eta_1+\eta_2)-0.12\eta_1\eta_2}{1-0.8(\eta_1+\eta_2)+0.6\eta_1\eta_2}} \tag{6.47}$$

For calculating beam stiffness coefficients K_{ij} that are functions of beam restraints, formulas of Table 6.5 have to be used.

If beams are subjected to large compression axial loads, stiffness values have to be reduced according to formulas of the Table 6.6.

For beams supporting reinforced concrete slabs, a higher stiffness has to be taken into account, using formulas of Table 6.7.

6.4.2 Design According to the US Approach

In accordance with the US practice, simplified methods to calculate the effective length factor K are admitted. For columns belonging to a moment frame system, alignment chart with sidesway inhibited (Figure 6.15) and with uninhibited sidesway (Figure 6.16) can be used. The parameter G governing the use of the alignment charts is defined as:

$$G = \frac{\sum \frac{E_c I_c}{L_c}}{\sum \frac{E_g I_g}{L_g}} \tag{6.48}$$

where I and L are the moment of inertia and the member length, respectively, E is the Young's modulus and subscripts c and g indicate columns and beams (girders), respectively.

The summation Σ is extended to all the members rigidly connected to the node of interest and subscripts A and B (Figures 6.15 and 6.16) identify the upper and lower joints of the considered column.

Therefore, in order to use these values for K, some adjustments are suggested by AISC 360-10 Commentary. In particular, for isolated columns, theoretical K values and recommended design values should be adopted as recommended in Table 6.8. Furthermore, for columns in frames with sidesway inhibited and sidesway uninhibited, the corrections herein described should be applied.

For sidesway inhibited frames, these adjustments for different beam end conditions are required:

- if the far end of a girder is fixed, multiply the flexural stiffness $(EI/L)g$ of the member by 2.0;
- if the far end of the girder is pinned, multiply the flexural stiffness $(EI/L)g$ of the member by 1.5.

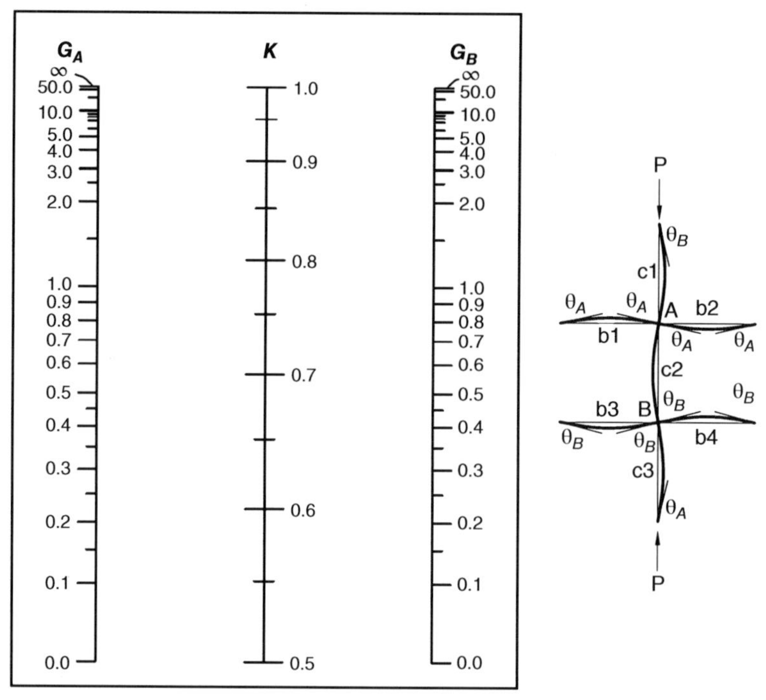

Figure 6.15 Alignment chart–sidesway inhibited (no-sway frames). From Figure C-A-7.1 of AISC 360-10.

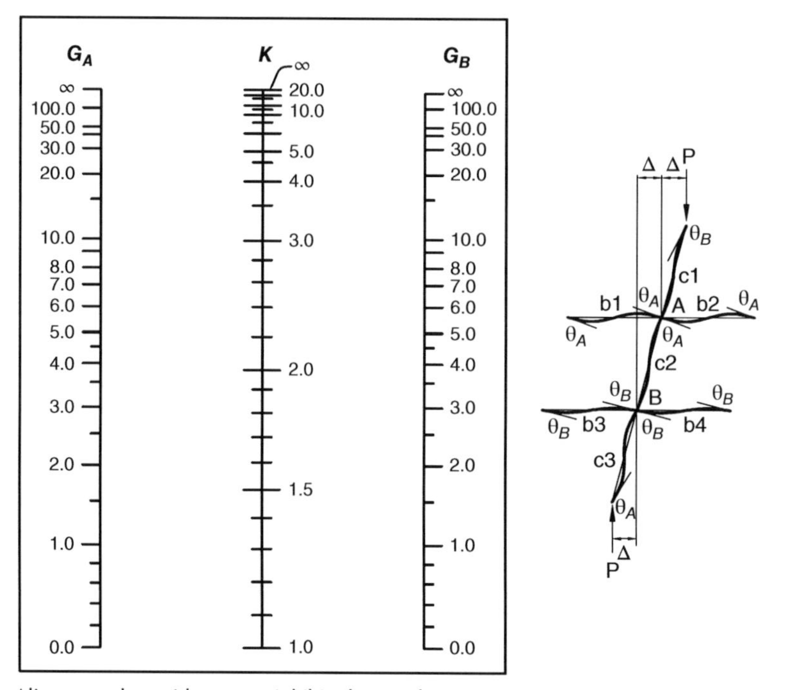

Figure 6.16 Alignment chart–sidesway uninhibited (sway frames). From Figure C-A-7.2 of AISC 360-10.

Table 6.8 Theoretical and recommended AISC values of the effective length factor (K) for isolated column.

		End condition restrains					
Top	End translation	Fixed	Fixed	Free	Fixed	Free	Free
	End rotation	Fixed	Free	Fixed	Free	Free	Fixed
Bottom	End translation	Fixed	Fixed	Fixed	Fixed	Fixed	Fixed
	End rotation	Fixed	Fixed	Fixed	Free	Fixed	Free
Theoretical K value		0.5	0.7	1.0	1.0	2.0	2.0
Recommended design value when ideal condition are approximated		0.85	0.8	1.20	1.0	2.1	2.0

For sidesway uninhibited frames and girders with different boundary conditions, the modified girder length, L'_g, should be used instead of the actual girder length, which is defined as:

$$L'_g = L_g\left(2 - \frac{M_F}{M_N}\right) \tag{6.49}$$

where M_F is the far end girder moment and M_N is the near end girder moment from a first-order analysis of the frame.

The ratio of the two moments M_F and M_N is positive if the girder is in reverse curvature. If M_F/M_N is more than 2.0, then L'_g becomes negative, in which case G is negative and the alignment chart equation must be used.

For sidesway uninhibited frames, these adjustments for different beam end conditions are admitted:

- if the far end of a girder is fixed, multiply the $(EI/L)g$ of the member by 2/3;
- if the far end of the girder is pinned, multiply the $(EI/L)g$ of the member by 0.5.

For girders with significant axial load, for both sidesway conditions, multiply the $(EI/L)g$ by the factor $1 - Q/Q_{cr}$ where Q is the axial load in the girder and Q_{cr} is the in-plane buckling load of the girder based on $K = 1.0$.

To account for inelasticity in columns, for both sidesway conditions, replace $(E_c\, I_c)$ with $\tau_b(E_c I_c)$ for all columns in the expression for G_A and G_B. The stiffness reduction factor, τ_b, has already been defined with reference to DAM approach presented in Eq. (12.12).

For columns with different end conditions, if they are supported by a foundation with a pinned connection, G should theoretically be infinity. Owing to the fact that usually pinned connections are not true friction-free pins, so G can be taken as 10 for practical design. Otherwise, if columns are rigidly connected to foundations, G should be 0, but it is prudent to take it as 1.0.

One important assumption in the use of the alignment charts is that all beam-column connections are fully restrained (FR connections). As seen previously, when the far end of a beam does not have an FR connection that behaves as assumed, an adjustment must be made, which implies a reduction of the beam stiffness. When a beam connection at the column of interest is a shear only connection (no moment can be transferred) then that beam cannot participate in the restraint of the column and it cannot be considered in the $\Sigma(EI/L)g$ term of the equation for G. Only FR connections can be used directly in the determination of G. Partially restrained (PR) connections with a well-known moment-rotation response can be utilized, but the $(EI/L)g$ of each beam must be adjusted to account for the connection flexibility.

6.5 Worked Examples

Example E6.1. HE-Shaped Column Available Strength Calculation According to EC3

Calculate the available strength of an EN 10025 S235 steel ($f_y = 235$ MPa (34 ksi)) HE 200 B. The column is 7.5 m (24.6 ft) long, pinned at its top and bottom in both axes, with a strong axis unbraced length of 7.5 m (24.6 ft) and a weak axis and torsional unbraced length of 3.75 m (12.3 ft).

HE 200 B properties

Height	h	200 mm (6.87 in.)
Flange width	b	200 mm (6.87 in.)
Flange thickness	t_f	15 mm (0.59 in.)
Web thickness	t_w	9 mm (0.35 in.)
Corner radius	r	18 mm (0.71 in.)
Area	A	78.1 cm^2 (12.11 in.^{2})
Moment of inertia about a strong axis	I_y	5696 cm^4 (136.8 in.^{4})
Moment of inertia about a weak axis	I_z	2003 cm^4 (48.12 in.^{4})

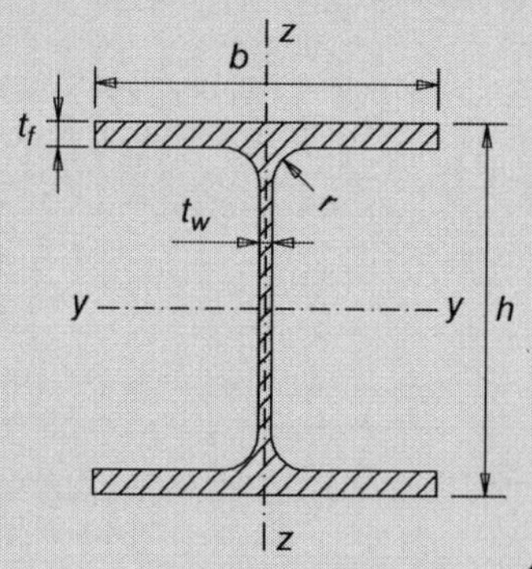

Design procedure

According to EC3, the column design curves are different for buckling about the y-axis and buckling about the z-axis. So it is not certain that the maximum compressive strength is associated with the minimum slenderness. Therefore, column strength calculation should be performed by taking into consideration member slenderness about both principal axes. The following steps have to be executed:

- cross-section classification;
- evaluation of elastic flexural buckling load about strong axis ($N_{cr,y}$);
- evaluation of relative slenderness about strong axis ($\bar{\lambda}_y$);
- choose the column design curve for flexural buckling about strong axis and computation of reduction factor for buckling about strong axis (χ_y);
- evaluation of elastic flexural buckling load about weak axis ($N_{cr,z}$);
- evaluation of relative slenderness about weak axis ($\bar{\lambda}_z$);
- choose the column design curve for flexural buckling about weak axis and computation of reduction factor for buckling about weak axis (χ_z);
- choose the minimum reduction factor ($\chi = \min(\chi_y, \chi_z)$) and calculate column available strength ($N_{b,Rd}$).

Section classification (Section 4.2.1). For S 235 steel grade: $\varepsilon = 1$

$$\text{Flange:}\quad (c/t_f) = [200 - 9 - (2\times 18)]/(2\times 15) = 5.2 \le 9 \qquad \text{Class 1}$$

$$\text{Web:}\quad (d/t_w) = [200 - (2\times 15) - (2\times 18)]/9 = 14.9 \le 33 \qquad \text{Class 1}$$

S 235 steel HE 200 B cross section is a class 1 section if subjected to axial load.

Calculate the elastic critical buckling stress about strong axis (Eq. (6.3)):

$$N_{cr,y} = \frac{\pi^2 EI_y}{L_{o,y}^{\ 2}} = \frac{\pi^2 \times 210000 \times (5696 \cdot 10^4)}{7500^2} \cdot 10^{-3} = 2098.78 \text{ kN} \ (471.8 \text{ kips})$$

Calculate the relative slenderness about strong axis (Eq. (6.26a)):

$$\bar{\lambda}_y = \sqrt{\frac{A \cdot f_y}{N_{cr,y}}} = \sqrt{\frac{7808 \times 235}{2098.78 \cdot 10^3}} = 0.935$$

Calculate the reduction factor χ_y: referring to curve b to be used for buckling about strong axis (y–y axis) in Table 6.3a of HE 200 B profile, reference has to be made (Table 6.4):

$\bar{\lambda}$	χ
0.9	0.6612
1.0	0.5970

Linear interpolation gives $\chi_y = 0.6386$.
Calculate elastic critical buckling stress about weak axis:

$$N_{cr,z} = \frac{\pi^2 EI_z}{L_{o,z}^{\ 2}} = \frac{\pi^2 \times 210000 \times (2003 \cdot 10^4)}{3750^2} \cdot 10^{-3} = 2952.1 \text{ kN} \ (663.6 \text{ kips})$$

Calculate the relative slenderness about weak axis (Eq. (6.26a)):

$$\bar{\lambda}_z = \sqrt{\frac{A \cdot f_y}{N_{cr,z}}} = \sqrt{\frac{7808 \times 235}{2952.1 \cdot 10^3}} = 0.788$$

Calculate the reduction factor χ_z: referring to curve c, to be used for buckling about weak axis (z–z axis) in Table 6.3a of HE 200 B profile, reference has to be made (Table 6.4):

$\bar{\lambda}$	χ
0.7	0.7247
0.8	0.6622

Linear interpolation gives $\chi_z = 0.6693$.
Calculate the available column strength (Eq. (6.24a)): let's consider as reduction factor χ the lower value between χ_y and χ_z:

$$N_{b,Rd} = \chi \cdot A \frac{f_y}{\gamma_{M1}} = 0.6387 \times 7808 \times \frac{235}{1.00} \cdot 10^{-3} \Rightarrow 1171.9 \text{ kN} \ (263.5 \text{ kips})$$

Example E6.2. W-Shape Column Available Strength Calculation According to AISC

Calculate the available strength of an ASTM A36 steel (F_y = 36 ksi (248.2 MPa)) W8 × 8 × 40 (W200 × 200 × 59). The column is 25 ft (7.62 m) long and is pinned on top and bottom in both directions, with a strong axis unbraced length of 25 ft (7.62 m) and a weak axis and torsional unbraced length of 12.5 ft (3.81 m).

W8 × 8 × 40 properties

Property	Symbol	Value
Height	d	8.25 in. (209.6 mm)
Flange width	b_f	8.07 in. (205 mm)
Flange thickness	t_f	0.56 in. (14.2 mm)
Web thickness	t_w	0.36 in. (9.1 mm)
Corner radius	r	0.69 in. (17.5 mm)
Area	A	11.7 in.^{2} (75.6 cm^2)
Moment of inertia about a strong axis	I_x	146 in.^{4} (6077 cm^4)
Moment of inertia about a weak axis	I_y	49.1 in.^{4} (2044 cm^4)
radius of gyration about a strong axis	i_x	3.53 in. (89.7 mm)
radius of gyration about a weak axis	i_y	2.04 in. (51.8 mm)

Design procedure

According to AISC 360-10, there is only one column design curve to be used for any kind of sections. So the maximum compressive strength is associated with the minimum slenderness. Therefore, the column strength calculation should be performed by taking into consideration member slenderness about both principal axes, performing the following steps:

- cross-section classification;
- check slenderness ratio about strong axis (KL_x/r_x);
- check slenderness ratio about weak axis (KL_y/r_y);
- choose the higher slenderness ratio;
- calculate the elastic critical buckling stress (F_e);
- calculate the flexural buckling stress (F_{cr});
- compute the nominal compressive strength (P_n);
- compute the available strength (LRFD: $\phi_c P_n$; ASD: P_n/Ω_c).

Section classification for local buckling (Section 4.2.1).

Flange:

$$b/t = (0.5 \times 8.07)/0.56 = 7.21 < 0.56\sqrt{E/F_y} = \quad \rightarrow \text{non-slender}$$
$$= 0.56 \times \sqrt{29000/36} = 15.89$$

Web:

$$h/t_w = (8.25 - 2 \times 0.56 - 2 \times 0.69)/0.56 = 15.97 < 1.49\sqrt{E/F_y} = \quad \rightarrow \text{non-slender}$$
$$= 1.49 \times \sqrt{29000/36} = 42.29$$

ASTM A36 steel W8 × 8 × 40, subjected to axial load, is a non-slender section ($Q = 1$).

Check the slenderness ratio about both axes (by assuming $K = 1.0$)

$$\frac{KL_x}{r_x} = \frac{1 \times (25.0 \cdot 12)}{3.53} = 84.9 \text{ governs}$$

$$\frac{KL_y}{r_y} = \frac{1 \times (12.5 \cdot 12)}{2.04} = 73.5$$

Calculate via Eq. (6.34) the elastic critical buckling stress (F_e).

$$F_e = \frac{\pi^2 E}{\left(\frac{KL_x}{r_x}\right)^2} = \frac{3.14^2 \times 29000\,\text{ksi}}{(84.9)^2} = 39.7\,\text{ksi}\,(273.2\,\text{MPa})$$

Calculate the flexural buckling stress (F_{cr}).

Check limit: $4.71\sqrt{\frac{E}{QF_y}} = 4.71 \times \sqrt{\frac{29000}{1 \times 36}} = 133.7 > 84.9$

Because $\frac{KL_x}{r_x} \leq 4.71\sqrt{\frac{E}{(QF_y)}}$ then, based on Eq. (6.32):

$$F_{cr} = \left[0.658^{\frac{(QF_y)}{F_e}}\right](QF_y) = \left[0.658^{\frac{1 \times 36}{39.7}}\right] \times (1 \times 36) = 24.61\,\text{ksi}\,(169.7\,\text{MPa})$$

Compute via Eq. (7.31) the nominal compressive strength (P_n).

$$P_n = F_{cr} A_g = 24.61 \times 11.7 = 286.9\,\text{kips}\,(1281\,\text{kN})$$

Compute the available strength.

$$\text{LFRD}: \ \phi_c P_n = 0.90 \times 287.9 = 259.1\,\text{kips}\,(1153\,\text{kN})$$

$$\text{ASD}: \ P_n / \Omega_c = 287.9/1.67 = 172.4\,\text{kips}\,(767.1\,\text{kN})$$

CHAPTER 7

Beams

7.1 Introduction

Members subjected to bending are also generally affected by shear forces, which have to be adequately considered in all the safety checks. Furthermore, the design of beams has to take into account both serviceability (mainly, check on deflections and dynamic effects) and ultimate limit states, including, in addition to resistance, stability verifications when relevant.

Preliminarily to the design rules of beams, in the following some key aspects related to the response of elements under flexure and shear are briefly discussed.

7.1.1 Beam Deformability

The deflection limits provided in any specification can be generally used only as a guide to the serviceability of the structure and may not be taken as an absolute guide to satisfactory performance in the cases of interest for routine design. It is the responsibility of the designer to verify that the limits used in the design are appropriate for the structure under consideration.

Maximum deflections (usually in the elastic range), v, can be evaluated on the basis of the elastic theory of structures and then must be compared with the standard limit, v_{Lim}, ensuring that:

$$v \leq v_{Lim} \tag{7.1}$$

The elastic beam deflection should always be considered to be the sum of two contributions, one associated with the flexural deformability, v_F, and one associated with the shear deformability, v_T, as:

$$v = v_F + v_T \tag{7.2}$$

Term v_F is usually reported in the designer manuals for the most common routine design cases. Term v_T is rarely offered in literature and can be evaluated using the principle of virtual work. In the case of isolated beam of length L, the following expression can be used:

$$v_T = \int_o^L \frac{\chi_T T(x)}{G \cdot A} \cdot T^1(x) \cdot \mathrm{d}x \tag{7.3}$$

Structural Steel Design to Eurocode 3 and AISC Specifications, First Edition. Claudio Bernuzzi and Benedetto Cordova.

where terms $T(x)$ and $T^1(x)$ represent the shear distribution on the beam associated with the load condition and the service condition (characterized by a unitary concentrated force applied where the displacement has to be evaluated), respectively, G is the shear modulus, A is the cross-section area and χ_T is the shear factor.

Shear factor, χ_T, is a dimensionless coefficient, depending on the shape of the cross-section. Its value is always greater than unity and it increases with the increase of the deviation of the shear stress distribution from a uniform distribution. A correct evaluation of χ_T can be obtained from the expression:

$$\chi_T = \frac{A}{I^2} \cdot \int_{y'}^{y''} \frac{S_i^2}{b_i} dy \tag{7.4}$$

where I is the moment of inertia, y' and y'' are the cross-section limits, S is the first moment of area of the part of the cross-section (below the chord with a width b_i with reference to the neutral axis) and A is the cross-section area.

As an example, in case of rectangular solid cross-sections the result is $\chi_T = 1.2$, while for I- and H-shaped European profiles (typically, IPE and HE profiles) the value of χ_T is always greater than 2 and in particular:

- for IPE profiles χ_T ranges between 2.2 and 2.6;
- for HEA and HEB profiles χ_T ranges between 2.1 and 4.7;
- for HEM profiles χ_T ranges between 2.1 and 4.4.

As alternative to the exact definition of the shear factor by Eq. (7.4), a simplified formula can be applied in cases of I- and H-shaped doubly symmetrical profiles under flexure in the web plane: an approximate estimation of χ_T is given by:

$$\chi_T = \frac{A}{A_w} \tag{7.5}$$

where A and A_w are the area of the cross-section and the area of the sole web, respectively.

The errors due to Eq. (7.5) are extremely limited: not more than 5% for the IPE profiles and always less than 9% for HE profiles.

The influence of the contribution to the overall deflection due to shear deformability depends on the load condition as well as on the beam slenderness, which can be defined as the ratio between the beam length (L) and its depth (H). As an example, in the following the results of a parametric analysis are presented with reference to the case of simply supported beam under uniformly distributed loads where all the IPE and HE profiles have been considered.

In case of a beam slenderness equal to 6 (i.e. $L = 6H$) the results are that:

- for IPE profiles v_T ranges from 24 to 30% of v_F;
- for HEA and HEB profiles v_T ranges from 23 to 58% of v_F;
- for HEM profiles v_T ranges from 23 to 49% of v_F.

In case of beam slenderness equal to 12 (i.e. $L = 12H$) the influence of v_T is significantly reduced and the results are that:

- for IPE profiles v_T ranges from 6 to 7% of v_F;
- for HEA and HEB profiles v_T ranges from 6 to 15% of v_F;
- for HEM profiles v_T ranges from 6 to 12% of v_F.

7.1.2 Dynamic Effects

In several design cases, dynamic effects need to be taken into account to ensure that vibrations do not impair the efficiency of the structure at the serviceability limit state. The vibrations of structures on which people can walk have to be limited to avoid significant discomfort to users and limits should be specified for each project and agreed with the client. To achieve a more than satisfactory vibration behaviour of buildings and of their structural main members under serviceability conditions, the following aspects, amongst others, have to be considered:

- the comfort of the user;
- the functioning of the structure or its structural members (e.g. cracks in partitions, damage to cladding, sensitivity of building contents to vibrations).

Other aspects should be considered for each project and agreed with the client: problems associated with dynamic effects can also be due to moving plant and machinery.

Referring to the literature for a more complete discussion on this topic, it should be noted that in case of free vibration of an isolated beam of length L, the natural (fundamental) beam frequency, f_0, expressed in hertz (i.e. cycles per second), can be estimated as:

$$f_0 = K \cdot \sqrt{\frac{EI}{(m \cdot L^4)}} \tag{7.6a}$$

where K is a coefficient accounting for the restraint conditions, E the dynamic elastic modulus of the material, I the moment of inertia and m represents the mass per unit length.

Term K can be evaluated as:

$$K = \frac{\alpha}{2\pi} \tag{7.6b}$$

For the most common cases of beam end restraints, the following values are proposed in literature:

- $\alpha = 9.869$ ($K = 1.57$) for a simply supported beam;
- $\alpha = 22.37$ ($K = 3.56$) for a fixed-end beam;
- $\alpha = 3.516$ ($K = 0.56$) for a cantilever beam;
- $\alpha = 14.538$ ($K = 2.45$) for a simply supported-fixed end beam.

With reference to the current design practice, generally, the direct calculation of f_0 is avoided, owing to the difficulties in the evaluation of the dynamic characteristics of the steel materials. It appears more preferable to use an approximated approach based on a direct evaluation of displacements. As an example, in case of a simply supported beam of length L with a uniform mass m, the displacement δ_m can be estimated neglecting the contribution due to shear deformability, as:

$$\delta_m = \frac{5 \cdot (mg) \cdot L^4}{384 \cdot E \cdot I} \tag{7.7}$$

where g is the acceleration due to gravity (conventionally assumed as $g = 9805.5\ \text{mm/s}^2$ (386 in./s^2), m is the mass per unit length, E the Young's modulus and I the moment of inertia.

Substituting term $m \cdot L^4$ in the previous equation and considering the maximum displacement δ_{max}, the frequency (expressed in hertz) can be approximated as:

$$f_0 = \frac{17.75}{\sqrt{\delta_{max}[\text{mm}]}} \approx \frac{18}{\sqrt{\delta_{max}[\text{mm}]}} \tag{7.6c}$$

Using the US system of units, the result is:

$$f_0 = \frac{3.54}{\sqrt{\delta_{max}[\text{in.}]}} \tag{7.6d}$$

For the serviceability limit state, the natural frequency of vibrations of the structure or of a structural member should be kept above appropriate values in order to avoid the phase (amplitude) resonance during normal use. Frequency limit depends upon the function of the building and the source of the vibration, and is agreed with the client and/or the relevant authority. If the natural frequency of vibrations of the structure obtained via this simplified approach is lower than the appropriate value, a more refined analysis of the dynamic response of the structure, including the consideration of damping, should be performed. The most severe load combination, with reference to the displacement δ_{max} should be considered in order to base the vibration verification on the lower values of frequency (fundamental frequency).

7.1.3 Resistance

In design practice, members generally have cross-sections with at least one axis of symmetry and one of the most frequent cases is the check for mono-axial bending, where one of the principal axes is also an axis of flexure. Furthermore, it should be noted that the more general case is the bi-axial bending, that is the bending axis does not coincide with the principal axis. With reference to Figure 7.1 the resulting bending moment M is given by the equation:

$$M = \sqrt{M_{y1}^2 + M_{z1}^2} \tag{7.8}$$

where M_{y1} and M_{z1} represent the moments acting on the generic planes y_1 and z_1.

The simplest method to analyse biaxial bending is to replace the moments M_{y1} and M_{z1} by their principal plane static equivalents M_y and M_z calculated from:

$$M_y = M_{y1} \cdot \cos\alpha + M_{z1} \cdot \text{sen}\alpha \tag{7.9a}$$

$$M_z = -M_{y1} \cdot \text{sen}\alpha + M_{z1} \cdot \cos\alpha \tag{7.9b}$$

in which α is the angle between the y_1- and z_1-axes and the principal y- and z-axes, as shown in Figure 7.1b.

From a practical point of view the superposition principle is frequently used and the mono-axial bending approach is extended to the case of bi-axial bending.

7.1.4 Stability

Open section beams bent in their stiffer principle plane are susceptible to a type of buckling deflecting sideways and twisting (Figure 7.2), the so-called *lateral instability, lateral-torsional* or *flexural-torsional instability*. In particular, this form of instability is due to the compression

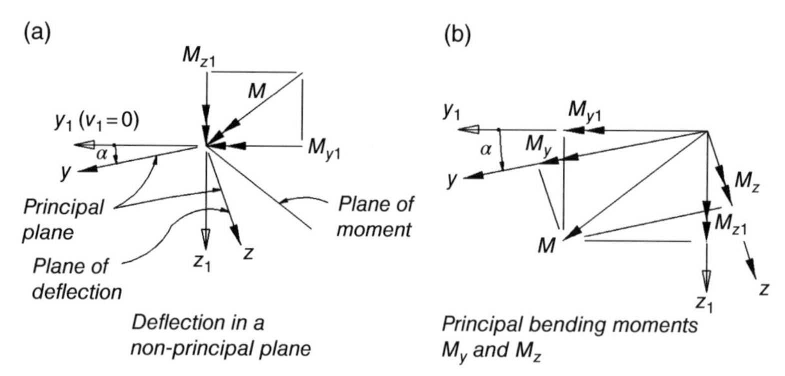

Figure 7.1 Bending in a non-principal plane: (a) deflection in a non-principal plane and (b) principal bending moments M_y and M_z.

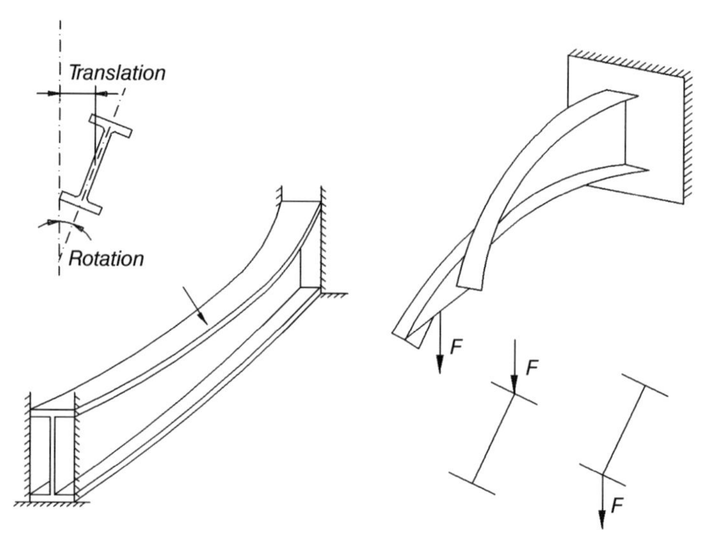

Figure 7.2 Deformed beam configurations associated with lateral torsional buckling. From Figure C-C2.3 of AISC 360-10.

force acting on a part of the profile causing instability with lateral deflection partially prevented by the tension part of the profile, which generates twist. Design standards consider lateral–torsional buckling as one of the ultimate limit states that must be checked for steel members in bending, when relevant. The buckling resistance assessment is usually based on appropriate buckling curves and requires the computation of the elastic critical moment, which is strongly dependent on several factors such as, the bending moment distribution, the restraints at the end supports and in correspondence of the load points, the beam cross-section, the distance between the load application point and the shear centre.

Due to the presence of both lateral and torsional deformations, a rigorous design approach is very complex but few simplifications are admitted. Beam design, taking into account lateral–torsional buckling, essentially consists of assessing the maximum moment that can safely be carried from knowledge of the section material and geometrical cross-section properties, the restraints provided and the arrangement of the applied loading. It should be noted that the eccentricity between the load application point and the shear centre plays an important role in the beam response. As an example, in case of I- or H-shaped cantilever beam in Figure 7.2, it can be noted that when the load is applied on the top flange or on the bottom flange, the load carrying capacity is significantly different, due to the fact that the load at the top flange has a negative effect (destabilizing load) with reference to lateral stability. Otherwise, load applied on the bottom flange stabilizes the beam response.

With reference to the design of civil and industrial structures, it should be noted that the floor slab could restraint efficiently the lateral buckling of the beams. However, buckling resistance has to be considered for the erection stage, in presence of load condition generally less severe than the one associated with the normal usage but without any restraint to lateral buckling. Furthermore, beam instability has to be considered in many cases for the roof beams, due to quite limited effect of metal sheeting in preventing the beam buckling. Verification criteria proposed for members in bending susceptible of lateral buckling are based on the value of the elastic critical moment (M_{cr}) and its evaluation, which in many cases is quite fairly complex, can be made by using numerical or theoretical approaches.

By exploiting the potentiality of finite elements (FEs) commercial analysis packages and their refined pre- and post-processors, it appears convenient, for design purposes, to generate complex three-dimensional refined meshes, especially in case of isolated beams. Available software pre-processors allow modelling exactly the shape of the profile with an accurate description (Figure 7.3a) of its components (web and flanges) by using shell or solid elements. As in case of the columns in the frames, an elastic buckling analysis or an incremental second order elastic analysis can be carried out on beams (Figure 7.3b,c) and attention has to be paid to the buckling deformed shape of interest for practical design, owing to the possibility that a simplified modelling of restraints or of the load introduction zones could lead to local buckling modes not relevant for design purposes (Figure 7.3d). In case of sub-frames or more complex framed systems, this modelling approach could lead to a large number of degrees of freedom as well as to an excessive number of elements in the zones where elements are connected and/or loaded.

Furthermore, beam formulations that are implemented in the most commonly used FE analysis packages neglect the warping of the cross-sections as well as all the associated effects and, as a consequence cannot be used to evaluate M_{cr} directly. As alternative, the well-established theoretical approach to evaluate directly the elastic critical moments M_{cr} can be used, which have been proposed in literature for most common cases of load cases and restraints, mainly with reference to profiles with bi- and mono-symmetrical I- and H-shaped cross-sections.

With reference to the more general case of mono-symmetrical I- or H-shaped unequal flange member (Figure 7.4), if the axis of symmetry is also axis of flexure and the moment distribution is constant (uniform) across the element (equal opposite moments applied at the beam ends), the critical elastic moment ($M_{cr,u}$) can be evaluated as:

$$M_{\mathrm{cr,u}} = \sqrt{\frac{\pi^2 EI_z}{(k_z L)^2}} \cdot \left\{ \sqrt{GI_t + \frac{\pi^2 EI_w}{(k_w L)^2} + \left[\frac{\beta_y}{2} \cdot \sqrt{\frac{\pi^2 EI_z}{(k_z L)^2}} \right]^2} + \frac{\beta_y}{2} \cdot \sqrt{\frac{\pi^2 EI_z}{(k_z L)^2}} \right\} \tag{7.10}$$

where L is the distance between two consecutive restrained cross-sections, E and G are the Young's and the tangential elasticity modulus of material, respectively, I_z is the moment of inertia along the weak axis, I_w and I_t are the warping and the torsion constant, respectively.

Terms k_w and k_z take into account the restraints of the cross-section.

Term k_w is an effective length factor accounting for warping end restraint, ranging from 0.5 (full fixity) to 1.0 (no fixity): $k_w = 0.7$ is recommended for one end fixed and the other end free.

Term k_z is an effective length factor accounting for rotation about y–y axis: it varies from 0.5 for full fixity to 1.0 for no fixity, with 0.7 for one end fixed and the other end free.

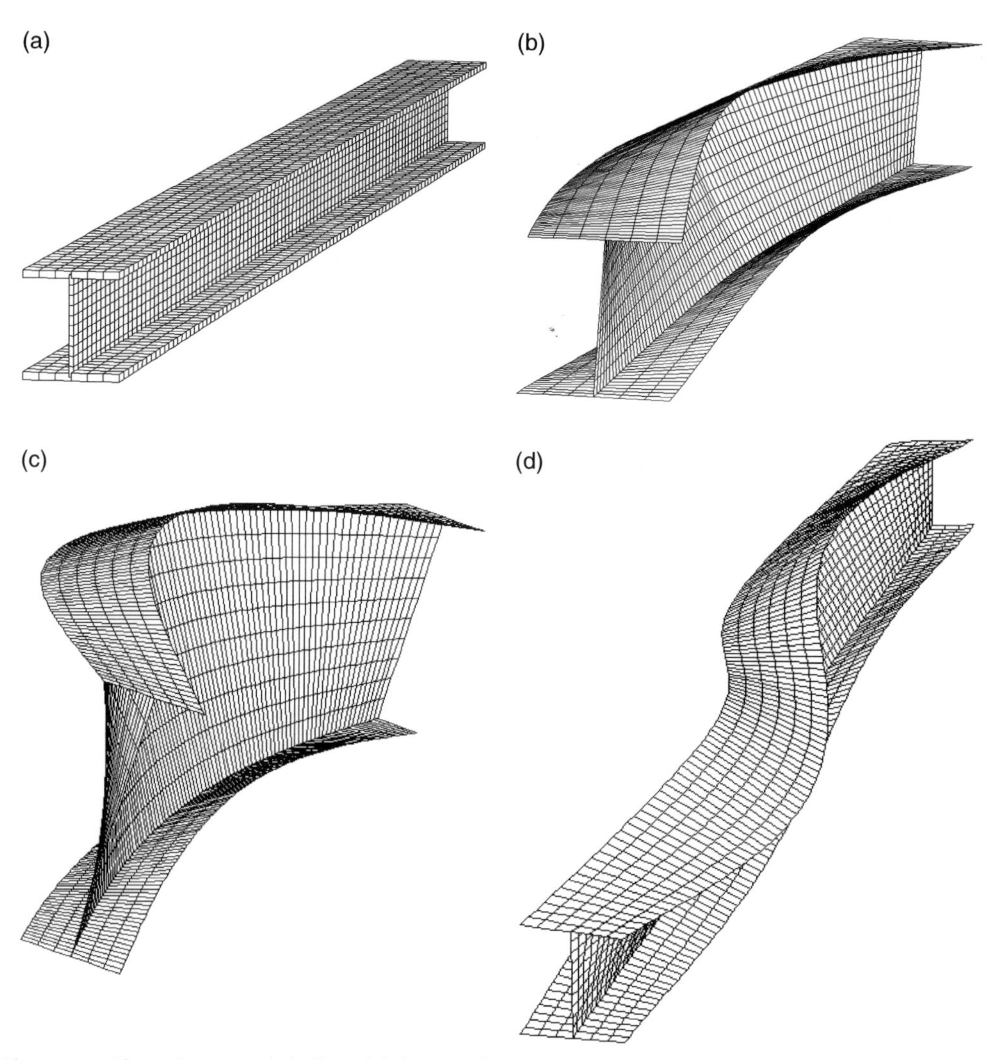

Figure 7.3 Three-dimensional shell models for a steel beams: the mesh (a) and typical critical deformed shapes obtained via buckling analysis in case of wide flange HE beam (b), standard I beam (c) and wide flange beam (d) for a high mode.

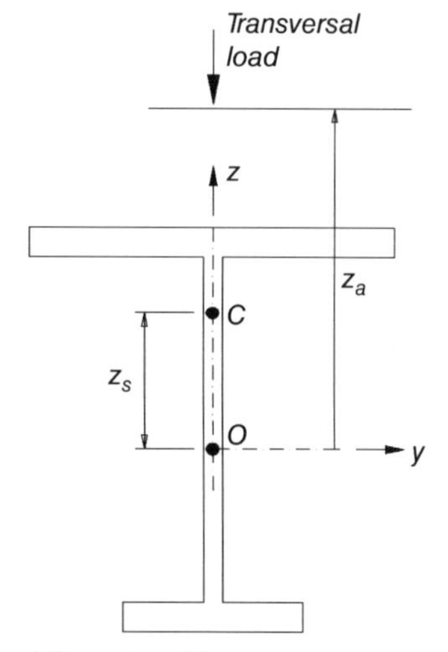

Figure 7.4 Mono-symmetrical unequal flange I profiles.

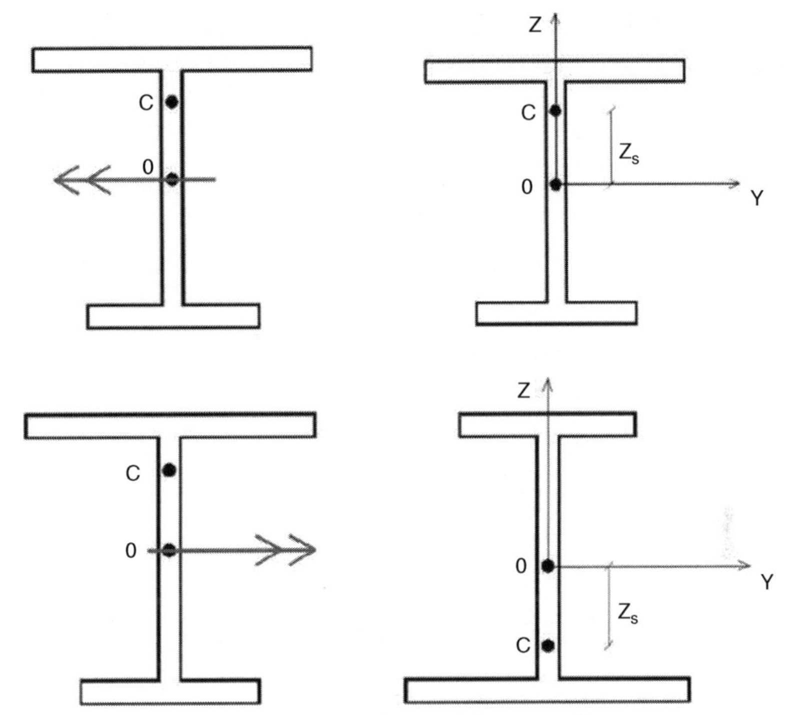

Figure 7.5 Bending load cases for mono-symmetrical unequal flange I profiles.

Term β_y is the Wagner coefficient, accounting for the eventual coincidence between shear centre and centroid. If the reference system has the origin in the centroid O (at a distance z_s of form centroid C), β_y is defined as:

$$\beta_y = 2 \cdot z_s - \frac{1}{I_y} \cdot \int_A \left(y^2 z + z^3\right) \mathrm{d}A \tag{7.11}$$

For the sake of simplicity, reference can be made to Figure 7.5 relating to the cases of simple bending for mono-symmetrical cross-section members.

This approach allows us to evaluate the elastic critical moment of the beam under uniform moment distribution, $M_{cr,u}$. Usually, bending moment distribution along the beam is not uniform and, as a consequence, standard codes propose the use of an equivalent uniform moment factor (EUMF) to be used to compute the elastic critical moment referred to the actual bending moment distribution by means of the expression:

$$M_{cr} = EUMF \cdot M_{cr,u} \tag{7.12}$$

Several approaches to evaluate the EUMF term are available nowadays. One of the most efficient ones was proposed by Serna *et al.*, who recently carried out a critical review of some of these approaches to evaluating EUMF coefficients recommended by modern steelwork standards. It has been shown that, whilst codes may lead to conservative values for simply supported beams, non-conservative values are obtained in the case of support types designed to restrict lateral bending and warping. The following general closed-form expression was hence proposed to assess EUMF coefficient:

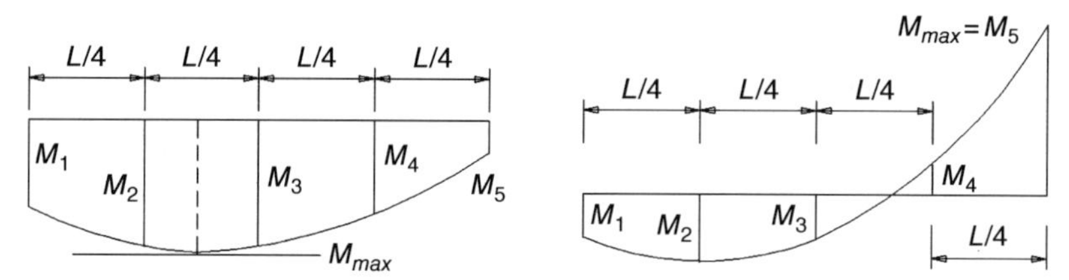

Figure 7.6 Moment diagrams and moment values for Eq. (7.14a,b).

$$EUMF = \frac{\sqrt{\sqrt{k}\cdot A_1 + \left[\frac{(1-\sqrt{k})}{2}A_2\right]^2} + \frac{(1-\sqrt{k})}{2}A_2}{A_1} \tag{7.13}$$

where terms A_1, A_2 and k are defined as:

$$A_1 = \frac{M_{\max}^2 + 9k\cdot M_2^2 + 16\cdot M_3^2 + 9k\cdot M_4^2}{[1+9k+16+9k]\cdot M_{\max}^2} \tag{7.14a}$$

$$A_2 = \left|\frac{M_{\max} + 4\cdot M_1 + 8\cdot M_2 + 12\cdot M_3 + 8\cdot M_4 + 4\cdot M_5}{37\cdot M_{\max}^2}\right| \tag{7.14b}$$

$$k = \sqrt{k_z\cdot k_w} \tag{7.14c}$$

It is worth mentioning that coefficient k depends on the same k_z and k_w coefficients already defined (Eq. (7.10)), accounting for both lateral bending (k_z) and warping (k_w) restraints at the end supports (see Eq. (7.14c)).

Bending moments M_1–M_5 are defined in Figure 7.6, which considers their absolute value.

For lateral bending and warping free at both end supports (i.e. $k_z = k_w = k = 1$), $EUMF$ coefficient results in a more simple expression as:

$$EUMF = \sqrt{\frac{35\cdot M_{\max}^2}{M_{\max}^2 + 9\cdot M_2^2 + 16\cdot M_3^2 + 9\cdot M_4^2}} \tag{7.15}$$

7.2 European Design Approach

Verification rules for beam elements are discussed in this sub-chapter, which are in accordance with the requirements reported in EN 1993-1-1.

7.2.1 Serviceability Limit States

7.2.1.1 Deformability

The current version of EN 1993-1-1 does not report any practical indications related to the deflection limits, which should be specified in the National Annex that each UE country has to develop. Reference is made to the general principles reported in EN 1990 – (basis of design) Annex A1.4, where no values are directly specified for limiting the vertical beam deflections. In the previous version of EN 1993-1-1, that is ENV 1993-1, suitable limits were proposed for the most common

Table 7.1 Recommended limiting values for vertical deflections from ENV 1993-1-1.

Conditions	Limits	
	$\delta_{max} = \delta_1 + \delta_2 - \delta_0$	δ_2
Roofs generally	$\frac{L}{200}$	$\frac{L}{250}$
Roofs frequently carrying personnel other than for maintenance	$\frac{L}{250}$	$\frac{L}{300}$
Floors generally	$\frac{L}{250}$	$\frac{L}{300}$
Floors and roofs supporting plaster or other brittle finish or non-flexible partitions	$\frac{L}{250}$	$\frac{L}{350}$
Floors supporting columns (unless the deflection has been included in the global analysis for the ultimate limit states)	$\frac{L}{400}$	$\frac{L}{500}$
Where δ_{max} can impair appearance of the building	$\frac{L}{250}$	—

δ_0 = precamber (hogging) of the beam in the unloaded state (state 0)
δ_1 = variation of the deflection of the beam due to permanent loads immediately after loading (state 1)
δ_2 = variation of the deflection of the beam due to variable loading plus any time dependant deformations due to permanent load (state 2)

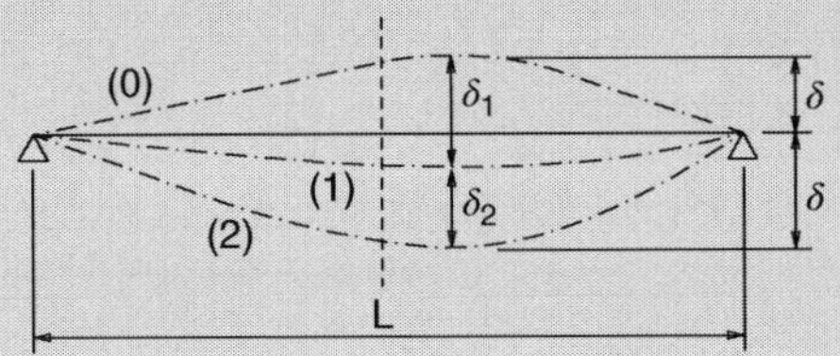

cases encountered in routine design. In addition to the contribution to deflection due to dead load (δ_1) and to the variable-live load (δ_2), the possible presence of a precamber (hogging) in the unloaded phase δ_0 is also considered, which should be required for manufacture in order to limit the total vertical deflection *in-service*.

Table 7.1 refers to simply supported beams and proposes the ENV deflection limits on both the sagging deflection in the final stage relative to the straight line joining the supports (δ_{max}) and the deflection due to variable-live load (δ_2) in service.

7.2.1.2 Vibrations

As for deflection limits, the current version of EN 1993-1-1 does not report any practical indication that is related to vibration checks, in this case also making reference to the contents of the National Annex or to the general principle of EN 1990 – Annex A1.4.4, where no values are recommended. As for beam deflection, the ENV 1993-1 proposes practical indications to be used for routine building design when the possibility of vibrations could cause discomfort to the users. In particular, it is required that:

- the fundamental frequency of floors in dwellings and offices should not be less than 3 cycles/s (i.e. $f_0 > 3$ Hz). This may be deemed to be satisfied when the sum of $\delta_1 + \delta_2$ (see Table 7.1) is less than 28 mm (1.1 in.);
- the fundamental frequency of floors used for dancing and gymnasium should not be less than 5 cycles/s (i.e. $f_0 > 5$ Hz). This may be deemed to be satisfied when the sum $\delta_1 + \delta_2$ (see Table 7.1) is less than 10 mm (0.39 in.).

7.2.2 Resistance Verifications

Flexural verifications have to take into account the presence of the shear force.

7.2.2.1 Bending Resistance

The design value of the bending moment, M_{Ed}, at each cross-section, must satisfy the condition:

$$M_{Ed} \le M_{c,Rd} \tag{7.16a}$$

where $M_{c,Rd}$ represents the bending resistance of cross-section determined by considering the eventual presence of fastener holes.

The design resistance $M_{c,Rd}$ for bending about one principal axis of a cross-section is determined, in case of absence of shear forces, on the basis of the cross-section class as described next:

- for class 1 or 2 cross-sections:

$$M_{c,Rd} = M_{pl,Rd} = W_{pl}\frac{f_y}{\gamma_{M0}} \tag{7.16b}$$

  where W_{pl} is the plastic section modulus, f_y is the yield strength and γ_{M0} is the partial safety factor.
- for class 3 cross-sections:

$$M_{c,Rd} = M_{el,Rd} = W_{el,\min}\frac{f_y}{\gamma_{M0}} \tag{7.16c}$$

  where $W_{el,\min}$ is the elastic section modulus related to the more stressed point.
- for class 4 cross-sections:

$$M_{c,Rd} = W_{\text{eff},\min}\frac{f_y}{\gamma_{M0}} \tag{7.16d}$$

  where $W_{\text{eff},\min}$ is the effective section modulus evaluated with reference to the effective cross-section, defined in accordance with the criteria summarized in Chapter 4.

Fastener holes in the tension flange may be ignored provided that the following condition for the tension flange is fulfilled:

$$\frac{A_{f,\text{net}}\cdot(0.9\cdot f_u)}{\gamma_{M2}} \ge \frac{A_f\cdot f_y}{\gamma_{M0}} \tag{7.17}$$

where A_f and $A_{f,\text{net}}$ represent the gross area and the effective area of the tension flange, respectively, and γ_{M0} and γ_{M2} are partial safety factors.

It should be noted that fastener holes in tension zone of the web are not allowed, provided that the limit given in Eq. (7.17) is satisfied for the complete tension zone, including the tension flange plus the tension zone of the web. Moreover, fastener holes except for oversize and slotted holes in compression zone of the cross-section are not allowed, provided that they are filled by fasteners.

7.2.2.2 Shear Resistance

The design value of the shear force V_{Ed} at each cross-section must not be greater than the design shear resistance, $V_{c,Rd}$, that is the following conditions must be fulfilled:

$$V_{Ed} \leq V_{c,Rd} \tag{7.18}$$

7.2.2.3 Plastic Design

For plastic design $V_{c,Rd}$ has to be assumed as the design plastic shear resistance, $V_{pl,Rd}$, which can be evaluated as:

$$V_{pl,Rd} = A_v \frac{f_y/\sqrt{3}}{\gamma_{M0}} \tag{7.19}$$

where A_v is the shear area, f_y is the yield strength and γ_{M0} is the partial safety factor.

The shear area A_v may be taken as follows:

- rolled I- and H-shaped sections, with load parallel to the web:

$$A_v = A - 2bt_f + (t_w + 2r)t_f \text{ with } A_v \geq \eta \cdot h \cdot t$$

- rolled channel sections, with load parallel to the web:

$$A_v = A - 2bt_f + (t_w + r)t_f$$

- rolled T-shaped section, with load parallel to the web:

$$A_v = A - bt_f + (t_w + 2r)\frac{t_f}{2}$$

- welded T-shaped section, with load parallel to the web:

$$A_v = t_w\left(h - \frac{t_f}{2}\right)$$

- welded I-, H-shaped and box sections, with load parallel to the web:

$$A_v = \eta\Sigma(h_w t_w)$$

- welded I-, H-shaped, channel and box sections, with load parallel to the flanges:

$$A_v = A - \Sigma(h_w t_w)$$

- rolled rectangular hollow sections of uniform thickness with load parallel to the depth:

$$A_v = Ah/(b + h)$$

- rolled rectangular hollow sections of uniform thickness with load parallel to the width:

$$A_v = Ab/(b + h)$$

- circular hollow sections and tubes of uniform thickness:

$$A_v = 2A/\pi$$

where A is the cross-section area, b and h are the overall width and depth, respectively, h_w is the depth of the web, r is the root radius, t is the thickness and subscripts f and w are related to the flange and the web, respectively.

Furthermore, it should be noted that in case of not constant web thickness, t_w has to be taken as the minimum thickness. The value of coefficient η is defined in EN 1993-1-5, which recommends $\eta = 1.2$ for S235 to S460 steel grades and $\eta = 1.0$ for steel grades over S460. Alternatively, this valued should be reported in the National Annexes implementing the Eurocodes but, however, it can be conservatively assumed equal to unity ($\eta = 1.0$).

7.2.2.4 Elastic Design

To verify the design shear resistance $V_{c,Rd}$ in Eq. (7.18), the following criterion for a critical point of the cross-section may be used unless the buckling verification in Section 5 of EN 1993-1-5 has to be fulfilled:

$$\tau_{Ed} \leq \frac{f_y}{\sqrt{3}\cdot\gamma_{M0}} \tag{7.20a}$$

where f_y is the yield strength and γ_{M0} is the partial safety factor.

Tangential stress τ_{Ed} due to shear force must be evaluated as:

$$\tau_{Ed} = \frac{V_{Ed}\cdot S}{I\cdot t} \tag{7.20b}$$

where V_{Ed} is the design value of the shear force, S is the first moment of the area above on either side of the examined point, I is the moment of inertia of the whole cross-section and t is the thickness at the examined point.

Shear buckling for stocky webs does not have to be considered in the following cases:

- for unstiffened webs:

$$\frac{h_w}{t_w} > 72\frac{\varepsilon}{\eta} \tag{7.21a}$$

- for transversely stiffened webs:

$$\frac{d}{t_w} > 31\frac{\varepsilon}{\eta}\sqrt{k_\tau} \tag{7.21b}$$

where h_w and t_w are the depth and the thickness of the web, respectively, $\varepsilon = \sqrt{235/f_y[MPa]}$, η is the coefficient already introduced for the evaluation of shear area and k_τ is the shear buckling coefficient.

In the case of absence of longitudinal stiffeners, defining a as the distance between two rigid transverse stiffeners, the following values are proposed in EN 1993-1-5:

- when $a \geq h_w$:

$$k_\tau = 5.34 + 4 \cdot \left(\frac{h_w}{a}\right)^2 \tag{7.22}$$

- when $a < h_w$:

$$k_\tau = 4 + 5.34 \cdot \left(\frac{h_w}{a}\right)^2 \tag{7.23}$$

7.2.2.5 Shear-Torsion Interaction

For combined shear force and torsional moment, the plastic shear resistance accounting for torsional effects should be reduced from $V_{pl,Rd}$ to $V_{pl,T,Rd}$ and the design shear force V_{Ed} must satisfy the condition:

$$V_{Ed} \leq V_{pl,T,Rd} \tag{7.24}$$

For the most common cases reduced shear resistance $V_{pl,T,,Rd}$ is defined as:

- for an I- or H-shaped section:

$$V_{pl,T,Rd} = V_{pl,Rd} \cdot \sqrt{1 - \frac{\tau_{t,Ed}}{\frac{1.25}{\gamma_{M0}} \cdot \left(\frac{f_y}{\sqrt{3}}\right)}} \tag{7.25a}$$

- for a channel section:

$$V_{pl,T,Rd} = V_{pl,Rd} \cdot \left[\sqrt{1 - \frac{\tau_{t,Ed}}{\frac{1.25}{\gamma_{M0}} \cdot \left(\frac{f_y}{\sqrt{3}}\right)}} - \frac{\tau_{w,Ed}}{\frac{1}{\gamma_{M0}} \cdot \left(\frac{f_y}{\sqrt{3}}\right)}\right] \tag{7.25b}$$

- for a structural hollow section:

$$V_{pl,T,Rd} = V_{pl,Rd} \cdot \left[1 - \frac{\tau_{t,Ed}}{\frac{1}{\gamma_{M0}} \cdot \left(\frac{f_y}{\sqrt{3}}\right)}\right] \tag{7.25c}$$

7.2.2.6 Bending and Shear Resistance

In case of shear force acting on beams, allowance must be made for its effects on the moment resistance. In particular, if the design shear force V_{Ed} is less than half of the plastic shear resistance $V_{pl,Rd}$ (i.e. $V_{Ed} < 0.5\ V_{pl,Rd}$), its effect on the moment resistance may be neglected, except when the shear buckling reduces the section resistance. Otherwise, the reduced moment resistance should be taken as the design resistance of the cross-section, based on a reduced yield strength, $f_{y,red}$, defined as:

$$f_{y,red} = (1-\rho)\cdot f_y \tag{7.26}$$

where f_y is the yield strength and ρ is a reduction factor defined as:

$$\rho = \left(\frac{2\cdot V_{Ed}}{V_{pl,Rd}} - 1\right)^2 \tag{7.27a}$$

In case of torsion (see Chapter 8), a suitably reduced shear plastic resistance $V_{pl,T,Rd}$ has to be considered instead of $V_{pl,Rd}$ when $V_{Ed} > 0.5\ V_{pl,T,Rd}$. Term ρ is consequently defined as:

$$\rho = \left(\frac{2\cdot V_{Ed}}{V_{pl,T,Rd}} - 1\right)^2 \tag{7.27b}$$

For I- and H-shaped cross-sections with equal flanges bent about the major axis, the reduced design plastic resistance moment allowing for the shear force, $M_{y,V,Rd}$, can alternatively be obtained as:

$$M_{y,V,Rd} = \left(W_{pl,y} - \frac{\rho\cdot A_w^2}{4\cdot t_w}\right)\cdot\frac{f_y}{\gamma_{M0}} \tag{7.28}$$

where A_w is the web area of the cross-section ($A_w = h_w\ t_w$).

In Figure 7.7 the resistant moment-shear (M–V) domain is proposed for doubly symmetrical H- and I-shaped profiles when loads are applied parallel to the web. By increasing the value of the shear force, the contribution of the bending moment transferred by the web decreases (M_v) up to the limit case of bending moment resistance due to the sole flanges (M_f).

7.2.3 Buckling Resistance of Uniform Members in Bending

Beams with sufficient restraints along the compression flange are not susceptible to lateral-torsional buckling. Furthermore, the beams with certain types of cross-sections, such as square or circular hollow sections, fabricated circular tubes or square box sections are also less susceptible to lateral-torsional buckling. On the contrary, in case of laterally unrestrained beam members subjected to major axis (y–y axis) bending, verification against this phenomenon is required. Defining

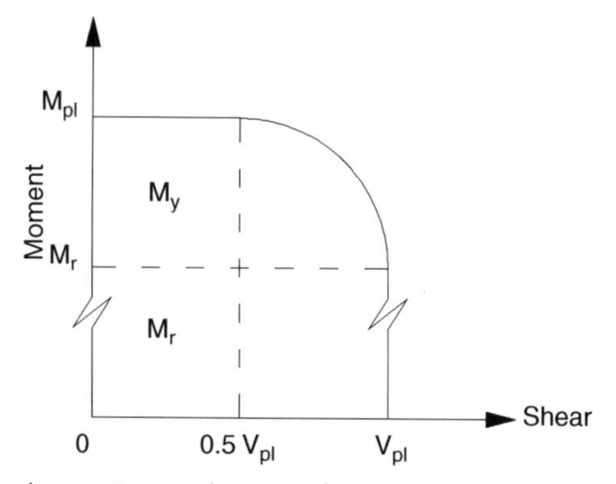

Figure 7.7 Bending moment-shear resistance domain.

M_{Ed} as the design value of the moment and $M_{b,Rd}$ the design buckling resistance moment, it must be guaranteed that:

$$M_{Ed} \leq M_{b,Rd} \tag{7.29}$$

It should be noted that the design approach proposed for beams subjected to lateral-torsional buckling is broadly similar to that used for compression members (columns in Chapter 6). In particular, the design buckling resistance moment $M_{b,Rd}$ of a laterally unrestrained beam is defined as:

$$M_{b,Rd} = \chi_{LT} \cdot W_y \cdot \frac{f_y}{\gamma_{M1}} \tag{7.30}$$

where W_y is the appropriate section modulus, depending on the class of cross-section, f_y is the yield strength, χ_{LT} is a reduction factor and γ_{M1} is the safety coefficient.

As to the section modulus, W_y, for class 1 or 2 cross-sections is the plastic modulus ($W_y = W_{pl,y}$), for class 3 the elastic modulus ($W_y = W_{el}$) and for class 4 the effective modulus ($W_y = W_{\mathrm{eff},y}$).

As to the reduction factor χ_{LT}, two different approaches are proposed: the general approach and the more refined approach for doubly symmetrical I- and H-shaped profiles.

7.2.3.1 The General Approach

The reduction factor for lateral-torsional buckling, χ_{LT}, is given by expression:

$$\chi_{LT} = \frac{1}{\phi_{LT} + \sqrt{\phi_{LT}^2 - \bar{\lambda}_{LT}^2}} \quad \text{with } \chi_{LT} \leq 1 \tag{7.31}$$

Term ϕ_{LT} is defined as:

$$\phi_{LT} = 0.5 \cdot \left[1 + \alpha_{LT}(\bar{\lambda}_{LT} - 0.2) + \bar{\lambda}_{LT}^2\right] \tag{7.32}$$

where α_{LT} is the imperfection factor corresponding to the appropriate buckling curve, which may be obtained from the National Annex or by considering the Eurocode recommended values in Table 7.2.

Buckling curve depends on the type of cross-section as well as on the ratio between the overall depth (h) and the overall width (b) of the beam, in accordance with the indications in Table 7.3.

Relative slenderness for lateral-torsional buckling, $\bar{\lambda}_{LT}$, is defined as:

$$\bar{\lambda}_{LT} = \sqrt{\frac{W_y \cdot f_y}{M_{cr}}} \tag{7.33}$$

where M_{cr} is the elastic critical moment for lateral-torsional buckling based on gross cross-sectional properties and taking into account the actual load condition, the real moment distribution and the lateral restraint, f_y, is the yield strength and W_y is the appropriate cross-section modulus already presented with reference to the Eq. (7.30).

Table 7.2 Eurocode recommended values for the imperfection factor α_{LT} for lateral torsional buckling curves.

Buckling curve	a	b	c	d
Imperfection factor α_{LT}	0.21	0.34	0.49	0.76

Table 7.3 Recommended values for the lateral torsional buckling curves using the general approach.

Cross-section	Limit	Stability curve
Rolled I-sections	$h/b \leq 2$	a
	$h/b > 2$	b
Welded I-sections	$h/b \leq 2$	c
	$h/b > 2$	d
Other cross-sections	—	d

Table 7.4 Recommendation for the lateral torsional buckling curve selection using the approach proposed for rolled sections or equivalent welded sections.

Cross-section	Limit	Stability curve
Rolled I-sections	$h/b \leq 2$	b
	$h/b > 2$	c
Welded I-sections	$h/b \leq 2$	c
	$h/b > 2$	d

7.2.3.2 The Method for I- or H-Shaped Profiles

For the lateral torsional buckling (LTB) verification of rolled or equivalent welded sections I- or H-shaped beams, the values of χ_{LT} for the appropriate relative slenderness can be determined as:

$$\chi_{LT} = \frac{1}{\phi_{LT} + \sqrt{\phi_{LT}^2 - \beta \cdot \bar{\lambda}_{LT}^2}}, \text{ with } \chi_{LT} \leq 1 \text{ and } \chi_{LT} \leq (1/\bar{\lambda}_{LT})^2 \tag{7.34}$$

Term ϕ_{LT} is expressed as:

$$\phi_{LT} = 0.5 \cdot \left[1 + \alpha_{LT}(\bar{\lambda}_{LT} - \bar{\lambda}_{LT,0}) + \beta \cdot \bar{\lambda}_{LT}^2\right] \tag{7.35}$$

The parameters β and $\bar{\lambda}_{LT,0}$ as well as any limitation of validity concerning the beam depth or h/b ratio may be given in the National Annex. For rolled sections or equivalent welded sections, $\bar{\lambda}_{LT,0} = 0.4$ (maximum value) and $\beta = 0.75$ (minimum value) are recommended to be used by selecting the buckling curve in accordance with Table 7.4.

In order to account for the moment distribution between the lateral restraints of the members, the reduction factor $\chi_{LT,mod}$ may be modified as follows:

$$\chi_{LT,\text{mod}} = \frac{\chi_{LT}}{f} \text{ with } \chi_{LT,\text{mod}} \leq 1 \tag{7.36}$$

The values of term f should be defined in the National Annex; the following minimum value is, however, recommended:

$$f = 1 - 0.5 \cdot (1 - k_c) \cdot \left[1 - 2(\bar{\lambda}_{LT} - 0.8)^2\right] \text{ with } f \leq 1 \tag{7.37}$$

where k_c is a correction factor according to Table 7.5.

A critical phase in the evaluation of the buckling bending resistance is the assessment of the elastic critical moment for lateral-torsional buckling, M_{cr}, which is the key parameter defining

Table 7.5 Correction factors k_c.

Moment distribution	k_c
$\psi = 1$	1.0
$-1 \leq \psi \leq 1$	$\frac{1}{1,33-0,33\psi}$
	0.94
	0.90
	0.91
	0.86
	0.77
	0.82

the relative slenderness $\bar{\lambda}_{LT}$ in Eq. (7.33). In the current edition of EN 1993-1-1 no practical indications are given for the evaluation of M_{cr}. It is declared that M_{cr} has to be evaluated on the basis of gross cross-sectional properties and by taking into account the load conditions, the effective moment distribution and the lateral restraints. Hence Equation (7.10) has very limited direct use for the evaluation of M_{cr}, owing to the fact that no practical indications are provided for routine design. The reasons for the omission of suitable formulations for routine design and the absence of practical guidance for designers seems to be associated with the complexity of the problem. As a reference, however, Eurocode 9 for aluminium structures (*EN 1999-1-1 Eurocode 9: Design of aluminium structures – Part 1-1: General structural rules*) proposes in its Annex I (*[informative] – Lateral torsional buckling of beams and torsional or torsional-flexural buckling of compressed members*) should be considered, being a very important and practical guidance for the evaluation of the elastic critical moment of beams.

Another important reference for the evaluation of M_{cr} should be Annex F of the previous version of EN 1993-1-1 (i.e. Annex F of ENV 1993-1-1), where expressions for the evaluation of M_{cr} have been proposed. With reference to a beam of uniform cross-section, symmetrical about its minor axis, for bending about the major axis, the elastic critical moment for lateral-torsional buckling (Figure 7.8) can be obtained as:

$$M_{cr} = C_1 \frac{\pi^2 EI_z}{(k_z L)^2} \cdot \left\{ \left[\sqrt{\left(\frac{k_z}{k_w}\right)^2 \frac{I_w}{I_z} + \frac{(k_z L)^2 GI_t}{\pi^2 EI_z} + \left(C_2 z_g - C_3 z_j\right)^2} \right] - \left(C_2 z_g - C_3 z_j\right) \right\} \tag{7.38}$$

It can be noted that several parameters affect the value of the elastic critical load. Term I_w is the warping constant, z_g is the distance between the load application point and the shear centre, that is $z_g = z_a - z_s$ (in general this term is positive when loads acting towards the shear centre, i.e. the gravity loads are applied above the shear centre, in accordance with Figure 7.5) and z_j is a parameter with units of length, which is equivalent to term β_j in Eq. (7.11) divided by 2, defined as:

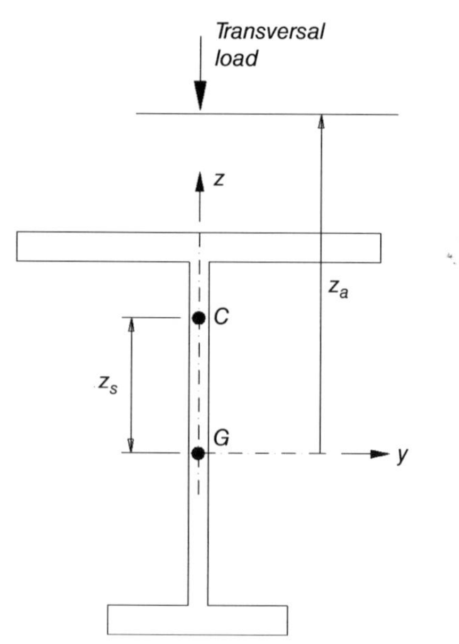

Figure 7.8 Mono-symmetrical cross-section (symmetry about the minor axis).

$$z_j = z_S - \frac{0.5}{I_y} \cdot \int_A \left(y^2 + z^2\right) \cdot z \, dA \tag{7.39}$$

Furthermore, it should be noted that parameter z_j reflects the degree of asymmetry of the cross-section with reference to the y–y axis; in case of I- and H-shaped profiles, the following values are assumed:

- $z_j = 0$ in case of doubly symmetrical cross-section;
- $z_j > 0$ if the flange with the largest moment of inertia about the z–z axis is compressed at the beam location with the maximum bending moment;
- $z_j < 0$ if the flange with the lowest moment of inertia about the z–z axis is compressed at the beam location with the maximum bending moment.

Terms k_w and k_z are the effective length factor dealing with warping end restraint and rotation about the y–y axis, respectively, already introduced with reference to Eq. (7.10).

Coefficients C_1, C_2 and C_3 depend on the shape of the bending moment diagram (i.e. by the load conditions), and on the support conditions. These coefficients are reported in Tables 7.6a and 7.6b for the most common design cases in accordance with those in Annex F of ENV 1993-1-1, which recent studies have demonstrated to be incorrect in few cases. Reference should be made to Tables 7.7a and 7.7b, proposing more correct values on the basis of the studies carried out by Boissonade *et al.* and published in 2006 in the ECCS document n.119 (*Rules for member stability in EN 1993-1-1*).

Furthermore, in case of I-shaped profiles with one axis of symmetry, a simplified formula is proposed to evaluate warping constant I_w on the basis of the height h_s of the profile (or, more correctly, as the distance between the shear centres of the flanges):

$$I_w = \beta_f \cdot \left(1 - \beta_f\right) \cdot I_z \cdot h_s^2 \tag{7.40}$$

Table 7.6a Coefficients C_1, C_2 and C_3 for beams with end moments (Annex F of ENV 1993-1-1).

Load and support conditions	Bending moment diagram	Value for k_z	C_1	C_2	C_3
			Value for coefficient		
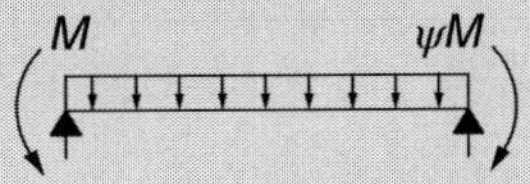 M, ψM	$\psi=+1$ 	1.0	1.000	—	1.000
		0.7	1.000		1.113
		0.5	1.000		1.144
	$\psi=+3/4$ 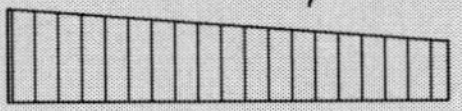	1.0	1.141	—	0.998
		0.7	1.270		1.565
		0.5	1.305		2.283
	$\psi=+1/2$ 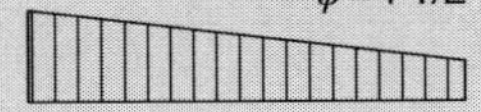	1.0	1.323	—	0.992
		0.7	1.473		1.556
		0.5	1.514		2.271
	$\psi=+1/4$ 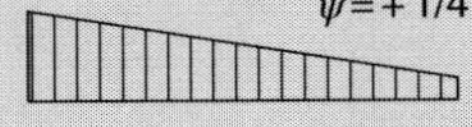	1.0	1.563	—	0.977
		0.7	1.739		1.531
		0.5	1.788		2.235
	$\psi=0$ 	1.0	1.879	—	0.939
		0.7	2.092		1.473
		0.5	2.150		2.150
	$\psi=-1/4$ 	1.0	2.281	—	0.855
		0.7	2.538		1.340
		0.5	2.609		1.957
	$\psi=-1/2$ 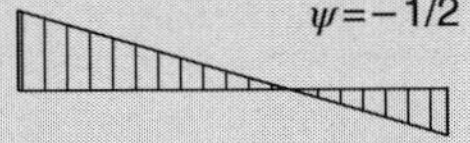	1.0	2.704	—	0.676
		0.7	3.009		1.059
		0.5	3.093		1.546
	$\psi=-3/4$ 	1.0	2.927	—	0.366
		0.7	3.009		0.575
		0.5	3.093		0.837
	$\psi=-1$ 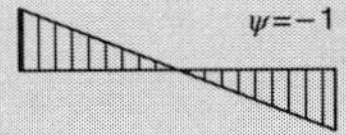	1.0	2.752	—	0.000
		0.7	3.063		0.000
		0.5	3.149		0.000

Table 7.6b Coefficients C_1, C_2 and C_3 for intermediate transverse load (Annex F of ENV 1993-1-1).

Load and support conditions	Bending moment diagram	Value for k_z	C_1	C_2	C_3
			Value for coefficient		
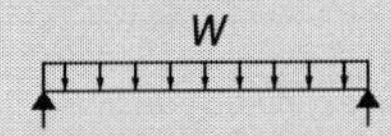 W (simply supported)		1.0	1.132	0.459	0.525
		0.5	0.972	0.304	0.980
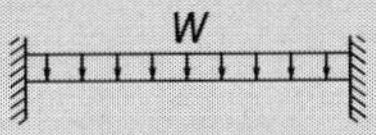 W (fixed ends)	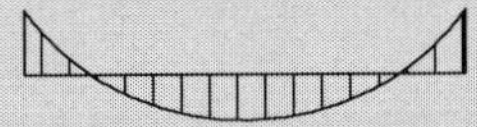	1.0	1.285	1.562	0.753
		0.5	0.712	0.652	1.070
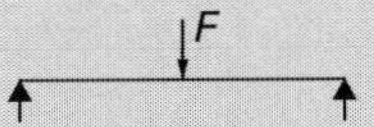 F (simply supported)	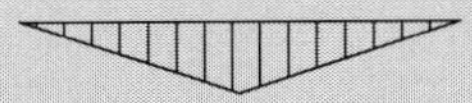	1.0	1.365	0.553	1.730
		0.5	1.070	0.432	3.050
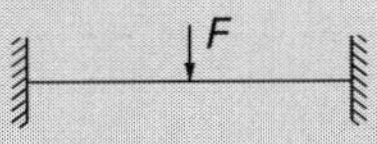 F (fixed ends)	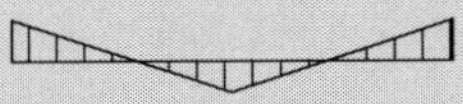	1.0	1.565	1.267	2.640
		0.5	0.938	0.715	4.800
 F, F at quarter points (= = = =)		1.0	1.046	0.430	1.120
		0.5	1.010	0.410	1.890

Table 7.7a Coefficients C_1, C_2 and C_3 for beams with end moments proposed by Boissonade *et al.* in the ECCS doc. No. 119.

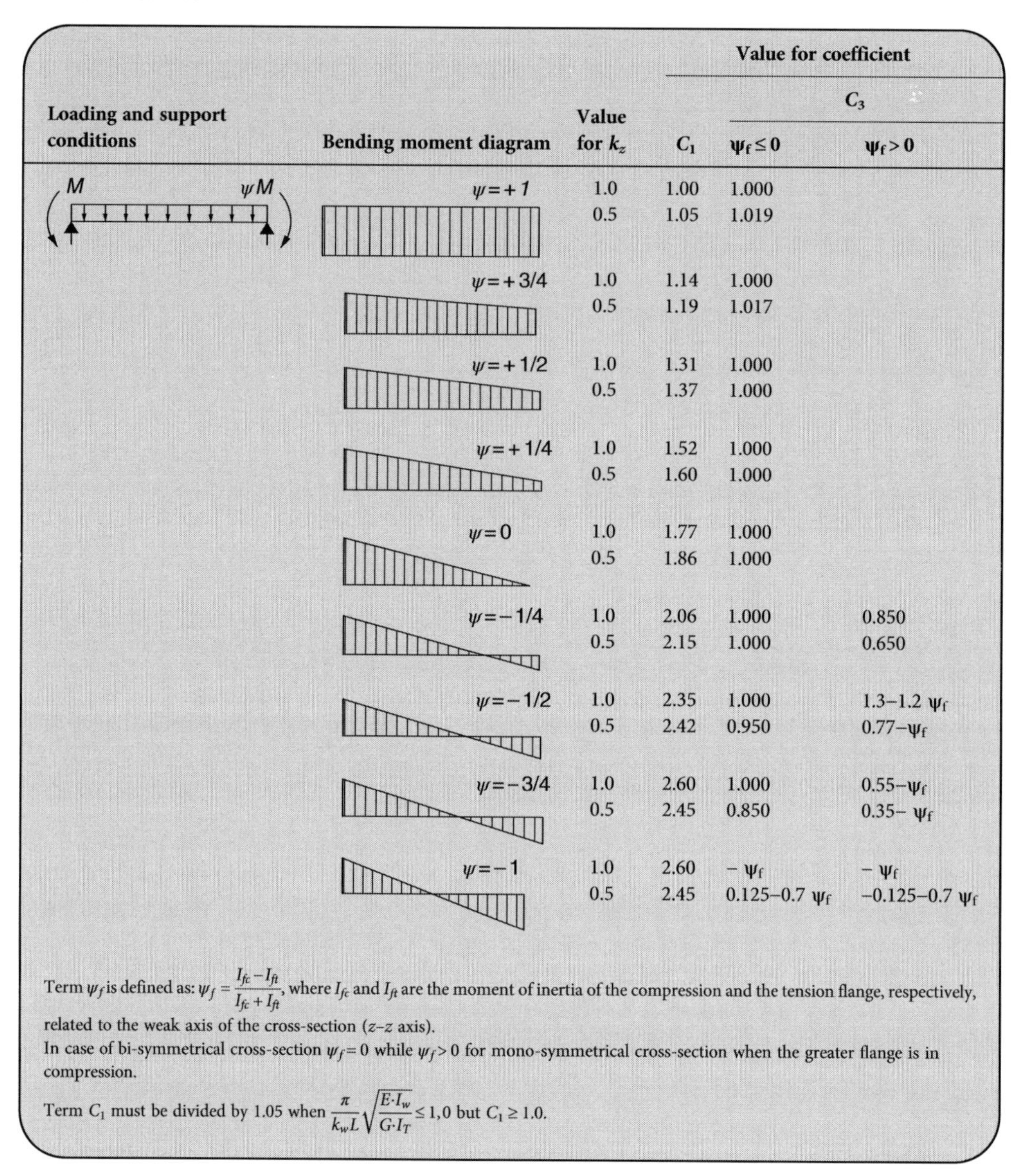

Loading and support conditions	Bending moment diagram	Value for k_z	Value for coefficient C_1	C_3: $\psi_f \le 0$	C_3: $\psi_f > 0$
M, ψM	$\psi = +1$	1.0	1.00	1.000	
		0.5	1.05	1.019	
	$\psi = +3/4$	1.0	1.14	1.000	
		0.5	1.19	1.017	
	$\psi = +1/2$	1.0	1.31	1.000	
		0.5	1.37	1.000	
	$\psi = +1/4$	1.0	1.52	1.000	
		0.5	1.60	1.000	
	$\psi = 0$	1.0	1.77	1.000	
		0.5	1.86	1.000	
	$\psi = -1/4$	1.0	2.06	1.000	0.850
		0.5	2.15	1.000	0.650
	$\psi = -1/2$	1.0	2.35	1.000	$1.3-1.2\,\psi_f$
		0.5	2.42	0.950	$0.77-\psi_f$
	$\psi = -3/4$	1.0	2.60	1.000	$0.55-\psi_f$
		0.5	2.45	0.850	$0.35-\psi_f$
	$\psi = -1$	1.0	2.60	$-\psi_f$	$-\psi_f$
		0.5	2.45	$0.125-0.7\,\psi_f$	$-0.125-0.7\,\psi_f$

Term ψ_f is defined as: $\psi_f = \dfrac{I_{fc}-I_{ft}}{I_{fc}+I_{ft}}$, where I_{fc} and I_{ft} are the moment of inertia of the compression and the tension flange, respectively, related to the weak axis of the cross-section (z–z axis).

In case of bi-symmetrical cross-section $\psi_f = 0$ while $\psi_f > 0$ for mono-symmetrical cross-section when the greater flange is in compression.

Term C_1 must be divided by 1.05 when $\dfrac{\pi}{k_w L}\sqrt{\dfrac{E \cdot I_w}{G \cdot I_T}} \le 1{,}0$ but $C_1 \ge 1.0$.

Table 7.7b Coefficients C_1, C_2 and C_3 in case of intermediate transverse loading proposed by Boissonade *et al.* in the ECCS doc. No. 119.

Loading and support conditions	Bending moment diagram	Value for k_z	Value for coefficient C_1	C_2	C_3
W		1.0	1.12	0.45	0.525
		0.5	0.97	0.36	0.478
F		1.0	1.35	0.59	0.411
		0.5	1.05	0.48	0.338
F, F (= = = =)		1.0	1.04	0.42	0.562
		0.5	0.95	0.31	0.539

where h_s is the distance between the shear centre of the webs and parameter β_f is defined as:

$$\beta_f = \frac{I_{fc}}{I_{fc} + I_{ft}} \tag{7.41}$$

where I_{fc} and I_{ft} are the moments of inertia referred to the weak axis related to the compression and tension flanges, respectively.

As to the evaluation of z_j, the following approximated equations can be used:

- if $\beta_f > 0.5$:

$$z_j = 0.8 \cdot \left(2\beta_f - 1\right) \cdot \frac{h_S}{2} \tag{7.42a}$$

- if $\beta_f < 0.5$:

$$z_j = 1.0 \cdot \left(2\beta_f - 1\right) \cdot \frac{h_S}{2} \tag{7.42b}$$

In the case of compression with a stiffened flange:

if $\beta_f > 0.5$:

$$z_j = 0.8 \cdot \left(2\beta_f - 1\right) \cdot \left(1 + \frac{h_L}{h}\right) \cdot \frac{h_s}{2} \tag{7.42c}$$

if $\beta_f < 0.5$:

$$z_j = 1.0 \cdot \left(2\beta_f - 1\right) \cdot \left(1 + \frac{h_L}{h}\right) \cdot \frac{h_s}{2} \tag{7.42d}$$

where h_L is the height of the stiffener.

In case of doubly-symmetrical I- or H-shaped profiles ($z_j = 0$) the expression of M_{cr} simplifies to:

$$M_{cr} = C_1 \frac{\pi^2 EI_z}{(k_z L)^2} \cdot \left[\sqrt{\left(\frac{k_z}{k_w}\right)^2 \frac{I_W}{I_z} + \frac{(k_z L)^2 GI_t}{\pi^2 EI_z} + \left(C_2 z_g\right)^2} - C_2 z_g\right] \tag{7.43}$$

Furthermore, in case of absence of end stiffeners, warping constant I_W can be evaluated as:

$$I_W = \frac{I_Z \cdot \left(h - t_f\right)^2}{4} \tag{7.44}$$

where h is the height of the profile and t_f is the thickness of the flanges.

If the load is applied directly on the shear centre ($z_g = 0$), from Eq. (7.38) the expression of the critical moment M_{cr} becomes:

$$M_{cr} = C_1 \frac{\pi^2 EI_z}{(k_z L)^2} \cdot \left[\sqrt{\left(\frac{k_z}{k_w}\right)^2 \frac{I_w}{I_z} + \frac{(k_z L)^2 GI_t}{\pi^2 EI_z}} \right] \tag{7.45}$$

In case of fixed ends, that is restraints to lateral translation as well as to rotation of the compressed flange ($k_z = k_w = 1$), Eq. (7.45) becomes:

$$M_{cr} = C_1 \frac{\pi^2 EI_z}{L^2} \cdot \left[\sqrt{\frac{I_w}{I_z} + \frac{L^2 GI_t}{\pi^2 EI_z}} \right] \tag{7.46}$$

Simplified methods for beams in buildings with restraints are available with great advantages to simplifying design. In case of members with discrete lateral restraint to the compression flange, lateral torsional buckling (LTB) can be neglected if the length L_c between restraints or the resulting relative slenderness $\bar{\lambda}_f$ of the equivalent compression flange satisfies:

$$\bar{\lambda}_f = \frac{k_c \cdot L_c}{i_{f,z} \cdot \lambda_1} \le \bar{\lambda}_{c0} \cdot \frac{M_{c,Rd}}{M_{y,Ed}} \tag{7.47}$$

where k_c is a slenderness correction factor for moment distribution between restraints (see Table 7.5), λ_1 is the material slenderness (Eq. (6.5)), $M_{c,Rd}$ is the moment resistance based on an appropriate section modulus corresponding to the compression flange and $M_{y,Ed}$ is the maximum design value of the bending moment within the restraint spacing.

Furthermore, term $i_{f,z}$ is the radius of gyration of the equivalent (effective) compression part of the cross-section composed of the compression flange plus 1/3 of the compressed part of the web area, about the minor axis of the cross-section:

$$i_{f,z} = \sqrt{\frac{I_{\text{eff},f}}{A_{\text{eff},f} + \frac{1}{3} A_{\text{eff},w,c}}} \tag{7.48a}$$

where $I_{\text{eff},f}$ is the effective moment of inertia of the compression flange about the minor axis of the section, $A_{\text{eff},f}$ is the effective area of the compression flange and $A_{\text{eff},w,c}$ is the effective area of the compressed part of the web.

The slenderness limit $\bar{\lambda}_{c0}$, which is related to the equivalent flange under compression, should, however, be given in the National Annex. In Eurocode 3 it is proposed:

$$\bar{\lambda}_{c,0} = \bar{\lambda}_{LT,0} + 0.1 \tag{7.48b}$$

where term $\bar{\lambda}_{LT,0}$ has been already introduced in Eq. (7.35).

If the slenderness of the compression flange $\bar{\lambda}_f$ exceeds the limit given in Eq. (7.47) the design buckling resistance moment may be evaluated with reference to an equivalent compression flange (isolated flange method). In particular, the buckling moment resistance, $M_{b,Rd}$, can be evaluated as:

$$M_{b,Rd} = k_{Fl} \cdot \chi \cdot M_{c,Rd} \tag{7.49}$$

where k_{Fl} is the modification factor accounting for the conservatism of the equivalent compression flange, which may be given in the National Annex, or alternatively can be assumed as 1.1 ($k_{Fl} = 1.1$).

Term χ is the reduction factor of the equivalent compression flange determined on the basis of relative slenderness $\bar{\lambda}_f$ evaluated with reference to the flange buckling, which could occur only in the flange plane, owing to the presence of the web restraining efficiently buckling along the weak axis of the flange.

Buckling curve c has to be adopted except for welded sections for which curve b has to be considered provided that:

$$\frac{h}{t_f} \leq 44\sqrt{\frac{235}{f_y[\mathrm{N/mm^2}]}} \tag{7.50}$$

where h is the overall depth of the cross-section and t_f is the thickness of the compression flange.

7.3 Design According to the US Approach

The main verification rules for beam elements are proposed in the following, which are mainly in accordance with the requirements reported in AISC 360-10 and in ASCE 7-10 Appendix C.

7.3.1 Serviceability Limit States

7.3.1.1 Deformability

Both AISC 360-10 chapter L and ASCE 7-10 Appendix C do not provide any detailed information about the allowable limits for beam deflections, due to the fact that such limits depend on the non-structural elements sustained by steel structures. Both codes report the historical (traditional) deflection limits used in US practice for designing steel beams, which are listed in Table 7.8.

7.3.1.2 Vibrations

As for deflection limits, both AISC 360-10 and ASCE 7-10 do not provide precise prescriptions. ASCE 7-10, in the commentary to Appendix C, states that:

> Many common human activities impact dynamic forces to a floor at frequencies (or harmonics) in the range of 2 to 6 Hz [...] As a general rule, the natural frequency of structural elements should be greater than 2,0 times the frequency of any steady-state excitation to which they are exposed unless vibration isolation is provided.

So, according to this point, at least a limit frequency of 3 Hz should be maintained for floors subjected to normal human activity (homes, offices, floors mainly subjected to walking), while for buildings hosting activities rhythmic in nature (dancing, aerobic exercise, etc.) higher limits should be adopted in design.

AISC 360-10 suggests, for a more detailed approach, to refer to the AISC publication Design Guide 11 (*Floor Vibration Due to Human Activity* by Murray *et al.*, 1997).

Table 7.8 Historical (traditional) limits for beam vertical deflections in accordance with AISC 360-10 and ASCE 7-10.

Conditions	Limits
Roof beams subjected to full nominal live load	$\frac{L}{240}$
Floor beams subjected to full nominal live load	$\frac{L}{360}$

7.3.2 Shear Strength Verification

Shear strength verification is addressed in chapter G of AISC 360-10 code.

LRFD approach	**ASD approach**
Design according to the provisions for *load and resistance factor design* (LRFD) satisfies the requirements of AISC Specification when the *design shear strength* $\Phi_v V_n$ equals or exceeds the *required shear strength* V_u, that is the maximum shear along the beam, determined on the basis of the LRFD load combinations. Design has to be performed in accordance with the following equation:	Design according to the provisions for *allowable strength design* (ASD) satisfies the requirements of AISC Specification when the *allowable shear strength* V_n equals or exceeds the *required shear strength* V_a, that is the maximum shear along the beam, determined on the basis of the ASD load combinations. Design has to be performed in accordance with the following equation:
$$V_u \le \phi_v V_n \quad (7.51)$$	$$V_a \le V_n/\Omega_v \quad (7.52)$$
where ϕ_v is the *shear resistance factor* (= 0.90 except in cases indicated in Table 7.9) and V_n represents the *nominal shear strength.*	where Ω_v is the *shear safety factor* (= 1.67 except in cases indicated in Table 7.9) and V_n represents the *nominal shear strength.*

Two methods are permitted for computing *nominal shear strength* V_n:

(1) the method based on the minimum shear strength between limit states of *shear yielding* and *shear buckling*;
(2) the method that also considers *post buckling strength* of web panels.

According to AISC code, there is no effect of shear stresses on flexural strength of a cross-section, so verifications for shear and bending are not connected or mutually influenced.

7.3.2.1 Method (1)

AISC 360-10 Section G2.1. This method applies to webs of singly or doubly symmetrical members and channels subjected to shear in the plane of a stiffened or unstiffened web. Conservatively, it can be used also in lieu of method 2, if the designer does not want to take advantage of the post-buckling increment of shear strength.

The *nominal shear strength* V_n is:

$$V_n = 0.6 F_y A_w C_v \tag{7.53}$$

where $A_w = d \cdot t_w$ is the area of web, the overall section depth (d) times the web thickness (t_w) and C_v is a coefficient that takes into account the *shear buckling*, which assumes the values indicated in Table 7.9.

Stiffeners are not required if:

(a) $h/t_w \le 2.46\sqrt{E/F_y}$
(b) the available shear strength computed with $k_v = 5$ is greater than the required shear strength (Table 7.10).

In Table 7.11 the values for ϕ_v, Ω_v and C_v for ASTM A6 W, M, S and HP profiles are listed for F_y = 50 ksi (345 MPa). It can be noted that many hot rolled cross-sections belong to typology (a) as

Table 7.9 C_v values for method 1.

Typology	Condition(s)	C_v
(a) Webs of hot-rolled I-shaped sections	$h/t_w \le 2.24\sqrt{E/F_y}$	1.0^a
(b) Webs of all other doubly-symmetrical and singly-symmetrical shapes (typically built-up welded I-shaped sections) and channels, except round HSS	$h/t_w \le 1.10\sqrt{k_v E/F_y}$	1.0
	$1.10\sqrt{k_v E/F_y} < h/t_w$ $h/t_w \le 1.37\sqrt{k_v E/F_y}$	$\frac{1.10\sqrt{k_v E/F_y}}{h/t_w}$
	$h/t_w > 1.37\sqrt{E/F_y}$	$\frac{1.51 \cdot k_v E}{(h/t_w)^2 F_y}$

[a] For this case, assume $\phi_v = 1.00$; $\Omega_v = 1.50$.

h for rolled shapes, the distance between flanges, minus corner radii; for built-up welded sections, the clear distance between flanges; for tee-shapes, the overall depth;

t_w is the web thickness;

k_v is a web plate buckling coefficient, to be determined as indicated in Table 7.10

Table 7.10 Evaluation of k_v.

Section type	Condition	k_v
Web without transverse stiffeners	$h/t_w < 260^a$	5
Web with transverse stiffeners	$a/h > 3.0$ or $a/h > \left[\frac{260}{h/t_w}\right]^2$	5
	$a/h \le 3{,}0$ or $a/h \le \left[\frac{260}{h/t_w}\right]^2$	$5 + \frac{5}{(a/h)^2}$
Tee-shaped sections	—	1.2

[a] All ASTM A6 W, S and HP shapes respect this condition.

a = clear distance between transverse stiffeners.

defined in Table 7.9 and just few of them belong to typology (b). C_v factor is in general equal to 1.00; values less than 1.00 are related to M profiles only.

7.3.2.2 Method (2)

Method 2 in AISC 360-10 Section G3.2 applies to I-shaped built-up members, with properly spaced thin webs and transverse stiffeners. The method takes into account the extra strength developed by web panels, bounded on top and bottom by flanges and on each side by stiffeners, after buckling. When buckling occurs, significant diagonal *tension field actions* form in the web panels while stiffeners act as vertical compressed members. The whole beam behaves as a truss girder: flanges are the chords, web panels the diagonal in tension and stiffeners the vertical element in compression. In order to account for the tension field actions, the following conditions must be respected:

(a) $a/h \le 3$ and $a/h \le [260/(h/t_w)]^2$

Table 7.11 ϕ_v, Ω_v and C_v for ASTM A6 hot rolled sections with $F_y = 50$ ksi.

Shape	Types	ϕ_v	Ω_v	C_v	Shape	Types	ϕ_v	Ω_v	C_v
W44 × 335-262	(a)	1.00	1.50	1.00	M12.5 × 12.4	(b)	0.90	1.67	0.81
W44 × 230	(b)	0.90	1.67	1.00	M12.5 × 11.6				0.81
W40 × 593-167	(a)	1.00	1.50	1.00	M12 × 11.8				0.96
W40 × 149	(b)	0.90	1.67	1.00	M12 × 10.8				0.87
W36 × 800-150	(a)	1.00	1.50	1.00	M12 × 10				0.80
W36 × 135	(b)	0.90	1.67	1.00	M10 × 9				1.00
W33 × 387-130	(a)	1.00	1.50	1.00	M10 × 8				0.94
W33 × 118	(b)	0.90	1.67	1.00	M10 × 7.5				0.84
W30 × 391-99	(a)	1.00	1.50	1.00	M8 × 6.5				1.00
W30 × 90	(b)	0.90	1.67	1.00	M8 × 6.2				1.00
W27 × 539-84	(a)	1.00	1.50	1.00	M6 × 4.4				1.00
W24 × 370-62					M6 × 3.7				1.00
W24 × 55	(b)	0.90	1.67	1.00	M5 × 17.9				1.00
W21 × 201-44	(a)	1.00	1.50	1.00	M4 × 6				1.00
W18 × 311-35					M4 × 4.08				1.00
W16 × 100-31					M4 × 3.45				1.00
W16 × 26	(b)	0.90	1.67	1.00	M4 × 3.2				1.00
W14 × 730-22	(a)	1.00	1.50	1.00	M3 × 2.9				1.00
W12 × 336-16					S (all)	(a)	1.00	1.50	1.00
W12 × 14	(b)	0.90	1.67	1.00	HP (all)	(a)	1.00	1.50	1.00
W10 × 112-12	(a)	1.00	1.50	1.00					
W8 × 67-10									
W6 × 25-7.5									
W5 × 19-16									
W4 × 13									

(b) $2A_w/\left(A_{fc}+A_{ft}\right) \le 2.5$
(c) $h/b_{fc} \le 6.0$ and $h/b_{ft} \le 6.0$

where A_{fc} is the area of compression flange, A_{ft} is the area of tension flange, b_{fc} is the width of compression flange and b_{ft} is the width of tension flange.

In addition, tension field action cannot be taken into account for end panels without stiffeners on their end side. Furthermore, there is also the following requirement on the moment of inertia of stiffeners (I_{st}):

$$I_{st} \ge bt_w^3 j \text{ with } j = \frac{2.5}{(a/h)^2} - 2 \ge 0.5 \tag{7.54}$$

where I_{st} is computed about an axis in the web centre for stiffener pairs, or about the face in contact with the web plate for single stiffeners and $b = \min(a; h)$.

This requirement for stiffeners is also valid for dimensioning stiffeners, if used when members are verified in accordance with the method 1.

If all these conditions are satisfied, the nominal shear strength V_n accounting for tension field action has to be taken in accordance with the symbols already defined:

$$V_n = 0.6F_yA_w \text{ if } h/t_w \le 1.10\sqrt{kvE/F_y} \tag{7.55}$$

$$V_n = 0.6F_yA_w\left[C_v + \frac{1-C_v}{1.15\sqrt{1+\left(\frac{a}{h}\right)^2}}\right] \text{ if } h/t_w > 1.10\sqrt{kvE/F_y} \tag{7.56}$$

With method (2), stiffeners cannot be avoided, which have to meet in addition to the requirement provided in Eq. (7.54) as well as the following:

(a) $(b/t)_{st} \le 0.56\sqrt{E/F_{yst}}$

(b) $I_{st} \ge I_{st1} + (I_{st2} - I_{st1})\left[\frac{V_r - V_{c1}}{V_{c2} - V_{c1}}\right]$

where $(b/t)_{st}$ is the width-to-thickness ratio of the stiffener and F_{yst} is the minimum yield stress of the stiffener material.

Other terms introduced in the previous equations are defined as:

$I_{st1} = bt_w^2 j$ (see Eq. (7.54));

$$I_{st2} = \frac{h^4 \rho_{st}^{1.3}}{40}\left(\frac{F_{yw}}{E}\right)^{1.5};$$

$$\rho_{st} = \max\left(\frac{F_{yw}}{F_{yst}};1.0\right);$$

where F_{yw} is the minimum yield stress of the web material, V_r is the larger of the required shear strength in the adjacent web panels, computed using LRFD or ASD load combinations, V_{c1} is the smaller of the available shear strength in the adjacent web panels, with V_n computed with method (1), V_{c2} is the smaller of the available shear strength in the adjacent web panels, with V_n computed with method (2).

Special provisions are given, in AISC specifications, for *rectangular hollow square section (HSS)* and *box-shaped members*. Verifications have to be performed according to method (1), with:

$$A_w = 2 \cdot h \cdot t$$

where h is the width resisting to the shear force, computed as the clear distance between flanges less the inside corners radius on each side, t is the design wall thickness (taken 0.93 time the nominal thickness for electric-resistance-welded (ERW) HSS, and equal to the nominal thickness for submerged-arc-welded (SAW) HSS), $t_w = t$ and $k_c = 5$.

Furthermore, for round HSS, the nominal shear strength is:

$$V_n = 0.5F_{cr}A_g \tag{7.57}$$

with:

$$F_{cr} = \max\left\{\frac{1.60E}{\sqrt{\frac{L_v}{D}} \cdot \left(\frac{D}{t}\right)^{\frac{5}{4}}};\frac{0.78E}{\left(\frac{D}{t}\right)^{\frac{3}{2}}}\right\} \le 0.6F_y \tag{7.58}$$

where A_g is the gross cross-section area of section, D is the outside diameter, L_v is the distance from maximum to zero shear force and t is the design wall thickness (taken equal to 0.93 time the nominal thickness for ERW HSS and equal to the nominal thickness for SAW HSS).

It should be noted that in Eq. (7.58), the value $0.6F_y$ represents the yielding limit state, while both the terms defining F_{cr} represent shear buckling limit state. For all standard sections, yielding limit state usually controls shear strength.

7.3.3 Flexural Strength Verification

Flexural strength verification is addressed in chapter F of AISC 360-10 specifications. The Code gives rules for determining the *nominal flexural strength* of the cross-section, M_n, as the minimum value among values computed for each applicable limit state.

AISC identifies the following limit states to be considered for beams in bending:

- Global section yielding, for doubly symmetrical sections (Y);
- Local buckling (LB);
- Compression flange yielding, for simply symmetrical sections (CFY);
- Tension flange yielding, for simply symmetrical sections (TFY);
- Lateral torsional buckling (LTB);
- Flange local buckling (FLB);
- Web local buckling (WLB)
- Leg local buckling, for single angles (LLB);
- Web leg local buckling, for double angles (DALB);
- Tee stem local buckling in flexural compression (TSLB).

The listed limit states take into account the three main collapse mechanisms:

- yielding of the cross-section;
- lateral-torsional buckling of the beam;
- local buckling of webs and/or flanges.

LRFD approach

Design according to the provisions for *load and resistance factor design* (LRFD) satisfies the requirements of AISC Specification when the *design flexural strength* $\phi_b M_n$ of each structural component equals or exceeds the *required flexural strength* M_u; that is, the maximum bending moment along the beam, determined on the basis of the LRFD load combinations. Design has to be performed in accordance with the following equation:

$$M_u \leq \phi_b M_n \tag{7.59}$$

where ϕ_b is the *flexural resistance factor* ($\phi_b = 0.90$) and M_n represents the *nominal flexural strength*

ASD approach

Design according to the provisions for *allowable strength design* (ASD) satisfies the requirements of AISC Specification when the *allowable flexural strength* M_n/Ω_b of each structural component equals or exceeds the *required flexural strength* M_a; that is, the maximum bending moment along the beam, determined on the basis of the ASD load combinations. Design has to be performed in accordance with the following equation:

$$M_a \leq M_n/\Omega_b \tag{7.60}$$

where Ω_b is the *flexural safety factor* ($\Omega_b = 1.67$) and M_n represents the *nominal flexural strength*

AISC specifications provide the expressions for the nominal flexural strength M_n in all the following cases (Table 7.12):

(a) *Doubly symmetrical compact I-shaped members and channels bent about their major axis;*
(b) *Doubly symmetrical compact I-shaped members with compact webs and non-compact or slender flanges bent about their major axis;*

(c) *Other I-shaped members with compact or non-compact webs, compact, non-compact or slender flanges, bent about their major axis;*
(d) *Doubly symmetrical and singly symmetrical I-shaped members with slender webs, compact, non-compact or slender flanges, bent about their major axis;*
(e) *I-shaped members and channels bent about their minor axis;*
(f) *Square and rectangular HSS and box-shaped members;*
(g) *Round HSS;*
(h) *Tees loaded in the plane of symmetry;*
(i) *Double angles loaded in the plane of symmetry;*
(j) *Single angles;*
(k) *Rectangular bar and rounds;*
(l) *Unsymmetrical shapes.*

Table 7.12 Appropriate limit states for flexural strength verification (from Table F1.1 of AISC 360-10).

Case	Section type and classification	AISC 360-10 Chapter F applicable section	Applicable limits states
(a)	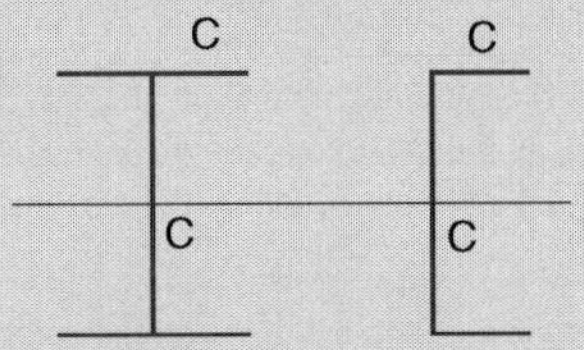 C C C C	F2	Y LTB
(b)	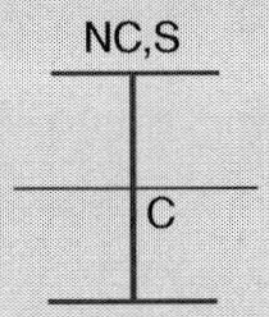 NC,S C	F3	LTB FLB
(c)	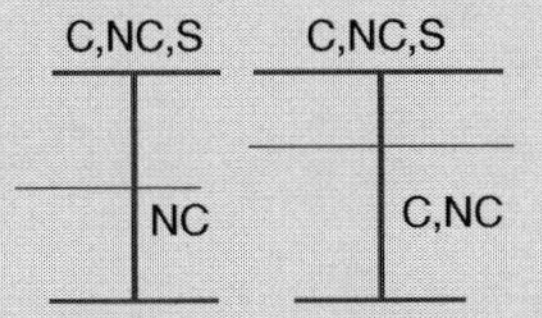 C,NC,S C,NC,S NC C,NC	F4	CFY LTB FLB TFY
(d)	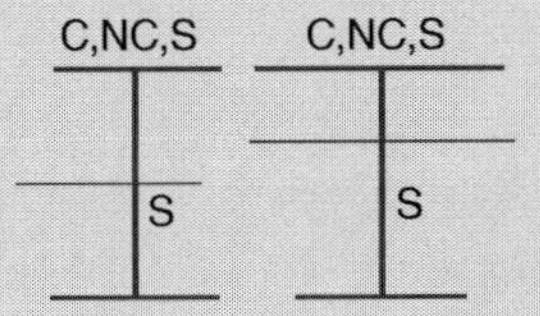 C,NC,S C,NC,S S S	F5	CFY LTB FLB FTY
(e)	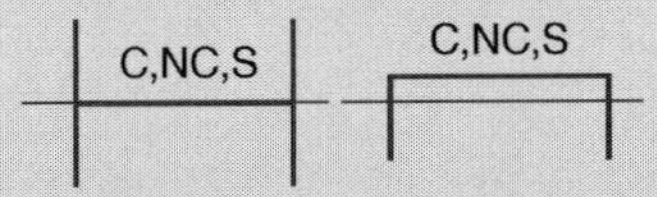 C,NC,S C,NC,S	F6	Y FLB
(f)	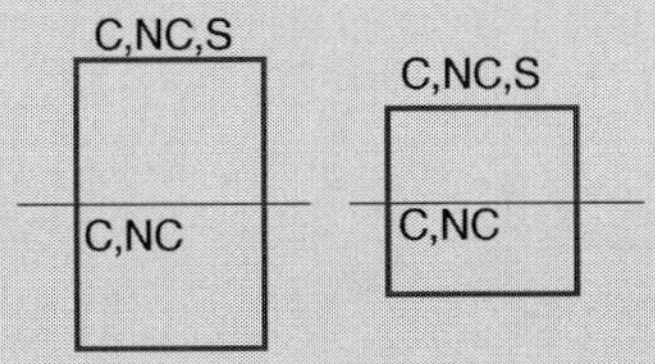 C,NC,S C,NC,S C,NC C,NC	F7	Y FLB WLB

(Continued)

Table 7.12 (Continued)

Case	Section type and classification	AISC 360-10 Chapter F applicable section	Applicable limits states
(g)		F8	Y LB
(h)	C,NC,S	F9	Y LTB FLB TSLB
(i)	C,NC,S	F9	Y LTB FLB DALB
(j)		F10	Y LTB LLB
(k)		F11	Y LTB
(l)	Unsymmetrical shapes, other than single angles	F12	All limit states

7.3.3.1 (a) Doubly Symmetrical Compact I-Shaped Members and Channels Bent about Their Major Axis

This case is applicable to almost all ASTM A6 W, S, M, C and MC shapes. The relevant limit states for compact I-shaped members and channels are:

- yielding of the whole section;
- lateral torsional buckling.

The nominal flexural strength M_n is the lowest value obtained according to the previously listed limit states. L_b is defined as the beam length between points that are either braced against lateral displacement of the compression flange or braced against twist of the cross-section and L_p is defined as:

$$L_p = 1.76 r_y \sqrt{\frac{E}{F_y}} \tag{7.61}$$

If $L_b \leq L_p$ then lateral-torsional buckling does not occur and the nominal strength M_n is determined by the yielding limit state (plastic moment):

$$M_n = M_p = F_y Z_x \tag{7.62}$$

If $L_b > L_p$, then lateral-torsional buckling governs the design verification. In this case beam should collapse for:

(a) inelastic lateral-torsional buckling, if $L_p < L_b \leq L_r$. The nominal flexural strength M_n is:

$$M_n = C_b \left[M_p - (M_p - 0.7F_y S_x) \left(\frac{L_b - L_p}{L_r - L_p} \right) \right] \leq M_p \tag{7.63}$$

(b) elastic lateral-torsional buckling, if $L_b > L_r$. The nominal flexural strength M_n is:

$$M_n = F_{cr} S_x \leq M_p \tag{7.64}$$

where:

$$F_{cr} = \frac{C_b \pi^2 E}{\left(\frac{L_b}{r_{ts}} \right)^2} \sqrt{1 + 0.078 \frac{J \cdot c}{S_x h_0} \left(\frac{L_b}{r_{ts}} \right)^2} \tag{7.65a}$$

The expression (7.65a) has to be applied for load on the centroid of the section. If the loads are applied on the top flange, the following alternative equation may be conservatively applied:

$$F_{cr} = \frac{C_b \pi^2 E}{\left(\frac{L_b}{r_{ts}} \right)^2} \tag{7.65b}$$

The length L_r is defined as:

$$L_r = 1.95 r_{ts} \frac{E}{0.7F_y} \sqrt{\frac{J \cdot c}{S_x h_0} + \sqrt{\left(\frac{J \cdot c}{S_x h_0} \right)^2 + 6.76 \left(\frac{0.7F_y}{E} \right)^2}} \tag{7.66}$$

where h_0 is the distance between the flange centroids, E is the modulus of elasticity, J is the torsional constant, S_x is the elastic section modulus taken about the x-axis, c is equal to the unity for doubly-symmetrical shapes or $c = \frac{h_0}{2} \sqrt{\frac{I_y}{C_w}}$ for channels and r_{ts} is defined as:

$$r_{ts} = \sqrt{\frac{\sqrt{I_y C_w}}{S_x}} \tag{7.67}$$

In Table 7.13 the decision-making process for computing M_n is summarized and in Figure 7.9 the typical M_n–L_p curve is plotted.

The C_b coefficient in Eqs. (7.63) and (7.65a) is a *modification factor that takes into account the non-uniform bending moment* diagrams in the beam. If the beam is subjected to a constant bending moment along the length L_b, then:

$$C_b = 1$$

The same value applies to cantilevers. In all the other cases:

$$C_b = \frac{12.5M_{\max}}{2.5M_{\max} + 3M_A + 4M_B + 3M_C} \tag{7.68}$$

where $M_{\max}$ is the absolute value of maximum moment in the unbraced length L_b, and M_A, M_B and M_C are the absolute values of moment at the quarter point of L_b, at the centre line of L_b and at the three-quarter point of L_b.

Table 7.13 Compute of M_n for case (a).

Compute M_p, L_p, L_r, C_b					
Is $L_b \le L_p$?	Yes	$M_n = M_p$			
	No	Is $L_b \le L_r$?	Yes	Compute $M_{n,LTB} = M_n$ of Eq. (7.63) M_n = minimum of $M_{n,LTB}$ and M_p	
			No	Loads applied at the centroid	Compute $M_{n,LTB} = M_n$ of Eqs. (7.64) and (7.65a) M_n = minimum of $M_{n,LTB}$ and M_p
				Loads applied at the top flange	Compute $M_{n,LTB} = M_n$ of Eqs. (7.64) and (7.65b) M_n = minimum of $M_{n,LTB}$ and M_p

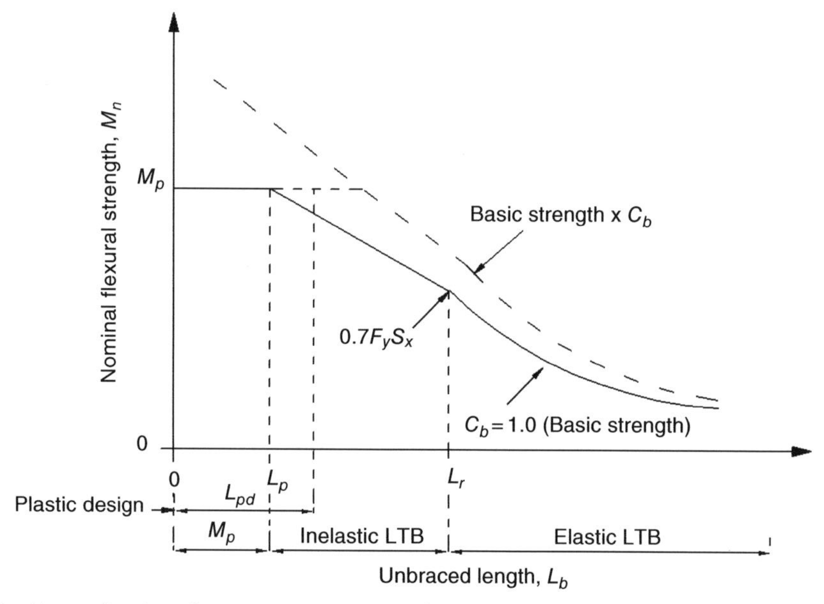

Figure 7.9 M_n as a function of L_b. From Figure C-F1.2 of AISC 360-10.

Table 7.14 Values of C_b in most common cases (From Table 3-1 of the *AISC* Manual).

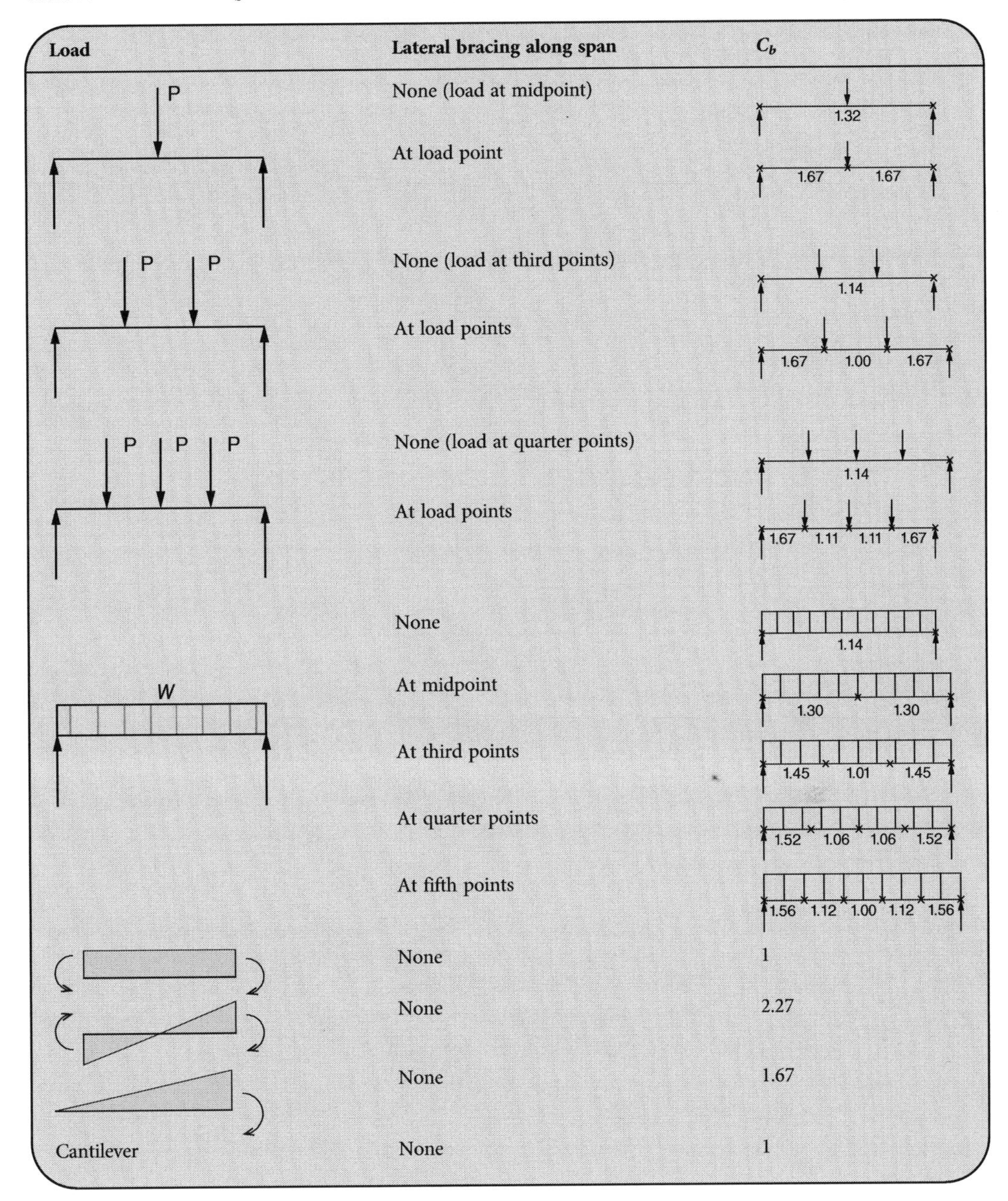

Load	Lateral bracing along span	C_b
P	None (load at midpoint)	1.32
	At load point	1.67 1.67
P P	None (load at third points)	1.14
	At load points	1.67 1.00 1.67
P P P	None (load at quarter points)	1.14
	At load points	1.67 1.11 1.11 1.67
W	None	1.14
	At midpoint	1.30 1.30
	At third points	1.45 1.01 1.45
	At quarter points	1.52 1.06 1.06 1.52
	At fifth points	1.56 1.12 1.00 1.12 1.56
	None	1
	None	2.27
	None	1.67
Cantilever	None	1

The most common values of isolated beams are reported in Table 7.14, which proposes the associated C_b coefficients.

For doubly symmetrical members with no transverse loading between brace points and moments at the ends equal to M_1 and M_2, C_b can be also computed with good approximation from the following equation, already contained in former editions of AISC specifications:

$$C_b = 1.75 + 1.05\left(\frac{M_1}{M_2}\right) + 0.3\left(\frac{M_1}{M_2}\right)^2 \tag{7.69}$$

7.3.3.2 (b) Doubly Symmetrical I-Shaped Members with Compact Webs and Non-Compact or Slender Flanges Bent about Their Major Axis

Just a few shapes have non-compact flanges for $F_y = 50$ ksi (345 MPa). They are: W21 × 48, W14 × 99, W14 × 90, W12 × 65, W10 × 12, W8 × 31, W8 × 10, W6 × 15, W6 × 9, W6 × 7.5 and M4 × 6. All other ASTM A6 W, S and M shapes with $F_y \leq 50$ ksi (345 MPa) have compact flanges, so they belong to case (a).

The relevant limit states for members belonging to this case are:

- lateral torsional buckling;
- compression flange local buckling.

The nominal flexural strength M_n is the lowest value obtained according to the previously listed limit states.

For the *lateral buckling limit state* M_n is computed as in case (a).

For the *compression FLB*, AISC specifications provide the following expressions for M_n:

(1) for cross-sections with non-compact flanges:

$$M_n = M_p - \left(M_p - 0.7F_yS_x\right)\left(\frac{\lambda - \lambda_{pf}}{\lambda_{rf} - \lambda_{pf}}\right) \tag{7.70}$$

(2) for cross-sections with slender flanges:

$$M_n = \frac{0.9Ek_cS_x}{\lambda^2} \tag{7.71}$$

where $\lambda = b_f/2t_f$, $\lambda_{pf} = 0.38\sqrt{E/F_y}$ is the limiting width-to-thickness ratio for a compact flange (see Section 4.3), $\lambda_{rf} = 1.0\sqrt{E/F_y}$ is the limiting width-to-thickness ratio for a non-compact flange (see Section 4.3), $k_c = 4/\sqrt{h/t_w}$ (with $0.35 \leq k_c \leq 0.76$) and the distance h defined in Section 4.3

In Table 7.15 the decision-making process for computing M_n is summarized and in Figure 7.10 the qualitative curve M_n–λ is outlined.

7.3.3.3 (c) Other I-Shaped Members with Compact or Non-Compact Webs, Compact, Non-Compact or Slender Flanges, Bent about Their Major Axis

This case applies to doubly symmetrical I-shaped members and singly symmetrical I-shaped members with a web attached to the mid-width of the flanges, bent about their major axis, with webs that are not slender. It applies mainly to welded I-shaped beams.

The relevant limit states for members belonging to this case are:

- compression flange yielding;
- compression flange local buckling (FLB);
- lateral torsional buckling;
- tension flange yielding.

The nominal flexural strength M_n is the lowest value obtained according to the previously listed limit states. For *compression flange yielding*, AISC specifications provide the following expressions for M_n:

$$M_n = R_{pc}M_{yc} \tag{7.72}$$

Table 7.15 Computation of M_n for case (b).

Compute M_p; classify flange (non-compact or slender)					
How is the flange classified?	Non-compact	Compute: $M_{n,FLB} = M_n$ of Eq. (7.70)			
	Slender	Compute: $M_{n,FLB} = M_n$ of Eq. (7.71)			
Compute L_p, L_r, C_b					
Is $L_b \le L_p$?	Yes	$M_n = M_{n,FLB}$			
	No	Is $L_b \le L_r$?	Yes	Compute $M_{n,LTB} = M_n$ of Eq. (7.63) M_n = minimum of $M_{n,LTB}$ and $M_{n,FLB}$	
			No	Loads applied at the centroid	Compute $M_{n,LTB} = M_n$ of Eqs. (7.64) and (7.65a) M_n = minimum of $M_{n,LTB}$ and $M_{n,FLB}$
				Loads applied at the top flange	Compute $M_{n,LTB} = M_n$ of Eqs. (7.64) and (7.65b) M_n = minimum of $M_{n,LTB}$ and $M_{n,FLB}$

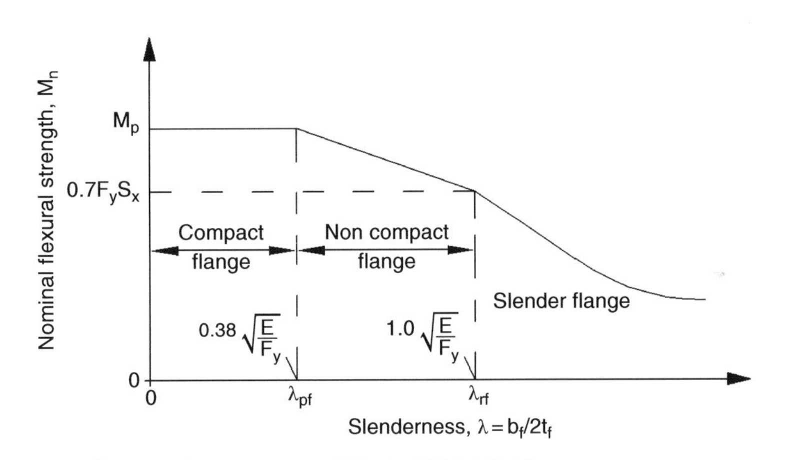

Figure 7.10 M_n as a function of λ. From Figure C-F1.1 of AISC 360-10.

where $M_{yc} = F_y S_{xc}$ is the value of the bending moment causing yield in the compression flange, S_{xc} is the elastic section modulus referred to compression flange and R_{pc} is the *web plastification factor* that takes into account the effect of inelastic bucking of the web (it varies from 1.0 to 1.6 and, conservatively, it can be assumed to equal 1.0).

A more accurate value of R_{pc} can be defined on the basis of the following conditions:

if $I_{yc}/I_y \le 0.23$:

$$R_{pc} = 1 \tag{7.73}$$

if $I_{yc}/I_y > 0.23$ and $\frac{h_c}{t_w} \le \lambda_{pw}$:

$$R_{pc} = \frac{M_p}{M_{yc}} \tag{7.74}$$

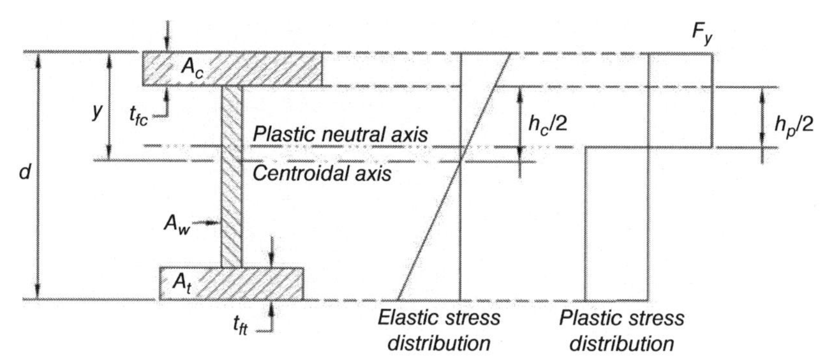

Figure 7.11 Elastic and plastic stress distribution for case (c). From Figure C-F411 of AISC 360-10.

if $I_{yc}/I_y > 0.23$ and $\dfrac{h_c}{t_w} > \lambda_{pw}$:

$$R_{pc} = \left[\frac{M_p}{M_{yc}} - \left(\frac{M_p}{M_{yc}} - 1\right)\left(\frac{\lambda - \lambda_{pw}}{\lambda_{rw} - \lambda_{pw}}\right)\right] \le \frac{M_p}{M_{yc}} \tag{7.75}$$

where $M_p = F_y Z_x \le 1.6 F_y S_{xc}$, $\lambda = h_c/t_w$, term $\lambda_{pw} = \dfrac{\dfrac{h_c}{h_p}\sqrt{\dfrac{E}{F_y}}}{\left[\dfrac{0.54 M_p}{M_y} - 0.09\right]^2} \le 5.70\sqrt{\dfrac{E}{F_y}}$ is the limiting slenderness ratio for a compact web (see Section 4.3) and term $\lambda_{rw} = 5.70\sqrt{\dfrac{E}{F_y}}$ is the limiting slenderness ratio for a non-compact web (see Section 4.3).

With reference to Figure 7.11, this results in:

$$h_c = 2(y - t_{fc});\ h_p = \frac{A - 2A_c}{t_w};\ S_{xc} = \frac{I_x}{y};\ S_{xt} = \frac{I_x}{d - y}$$

For *compression FLB*, AISC specifications provide the following expressions for M_n that are slightly different from those obtained via Eqs. (7.70) and (7.71) of case (b):

(1) For sections with non-compact flanges:

$$M_n = R_{pc} M_{yc} - (R_{pc} M_{yc} - F_L S_{xc})\left(\frac{\lambda - \lambda_{pf}}{\lambda_{rf} - \lambda_{pf}}\right) \tag{7.76}$$

(2) For sections with slender flanges:

$$M_n = \frac{0.9 E k_c S_{xc}}{\lambda^2} \tag{7.77}$$

where $\lambda = b_{fc}/2t_{fc}$, $\lambda_{pf} = 0.38\sqrt{E/F_y}$ is the limiting width-to-thickness ratio for a compact flange (see Section 4.3), $\lambda_{rf} = 1.0\sqrt{E/F_y}$ is the limiting width-to-thickness ratio for a non-compact flange (see Section 4.3), $k_c = 4/\sqrt{h/t_w}$ (with $0.35 \le k_c \le 0.76$) and the distance h defined in Section 4.3).

The stress F_L is determined as follows:

(1) if $S_{xt}/S_{xc} \geq 0.7$:

$$F_L = 0.7F_y \tag{7.78}$$

(2) if $S_{xt}/S_{xc} < 0.7$:

$$F_L = \frac{S_{xt}}{S_{xc}} F_y \geq 0.5F_y \tag{7.79}$$

For cross-sections with compact flange this limit state does not apply, obviously.
For *lateral-torsional buckling*, AISC provisions are very close to those in case (a).

(1) If $L_b \leq L_p$, then lateral-torsional buckling does not occur, L_p being different than in case (a), defined as:

$$L_p = 1.1 r_t \sqrt{\frac{E}{F_y}} \tag{7.80}$$

where r_t is the effective radius of gyration for lateral torsional buckling, defined as:

$$r_t = \frac{b_{fc}}{\sqrt{12\left(\frac{h_0}{d} + \frac{1}{6} a_w \frac{h^2}{h_0 d}\right)}} \tag{7.81a}$$

with:

$$a_w = \frac{h_c t_w}{b_{fc} t_{fc}} \tag{7.81b}$$

where b_{fc} is the width of compression flange and t_{fc} is the thickness of compression flange.

(2) if $L_p < L_b \leq L_r$:

$$M_n = C_b \left[R_{pc} M_{yc} - \left(R_{pc} M_{yc} - F_L S_{xc} \right) \left(\frac{L_b - L_p}{L_r - L_p} \right) \right] \leq R_{pc} M_{yc} \tag{7.82}$$

(3) if $L_b > L_r$:

$$M_n = F_{cr} S_{xc} \leq R_{pc} M_{yc} \tag{7.83}$$

where:

$$M_{yc} = F_y S_{xc} \tag{7.84}$$

$$F_{cr} = \frac{C_b \pi^2 E}{\left(\frac{L_b}{r_t}\right)^2} \sqrt{1 + 0.078 \frac{J}{S_{xc} h_0} \left(\frac{L_b}{r_t}\right)^2} \tag{7.85}$$

If $I_{yc}/I_y \leq 0.23$ then $J = 0$, with I_{yc} representing the moment of inertia of compression flange about the y-axis. The length L_r is defined as:

$$L_r = 1.95 r_t \frac{E}{F_L} \sqrt{\frac{J}{S_{xc} h_0} + \sqrt{\left(\frac{J}{S_{xc} h_0}\right)^2 + 6.76 \left(\frac{F_L}{E}\right)^2}} \tag{7.86}$$

For *tension flange yielding*:

(1) if $S_{xt} \geq S_{xc}$, the limit state of tension flange yielding does not apply.
(2) if $S_{xt} < S_{xc}$:

$$M_n = R_{pt} M_{yt} \tag{7.87}$$

where $M_{yt} = F_y S_{xt}$ is the moment that causes yield in the tension flange, S_{xt} is the elastic section modulus referred to tension flange and R_{pt} is the *web plastification factor* corresponding to tension flange yielding limit state, which is defined as:

if $\frac{h_c}{t_w} \leq \lambda_{pw}$:

$$R_{pt} = \frac{M_p}{M_{yt}} \tag{7.88}$$

if $\frac{h_c}{t_w} > \lambda_{pw}$:

$$R_{pt} = \left[\frac{M_p}{M_{yt}} - \left(\frac{M_p}{M_{yt}} - 1\right)\left(\frac{\lambda - \lambda_{pw}}{\lambda_{rw} - \lambda_{pw}}\right)\right] \leq \frac{M_p}{M_{yt}} \tag{7.89}$$

where $M_p = F_y Z_x \leq 1.6 F_y S_{xc}$, $\lambda = h_c/t_w$, $\lambda_{pw} = \dfrac{\frac{h_c}{h_p}\sqrt{\frac{E}{F_y}}}{\left[\frac{0.54 M_p}{M_y} - 0.09\right]^2} \leq 5.70\sqrt{\frac{E}{F_y}}$ is the limiting slenderness ratio for a compact web (see Section 4.3), $\lambda_{rw} = 5.70\sqrt{\frac{E}{F_y}}$ is the limiting slenderness ratio for a non-compact web (see Section 4.3).

In Table 7.16 the decision-making process for computing M_n is summarized.

Table 7.16 Computation of M_n for case (c) (from Figure C-F10.1 of AISC 360-10).

Compute M_{yc}, S_{xc}, R_{pc}				
Classify flange (compact, non-compact or slender)				
How is the flange classified?	Compact			Compute: $M_{n,CF} = M_{n,CFY} = M_n$ of Eq. (7.72)
	Non-compact			Compute: $M_{n,CF} = M_{n,FLB} = M_n$ of Eq. (7.76)
	Slender			Compute: $M_{n,CF} = M_{n,FLB} = M_n$ of Eq. (7.77)
Compute S_{xt}				
Is $S_{xt} < S_{xc}$?		Yes		Compute R_{pt} Compute: $M_{n,TF\ Y} = M_n$ of Eq. (7.87)
		No		Tension flange yielding is not applicable
Compute L_p, L_r, C_b				
Is $L_b \le L_p$?	Yes	Choose the minimum of $M_{n,CF}$ and $M_{n,TFY}$ (if applicable)		
	No	Is $L_b \le L_r$?	Yes	Compute $M_{n,LTB} = M_n$ of Eq. (7.82); Choose the minimum of $M_{n,LTB}$, $M_{n,CF}$ and $M_{n,TFY}$ (if applicable)
			No	Compute $M_{n,LTB} = M_n$ of Eq. (7.83); Choose the minimum of $M_{n,LTB}$, $M_{n,CF}$ and $M_{n,TFY}$ (if applicable)

7.3.3.4 (d) Doubly Symmetrical and Singly Symmetrical I-Shaped Members with Slender Webs, Compact, Non-Compact or Slender Flanges, Bent about Their Major Axis

This case applies to doubly symmetrical I-shaped members and singly symmetrical I-shaped members with a slender web attached to the mid-width of the flanges, bent about their major axis. It applies mainly to welded I-shaped beams. Case (d) is then similar to case (c), with the difference that the webs are slender, while in case (c) they are compact or non-compact. The relevant limit states for this case are:

- compression flange yielding;
- compression flange local buckling;
- lateral torsional buckling;
- tension flange yielding.

The nominal flexural strength M_n is the lower value obtained according to the previously listed limit states.

For *compression flange yielding*, AISC specifications provide the following expressions for M_n:

$$M_n = R_{pg} M_{yc} \tag{7.90}$$

where $M_{yc} = F_y S_{xc}$ is the moment that causes yield in the compression flange, S_{xc} is the elastic section modulus referred to compression flange and R_{pg} is the *bending strength reduction factor*, defined as:

$$R_{pg} = 1 - \frac{a_w}{1200 + 300a_w}\left(\frac{h_c}{t_w} - 5.7\sqrt{\frac{E}{F_y}}\right) \le 1.0 \tag{7.91}$$

where a_w has been defined previously, Eq. (7.81b), but in this case $a_w \le 1.0$

For *compression FLB*, AISC specifications provide the following expressions for M_n:

(1) for sections with non-compact flanges:

$$M_n = R_{pg}\left[F_y - (0.3F_y)\left(\frac{\lambda - \lambda_{pf}}{\lambda_{rf} - \lambda_{pf}}\right)\right]S_{xc} \tag{7.92}$$

(2) for sections with slender flanges:

$$M_n = R_{pg}\frac{0.9Ek_c}{\left(\frac{b_f}{2t_f}\right)^2}S_{xc} \tag{7.93}$$

where $\lambda = b_{fc}/2t_{fc}$, $\lambda_{pf} = 0.38\sqrt{E/F_y}$ is the limiting width-to-thickness ratio for a compact flange (see Section 4.3), $\lambda_{rf} = 1{,}0\sqrt{E/F_y}$ is the limiting width-to-thickness ratio for a non-compact flange (see Section 4.3), $k_c = 4/\sqrt{h/t_w}$ (where $0.35 \le k_c \le 0.76$) and h is the distance defined in Section 4.3.

For sections with compact flange this limit state does not apply, obviously.
For *lateral-torsional buckling*, AISC provisions are very close to those of case (a).

(1) if $L_b \le L_p$, then lateral-torsional buckling does not occur, with L_p defined by Eq. (7.80).
(2) if $L_p < L_b \le L_r$:

$$M_n = C_b R_{pg}\left[F_y - (0.3F_y)\left(\frac{L_b - L_p}{L_r - L_p}\right)\right]S_{xc} \le R_{pg}M_{yc} \tag{7.94}$$

(3) if $L_b > L_r$:

$$M_n = R_{pg}F_{cr}S_{xc} \le R_{pg}M_{yc} \tag{7.95}$$

where:

$$M_{yc} = F_y S_{xc} \tag{7.96}$$

$$F_{cr} = \frac{C_b \pi^2 E}{\left(\frac{L_b}{r_t}\right)^2} \tag{7.97}$$

where I_{yc} is the moment of inertia of compression flange about the y-axis.

It should be noted that the Eq. (7.97) to evaluate F_{cr} is equal to that of case (c) Eq. (7.85) with $J = 0$. This causes a discontinuity in transition between cases (c) and (d). So, in case of a welded I-shaped beam with $F_y = 50$ ksi (345 MPa) and a web slenderness $h/t_w = 137$ (non-compact web), case (c) has to be used for verification; if $h/t_w = 138$ (slender web) then it is necessary to switch verification to case (d). In any case, the differences are generally small and acceptable from an engineering point of view.

The length L_r is defined as:

$$L_r = \pi r_t \sqrt{\frac{E}{0.7F_y}} \tag{7.98}$$

where r_t is the effective radius of gyration for lateral torsional buckling, defined by Eq. (7.81a).

Table 7.17 Computation of M_n for case (d).

Compute M_{yc}, S_{xc}, R_{pg}				
Classify flange (compact, non-compact or slender)				
How is the flange classified?	Compact			Compute: $M_{n,CF} = M_{n,CFY} = M_n$ of Eq. (7.90)
	Non-compact			Compute: $M_{n,CF} = M_{n,FLB} = M_n$ of Eq. (7.92)
	Slender			Compute: $M_{n,CF} = M_{n,FLB} = M_n$ of Eq. (7.93)
Compute S_{xt}				
Is $S_{xt} < S_{xc}$?	Yes			Compute: $M_{n,TFY} = M_n$ of Eq. (7.99)
	No			Tension flange yielding is not applicable
Compute L_p, L_r, C_b				
Is $L_b \leq L_p$?	Yes			Choose the minimum of $M_{n,CF}$ and $M_{n,TFY}$ (if applicable)
	No	Is $L_b \leq L_r$?	Yes	Compute $M_{n,LTB} = M_n$ of Eq. (7.94) Choose the minimum of $M_{n,LTB}$, $M_{n,CF}$ and $M_{n,TFY}$ (if applicable)
			No	Compute $M_{n,LTB} = M_n$ of Eq. (7.95) Choose the minimum of $M_{n,LTB}$, $M_{n,CF}$ and $M_{n,TFY}$ (if applicable)

Finally, for *tension flange yielding*:

(1) if $S_{xt} \geq S_{xc}$, the limit state of tension flange yielding does not apply.
(2) if $S_{xt} < S_{xc}$:

$$M_n = M_{yt} = F_y S_{xt} \tag{7.99}$$

Equation (7.99) is identical to Eq. (7.89) computed with $\lambda \geq \lambda_{pw}$.
In Table 7.17 the decision-making process for computing M_n is summarized.

7.3.3.5 (e) I-Shaped Members and Channels Bent about Their Minor Axis

The relevant limit states for members belonging to this case are:

- yielding of the whole section;
- flange local buckling.

The nominal flexural strength M_n is the lowest value obtained according to the previously listed limit states. The majority of sections belonging to ASTM A6 S, M, C and MC shapes have compact flanges at $F_y = 50$ ksi (345 MPa). So for them the only limit state to consider is yielding.

FLB has to be considered just for W21 × 48, W14 × 99, W14 × 90, W12 × 65, W10 × 12, W8 × 31, W8 × 10, W6 × 15, W6 × 9, W6 × 8.5 and W4 × 6.

The nominal strength M_n associated with the *yielding limit state* (plastic moment) is:

$$M_n = M_p = F_y Z_y \leq 1.6 F_y S_y \tag{7.100}$$

where Z_y is the plastic modulus of the section taken about the minor axis and S_y is the elastic modulus of the section taken about the minor axis.

For *compression FLB*, AISC specifications provide the following expressions for M_n:

(1) For sections with non-compact flanges:

$$M_n = M_p - \left(M_p - 0.75F_yS_y\right)\left(\frac{\lambda - \lambda_{pf}}{\lambda_{rf} - \lambda_{pf}}\right), \tag{7.101}$$

(2) For sections with slender flanges:

$$M_n = \frac{0.69ES_y}{\lambda^2} \tag{7.102}$$

where $\lambda = b/t_f$, b is the length of the outstand part of the flange (for flanges of I-shaped members $b = b_f/2$ and for channels, b is the full nominal dimension of the flange), $\lambda_{pf} = 0.38\sqrt{E/F_y}$ is the limiting width-to-thickness ratio for a compact flange (see Section 4.3), $\lambda_{rf} = 1.0\sqrt{E/F_y}$ is the limiting width-to-thickness ratio for a non-compact flange (see Section 4.3) and for I-shaped members and S_y is the elastic modulus taken about the y-axis (for a channel, it is the minimum section modulus).

For sections with compact flange this limit state does not apply, obviously.

7.3.3.6 (f) Square and Rectangular HSS and Box-Shaped Members

This case applies to members bent about either axis, having compact or non-compact webs and compact, non-compact or slender flanges. The relevant limit states for members belonging to this case are:

- yielding of the whole section;
- flange local buckling;
- web local buckling.

Square and rectangular HSS are not subjected to lateral-torsional buckling, due to their high torsional resistance, so lateral-torsional buckling is not a relevant limit state for these sections. The nominal flexural strength M_n is the lowest value obtained according to the previously listed limit states.

The nominal strength M_n associated with the *yielding limit state* (plastic moment) is:

$$M_n = M_p = F_yZ \tag{7.103}$$

where Z is the plastic modulus of the section taken about the flexural axis.

For *compression FLB*, AISC specifications provide the following expressions for M_n:

for sections with non-compact flanges:

$$M_n = M_p - \left(M_p - F_yS\right)\left(3.57\frac{b}{t_f}\sqrt{\frac{F_y}{E}} - 4.0\right) \le M_p \tag{7.104}$$

where S is the elastic modulus of the section taken about the flexural axis.

for sections with slender flanges:

$$M_n = F_y S_e \tag{7.105}$$

where S_e is the effective modulus of the section taken about the flexural axis, and determined with the *effective width*, b_e, of the compression flange, computed as:

$$b_e = 1.92 t_f \sqrt{\frac{E}{F_y}} \left[1 - \frac{0.38}{b/t_f} \sqrt{\frac{E}{F_y}}\right] \le b \tag{7.106}$$

For sections with compact flanges this limit state does not apply, obviously.

As to the *web local buckling*, if webs are compact, this limit state is not applicable. If they are non-compact AISC specifications provide the following expression for M_n:

$$M_n = M_p - (M_p - F_y S_x)\left(0.305 \frac{h}{t_w} \sqrt{\frac{F_y}{E}} - 0.738\right) \le M_p \tag{7.107}$$

7.3.3.7 (g) Round HSS

This case applies to round HSS (hot-formed seamless pipes, ERW pipes and fabricated tubing) for which the following condition is verified:

$$D/t < 0.45 \frac{E}{F_y}$$

The relevant limit states for round HSS are:

- yielding;
- local buckling.

As in case of square and rectangular HSS, also round HSS are not subjected to lateral-torsional buckling. The nominal flexural strength M_n is the lower value obtained according to the previously listed limit states. The nominal strength M_n associated with the *yielding limit state* (plastic moment) is:

$$M_n = M_p = F_y Z \tag{7.108}$$

where Z is the plastic modulus of the section.

For *local buckling*, AISC specifications provide the following expressions for M_n:

(1) for non-compact sections $\left(0.07 \frac{E}{F_y} < D/t \le 0.31 \frac{E}{F_y}\right)$:

$$M_n = \left[\frac{0.021E}{D/t} + F_y\right] S \tag{7.109}$$

where S is the elastic modulus of the section, D is the diameter of the section and t is the thickness of the cross-section.

Table 7.18 Values of M_n for round HSS, case (g).

Section classification	D/t	M_n	Limit state
Compact	$<0.07\frac{E}{F_y}$	M_P	Y
Non-compact	From $0.07\frac{E}{F_y}$	$1.3M_Y$	LB
	To $0.31\frac{E}{F_y}$	$1.06M_Y$	LB
Slender	From $0.31\frac{E}{F_y}$	$1.06M_Y$	LB
	To $0.45\frac{E}{F_y}$	$0.73M_Y$	LB

(2) for slender sections $\left(0.31\frac{E}{F_y} < D/t < 0.45\frac{E}{F_y}\right)$:

$$M_n = \left[\frac{0.33E}{D/t}\right]S \tag{7.110}$$

For compact sections this limit state does not apply.
Significant values of M_n for round HSS are listed in Table 7.18.

7.3.3.8 (h) Tees Loaded in the Plane of Symmetry

The relevant limit states for tees are:

- yielding;
- lateral-torsional buckling;
- flange leg local buckling;
- Stem local buckling.

The nominal flexural strength M_n is the lowest value obtained according to the previously listed limit states.

The nominal strength M_n associated with the *yielding limit state* (plastic moment) is:

(1) if the stem is in tension:

$$M_n = M_p = F_y Z_x \leq 1.6M_y \tag{7.111}$$

(2) if the stem is in compression:

$$M_p = M_p = F_y Z_x \leq M_y \tag{7.112}$$

where $M_y = F_y S_x$.

For *lateral torsional buckling*, nominal flexural strength is computed as follows.

(1) if the stem is in tension:

$$M_n = M_{cr} = \frac{\pi}{L_b}\sqrt{EI_yGJ}\left[+2.3\frac{d}{L_b}\sqrt{\frac{I_y}{J}} + \sqrt{1+\left(2.3\frac{d}{L_b}\sqrt{\frac{I_y}{J}}\right)^2}\right] \tag{7.113}$$

(2) If the stem is in compression:

$$M_n = M_{cr} = \frac{\pi}{L_b}\sqrt{EI_yGJ}\left[-2.3\frac{d}{L_b}\sqrt{\frac{I_y}{J}} + \sqrt{1+\left(2.3\frac{d}{L_b}\sqrt{\frac{I_y}{J}}\right)^2}\right] \tag{7.114}$$

For *tee FLB*, AISC specifications provide the following expressions for M_n:

(1) for tees with a non-compact flange in flexural compression:

$$M_n = M_p - \left(M_p - 0.75F_yS_{xc}\right)\left(\frac{\lambda-\lambda_{pf}}{\lambda_{rf}-\lambda_{pf}}\right) \le 1.6M_y \tag{7.115}$$

(2) for tees with a slender flange in flexural compression:

$$M_n = \frac{0.7ES_{xc}}{\lambda^2} \tag{7.116}$$

where $\lambda = b_f/2t_f$, $\lambda_{pf} = 0.38\sqrt{E/F_y}$ is the limiting width-to-thickness ratio for a compact flange (see Section 4.3) and $\lambda_{rf} = 1.0\sqrt{E/F_y}$ is the limiting width-to-thickness ratio for a non-compact flange (see Section 4.3).

For sections with compact flange this limit state does not apply, obviously.

Finally, if reference is made to the *local buckling of stem in flexural compression*, for this limit state:

(1) for a compact stem $\left(d/t_w \le 0.84\sqrt{\frac{E}{F_y}}\right)$:

$$M_n = F_yS_x \tag{7.117}$$

(2) for a non-compact stem $\left(0.84\sqrt{\frac{E}{F_y}} < d/t_w \le 1.03\sqrt{\frac{E}{F_y}}\right)$:

$$M_n = \left[2.55 - 1.84\frac{d}{t_w}\sqrt{\frac{F_y}{E}}\right]F_yS_x \tag{7.118}$$

Table 7.19 Values of M_n for tees, case (h).

Stem in tension		Compute $M_{n,Y} = M_n$ of Eq. (7.111)	Y
		Compute $M_{n,LTB} = M_n$ of Eq. (7.113)	LTB
	Compact flange	—	FLB
	Non-compact flange	Compute $M_{n,FLB} = M_n$ of Eq. (7.115)	
	Slender flange	Compute $M_{n,FLB} = M_n$ of Eq. (7.116)	
	$M_n = \min\{M_{n,Y}; M_{n,LTB}; M_{n,FLB}\}$		
Stem in compression		Compute $M_{n,Y} = M_n$ of Eq. (7.112)	Y
		Compute $M_{n,LTB} = M_n$ of Eq. (7.114)	LTB
	Compact stem	Compute $M_{n,TSLB} = M_n$ of Eq. (7.117)	TSLB
	Non-compact stem	Compute $M_{n,TSLB} = M_n$ of Eq. (7.118)	
	Slender stem	Compute $M_{n,TSLB} = M_n$ of Eq. (7.119)	
	$M_n = \min\{M_{n,Y}; M_{n,LTB}; M_{n,TSLB}\}$		

(3) for a slender stem $\left(d/t_w > 1.03\sqrt{\frac{E}{F_y}}\right)$:

$$M_n = \left[\frac{0.69E}{\left(\frac{d}{t_w}\right)^2}\right] S_x \tag{7.119}$$

In Table 7.19 the decision-making process for computing M_n is summarized.

7.3.3.9 (i) Double Angles Loaded in the Plane of Symmetry

This case applies to double angles loaded in the plane of symmetry. The relevant limit states for double angles are:

- yielding;
- lateral-torsional buckling;
- double angle flange legs local buckling;
- double angle web legs local buckling.

The nominal flexural strength M_n is the lowest value obtained according to the previously listed limit states. The nominal strength M_n associated with the *yielding limit state* (plastic moment) is:

(1) if the web legs are in tension:

$$M_n = M_p = F_y Z_x \le 1.6 M_y \tag{7.120}$$

(2) if the web legs are in compression:

$$M_p = M_p = F_y Z_x \le M_y \tag{7.121}$$

where $M_y = F_y S_x$

For *lateral torsional buckling*, nominal flexural strength is computed as follows.

(1) if the web legs are in tension:

$$M_n = M_{cr} = \frac{\pi}{L_b}\sqrt{EI_yGJ}\left[+2.3\frac{d}{L_b}\sqrt{\frac{I_y}{J}} + \sqrt{1 + \left(2.3\frac{d}{L_b}\sqrt{\frac{I_y}{J}}\right)^2}\right] \tag{7.122}$$

(2) if the web legs are in compression:

$$M_n = M_{cr} = \frac{\pi}{L_b}\sqrt{EI_yGJ}\left[-2.3\frac{d}{L_b}\sqrt{\frac{I_y}{J}} + \sqrt{1 + \left(2.3\frac{d}{L_b}\sqrt{\frac{I_y}{J}}\right)^2}\right] \tag{7.123}$$

For *flange leg local buckling*, AISC specifications provide the following expressions for M_n:

(1) for angles with non-compact legs $\left(0.54\sqrt{\frac{E}{F_y}} < \frac{b}{t} \le 0.91\sqrt{\frac{E}{F_y}}\right)$:

$$M_n = \left(2.43 - 1.72\lambda\sqrt{\frac{F_y}{E}}\right)F_yS_c \le 1.5F_yS_c \tag{7.124}$$

(2) for angles with slender legs $\left(\frac{b}{t} > 0.91\sqrt{\frac{E}{F_y}}\right)$:

$$M_n = \frac{0.71E}{\lambda^2}S_c \tag{7.125}$$

where $\lambda = b/t$, b is the full width of the flange legs in compression, t is the thickness of angle and S_c is the elastic section modulus of the leg in compression.

For sections with a compact flange this limit state does not apply. Finally, with reference to the *local buckling of web legs in flexural compression*:

(1) for angles with non-compact legs $\left(0.54\sqrt{\frac{E}{F_y}} < \frac{b}{t} \le 0.91\sqrt{\frac{E}{F_y}}\right)$:

$$M_n = \left(2.43 - 1.72\lambda\sqrt{\frac{F_y}{E}}\right)F_yS_c \le 1.5F_yS_c \tag{7.126}$$

(2) for angles with slender legs $\left(\frac{b}{t} > 0.91\sqrt{\frac{E}{F_y}}\right)$:

Table 7.20 Values of M_n for double angles, case (i).

Web legs in tension		Compute $M_{n,Y} = M_n$ of Eq. (7.120)	Y
		Compute $M_{n,LTB} = M_n$ of Eq. (7.122)	LTB
	Compact flange leg	—	FLB
	Non-compact flange leg	Compute $M_{n,FLB} = M_n$ of Eq. (7.124)	
	Slender flange leg	Compute $M_{n,FLB} = M_n$ of Eq. (7.125)	
	$M_n = \min\{M_{n,Y}; M_{n,LTB}; M_{n,FLB}\}$		
Web legs in compression		Compute $M_{n,Y} = M_n$ of Eq. (7.121)	Y
		Compute $M_{n,LTB} = M_n$ of Eq. (7.123)	LTB
	Compact web leg	—	DALB
	Non-compact web leg	Compute $M_{n,DALB} = M_n$ of Eq. (7.126)	
	Slender web leg	Compute $M_{n,DALB} = M_n$ of Eq. (7.127)	
	$M_n = \min\{M_{n,Y}; M_{n,LTB}; M_{n,DALB}\}$		

$$M_n = \frac{0.71E}{\lambda^2} S_c \tag{7.127}$$

where $\lambda = b/t$, b is the full width of web legs in flexural compression, t is the thickness of angle and S_c is the elastic section modulus of the toe in compression.

For sections with compact flange this limit state does not apply.

In Table 7.20 the decision-making process for computing M_n is summarized.

7.3.3.10 (j) Single Angles

This case applies to single equal-leg or unequal-leg angles, with or without continuous lateral-torsional restraint along their length. If single angles are not laterally restrained along their length, the relevant limit states are:

- yielding;
- lateral-torsional buckling;
- local leg buckling.

The nominal flexural strength M_n is the lowest value obtained according to the previously listed limit states. The nominal strength M_n associated with the *yielding limit state* is:

$$M_n = 1.5M_y \tag{7.128}$$

where M_y is the yield moment about the bending axis.

As can be seen, the strength is not referred to plastic moment but limited to a shape factor of 1.5 applied to the yield moment. Shape factors for angles range from 1.73 to 1.96 actually, so Eq. (7.128) is intended to be quite conservative.

For *lateral torsional buckling*, AISC specifications consider different cases:

7.3.3.10.1 Laterally Unrestrained Unequal-Leg Single Angle

(1) *subjected to bending moment about the major principal axis*

(a) If $M_e \le M_y$:

$$M_n = \left(0.92 - \frac{0.17M_e}{M_y}\right) M_e \tag{7.129}$$

Table 7.21 β_w values for angles.

Angle size in. (mm)	β_w in. (mm)
8 × 6 (203 × 152)	3.31 (84.1)
8 × 4 (203 × 102)	5.48 (139)
7 × 4 (178 × 102)	4.37 (111)
6 × 4 (152 × 102)	3.14 (79.8)
6 × 3½ (152 × 89)	3.69 (93.7)
5 × 3½ (127 × 89)	2.40 (61)
5 × 3 (127 × 76)	2.99 (75.9)
4 × 3½ (102 × 89)	0.87 (22.1)
4 × 3 (102 × 76)	1.65 (41.9)
3½ × 3 (89 × 76)	0.82 (22.1)
3½ × 2½ (89 × 64)	1.62 (41.1)
3 × 2½ (76 × 64)	0.86 (21.8)
3 × 2 (76 × 51)	1.56 (39.6)
2½ × 2 (64 × 51)	0.85 (21.6)
2½ × 1½ (64 × 38)	1.49 (37.8)
Equal leg	0

(b) If $M_e < M_y$:

$$M_n = \left(1.92 - 1.17\sqrt{\frac{M_y}{M_e}}\right) M_y \le 1.5 M_y \tag{7.130}$$

M_e is evaluated as:

$$M_e = \frac{4.9 E I_z C_b}{L_b^2}\left[\sqrt{\beta_w^2 + 0.052\left(\frac{L_b t}{r_z}\right)^2} + \beta_w\right] \tag{7.131}$$

where M_e is the elastic lateral-torsional buckling moment, C_b is computed using Eq. (7.68), but $C_b \le 1.5$; M_y is the yield moment about the bending axis, I_z is the minor principal axis moment of inertia, r_z is the radius of gyration about the minor principal axis, t is the thickness of the angle leg and term β_w represents a section property, to be chosen from Table 7.21 with positive sign for short leg in compression and negative sign for long leg in compression.

(2) *subjected to a generically oriented bending moment*

(a) Resolve moment into components along both principal axes.

(b) For component along the major principal axis, compute M_n as in (1).

(c) Verify the angle for *biaxial bending*, according to Section 9.3.

7.3.3.10.2 Laterally Unrestrained Equal-Leg Single Angle

(1) *subject to a bending moment about the major principal axis*

(a) If $M_e \le M_y$:

$$M_n = \left(0.92 - \frac{0.17 M_e}{M_y}\right) M_e \tag{7.132}$$

(b) If $M_e < M_y$:

$$M_n = \left(1.92 - 1.17\sqrt{\frac{M_y}{M_e}}\right) M_y \le 1.5 M_y \quad (7.133)$$

where C_b is computed using Eq. (7.68), with the limitation $C_b \le 1.5$, M_y is the yielding moment about the bending axis, t is the thickness of the angle leg, b is the full width of leg and M_e is the elastic lateral-torsional buckling moment defined as:

$$M_e = \frac{0.46 E b^2 t^2 C_b}{L_b} \quad (7.134)$$

It should be noted that Eqs. (7.132) and (7.133) are identical to Eqs. (7.129) and (7.130).

(2) *subjected to a generically oriented bending moment*
  (a) resolve the moment into components along the two principal axes.
  (b) for components along the major principal axis, compute M_n as in (1a).
  (c) verify the angle for *biaxial bending*, according to Section 9.3.

The same procedure as for unequal-leg angles has to be adopted.

(3) *subjected to bending moment about one of the geometric axes of the angle* (alternative to method 2b)
  (a) if $M_e \le M_y$:

$$M_n = \left(0.92 - \frac{0.17 M_e}{(0.8 M_y)}\right) M_e \quad (7.135)$$

  (b) if $M_e < M_y$:

$$M_n = \left(1.92 - 1.17\sqrt{\frac{M_y}{M_e}}\right)(0.8 M_y) \le 1.5(0.8 M_y) \quad (7.136)$$

where M_y is the yield moment about the geometric axis, M_e is the elastic lateral-torsional buckling moment, computed as follows:
  (c) with maximum compression at the toe:

$$M_e = \frac{0.66 E b^4 t C_b}{L_b^2}\left[\sqrt{1 + 0.78\left(\frac{L_b t}{b^2}\right)^2} - 1\right] \quad (7.137)$$

  (d) with maximum tension at the toe:

$$M_e = \frac{0.66 E b^4 t C_b}{L_b^2}\left[\sqrt{1 + 0.78\left(\frac{L_b t}{b^2}\right)^2} + 1\right] \quad (7.138)$$

If there is a lateral-torsional restraint at the point of maximum moment, Eqs. (7.135) and (7.136) shall be used substituting $0.8M_y$ with M_y, and (Eqs. (7.137) and (7.138)) become:

$$M_e = 1.25\frac{0.66Eb^4tC_b}{L_b^2}\left[\sqrt{1+0.78\left(\frac{L_bt}{b^2}\right)^2}-1\right] \tag{7.139}$$

$$M_e = 1.25\frac{0.66Eb^4tC_b}{L_b^2}\left[\sqrt{1+0.78\left(\frac{L_bt}{b^2}\right)^2}+1\right] \tag{7.140}$$

Method (2c) for equal-leg angles is a simplification of the more general method (2b).

When bending is applied about one leg of an unrestrained single angle, the angle will deflect not only in the bending direction but also laterally, and the maximum stress at the angle tip will be approximately 25% greater than the calculated stress using the geometric axis section modulus.

Finally, the limit state of *local buckling* has to be considered when the toe of the leg is in compression and the section is not compact; AISC provides the following expressions for M_n:

(1) for cross-sections with non-compact legs:

$$M_n = \left[2.43-1.72\left(\frac{b}{t}\right)\sqrt{\frac{F_y}{E}}\right]F_yS_c \tag{7.141}$$

(2) for cross-sections with slender legs:

$$M_n = \left[\frac{0.71E}{\left(\frac{b}{t}\right)^2}\right]S_c \tag{7.142}$$

where S_c is the elastic section modulus of the toe in compression relative to the axis of bend. For bending about one of the geometric axes of equal-leg axes with no lateral-torsional restraint, S_c is taken equal to 80% of the geometric axis section modulus.

7.3.3.11 (k) Rectangular Bar and Rounds

This case applies to solid bars with a rectangular and round cross-section. The relevant limit states are:

- yielding;
- lateral-torsional buckling.

Lateral-torsional buckling occurs only for rectangular bars, when depth is larger than width; otherwise the only limit state is the attainment of a full plastic moment.

The nominal flexural strength M_n is the lower value obtained according to the previously listed limit states.

The nominal strength M_n associated with the *yielding limit state* (plastic moment) is:

$$M_n = M_p = F_yZ \le 1.6M_y \tag{7.143}$$

For *lateral-torsional buckling* AISC prescribes:

(1) for rectangular bars with $\frac{L_b d}{t^2} \leq \frac{0.08E}{F_y}$ bent about their major axis, this limit state does not occur.

(2) for rectangular bars with $\frac{0.08E}{F_y} < \frac{L_b d}{t^2} \leq \frac{1.9E}{F_y}$ bent about their major axis:

$$M_n = C_b \left[1.52 - 0.274 \left(\frac{L_b d}{t^2}\right) \frac{F_y}{E}\right] M_y \leq M_p \tag{7.144}$$

(3) for rectangular bars with $\frac{L_b d}{t^2} > \frac{1.9E}{F_y}$ bent about their major axis:

$$M_n = \left[\frac{1.9EC_b}{\left(\frac{L_b d}{t^2}\right)}\right] S_x \leq M_p \tag{7.145}$$

(4) for rectangular bars bent about their minor axis and for round bars this limit state does not occur.

7.3.3.12 (l) Unsymmetrical Shapes

In this case all unsymmetrical shapes (except angles) are grouped. AISC prescriptions are quite generic. The limit states to be considered are:

- yielding;
- lateral-torsional buckling;
- local buckling.

Critical stresses for the last two limit states are to be defined by textbooks, handbooks or by FE analysis.

7.4 Design Rules for Beams

With reference to the beam design, the member of an appropriate cross-section must be selected by considering the need to fulfil all the specific requirements related to the service condition as well resistance and ultimate stability limit-state. As a consequence, this choice can be based on the designer experience and/or on the use of suitable rules related to the application of the criteria to verify for safety checks.

In case of uniform doubly-symmetrical I- or H-shaped beams, appropriate equations can be easily obtained to define the minimum value of the moment of inertia (I_{min}) as well as the section modulus (W_{min}) required to guarantee the safety of the profile.

With reference to a simply supported beam having a span L, with a uniformly distributed load, comprising of dead (g) and live load (q) contribution, the following limit conditions can be considered:

$$\frac{5}{384} \cdot \frac{(g+q) \cdot L^4}{E \cdot I_{min}} = \delta_{Lim} \tag{7.146a}$$

$$\frac{3}{2} \cdot \frac{(g+q) \cdot L^2}{8 \cdot W_{min}} = f_d \tag{7.146b}$$

Table 7.22 Indications for the minimum geometrical characteristics of cross-sections.

Load condition	I_{min} Minimum moment of inertia	W_{min} Minimum section modulus
g,q (uniform load, span L)	$I_{min}=\frac{5}{384}\cdot\frac{(g+q)\cdot L^4}{E\cdot\delta_{Lim}}$	$W_{min}=\frac{3}{2}\cdot\frac{(g+q)\cdot L^2}{8\cdot f_d}$
P (midspan point load), L	$I_{min}=\frac{1}{48}\cdot\frac{(P_g+P_q)\cdot L^3}{E\cdot\delta_{Lim}}$	$W_{min}=\frac{3}{2}\cdot\frac{(P_g+P_q)\cdot L}{4\cdot f_d}$
P, P (two point loads), x, x, x	$I_{min}=\frac{23}{648}\cdot\frac{(P_g+P_q)\cdot L^3}{E\cdot\delta_{Lim}}$	$W_{min}=\frac{3}{2}\cdot\frac{(P_g+P_q)\cdot L}{3\cdot f_d}$

where E is the modulus of elasticity of the material, $\delta_{\rm Lim}$ is the maximum displacement compatible with the beam use and f_d is the design strength for the material.

These limit conditions regard the deflection and resistance criteria for the beam, respectively, for which the associated load safety factors in accordance with the European practice can be taken in this pre-sizing phase equal to 1.5 (i.e. $\gamma_{g1}=\gamma_{g2}=\gamma_q=1.5$). It should be noted that on the basis of the computed value of I_{min} and W_{min} two different cross-section types are generally identified and the greater has to be adopted for verification checks.

For three different load conditions common in simply supported beams, Table 7.22 proposes the equation to be used to evaluate I_{min} and W_{min}.

With reference to I- or H-shaped bi-symmetrical profiles, moment of inertia (I) and section modulus (W) are directly connected to each other via the beam depth (H) by means of the relationship:

$$W=\frac{2\cdot I}{H} \tag{7.147}$$

The displacement limit $\delta_{\rm Lim}$ can be considered to be the displacement ($\delta_{\rm Lim,tot}$) associated with the total load or the displacement ($\delta_{\rm Lim,2}$) due to sole variable load. For practical design purposes, tables of immediate applicability for the selection of the depth of the profile can easily be developed. As an example, Table 7.23 deals with two common load cases: for each of them and for different steel grades, with reference both to floor and roof beams two values of $H_{\rm min}$ are proposed, differing for the considered displacement limit, the first one associated with total loads $\delta_{\rm Lim,tot}$ and the second one with variable loads $\delta_{\rm Lim,2}$. In particular the following limits have been considered:

- floor beam: $\delta_{\rm Lim,tot}=L/250$ and $\delta_{\rm Lim,2}=L/300$;
- roof beam: $\delta_{\rm Lim,tot}=L/200$ and $\delta_{\rm Lim,2}=L/250$.

As an example related to the main steps to evaluate expressions presented in Table 7.23, the case of beam with uniform load is considered with reference to the displacement limit $\delta_{\rm Lim,tot}$. From Eq. (7.147), substituting the expressions in Table 7.22 we can obtain:

$$W_{\rm min}=\frac{3}{2}\cdot\frac{(g+q)\cdot L^2}{8\cdot f_d}=\frac{2\cdot I}{H_{\rm min}}=\frac{2}{H_{\rm min}}\cdot\frac{5}{384}\cdot\frac{(g+q)\cdot L^4}{E\cdot\delta_{Lim,tot}} \tag{7.148}$$

Table 7.23 Value of H_{min}: $\alpha = q/(g+q)$; $\beta = P_q/(P_g+P_q)$ for European steel grades.

	Steel grade				
	S 235	S 275	S 355	S 420	S 460
Floor beam (H_{min} = minimum beam depth)					
g, q, L	$\frac{L}{27}$ $\alpha\frac{2L}{45}$	$\frac{L}{23}$ $\alpha\frac{L}{19}$	$\frac{2L}{35}$ $\alpha\frac{2L}{29}$	$\frac{L}{15}$ $\alpha\frac{2L}{25}$	$\frac{2L}{27}$ $\alpha\frac{2L}{23}$
P_q, P_g, L	$\frac{2L}{67}$ $\beta\frac{L}{28}$	$\frac{2L}{57}$ $\beta\frac{L}{24}$	$\frac{L}{22}$ $\beta\frac{2L}{37}$	$\frac{2L}{37}$ $\beta\frac{2L}{31}$	$\frac{L}{17}$ $\beta\frac{L}{14}$
Roof beam (H_{min} = minimum beam depth)					
g, q, L	$\frac{2L}{67}$ $\alpha\frac{L}{27}$	$\frac{2L}{57}$ $\alpha\frac{2L}{23}$	$\frac{L}{22}$ $\alpha\frac{2L}{35}$	$\frac{2L}{37}$ $\alpha\frac{L}{15}$	$\frac{L}{17}$ $\alpha\frac{2L}{27}$
P_q, P_g, L	$\frac{L}{42}$ $\beta\frac{2L}{67}$	$\frac{L}{36}$ $\beta\frac{2L}{57}$	$\frac{2L}{55}$ $\beta\frac{L}{22}$	$\frac{2L}{47}$ $\beta\frac{2L}{37}$	$\frac{2L}{43}$ $\beta\frac{L}{17}$

As a consequence, the minimum beam depth, H_{min}, can be expressed as:

$$H_{\min} = 2\cdot\left(\frac{5}{384}\cdot\frac{L^2}{E\cdot\delta_{Lim,tot}}\right)\cdot\frac{2\cdot 8\cdot f_d}{3} \tag{7.149}$$

As an example, considering the value of elastic modulus in accordance with the European practice (i.e. $E = 210\,000$ N/mm^2), $H_{\min}$ expressed in millimetres, can be evaluated as:

$$H_{\min} = \frac{f_d\cdot L^2}{1512000\cdot\delta_{Lim,tot}} \tag{7.150}$$

In the same way, by considering the displacement limit δ_2 associated with the sole live load, we can obtain:

$$W_{\min} = \frac{3}{2}\cdot\frac{(g+q)\cdot L^2}{8\cdot f_d} = \frac{2\cdot I}{H_{\min}} = \frac{2}{H_{\min}}\cdot\frac{5}{348}\cdot\frac{q\cdot L^4}{E\cdot\delta_{\text{Lim},2}} \tag{7.151}$$

Term $H_{\min}$ is obtained from:

$$H_{\min} = 2\cdot\left(\frac{5}{384}\cdot\frac{qL^2}{(g+q)\cdot E\cdot\delta_{\text{Lim},2}}\right)\cdot\frac{2\cdot 8\cdot f_d}{3} \tag{7.152}$$

Table 7.24 Indications for the minimum geometrical characteristics of cross-sections (AISC-ASD).

Load condition	I_{min} – Minimum moment of inertia	W_{min} – Minimum section modulus
g,q (uniform load)	$I_{min}=\frac{5}{384}\cdot\frac{\alpha(g+q)L^4}{E\cdot\delta_{Lim}}$	$Z_{min}=\Omega_b\frac{(g+q)L^4}{8F_y}$
P (one point load, span L)	$I_{min}=\frac{1}{48}\cdot\frac{\beta(P_g+P_q)L^3}{E\cdot\delta_{Lim}}$	$Z_{min}=\Omega_b\frac{(P_g+P_q)L}{4F_y}$
P, P (two point loads, spacings x, x, x)	$I_{min}=\frac{23}{648}\cdot\frac{\beta(P_g+P_q)L^3}{E\cdot\delta_{Lim}}$	$Z_{min}=\Omega_b\frac{(P_g+P_q)L}{3F_y}$

$\alpha=q/(g+q)$; $\beta=P_q/(P_g+P_q)$; $\Omega_b=1,67$

As to units, if forces are measured in newton and dimensions in millimetres, by substituting directly the value of E, we can obtain:

$$H_{min}=\frac{f_d}{1512000\cdot\delta_{Lim,2}}\cdot\frac{qL^2}{(g+q)} \tag{7.153}$$

It should be noted from Table 7.23 that H_{min} increases with the increase of the steel grade. This apparent nonsense is due to the fact that the choice of a better quality of steel, which corresponds to a deeper beam with respect to a beam selected with a lower steel grade, is associated with a higher distance between parallel beams; that is with a greater load carrying capacity.

With reference to the AISC 360-10 code, Eqs. (7.146a) and (7.146b) can be rewritten as:

$$\frac{5}{384}\cdot\frac{qL^4}{EI_{min}}=\frac{5}{384}\cdot\frac{\alpha(g+q)L^4}{EI_{min}}=\delta_{Lim} \tag{7.154a}$$

$$\frac{(g+q)L^2}{8}=\frac{F_yZ_{min}}{\Omega_b} \tag{7.154b}$$

where $\alpha=q/(g+q)$ is the displacement computed with reference to live load only.

Equation (7.154b) is written with reference to ASD. Table 7.24 represents the equation to be used to evaluate I_{min} and Z_{min} for three different load conditions common in simply supported beams.

Table 7.25 corresponds to the 'translation' of Table 7.23 into AISC code. The main differences are:

(1) consider one value for δ_{Lim}, computed for live loads only. In particular the following data have been assumed:
Floor beam: $\delta_{Lim}=L/360$;
Roof beam: $\delta_{Lim}=L/240$.
(2) consider the following steel grades 36, 42, 46 and 50 (F_y [ksi]).
(3) refer to ASD for calculations.

It should be noted that, in accordance with AISC code notation, term d is used instead of H to identify the section height.

Table 7.25 Value of d_{min}: $\alpha = q/(g+q)$; $\beta = P_q/(P_g+P_q)$ for ASTM steel grades.

	Steel grade (F_y (ksi))			
	36	42	46	50
Floor beam (d_{min} = minimum beam depth)				
Uniform load g, q; span L	$\alpha\frac{2L}{36}$	$\alpha\frac{2L}{31}$	$\alpha\frac{2L}{28}$	$\alpha\frac{2L}{26}$
Concentrated load P_q, P_g; span L	$\beta\frac{L}{22}$	$\beta\frac{L}{19}$	$\beta\frac{2L}{35}$	$\beta\frac{2L}{32}$
Roof beam (d_{min} = minimum beam depth)				
Uniform load g, q; span L	$\alpha\frac{L}{27}$	$\alpha\frac{L}{23}$	$\alpha\frac{2L}{42}$	$\alpha\frac{2L}{39}$
Concentrated load P_q, P_g; span L	$\beta\frac{2L}{67}$	$\beta\frac{2L}{58}$	$\beta\frac{L}{26}$	$\beta\frac{2L}{48}$

As a consequence, the rewritten Eq. (7.147), using AISC symbols, is:

$$Z = \frac{2I}{d} \tag{7.155}$$

The approach to evaluate the expressions is presented in Table 7.25. Also in this case, a beam with uniform load is considered with reference to the displacement limit δ_{Lim}. From Eq. (7.155), by substituting the expressions in Table 7.24 we can obtain:

$$Z_{min} = \Omega_b \frac{(g+q)L^2}{8F_y} = \frac{2 \cdot I}{d_{min}} = \frac{2}{d_{min}} \cdot \frac{5}{384} \cdot \frac{\alpha(g+q)L^4}{E \cdot \delta_{Lim}} \tag{7.156}$$

As a consequence, d_{min} can be expressed as:

$$d_{min} = 2 \cdot \left(\frac{5}{384} \cdot \frac{L^2}{E\delta_{Lim}} \right) \cdot \frac{8\alpha F_y}{\Omega_b} \tag{7.157}$$

By considering the value of elastic modulus in accordance with the US practice (i.e. $E = 29\,000$ ksi), the minimum beam depth $d_{\min}$ is:

$$d_{\min} = \frac{\alpha F_y \cdot L^2}{232450 \cdot \delta_{Lim}} \tag{7.158}$$

If reference is made to $\delta_{\text{Lim}} = L/360$ for the floor beams and $\delta_{\text{Lim}} = L/240$ for the roof beams, we can obtain:

- for floor beams: $d_{\min} = \dfrac{\alpha \cdot L}{(646/F_y)}$
- for roof beams: $d_{\min} = \dfrac{\alpha \cdot L}{(969/F_y)}$

For beams with a concentrated load at midspan, from Eq. (7.155), by substituting the expressions in Table 7.24 we can obtain:

$$Z_{\min} = \Omega_b \frac{(P_g + P_q)L}{4F_y} = \frac{2 \cdot I}{d_{\min}} = \frac{2}{d_{\min}} \cdot \frac{1}{48} \cdot \frac{\beta (P_g + P_q) L^3}{E \cdot \delta_{Lim}} \tag{7.159}$$

As a consequence, $d_{\min}$ can be expressed as:

$$d_{\min} = 2 \cdot \left(\frac{1}{48} \cdot \frac{L^2}{E \delta_{Lim}} \right) \cdot \frac{4 \beta F_y}{\Omega_b} \tag{7.160}$$

By considering the US value of elastic modulus again (i.e. $E = 29\,000$ ksi), the minimum beam depth $d_{\min}$ is:

$$d_{\min} = \frac{\beta F_y \cdot L^2}{290580 \cdot \delta_{Lim}} \tag{7.161}$$

If δ_{Lim} is $L/360$ for floor beams and $L/240$ for roof beams, we can obtain:

- for floor beams: $d_{\min} = \dfrac{\alpha \cdot L}{(807/F_y)}$
- for roof beams: $d_{\min} = \dfrac{\alpha \cdot L}{(1210/F_y)}$

7.5 Worked Examples

Example E7.1 Beam Design in Accordance with the EU Approach

Verify a S275 IPE 300 simply supported beam (Figure E7.1.1) in accordance with EC3, which is subjected to a uniform dead load of 5.0 kN/m (0.343 kip/ft) and a uniform live load of 10.0 kN/m (0.685 kip/ft). Two cases are considered that differ for the load condition: loads are applied at the shear centre or on the top flange. The beam is not braced against lateral-torsional buckling along its entire length. Warping and lateral rotation are free at both ends; lateral displacement and torsion are prevented at both ends.

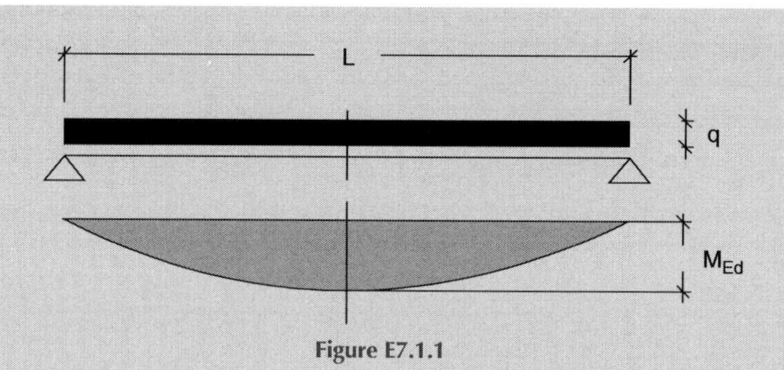

Figure E7.1.1

Geometrical properties:

$$H = 300\ \text{mm}(11.8\ \text{in.})\quad J_y = 8356\ \text{cm}^4(200.8\ \text{in.}^4)$$
$$b_f = 150\ \text{mm}(5.91\ \text{in.})\quad J_z = 603.8\ \text{cm}^4(14.51\ \text{in.}^4)$$
$$t_f = 10.7\ \text{mm}(0.42\ \text{in.})\quad I_t = 20.1\ \text{cm}^4(0.483\ \text{in.}^4)$$
$$t_w = 7.1\ \text{mm}(0.28\ \text{in.})\quad I_w = 125934.1\ \text{cm}^6(469\ \text{in.}^6)$$
$$r = 15\ \text{mm}(0.59\ \text{in.})\quad W_{pl,y} = 628.4\ \text{cm}^3(37.35\ \text{in.}^3)$$
$$L = 5\ \text{m}(16.4\ \text{ft})\quad W_{el,y} = 557.1\ \text{cm}^3(34.0\ \text{in.}^3)$$
$$L_{cr,LT} = 5\ \text{m}\quad A = 53.8\ \text{cm}^2(7.34\ \text{in.}^2)$$

Material properties:

$$\text{Steel}: \text{S275}\quad f_y = 275\ \text{MPa}(34.08\ \text{ksi})\quad f_u = 430\ \text{MPa}(62.37\ \text{ksi})$$

$$\text{Flange}: (c/t_f) = [150 - 7.1 - (2 \times 15)]/(2 \times 10.7) = 5.3 \leq 8.3\quad \text{Class 1}$$
$$\text{Web}: (d/t_w) = [300 - (2 \times 10.7) - (2 \times 15)]/7.1 = 35.0 \leq 66\quad \text{Class 1}$$
$$\text{Section}: \quad \text{Class 1}$$

Loads:

$$\text{Dead load}: q_p = 5.0\ \text{kN/m}(0.343\ \text{kip/ft})$$
$$\text{Live load}: q_s = 10.0\ \text{kN/m}(0.685\ \text{kip/ft})$$
$$\text{Factorized load}: q = 1.35 \times 5.0 + 1.5 \times 10.0 = 21.75\ \text{kN/m}(1.49\ \text{kip}/ft)$$
$$\text{Maximum total load deflection permitted}: f_p = L/400$$

Shear strength verification.
Maximum design shear force:

$$V_{Ed} = 21.75 \times 5/2 = 54.4\ \text{kN}(12.2\ \text{kips})$$

Shear area:

$$A_v = A - 2bt_f + (t_w + 2r)t_f = 53.8 \cdot 10^2 - 2 \times 150 \times 10.7 + (7.1 + 2 \times 15) \times 10.7$$
$$= 2567\ \text{mm}^2 = 25.7\ \text{cm}^2(4.0\ \text{in.}^2)$$

Design shear resistance:

$$V_{c,Rd} = \frac{A_v f_y}{\sqrt{3}\cdot\gamma_{M0}} = \frac{2567\times 275}{\sqrt{3}\times 1.0}\cdot 10^{-3} = 407.6\,\text{kN}\,(91.6\,\text{kips})$$

$$V_{Ed}/V_{c,Rd} = 54.4/407.6 = 0.13 < 0.50 \rightarrow \text{no influence on design resistance for bending}$$

Flexural strength verification.
Maximum design bending moment:

$$M_{Ed} = 21.75 \times 5^2/8 = 68.0\ \text{kNm}\ (50.2\ \text{kip-ft})$$

Design resistance for bending:

$$M_{c,Rd} = W_y f_y/\gamma_{M0} = (628.4 \times 275\,/\,1.00)\cdot 10^{-3} = 172.8\,\text{kNm} > 68.0\,\text{kNm}\ \ \text{OK}$$

$$(127.5\,\text{kips-ft} > 50.2\,\text{kip-ft})$$

Deflection verification.
Computed deflection at midspan:

$$f = \frac{5}{384}\frac{qL^4}{EJ_y} = \frac{5}{384}\frac{[(5.0+10.0)\cdot 10^{-2}]\times 500^4}{21000\times 8356} = 0.69\ \text{cm}\ (0.272\ \text{in.}) < f_p$$

$$= 500/400 = 1.25\ \text{cm}\,(0.492\ \text{in.})\ \ \text{OK}$$

Lateral-torsional buckling (LTB) verification.

(a) Verification According to EC3 – General Approach
Compute critical moment using Eq. (7.43) (with $C_3\ z_j = 0$ because section is symmetrical and then $z_j = 0$) considering the load applied to the shear centre:

$$M_{cr} = C_1\frac{\pi^2 EJ_z}{(k_z L)^2}\left\{\sqrt{\left(\frac{k_z}{k_w}\right)^2\frac{I_w}{J_z} + \frac{(kL)^2 GI_t}{\pi^2 EJ_z} + (C_2 z_g)^2} - C_2 z_g\right\}$$

$$= 1.132 \times \frac{3{,}14^2\times 21000\times 603{,}8}{(1\times 500)^2}$$

$$\times\left\{\sqrt{\left(\frac{1}{1}\right)^2\frac{125934.1}{603.8} + \frac{(1\times 500)^2\times 8077\times 20{,}1}{3.14^2\times 21000\times 603.8} + (0.459\times 0)^2} - 0.459\times 0\right\}10^{-2} = 130.8\,\text{kNm}$$

$$(96.5\,\text{kip-ft})$$

Assume:

$k_z = k_w = 1$ (free rotation in horizontal plane and warping at both end)
$z_g = 0$ (load applied at the shear centre)
$C_1 = 1.132$ (see Table 7.6b); $C_2 = 0.459$;
$E = 21\,000\ \text{kN/cm}^2$ (30 460 ksi); $G = 8077\ \text{kN/cm}^2$ (11710 ksi).

Compute relative slenderness (Eq. (7.33)):

$$\bar{\lambda}_{LT} = \sqrt{\frac{W_y f_y}{M_{cr}}} = \sqrt{\frac{628.4 \times 27.50}{(130.8 \cdot 10^2)}} = 1.15$$

Choose, according to Tables 7.2 and 7.3, the imperfection factor α_{LT}:

$$\alpha_{LT} = 0.21 (\text{rolled I-sections with H/b} \leq 2)$$

Hence:

$$\Phi_{LT} = 0.5\left[1 + \alpha_{LT}(\bar{\lambda}_{LT} - 0.2) + \bar{\lambda}_{LT}^2\right] = 0.5\left[1 + 0.21(1.15 - 0.2) + 1.15^2\right] = 1.26$$

$$\chi_{LT} = \frac{1}{\Phi_{LT} + \sqrt{\Phi_{LT}^2 - \bar{\lambda}_{LT}^2}} = \frac{1}{1.26 + \sqrt{1.26^2 - 1.15^2}} = 0.563\ (\leq 1)$$

And finally:

$$M_{b.Rd} = \chi_{LT} \frac{W_y f_y}{\gamma_{M1}} = 0.563 \times \frac{628.4 \times 27.50}{1.00} \cdot 10^{-2} = 97.3\,\text{kNm}\ (71.8\ \text{kip-ft})$$

LTB check will then be:

$$M_{Ed} = 68.0\ \text{kNm} \leq M_{b,Rd} = 97.3\ \text{kNm}\quad \text{OK}$$

Consider now the load applied on top flange of the beam. The distance from the shear centre to the top of beam flange is:

$$z_g = 150\ \text{mm}\ (5.9\ \text{in.})$$

Critical moment varies as follows:

$$M_{cr} = 1.132 \times \frac{3.14^2 \times 21000 \times 603.8}{(1 \times 500)^2}$$
$$\times \left\{ \sqrt{\left(\frac{1}{1}\right)^2 \frac{125934.1}{603.8} + \frac{(1 \times 500)^2 \times 8077 \times 20.1}{3.14^2 \times 21000 \times 603.8} + (0.459 \times 15)^2} - 0.459 \times 15 \right\} 10^{-2} = 97.5\,\text{kNm}$$

(71.9 kip-ft)

Hence:

$$\bar{\lambda}_{LT} = \sqrt{\frac{W_y f_y}{M_{cr}}} = \sqrt{\frac{628.4 \times 27.50}{(97.5 \times 100)}} = 1.33$$

$$\Phi_{LT} = 0.5\left[1 + \alpha_{LT}(\bar{\lambda}_{LT} - 0.2) + \bar{\lambda}_{LT}^2\right] = 0.5\left[1 + 0.21(1.33 - 0.2) + 1.33^2\right] = 1.505$$

$$\chi_{LT} = \frac{1}{\Phi_{LT} + \sqrt{\Phi_{LT}{}^2 - \bar{\lambda}_{LT}^2}} = \frac{1}{1.503 + \sqrt{1.503^2 - 1.33^2}} = 0.453\,(\leq 1)$$

$$M_{b,Rd} = \chi_{LT}\frac{W_y f_y}{\gamma_{M1}} = 0.454 \times \frac{628.4 \times 27.50}{1.00} \cdot 10^{-2} = 78.3\,\text{kNm}\,(57.6\text{ kip-ft})$$

LTB check will then be:

$$M_{Ed} = 68.0\text{ kNm} \leq M_{b,Rd} = 78.3\text{ kNm} \quad \text{OK}$$

(b) Verification According to EC3 – Method for I- or H-Shaped Profiles (see Section 6.3.2.3)
Consider loads applied to the shear centre ($z_g = 0$).
Critical moment and relative slenderness are the same as in case (a). Hence:

$$M_{cr} = 130.8\,\text{kNm}\,(96.5\,\text{kip-ft}); \bar{\lambda}_{LT} = 1.15$$

Compute f factor (Eq. (7.37)):
Assume correction factor $k_c = 0.94$ (from Table 7.7)

$$\bar{\lambda}_{LT,0} = 0.4; \beta = 0.75$$

$$f = 1 - 0.5(1 - k_c)\left[1 - 2.0(\bar{\lambda}_{LT} - 0.8)^2\right] = 1 - 0.5(1 - 0.94)\left[1 - 2.0(1.15 - 0.8)^2\right] = 0.977$$

f shall be ≤ 1.
Assume $\alpha_{LT} = 0.34$ (From Tables 7.2 and 7.4 (note: Table 7.4 and not Table 7.3, because this leads to a different value for α_{LT}).
Hence:

$$\Phi_{LT} = 0.5\left[1 + \alpha_{LT}(\bar{\lambda}_{LT} - \bar{\lambda}_{LT,0}) + \beta\bar{\lambda}_{LT}^2\right]$$
$$= 0.5\left[1 + 0.34(1.15 - 0.4) + 0.75 \times 1.15^2\right] = 1.123$$

$$\chi_{LT} = \frac{1}{\Phi_{LT} + \sqrt{\Phi_{LT}{}^2 - \beta\bar{\lambda}_{LT}^2}} = \frac{1}{1.123 + \sqrt{1.123^2 - 0.75 \times 1.15^2}} = 0.609$$

It results in:

$$\chi_{LT} = 0.609 \leq 1; \chi_{LT} = 0.609 \leq \frac{1}{\bar{\lambda}_{LT}^2} = \frac{1}{1.15^2} = 0.756; \frac{\chi_{LT}}{f} = \frac{0.609}{0.977} = 0.623 \leq 1$$

And finally:

$$M_{b,Rd} = \frac{\chi_{LT}}{f}\frac{W_y f_y}{\gamma_{M1}} = 0.623 \times \frac{628.4 \times 27.50}{1.00} \cdot 10^{-2} = 107.7\text{ kNm}\,(79.4\text{ kip-ft})$$

The LTB check will then be:

$$M_{Ed} = 68.0\text{ kNm} \leq M_{b,Rd} = 107.7\text{ kNm} \quad \text{OK}$$

The value of $M_{b,Rd}$ is greater than that computed using the general method.
Consider now loads applied on top flange ($z_g = 150$ mm).
$M_{cr} = 97.5$ kNm (71.9 kip-ft); $\bar{\lambda}_{LT} = 1.33$ (unchanged values respect to case (a).
Compute f factor (Eq. (7.37)):
Assume correction factor $k_c = 0.94$ (from Table 7.5)

$$\bar{\lambda}_{LT,0} = 0.4; \beta = 0.75$$

$$f = 1 - 0.5(1 - k_c)\left[1 - 2.0(\bar{\lambda}_{LT} - 0.8)^2\right] = 1 - 0.5(1 - 0.94)\left[1 - 2.0(1.33 - 0.8)^2\right] = 0.987$$

f shall be ≤ 1.
Assume $\alpha_{LT} = 0.34$ (From Tables 7.2 and 7.4 (note: Table 7.4 and not Table 7.3, because this leads to a different value for α_{LT}).
Hence:

$$\Phi_{LT} = 0.5\left[1 + \alpha_{LT}(\bar{\lambda}_{LT} - \bar{\lambda}_{LT.0}) + \beta\bar{\lambda}_{LT}^2\right] = 0.5\left[1 + 0.34(1.33 - 0.4) + 0.75 \times 1.33^2\right] = 1.322$$

$$\chi_{LT} = \frac{1}{\Phi_{LT} + \sqrt{\Phi_{LT}^2 - \beta\bar{\lambda}_{LT}^2}} = \frac{1}{1.322 + \sqrt{1.322^2 - 0.75 \times 1.33^2}} = 0.507$$

It results in:

$$\chi_{LT} = 0.507 \leq 1; \chi_{LT} = 0.507 \leq \frac{1}{\bar{\lambda}_{LT}^2} = \frac{1}{1.33^2} = 0.565; \frac{\chi_{LT}}{f} = \frac{0.507}{0.987} = 0.514 \leq 1$$

And finally:

$$M_{b,Rd} = \frac{\chi_{LT}}{f}\frac{W_y f_y}{\gamma_{M1}} = 0.514 \times \frac{628.4 \times 27.50}{1.00} \cdot 10^{-2} = 88.8\,\text{kNm (65.5 kip-ft)}$$

The LTB check will then be:

$$M_{Ed} = 68.0 \text{ kNm} \leq M_{b,Rd} = 88.8 \text{ kNm} \quad \text{OK}$$

Also with loads applied on top flange, the value of $M_{b,Rd}$ is greater than that computed using the general method.

In Table E7.1.1 the different values computed for both EC3 methods and for loads applied at the shear centre and on top flange are reported.

Table E7.1.1 LTB results.

Code	$M_{b,Rd}$ (kNm)	
	$z_g = 0$ mm	$z_g = 150$ mm
(a) General approach (EC3 – Section 6.3.2.2)	97.3	78.3
(b) Approach for I- and H-shaped profiles (EC3 – Section 6.3.2.3)	107.7	88.8

Example E7.2 Beam Design in Accordance with the US Approach

Verify an ASTM A992 W12 × 30 simply supported beam (Figure E7.2.1) in accordance with AISC, loaded by a uniform dead load of 0.40 kip/ft (5.8 kN/m) (first load case) and a uniform live load (second load case) of 0.70 kip/ft (10.2 kN/m). Loads are applied at the shear centre or on the top flange. The beam is not braced against lateral-torsional buckling along its entire length. Warping and lateral rotation free at both ends; lateral displacement and torsion prevented at both ends.

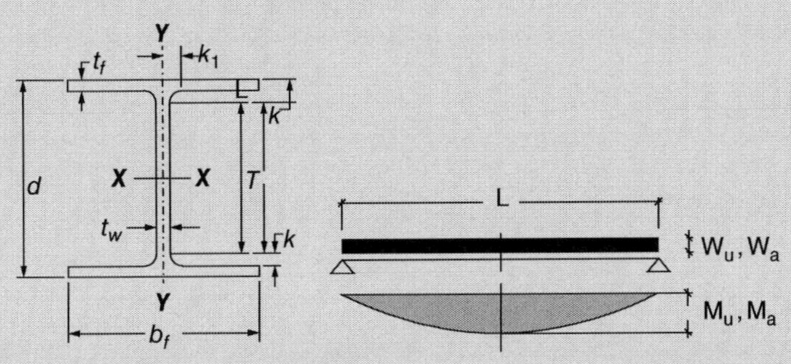

Figure E7.2.1

Geometrical properties:

$$d = 12.3 \text{ in.}(312 \text{ mm}) \quad A_g = 7.79 \text{ in.}^2(56.7 \text{ cm}^2)$$

$$b_f = 6.52 \text{ in.}(166 \text{ mm}) \quad Z_x = 43.1 \text{ in.}^3(706.3 \text{ cm}^3)$$

$$t_f = 0.44 \text{ in.}(11.2 \text{ mm}) \quad S_x = 37.6 \text{ in.}^3(632.5 \text{ cm}^3)$$

$$t_w = 0.26 \text{ in.}(6.6 \text{ mm}) \quad Z_y = 9.56 \text{ in.}^3(156.7 \text{ cm}^3)$$

$$k = 0.74 \text{ in.}(18.8 \text{ mm}) \quad S_y = 6.24 \text{ in.}^3(102.3 \text{ cm}^3)$$

$$L = 17 \text{ ft}(5.18 \text{ m}) \quad I_x = 238 \text{ in.}^4(9906 \text{ cm}^4)$$

$$L_x = 17 \text{ ft} \quad I_y = 20.3 \text{ in.}^4(845 \text{ cm}^4)$$

$$L_y = 17 ft \quad J = 0.457 \text{ in.}^4(19 \text{ cm}^4)$$

$$L_b = 17 \text{ ft} \quad C_w = 720 \text{ in.}^6(193300 \text{ cm}^6)$$

$$r_x = 5.21 \text{ in.}(13.2 \text{ cm})$$

$$r_y = 1.52 \text{ in.}(3.9 \text{ cm})$$

Material properties:

$$\text{Steel: ASTM A992} \quad F_y = 50 \text{ ksi}(345 \text{ MPa}) \quad F_u = 65 \text{ ksi}(448 \text{ MPa})$$

Limit the live load deflection to $L/360$.
Loads:
$w_D = 0.40$ kip/ft (5.8 kN/m); $w_L = 0.70$ kip/ft (10.2 kN/m)

$$\text{LRFD: } w_u = 1.2 \times 0.40 + 1.6 \times 0.70 = 1.6 \text{ kip/ft (23.4 kN/m)}$$

$$\text{ASD: } w_a = 0.40 + 0.70 = 1.1 \text{ kip/ft (16.1 kN/m)}$$

The required shear strength is:

$$\text{LRFD: } V_u = w_u L/2 = 1.6 \times 17/2 = 13.6 \text{ kips (60.5 kN)}$$
$$\text{ASD: } V_a = w_a L/2 = 1.1 \times 17/2 = 9.4 \text{ kips (41.8 kN)}$$

The required flexural strength is:

$$\text{LRFD}: M_u = \frac{w_u L^2}{8} = \frac{1.6 \times 17^2}{8} = 57.8 \text{ kip-ft (78.4 kNm)}$$
$$\text{ASD}: M_a = \frac{w_a L^2}{8} = \frac{1.1 \times 17^2}{8} = 39.7 \text{ kip-ft (58.9 kNm)}$$

Section classification for local buckling.
Flange:

$$b/t = (0.5 \times 6.52)/0.44 = 7.41 < 0.38\sqrt{E/F_y} = 0.38 \times \sqrt{29000/50} = 9.15 \rightarrow \text{compact}$$

Web:

$$h/t_w = (12.3 - 2 \times 0.74)/0.26 = 41.6 < 3.76\sqrt{E/F_y} = 3.76 \times \sqrt{29000/50} = 90.6 \rightarrow \text{compact}$$

ASTM A992 steel W12 × 30, subjected to bending moment, is a compact section.

Check for deflection.

Compute deflection at midspan:

$$f = \frac{5}{384}\frac{w_L L^4}{EI_x} = \frac{5}{384}\frac{0.7/12 \times (17 \times 12)^4}{29000 \times 238} = 0.19 \text{ in. (0.49 cm)} <$$
$$= 17 \times 12/360 = 0.57 \text{ in. (1.45 cm) OK}$$

Verification of shear.

Shear area:

$$A_w = d \cdot t_w = 12.3 \times 0.26 = 3.2 \text{ in.}^2 \left(20.6 \text{ cm}^2\right)$$
$$h/t_w = (12.3 - 2 \times 0.74)/0.26 = 41.6 \leq 2.24\sqrt{E/F_y} = 2.24 \times \sqrt{29000/50} = 53.9$$

Apply Method 1 (AISC 360-10 Section G2.1).
$C_v = 1.0$
Nominal shear strength V_n:

$$V_n = 0.6 F_y A_w C_v = 0.6 \times 50 \times 3.2 \times 1.0 = 96.0 \text{ kips (427 kN)}$$

Compute the available shear strength.

$$\text{LRFD: } \Phi_b M_n = 1.00 \times 96.0 = 96.0 \text{ kips (427 kN)} > V_u = 13.6 \text{ kips}$$

$$\text{ASD: } V_n/\Omega_v = 96.0/1.50 = 64.0 \text{ kips (285 kNm)} > V_a = 9.4 \text{ kips}$$

Verification of bending.

Apply verifications of case (a):

(a) Doubly symmetrical compact I-shaped members and channels bent about their major axis.

Compute plastic moment:

$$M_p = F_y \times Z_x = 50 \times 43.1/12 = 179.6 \text{ kip-ft}(243.5 \text{ kNm})$$

Flexural strength corresponding to lateral torsional buckling limit state M_{LT}):

Compute the modification factor C_b.

$$M_{max} = M_B = 57.8 \text{ kip-ft}$$

$$M_A = M_C = \left(\frac{wL}{2}\right)\cdot\frac{L}{4} - \left(\frac{wL}{4}\right)\left(\frac{1}{2}\cdot\frac{L}{4}\right) = \frac{3}{32}wL^2 = \frac{3}{32}\times 1.6 \times 17^2 = 43.4 \text{ kip-ft (58.8 kNm)}$$

$$C_b = \frac{12.5M_{max}}{2.5M_{max} + 3M_A + 4M_B + 3M_C} = \frac{12.5 \times 57.8}{2.5 \times 57.8 + 3 \times 43.4 + 4 \times 57.8 + 3 \times 43.4} = 1.136$$

$$L_p = 1.76 r_y \sqrt{\frac{E}{F_y}} = 1.76 \times 1.52 \times \sqrt{\frac{29000}{50}} = 64.4 \text{ in.}/12 = 5.37 \text{ ft (1.64 m)}$$

$$r_{ts} = \sqrt{\frac{\sqrt{I_y C_w}}{S_x}} = \sqrt{\frac{\sqrt{20.3 \times 720}}{38.6}} = 1.770 \text{ in. (4.5 cm)}$$

$c = 1$ (doubly symmetrical I-shape)

$$h_o = d - t_f = 12.3 - 0.44 = 11.86 \text{ in. (301.2 mm)}$$

$$L_r = 1.95 r_{ts}\frac{E}{0.7F_y}\sqrt{\frac{J\cdot c}{S_x h_0} + \sqrt{\left(\frac{J\cdot c}{S_x h_0}\right)^2 + 6.76\left(\frac{0.7F_y}{E}\right)^2}}$$

$$= 1.95 \times 1.770 \times \frac{29000}{0.7 \times 50}\sqrt{\frac{0.457 \times 1}{38.6 \times 11.86} + \sqrt{\left(\frac{0.457 \times 1}{38.6 \times 11.86}\right)^2 + 6.76 \times \left(\frac{0.7 \times 50}{29000}\right)^2}}$$

$$= 187.3 \text{ in.}/12 = 15.6 \text{ ft (4.76 m)}$$

$$L_b = 17 \text{ ft} > L_r = 15.6 \text{ ft.}$$

Hence:

(a) Consider loads applied at the shear centre.

$$F_{cr} = \frac{C_b \pi^2 E}{\left(\frac{L_b}{r_{ts}}\right)^2}\sqrt{1 + 0.078\frac{J\cdot c}{S_x h_0}\left(\frac{L_b}{r_{ts}}\right)^2}$$

$$= \frac{1.136 \times \pi^2 \times 29000}{\left(\dfrac{17 \cdot 12}{1.770}\right)^2} \sqrt{1 + 0.078 \frac{0.457 \times 1}{38.6 \times 11.86} \left(\frac{17 \cdot 12}{1.770}\right)^2} = 34.9 \text{ ksi} \, (241 \text{ MPa})$$

$$M_{LTB} = F_{cr} S_x = 34.9 \times 37.6/12 = 112.3 \text{ kip-ft} \, (152.3 \text{ kNm})$$

Compute the nominal flexural strength (M_n).

$$M_n = \min(M_p; M_{LTB}) = \min(179.6; 112.3) = 112.3 \text{ kip-ft}(152.3 \text{ kNm})$$

Compute the available strength.

$$\text{LRFD: } \Phi_b M_n = 0.90 \times 112.3 = 101.1 \text{ kip-ft} \, (137.1 \text{ kN})$$

$$\text{ASD: } M_n/\Omega_b = 112.3/1.67 = 67.3 \text{ kip-ft} \, (91.2 \text{ kN})$$

(b) Consider now loads applied on the top flange.

$$F_{cr} = \frac{C_b \pi^2 E}{\left(\dfrac{L_b}{r_{ts}}\right)^2} = \frac{1.136 \times \pi^2 \times 29000}{\left(\dfrac{17 \cdot 12}{1.770}\right)^2} = 24.5 \text{ ksi} \, (169 \text{ MPa})$$

$$M_{LTB} = F_{cr} S_x = 24.5 \times 38.6/12 = 77.7 \text{ kip-ft} \, (106.8 \text{ kNm})$$

Compute the nominal flexural strength (M_n).

$$M_n = \min(M_p; M_{LTB}) = \min(179.6; 77.7) = 77.7 \text{ kip-ft} \, (106.8 \text{ kNm})$$

Compute the available strength.

$$\text{LRFD: } \Phi_b M_n = 0.90 \times 78.7 = 70.9 \text{ kip-ft} \, (96.1 \text{ kNm})$$

$$\text{ASD: } M_n/\Omega_b = 78.7/1.67 = 47.1 \text{ kip-ft} \, (63.9 \text{ kNm})$$

Table E7.2.1 Values for design and allowable flexural strength.

Code	$\Phi_b M_n$, M_n/Ω_b (kip-ft)	
	Loads at shear centre	**Loads on top flange**
AISC 360-10 Section F2 – LRFD	101.1	70.9
AISC 360-10 Section F2 – ASD	67.3	47.1

In Table E7.2.1 computed values for design and allowable flexural strength are summarized.

CHAPTER 8
Torsion

8.1 Introduction

Sole pure torsion very rarely acts in steel structures. Most commonly, torsion occurs in combination with shear forces and bending moments. Although torsion does not generally have predominant effects on stress distribution in steel structures (compared to those associated with bending moments, shear or axial forces), the response of steel members under torsion is quite complex to predict and, if possible, the design has to be developed in order to reduce the effects associated with torsion. A fundamental role in the study of the torsional response of steel members is played by the *shear centre*. If this point lies on the line of application of an external load, no rotation of the cross-section occurs. Otherwise, the cross-section rotates with respect to an axis through this point, parallel to the longitudinal member axis; that is, torsional moments result from any applied force that does not pass through the shear centre.

In some cases, the shear centre can be directly determined. If a cross-section has two axes of symmetry, the shear centre coincides with its centroid as it does when the cross-section has a point of symmetry (typically unstiffened Z-shaped and stiffened Z-shaped members). In the case of L-, V- and T-shaped members – that is cross-sections composed by thin rectangular elements that intersect at a common point – this point is the shear centre. By neglecting the general case of cross-sections without any axis of symmetry, which is extremely unusual in steel construction practice, when a cross-section has one axis of symmetry the shear centre lies on this axis and its position can be determined on the basis of the traditional approaches of the theory of structures. If reference is made to thin-walled open cross-sections with constituent elements of equal thickness t, Table 8.1 can be considered for the most commonly used cross-section. It reports, for each of them, the expression of the eccentricity between shear centre and centroid.

The resistance of a structural member to torsional moment, T, may be considered to be the sum of two components: *pure torsional moment*, T_t, also identified as *St Venant's torsion* or the *uniform torsional moment* and *warping torsional moment*, T_ω or *non-uniform torsional moment*. From an equilibrium condition, this results in:

$$T = T_t + T_\omega \tag{8.1}$$

Pure torsion assumes that a cross-section that is plane in absence of torsion remains plane and only rotation occurs. As an example, a circular shaft subjected to torsion presents a situation where pure torsion exists as the only torsion action. Warping torsion is characterized by the

Structural Steel Design to Eurocode 3 and AISC Specifications, First Edition. Claudio Bernuzzi and Benedetto Cordova.

Table 8.1 Position of the shear centre C (point O identifies the centroid of cross-section).

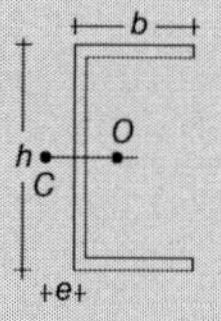

Cross-section	Shear centre position	Cross-section	Shear centre position
Channel (b, h, e, C, O)	$e=\frac{3b^2}{h+6b}$	Lipped channel (b, b₁, h, e, C, O)	$e=b\cdot\frac{3h^2b+6h^2b_1-8b_1^3}{h^3+6h^2b+6h^2b_1+8b_1^3-12hb_1^2}$
Mono-symmetric I (b₂, h, e, C, O)	$e=\frac{b_1^3h}{b_1^3+b_2^3}$	Hat section (b, b₁, h, e, C, O)	$e=b\frac{3h^2b+6h^2b_1-8b_1^3}{h^3+6h^2b+6h^2b_1+8b_1^3+12hb_1^2}$
Angle (a, e, C, O)	$e=0.707a$	Lipped angle (a, b, e, C, O)	$e=0.707ab^2\frac{3a-2b}{2a^3-(a-b)^3}$
Tee (b, h, e, C, O)	$e=0.5\frac{h^2}{h+b}$	Circular arc (α, r, e, C, O)	$e=2r\frac{\sin\alpha-\alpha\cos\alpha}{\alpha-\sin\alpha\cos\alpha}$

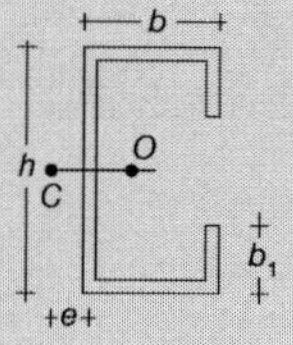

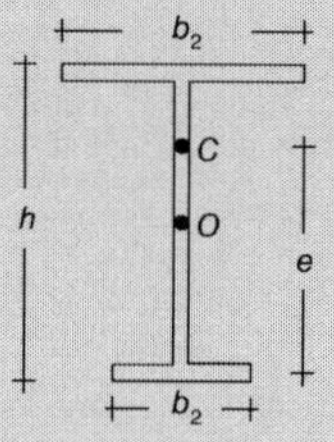

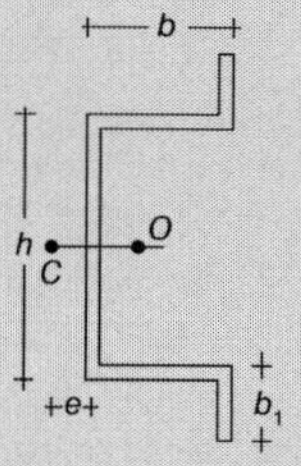

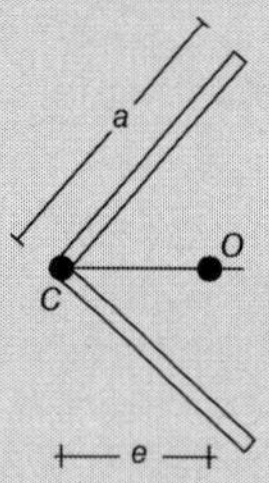

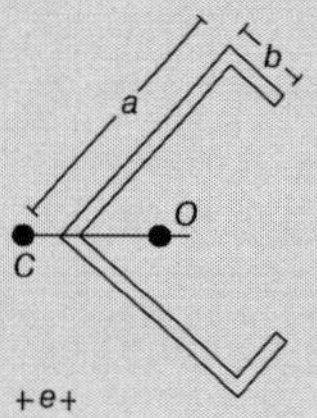

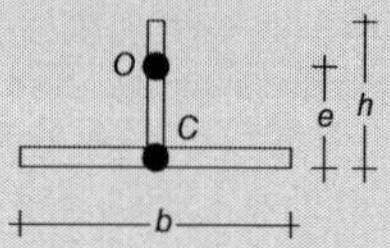

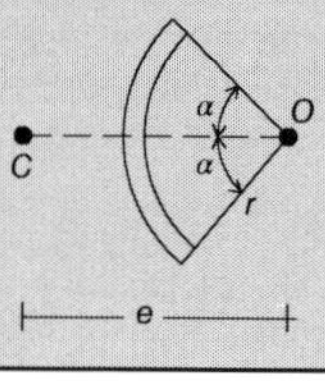

out-of-plane effect that arises when the flanges are laterally displaced during twisting, analogous to bending from laterally applied loads. Hence, the warping concept is strictly dependent on rejecting the assumption of planarity of the cross-section, which can be easily understood with reference to the beam-to-column rigid joint presented in Figure 8.1. In case of beam flanges free to deform in their own plane (i.e. cross-section planarity is not respected) warping occurs without the development of any warping torsional moment. Otherwise, if appropriate warping restraints, such as horizontal and/or diagonal stiffeners, are placed at the joint location, warping is prevented and a non-uniform torsional moment acts on the joint.

In steel structures, thin-walled open cross-sections are frequently used, which are composed by plates of three geometric dimensions (length, width and thickness) with an order of magnitude difference between them. In these cases St Venant's theory underestimates the resistance of the section and, therefore, the design phase requires the use of more sophisticated approaches, such as the ones based on the studies related to open thin-walled beams developed by Vlasov.

In cases of closed solid or boxed cross-section, pure torsion dominates the torsional response of the members and warping torsion can be neglected ($T_t \gg T_\omega$) for routine design. Otherwise, in case of open cross-sections, as in case of channels, I- and H-shaped profiles, warping torsion is relevant and in many design cases the contribution of pure torsion can be neglected ($T_\omega \gg T_t$).

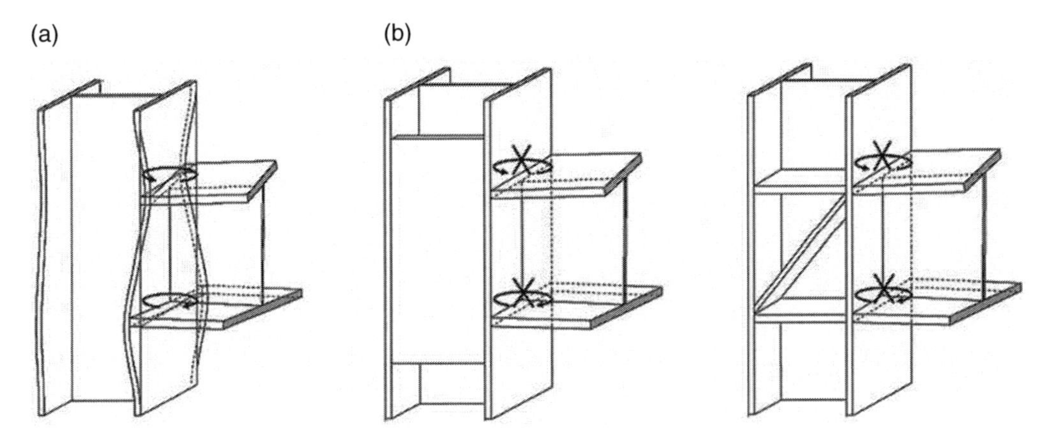

Figure 8.1 Free warping (a) and restrained warping and (b) in a beam-to-column rigid joint.

8.2 Basic Concepts of Torsion

On the basis of St Venant's theory, uniform torsion induces distortion that is caused by the rotation of the cross-sections around the longitudinal axis and the angle of rotation per unit length θ_L; that is, the rate of twist, which can be expressed as:

$$\theta_L = \frac{d\theta}{dx} = \frac{\theta}{L} = \text{constant} \tag{8.2}$$

where L is the member length and θ represents the relative rotation between two cross-sections at the longitudinal distance of x, or equivalently is the difference between the rotation of the cross-section of abscissa x and the one of abscissa zero (i.e. $x = 0$).

The angle of rotation per unit length can be associated with the moment of pure torsion, T_t, through the following equation:

$$T_t = G{\cdot}I_t{\cdot}\frac{d\theta}{dx} \tag{8.3}$$

where G is the shear modulus and I_t is the torsion constant (term $G{\cdot}I_t$ is the torsional rigidity of the members).

The shear stress distribution due to uniform torsion (Figure 8.2) can be obtained according to different methodologies, depending on the shape of the cross-section.

Circular cross-section: For members with solid or hollow circular cross-section, the shear stresses vary linearly with the distance from the shear centre (Figure 8.2a) and the maximum shear stress is:

$$\tau_{t,\max} = \frac{T_t{\cdot}R}{I_p} \tag{8.4a}$$

where R is the external cross-section radius and I_p is the polar moment of inertia.

In case of solid cross-sections of radius R, the polar moment of inertia is $I_p = \pi{\cdot}R^4/2$ while in case of a circular hollow cross-section, if R and R_i identify the external and internal radius, respectively, it results in $I_p = \pi(R^4 - R_i^{\,4})/4$.

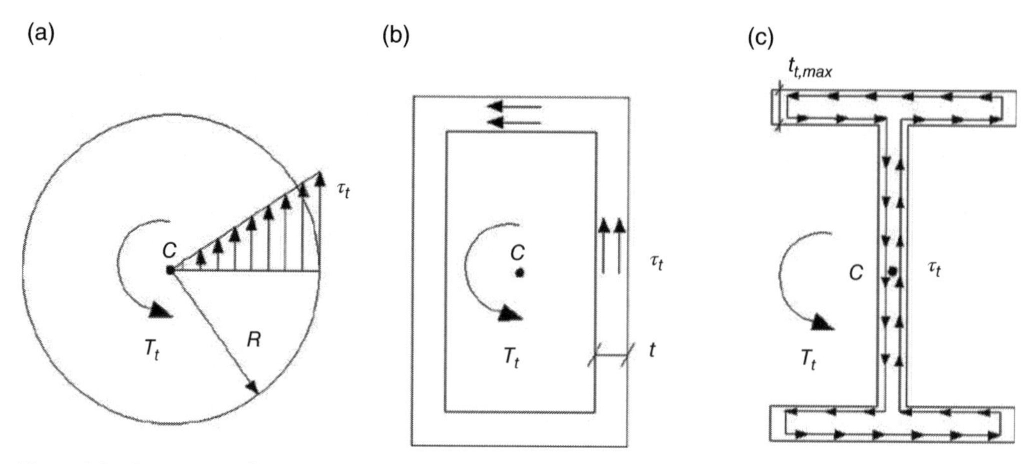

Figure 8.2 Pure torsion shear stresses for circular (a), rectangular hollow (b) and open cross-section (c).

Hollow closed cross-section: In the case of thin-walled members with hollow closed cross-sections, such as square or rectangular (Figure 8.2b), Bredt's theory can be used. Shear stresses distribution varies along the cross-section such that the shear flow ($\tau \cdot t$) is constant and the maximum shear stress, $\tau_{t,\max}$, is:

$$\tau_{t,\max} = \frac{T_t}{2 \cdot \Omega \cdot t} \tag{8.4b}$$

where Ω is the area defined by the middle line of the closed cross-section and t is the thickness.

Open cross-section: In the case of thin-walled open cross-sections, that is cross-sections composed by plates with the width-to-thickness ratio (b/t) approximately greater than 10, the maximum shear stress, $\tau_{t,\max}$, in the plate of maximum thickness t can be evaluated as:

$$\tau_{t,\max} = \frac{T_t \cdot t}{I_t} \tag{8.4c}$$

By neglecting the presence of the re-entrant corners between the plates constituting the cross-section (in case of I- or H-shaped members, in correspondence of the intersection between the web and the flange), that is neglecting the presence of the fillet regions where cross-section components are joined, I_t, can be evaluated as:

$$I_t = \sum_{i=1}^{n} \frac{b_i \cdot t_i^3}{3} \tag{8.5a}$$

where n is the number of plates of the cross-section and, for each of them, b_i and t_i indicate the width and the thickness, respectively of the i-plate.

Furthermore, the re-entrant corners between the beam flanges and the web for hot-rolled profiles and the welding fillets in welded beams can significantly increase the value of I_t obtained from Eq. (8.5a). In case of connected elements of the same thickness t, the contribution ΔI_t to the total torsional inertia of each corner can be estimated as:

$$\Delta I_t = [(p + q \cdot N) \cdot t]^4 \tag{8.5b}$$

where N is the ratio between the height of the corner and the thickness of the connected elements and p and q are empirical constants, the values of which are proposed in technical literature (usually, for hot-rolled angles $p = 0.99$ and $q = 0.22$ are frequently adopted).

The theoretical approaches to evaluate the stress distribution in the cross-section as applied to plates become complex, as both normal and tangential stresses are affected by warping as explained in the following. Though essentially the angle of twist is unaffected, the maximum shear stress, $\tau_{t,\max}$, can be evaluated as:

$$\tau_{t,\max} = \alpha \frac{T_t}{a \cdot b^2} \tag{8.5c}$$

where α depends on the ratio between the depth-to-width ratio (a/b), which can be evaluated directly from Table 8.2.

If equal angles are considered, with legs of length a and thickness t, term I_t can be evaluated as:

$$I_t = \frac{2[a-(2+N)t]t^3}{3} + [(0.99+0.22N)t]^4 \tag{8.5d}$$

Alternative to this equation, a more accurate expression of I_t is:

$$\begin{aligned} I_t = at^3 \cdot \left[\frac{1}{3} - 0.21\frac{t}{a}\left(1 - \frac{t^4}{12a^4}\right)\right] + (a-t)t^3 \left\{\frac{1}{3} - 0.105\frac{t}{(a-t)}\left[1 - \frac{t^4}{192(a-t)^4}\right]\right\} + \\ + \left(0.07 + 0.076\frac{r}{t}\right)\left[2 \cdot \left(2t + 3r - \sqrt{2 \cdot (2r+t)^2}\right)\right]^4 \end{aligned} \tag{8.5e}$$

where r is the root radius (fillet radius) between the legs with the limitation of $t < 2r$.

Warping torsion, which typically occurs in thin-walled open cross-sections (e.g. I- and H-profiles and channels), is much more complex to deal with. In order to better understand the effects associated with warping torsion, reference can be made to the cantilever beam presented in Figure 8.3, which is loaded by a transversal force F parallel to the flanges and applied to the top flange. This load condition can be correctly considered as the sum of a symmetrical (Figure 8.3b) and a hemi-symmetrical load case, described in part (b) and (c), respectively, of the figure.

Equal loads $F/2$ applied to the beam flanges inflect the beam along the weak axis and cross-sections keep their planarity when the symmetrical al load condition is considered.

On the basis of St Venant's bending theory, flanges are affected by a linear distribution of longitudinal normal stresses, σ_w, and by a parabolic distribution of shear stresses, τ_w. Otherwise, when flanges are affected by opposite horizontal forces, each flange bends along its plane with rotation and displacements opposite to the ones of the other flange. Also for this hemi-symmetrical load condition, flanges are affected by a linear distribution of longitudinal normal stresses, σ_w, and by a parabolic distribution of shear stresses, τ_w, but the cross-sections do not remain plane because of the warping occurrence. Furthermore, due to torsion, additional shear stresses τ_T act on the cross-section. The resulting stress distribution is hence quite complex, as is presented in Figure 8.4 where the distribution of the shear stresses due to pure

Table 8.2 Value of α for a rectangular cross-section (with a > b).

a/b	1.0	1.2	1.5	2.0	2.5	3.0	4.0	5.0	∞
α	4.81	4.57	4.33	4.07	3.88	3.75	3.55	3.44	3

torsion τ_t (Figure 8.4a), is shown together with the distributions of both shear stresses τ_ω (Figure 8.4b) and normal stresses $\sigma_{x,\omega}$ (Figure 8.4c), due to non-uniform torsion.

Function $w(x)$, which describes the warping effects, that is the field of displacements of the midline of the cross-section in the x-direction (longitudinal axis of the member), can be expressed as:

$$w(x) = \omega \cdot \frac{d\theta}{dx} \tag{8.6}$$

where ω is the sectorial area defined as:

$$\omega = \omega(s) = \int_0^N r_t(s) ds \tag{8.7}$$

The sectorial area is the double of the area swept by the radius $r_t(s)$, which moves along the midline of the cross-section (Figure 8.5) from the point $s = 0$ to the point under consideration (the swept area is generally taken to be positive when radius $r_t(s)$ rotates in the positive direction). The term $r_t(s)$ is generally assumed to be the distance between the shear centre (point C) and the axis tangent to the cross-section in point s.

Making reference to the compatibility conditions, longitudinal strain $\varepsilon_{x,\omega}$ takes place due to displacement w along the longitudinal member (x-axis), which can be expressed as:

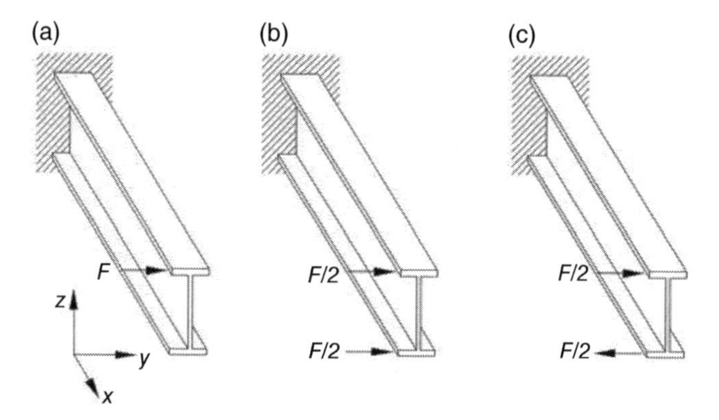

Figure 8.3 Torsion in an open cross-section member: the loading condition (a) considered as the sum of a symmetrical (b) and a hemi-symmetrical loading condition (c).

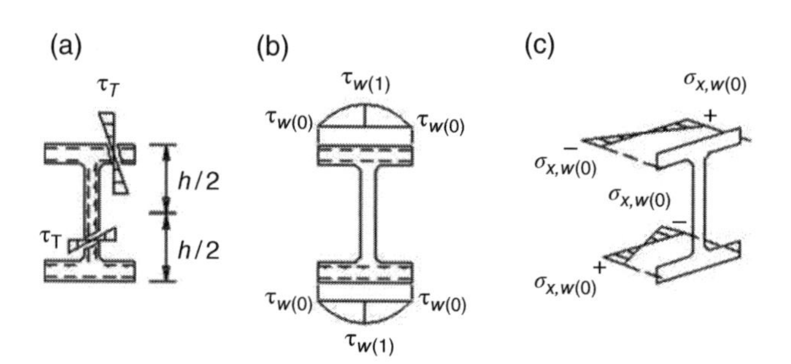

Figure 8.4 Distribution of stresses in the cantilever beam of Figure 8.3 due to pure torsion (shear stresses τ_t in (a)) and to the non-uniform torsion shear stresses τ_ω in (b) and normal stresses $\sigma_{x,\omega}$ in (c).

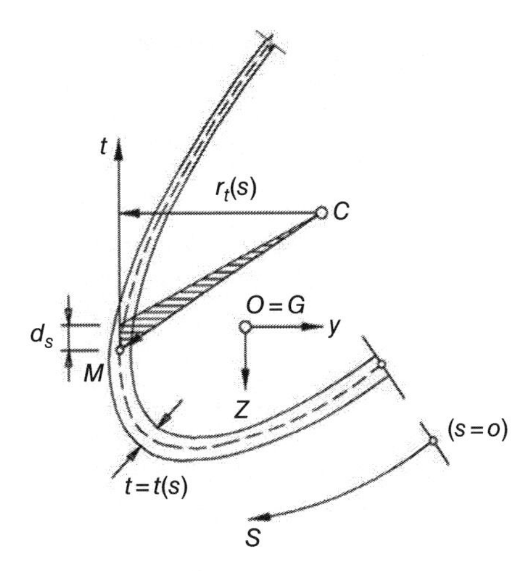

Figure 8.5 Shaded area to evaluate the sectorial area.

$$\varepsilon_{x,\omega} = \frac{dw(x)}{dx} \tag{8.8a}$$

As a consequence, normal stress $\sigma_{x,\omega}$, due to the prevention of warping, can be obtained directly by Hooke's law as:

$$\sigma_{x,\omega} = E \cdot \omega \cdot \frac{d^2\theta}{dx^2} \tag{8.8b}$$

The values of both strain $\varepsilon_{x,\omega}$ and stress $\sigma_{x,\omega}$ depend strictly on the considered point, owing to the definition of the sectorial area ω. As anticipated, for the case of non-uniform torsion, in addition to normal stress $\sigma_{x,\omega}$, warping also generates tangential shear τ_ω (Figure 8.4c) whose value can be obtained from:

$$\tau_\omega = -\frac{E \cdot \int_A \omega \cdot dA}{t} \frac{d^3\theta}{dx^3} = -\frac{E \cdot S_\omega}{t} \frac{d^3\theta}{dx^3} \tag{8.9}$$

where t is the thickness of the part of the cross-section under consideration, A is the area of the cross-section and S_ω is the first moment of area of the sectorial area (static moment of the sectorial area), defined as:

$$S_\omega = \int_A \omega \cdot dA \tag{8.10}$$

Non-uniform torsional moment T_ω is given by the expression:

$$T_\omega = -E \cdot I_\omega \cdot \frac{d^3\theta}{dx^3} \tag{8.11}$$

where I_ω is the moment of inertia of the sectorial area (sectorial moment of inertia) defined as:

$$I_\omega = \int_A \omega^2 \cdot dA \tag{8.12}$$

By substituting previous equations, non-uniform shear stress τ_ω can be expressed as:

$$\tau_\omega = \frac{T_\omega \cdot S_\omega}{I_\omega \cdot t} \tag{8.13}$$

Generally, with reference to non-uniform torsion, a new variable is conveniently introduced, which is identified as *bimoment* (*B*).

With reference to the cantilever beam in Figure 8.3, beam flanges are forced to bend in their plane. This bending generates clockwise rotation in one flange and anti-clockwise rotation in the other one so as resulting effect two equal and opposite bending moments and rotations are generated. This force system, which is induced in the flanges by warping restraint, that is the *bimoment*, is usually identified with symbol *B*, and is expressed as:

$$B = B(x) = \int_0^x T_\omega(c) \cdot dc = -E \cdot I_\omega \cdot \frac{d^2\theta}{dx^2} \tag{8.14a}$$

Bimoment *B* has the measurement unit of force × length^2 (*moment* × *distance*) and in the case of the cantilever beam of Figure 8.3 can be expressed as:

$$B = 2 \cdot \frac{F}{2} \cdot \frac{h}{2} \cdot (L - x) \tag{8.14b}$$

Normal stress ($\sigma_{x,\omega}$) associated with the warping torsion can be also expressed as:

$$\sigma_{x,\omega} = \frac{B}{I_\omega} \omega \tag{8.15}$$

8.2.1 I- and H-Shaped Profiles with Two Axes of Symmetry

As previously mentioned, for I- and H-shaped cross-sections with two axes of symmetry, the shear centre coincides with the centroid. By identifying with b and h the flange width and the beam depth, respectively, alternative to Eq. (8.5a), torsional constant can be evaluated more accurately as:

$$I_t = \frac{2(b - 0.63 t_b) t_f^3}{3} + \frac{(h - t_f) t_w^3}{3} + \frac{2 t_w}{t_f}\left(0.145 + 0.1 \frac{r}{t_f}\right)\left[\frac{(r + t_w/2)^2 + (r + t_f)^2 - r^2}{2r + t_f}\right]^4 \tag{8.16}$$

where r is the fillet radius and t_f and t_w are the thickness of the flange and of the web, respectively.

The sectorial area (ω) associated with each half flange, can be approximated as:

$$\omega = \omega(s) = \frac{(h - t_f)}{2} \int_0^s ds = \frac{(h - t_f) \cdot s}{2} \tag{8.17a}$$

Functions S_ω and I_ω are given, respectively, by expressions:

$$S_\omega = \int_0^s \frac{(h-t_f)\cdot s}{2}\cdot(t_f\cdot ds) = \frac{[(h-t_f)\cdot t_f]\cdot s^2}{4} \tag{8.18a}$$

$$I_\omega = \int_A \omega^2\cdot dA = \int_0^s \left[\frac{(h-t_f)\cdot s}{2}\right]^2 (t_f\cdot ds) = \frac{(h-t_f)^2\cdot t_f\cdot s^3}{12} \tag{8.19a}$$

Figure 8.6 proposes the distribution of ω and S_ω functions for the doubly symmetrical I- and H-shaped profiles.

For design purposes, reference has to be made to the maximum value of these functions in order to base the design on more severe verification conditions. By substituting with the variable s the values of the relevant coordinates, which are referred to the midline of the cross-section ($s = b$), we can obtain:

$$\omega_{\max} = b\cdot\frac{(h-t_f)}{4} \tag{8.17b}$$

$$S_{\omega,\max} = b^2\cdot\frac{t_f\cdot(h-t_f)}{16} \tag{8.18b}$$

$$I_\omega = 4\cdot\frac{(h-t_f)^2\cdot t_f\cdot b^3}{12\cdot 8} \approx I_z\left(\frac{h-t_f}{2}\right)^2 \tag{8.19b}$$

where I_z is the moment of inertia along the weak axis of the cross-section.

By re-considering the example of the cantilever beam in Figure 8.3, the stress distribution and the maximum value of $\sigma_{x,\omega}$ and τ_ω can be evaluated by simple considerations. In particular, torsional moment T is due to the horizontal force and can expressed as $T = (F/2)\cdot(h-t_f)$, as it results from the load condition in Figure 8.3c.

Bimoment B acting at the restrained cross-section end of the cantilever ($x = L$), where warping is totally prevented, assumes the value $B = T\cdot L$ and the maximum normal stress ($\sigma_{x,\omega,max}$) acting at the flange boundary can be evaluated as:

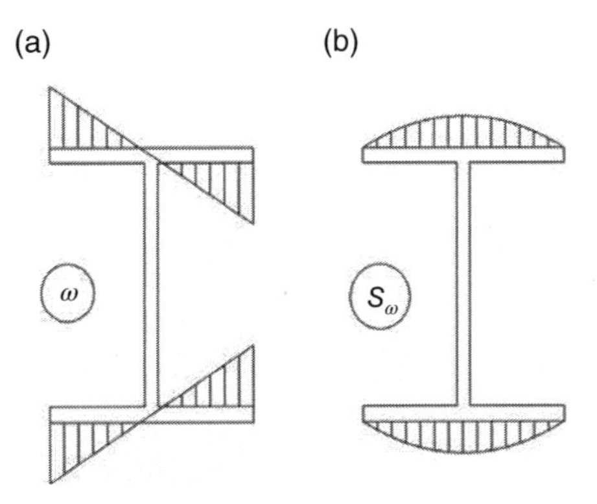

Figure 8.6 Distribution of ω (a) and S_ω (b) in case of a bi-symmetrical I- and H-shaped profile.

$$\sigma_{x,\omega,\max}(x)=\frac{B}{I_\omega}\cdot\omega=\frac{T\cdot L}{\left[I_z\left(\frac{h-t_f}{2}\right)^2\right]}\cdot\left[b\cdot\frac{(h-t_f)}{4}\right]=\frac{T\cdot L}{I_z}\cdot\frac{b}{h-t_f} \tag{8.20}$$

Maximum shear stress ($\tau_{\omega,max}$) corresponds with the centroid of the flange at the fixed cantilever end (where warping is totally prevented, being $T=T_\omega$) and assumes the value:

$$\tau_{\omega,\max}=\frac{T_\omega\cdot S_\omega}{I_\omega\cdot t}=\frac{T\cdot\left[b^2\cdot\frac{t_f\cdot(h-t_f)}{16}\right]}{\left[I_z\left(\frac{h-t_f}{2}\right)^2\right]\cdot t_f} \tag{8.21}$$

It should be noted that the design value of both $\sigma_{x,\omega}$ and τ_ω should also be obtained by applying the bending beam theory to the beam flanges. With reference to the case in Figure 8.3, by approximating the moment of inertia of the beam flange in its plane I_{zf} as $I_{zf}=t_f\cdot b^3/12$ and neglecting the presence of the corners between the beam flanges and the web, maximum normal stress ($\sigma_{x,\omega,\max}$) can be obtained as:

$$\sigma_{x,\omega,\max}(x)=\frac{(F\cdot L)\cdot\frac{b}{2}}{I_{zf}}=\cdot\frac{\left(\frac{T}{(h-t_f)}\cdot L\right)\cdot\frac{b}{2}}{\left(\frac{t_f b^3}{12}\right)}=6\cdot\frac{T\cdot L}{(h-t_f)\cdot t_f\cdot b^2} \tag{8.22}$$

In a similar way, by considering the shear distribution of a rectangular cross-section based on the Jourawsky's approach for the flange, the definition of the maximum shear stress ($\tau_{\omega,\max}$) is:

$$\tau_{\omega.\max}=\frac{3}{2}\frac{T}{(h-t_f)}\frac{1}{b\cdot t_f} \tag{8.23}$$

It is worth mentioning that Eq. (8.22) coincides with Eq. (8.20) if I_z is evaluated neglecting the web contribution (i.e. $I_z=b^3t_f/6$). Under this assumption, Eqs. (8.21) and (8.23) are also coincident.

8.2.2 Mono-symmetrical Channel Cross-Sections

The case of a channel cross-section with one axis of symmetry is considered here with reference to flanges and webs of different thickness (Figure 8.7). It can be convenient to evaluate, at first, the shear centre location (point C), usually measured by the distance e from the midline of the web, on the basis of the distribution of the shear stresses. By considering Jourawski's theory, a shear force V_z, applied to the cross-section in a direction parallel to the web, is balanced by a shear flow distribution ($\tau\cdot t$), which is obtained by the product between the shear stress (τ) and the thickness (t). As to $\tau\cdot t$, a parabolic distribution acts on the web of resultant V_w and parallel to the web, and a linear distribution is in each flange of resultant V_f and parallel to the flange. By considering the equilibrium conditions, external force (V_z) is balanced by the shear stresses resultant (V_w), that is $V_w=V_z$, and the resulting force on each flange generates a torsional moment.

As previously mentioned, the shear centre (C) is located where no torsion occurs when flexural shears act in planes passing through that location. This definition is used to identify the position of

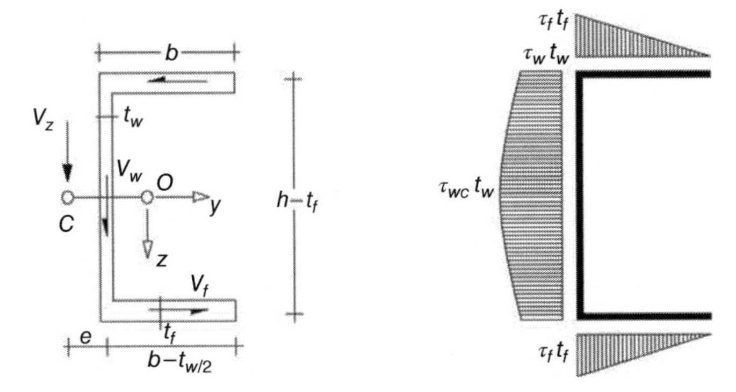

Figure 8.7 Distribution of the shear stresses in a channel loaded on the shear centre.

the shear centre. In particular, the equilibrium condition with reference to the rotation of the cross-section is satisfied if:

$$V_z \cdot e = V_w \cdot e = V_f \cdot (h - t_f) \tag{8.24a}$$

As a consequence, distance e can be obtained directly as:

$$e = \frac{V_f \cdot (h - t_f)}{V_w} \tag{8.24b}$$

If I_y identified the moment of inertia with reference to the y–y axis of the cross-section, shear stresses can be directly obtained by means of the Jourawsky approach. In particular, at the intersection between the flange and the web, due to the constant values of the shear flow $(\tau_j \cdot t_j)$, it can be assumed:

$$\tau_f \cdot t_f = \tau_\omega \cdot t_w = \frac{V_z}{I_y} \cdot \frac{\left[(b - t_w/2) \cdot t_f\right] \cdot (h - t_f)}{2} \tag{8.25a}$$

where t is the thickness, subscripts f and w are related to the flange and the web, respectively, and h and b are the width and the height of the channel, respectively.

At the centre of the web, in correspondence with the symmetry axis, the shear stress τ_{wc} is given by expression:

$$\tau_{wc} \cdot t_w = \frac{V_z}{I_y} \cdot \frac{\left[(b - t_w/2) \cdot t_f\right] \cdot (h - t_f)}{2} + \frac{V_z}{I_y} \cdot \frac{\left[(h - t_f)^2 \cdot t_w\right]}{8} \tag{8.25b}$$

Resulting forces V_f and V_w can be expressed as

$$V_f = \frac{(\tau_f \cdot t_f) \cdot (b - t_w/2)}{2} \tag{8.26}$$

$$V_w = (\tau_f \cdot t_f) \cdot (h - t_f) + \frac{2}{3}\left[\left((\tau_{wc} - \tau_w) \cdot t_w\right] \cdot (h - t_f)\right)\right] \tag{8.27}$$

By substituting the expressions of the resulting forces in Eq. (8.24b) the shear centre position can be identified on the basis of the sole geometry of the cross-section as:

$$e = \frac{3(b - t_w/2)^2 \cdot t_f}{6(b - t_w/2) \cdot t_f + (h - t_f) \cdot t_w} \tag{8.28}$$

As for the case of I- and H-shaped profiles, torsional properties can be easily evaluated for channel cross-sections. In Figure 8.8a,b the distribution of the sectorial area, ω, and the first moment of the sectorial area, S_ω, are indicated, which can be qualitatively associated with the distribution of the normal stresses $\sigma_{x,\omega}$ and shear stresses τ_ω.

As to the distribution of the sectorial area in the key points of the cross-section, the following values have to be considered:

$$\omega_A = \frac{(h - t_f)}{2}\left(b - \frac{t_w}{2} - e\right) \tag{8.29a}$$

$$\omega_B = -e \cdot \frac{(h - t_f)}{2} \tag{8.29b}$$

As to the local values defining the distribution of the first moment of the sectorial area, the result is:

$$S_{\omega B} = \frac{t_f \cdot \left(b - \frac{t_w}{2}\right) \cdot (h - t_f)}{2}\left(\frac{b}{2} - \frac{t_w}{4} - e\right) \tag{8.30a}$$

$$S_{\omega 1} = S_{\omega B} - \frac{t_w \cdot e \cdot (h - t_f)^2}{8} \tag{8.30b}$$

$$S_{\omega 2} = \frac{t_f \cdot (h - t_f)}{4}\left[\left(b - \frac{t_w}{2}\right) - e\right]^2 \tag{8.30c}$$

Second moment of sectorial area (warping constant) can be expressed as:

$$I_\omega = \frac{\left(b - \frac{t_w}{2}\right)^3 (h - t_f)^2 t_f}{12}\left[\frac{2(h - t_f)t_w + 3\left(b - \frac{t_w}{2}\right)t_f}{(h - t_f)t_w + 6\left(b - \frac{t_w}{2}\right)t_f}\right] \tag{8.31}$$

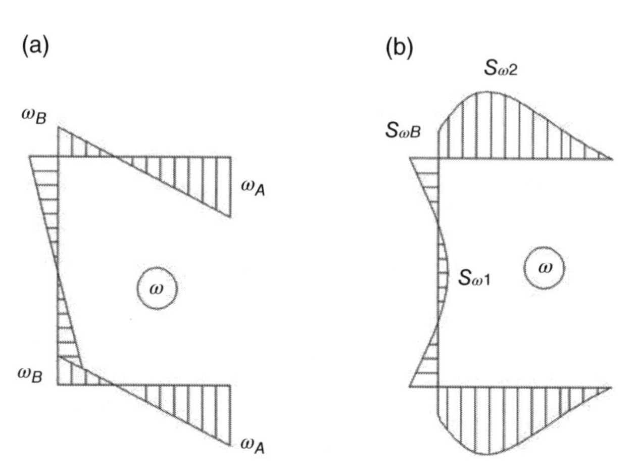

Figure 8.8 Distribution for a channel section of: (a) sectorial area (ω) and (b) first moment of area of the sectorial area (S_ω).

8.2.3 Warping Constant for Most Common Cross-Sections

As already mentioned, in case of L-, T- and V-shaped profiles, that is a member with cross-sections composed by thin plates having a common point of intersection that is the shear centre, the warping constant is always zero ($I_\omega = 0$); otherwise it has to be computed. Warping constants are reported in Table 8.3 for some of the most common shapes of cross-section in which the centroid coincides with the shear centre, assuming that all the constituent plates have equal thickness t.

Table 8.4 proposes the warping constant I_ω for some of the most common cross-sections with one axis of symmetry, under assumption of constant thickness of plates forming cross-section.

Furthermore, with reference to the more general cases of cross-section composed of plates of different thickness, a simplified procedure can be used to evaluate cross-section constants, which requires division of the cross-section into n plates, each of them identified with a progressive number (from 1 to n). Nodes are inserted between the parts, which are numbered from 0 to n (Figure 8.9). As a consequence the generic plate i is defined by nodes $i-1$ and i. Each node has coordinates y_i and z_i and each part has a thickness t_i, which is constant for the plate. In the following the expression of the main geometrical properties relevant for torsional design are proposed:

- Area A is the sum of the area of plates forming the cross-section:

$$A = \sum_{i=1}^{n} dA_i = \sum_{i=1}^{n} t_i \sqrt{(\bar{y}_i - \bar{y}_{i-1})^2 + (\bar{z}_i - \bar{z}_{i-1})^2} \tag{8.32}$$

- Moments of inertia $S_{\bar{y}0}$, and $S_{\bar{z}0}$ are defined with respect to original $\bar{y}0$ - and $\bar{z}0$ -axis:

$$S_{\bar{y}0} = \sum_{i=1}^{n} [\bar{z}_i + \bar{z}_{i-1}] \cdot \frac{dA_i}{2} \tag{8.33a}$$

$$S_{\bar{z}0} = \sum_{i=1}^{n} [\bar{y}_i + \bar{y}_{i-1}] \cdot \frac{dA_i}{2} \tag{8.33b}$$

- Coordinates z_{gc} and y_{gc} of the centroid:

$$\bar{z}_{gc} = \frac{S_{y0}}{A} \tag{8.34a}$$

$$\bar{y}_{gc} = \frac{S_{z0}}{A} \tag{8.34b}$$

Table 8.3 Warping constants when the centroid is coincident with the shear centre.

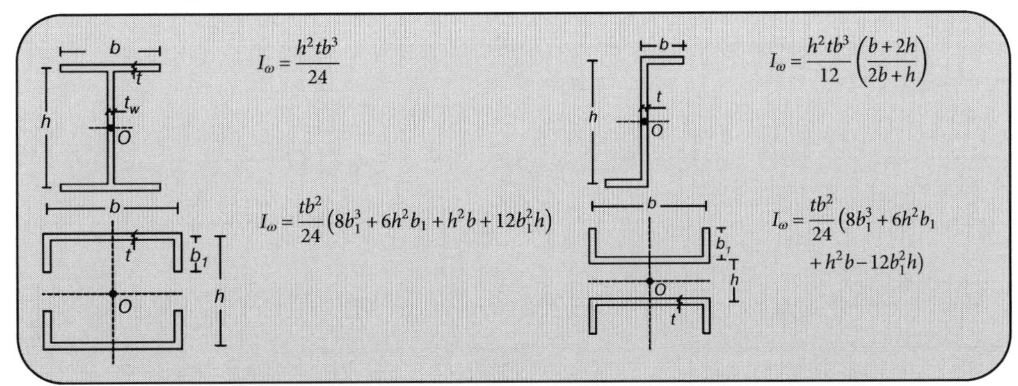

Table 8.4 Warping constants for mono-symmetrical cross-sections.

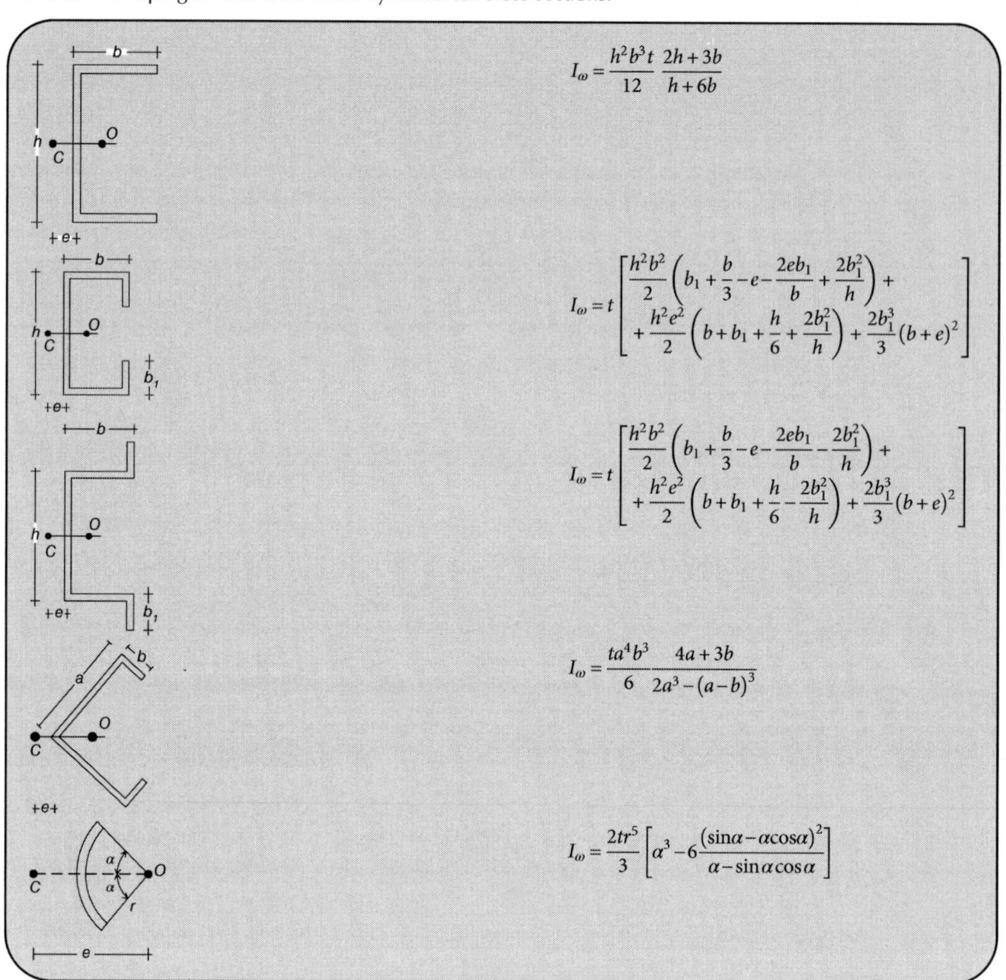

Cross-section	Warping constant
Channel (h, b, e, C, O)	$I_\omega = \dfrac{h^2 b^3 t}{12}\dfrac{2h+3b}{h+6b}$
Lipped channel (h, b, b_1, e, C, O)	$I_\omega = t\left[\dfrac{h^2b^2}{2}\left(b_1+\dfrac{b}{3}-e-\dfrac{2eb_1}{b}+\dfrac{2b_1^2}{h}\right)+\dfrac{h^2e^2}{2}\left(b+b_1+\dfrac{h}{6}+\dfrac{2b_1^2}{h}\right)+\dfrac{2b_1^3}{3}(b+e)^2\right]$
Channel with outward lips (h, b, b_1, e, C, O)	$I_\omega = t\left[\dfrac{h^2b^2}{2}\left(b_1+\dfrac{b}{3}-e-\dfrac{2eb_1}{b}-\dfrac{2b_1^2}{h}\right)+\dfrac{h^2e^2}{2}\left(b+b_1+\dfrac{h}{6}-\dfrac{2b_1^2}{h}\right)+\dfrac{2b_1^3}{3}(b+e)^2\right]$
Lipped angle (a, b, e, C, O)	$I_\omega = \dfrac{ta^4b^3}{6}\dfrac{4a+3b}{2a^3-(a-b)^3}$
Circular arc (r, α, e, C, O)	$I_\omega = \dfrac{2tr^5}{3}\left[\alpha^3-6\dfrac{(\sin\alpha-\alpha\cos\alpha)^2}{\alpha-\sin\alpha\cos\alpha}\right]$

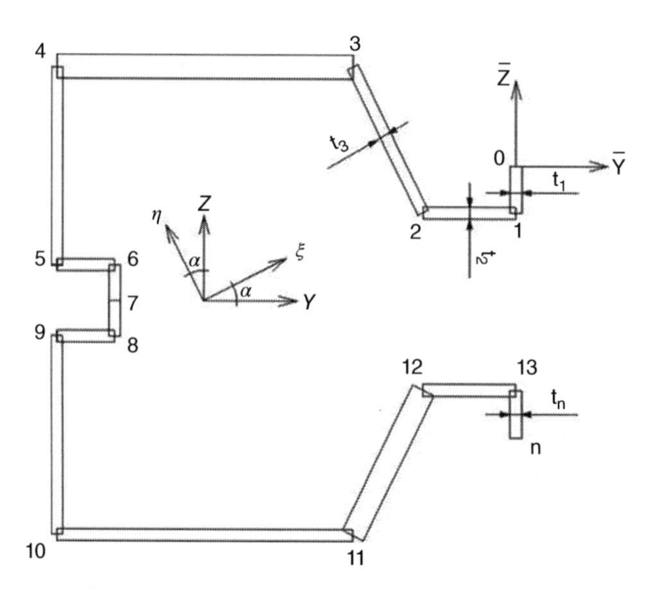

Figure 8.9 Cross-section nodes.

- Moments of inertia $I_{\bar{y}0}$, $I_{\bar{z}0}$ and $I_{\bar{y}\bar{z}0}$ defined with respect to the original $\bar{y}0$ - and $\bar{z}0$ -axis:

$$I_{\bar{y}0} = \sum_{i=1}^{n} \left[(\bar{z}_i)^2 + (\bar{z}_{i-1})^2 + \bar{z}_i \cdot \bar{z}_{i-1}\right] \cdot \frac{dA_i}{3} \tag{8.35a}$$

$$I_{\bar{z}0} = \sum_{i=1}^{n} \left[(\bar{y}_i)^2 + (\bar{y}_{i-1})^2 + \bar{y}_i \cdot \bar{y}_{i-1}\right] \cdot \frac{dA_i}{3} \tag{8.35b}$$

$$I_{\bar{y}\bar{z}0} = \sum_{i=1}^{n} \left(2 \cdot \bar{y}_{i-1} \cdot \bar{z}_{i-1} + 2\bar{y}_i \cdot \bar{z}_i + \bar{y}_{i-1} \cdot \bar{z}_i + \bar{y}_i \cdot \bar{z}_{i-1}\right) \cdot \frac{dA_i}{6} \tag{8.35c}$$

- Moments of inertia, I_y, I_{zo} and I_{yz} with respect to the y- and z-axis passing through the centroid:

$$I_y = I_{\bar{y}0} - A \cdot \bar{z}_{gc}^2 \tag{8.36a}$$

$$I_z = I_{\bar{z}0} - A \cdot \bar{y}_{gc}^2 \tag{8.36b}$$

$$I_{yz} = I_{\bar{y}\bar{z}0} - \frac{S_{\bar{y}0} \cdot S_{\bar{z}0}}{A} \tag{8.36c}$$

- Principal axes:

$$\alpha = \frac{1}{2} \arctan\left(\frac{2I_{yz}}{I_z - I_y}\right) \quad \text{if } (I_z - I_y) \neq 0 \quad \text{otherwise } \alpha = 0 \tag{8.37}$$

$$I_\xi = \frac{1}{2}\left[I_y + I_z - \sqrt{(I_z - I_y)^2 + 4 \cdot I_{yz}^2}\right] \tag{8.38a}$$

$$I_\eta = \frac{1}{2}\left[I_y + I_z + \sqrt{(I_z - I_y)^2 + 4 \cdot I_{yz}^2}\right] \tag{8.38b}$$

- Sectorial coordinates:

$$\omega_0 = 0 \tag{8.39a}$$

$$\omega_{0i} = \bar{y}_{i-1}\bar{z}_1 - \bar{y}_i\bar{z}_{1-1} \tag{8.39b}$$

$$\omega_i = \omega_{i-1} + \omega_0 \tag{8.39c}$$

- Mean values of the sectorial coordinate:

$$I_\omega = \sum_{i=1}^{n} (\omega_{i-1} + \omega_i) \cdot \frac{dA_i}{2} \tag{8.40}$$

- Sectorial constants:

$$\begin{aligned} I_{y\omega} &= I_{y\omega 0} - \frac{S_{z0} \cdot I_\omega}{A} = \\ &= \sum_{i=1}^{n} \left(2 \cdot \bar{y}_{i-1} \cdot \omega_{i-1} + 2\bar{y}_i \cdot \omega_i + \bar{y}_{i-1} \cdot \omega_i + \bar{y}_i \cdot \omega_{i-1}\right) \cdot \frac{dA_i}{6} - \frac{S_{z0} \cdot I_\omega}{A} \end{aligned} \tag{8.41a}$$

$$I_{z\omega} = I_{z\omega 0} - \frac{S_{y0} \cdot I_{\omega}}{A} = \\ = \sum_{i=1}^{n} (2 \cdot \bar{z}_{i-1} \cdot \omega_{i-1} + 2\bar{z}_i \cdot \omega_i + \bar{z}_{i-1} \cdot \omega_i + \bar{z}_i \cdot \omega_{i-1}) \cdot \frac{dA_i}{6} - \frac{S_{y0} \cdot I_{\omega}}{A} \tag{8.41b}$$

$$I_{\omega\omega} = I_{\omega\omega 0} - \frac{I_{\omega}^2}{A} = \sum_{i=1}^{n} \left((\omega_i)^2 + (\omega_{i-1})^2 + \omega_i \cdot \omega_{i-1} \right) \cdot \frac{dA_i}{3} - \frac{I_{\omega}^2}{A} \tag{8.42}$$

- Shear centre coordinates $\left(I_y I_z - I_{yz}^2 \right) \neq 0$:

$$\bar{y}_{sc} = \frac{I_{z\omega} I_z - I_{y\omega} I_{yz}}{I_y I_z - I_{yz}^2} \tag{8.43a}$$

$$\bar{z}_{sc} = \frac{-I_{y\omega} I_y - I_{z\omega} I_{yz}}{I_y I_z - I_{yz}^2} \tag{8.43b}$$

- Warping constant (I_w):

$$I_w = I_{\omega\omega} + \bar{z}_{sc} \cdot I_{y\omega} - \bar{y}_{sc} \cdot I_{z\omega} \tag{8.44}$$

- Torsion constants (I_t):

$$I_t = \sum_{i=1}^{n} dA_i \cdot \frac{t_i^2}{3} \tag{8.45}$$

8.3 Member Response to Mixed Torsion

Member response to mixed torsion, in both statically determinate and indeterminate structures, depends strictly on the torsional restraints at its end. Traditional ideal restraints of fixed ends, typically used for members in bending, can be differently classified when torsion is considered. In Figure 8.10 two types of torsional restraints are presented:

(a) *simple torsional restraint* (identified, for sake of simplicity as STR), which can absorb the torsional end moment but cannot prevent warping and hence is completely free (i.e. no planarity of cross-section is guaranteed by this restraint);
(b) *fixed torsional restraint* (FTR), which can absorb torsional end moment and prevent warping completely.

As already mentioned, the applied torsional moment is resisted by a combination of uniform and warping torsion. As results from Eq. (8.1), by substituting the definitions given by Eqs. (8.3) and (8.11), we can obtain:

$$T = G \cdot I_t \cdot \frac{d\theta}{dx} - E \cdot I_{\omega} \cdot \frac{d^3\theta}{dx^3} \tag{8.46a}$$

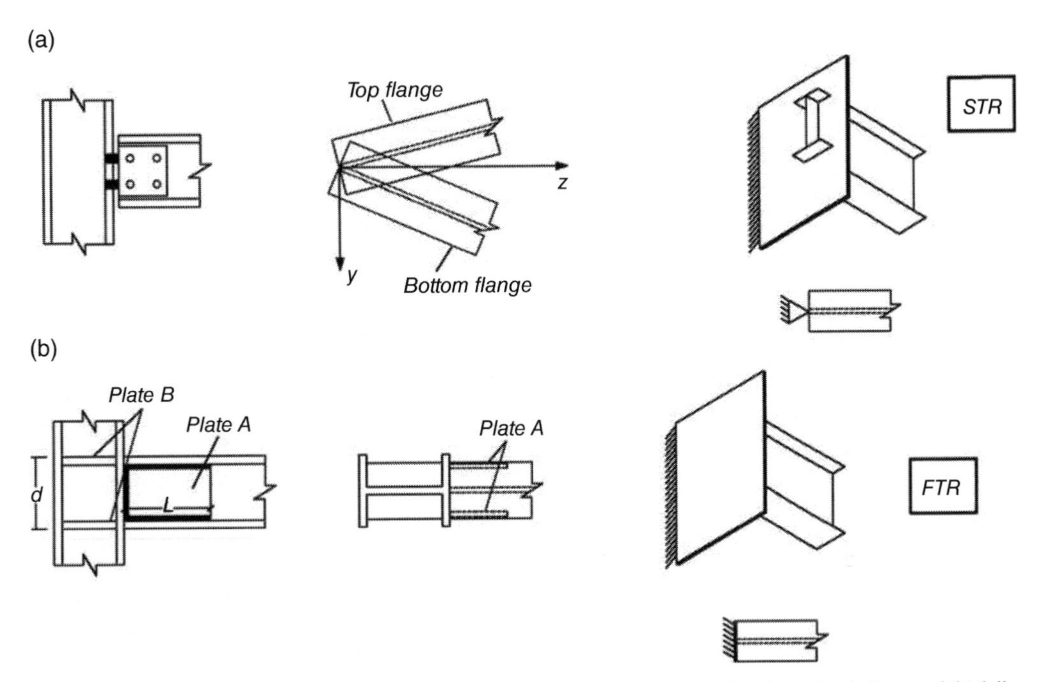

Figure 8.10 Examples of torsional restraints: (a) simple support restraining torsion (warping is free) and (b) fully fixed restraint to torsion (warping is prevented).

Reference can be made suitably to the torsion parameter λ_T, defined as $\lambda_T = \sqrt{G{\cdot}I_t/E{\cdot}I_\omega}$, which indicates the dominant type of torsion. In case of uniform torsion, λ_T is very large, as for thin-walled closed-section members whose torsional rigidities are very high (members with narrow rectangular sections, angle and tee-sections, whose warping rigidities are negligible). On the other hand, if the second component of resistance due to torsional loading completely dominates with respect to the first, the member is in a limiting state of non-uniform torsion referred to as *warping torsion*. This may occur when the torsion parameter λ_T is very limited, which is the case for some very thin-walled open sections (such as light gauge cold-formed sections) whose torsional rigidities are very small.

By introducing the λ_T term, Eq. (8.46a) can be re-written as:

$$\frac{T}{G{\cdot}I_t} = \cdot\frac{d\theta}{dx} - \frac{1}{\lambda_T^2}\cdot\frac{d^3\theta}{dx^3} \tag{8.46b}$$

The differential equation permits the following general solution:

$$\theta(x) = a + b{\cdot}\sinh(\lambda_T x) + c{\cdot}\cosh(\lambda_T x) + \theta_p \tag{8.47}$$

where a, b and c are constants depending on the boundary conditions and θ_p is the particular solution, associated with both loading and restraints conditions.

In case of the cantilever beam of length L represented in Figure 8.11, by considering a concentrate torsional moment applied at the free end ($x = L$) and the fixed end ($x = 0$) able to totally prevent the restraint, the particular solution θ_p is:

$$\theta_p = \frac{T}{G{\cdot}I_t}x \tag{8.48}$$

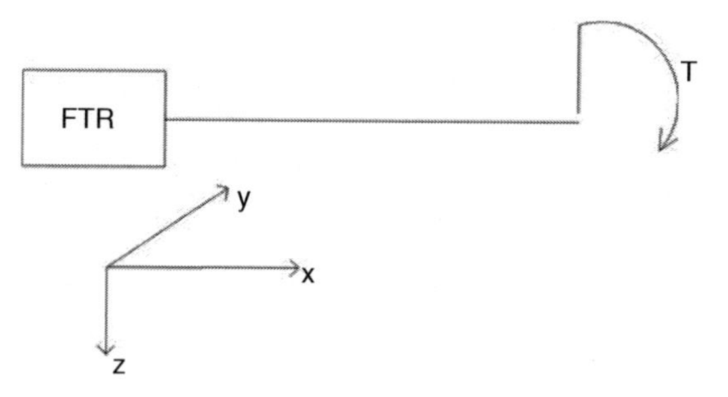

Figure 8.11 Cantilever beam loaded by a torsional moment at the free end ($x = L$) with torsional restraint preventing warping at the fixed end ($x = 0$).

Boundary conditions for this case are listed next:

- at the restrain location ($x = 0$), rotation is zero ($\theta = 0$) and warping is totally prevented $\left(\frac{d\theta}{dx} = 0\right)$; by considering the general solution expressed by Eq. (8.47), we can obtained:

$$a + c = 0 \tag{8.49a}$$

$$\lambda_T b + \frac{T}{G \cdot I_t} = 0 \tag{8.49b}$$

- at the free end, where the load is applied ($x = L$), the bimoment is zero $\left(\frac{d^2\theta}{dx^2} = 0\right)$ and hence, by deriving the Eq. (8.47) twice, it results in:

$$b \cdot \lambda_T^2 \cdot \sinh(\lambda_T L) + c \cdot \lambda_T^2 \cosh(\lambda_T L) = 0 \tag{8.49c}$$

The constants assume the following values:

$$a = -c = -\frac{T}{\lambda_T G I_t} \tanh(\lambda_T L) \tag{8.50}$$

$$b = -\frac{T}{\lambda_T G I_t} \tag{8.51}$$

Rotation is expressed as:

$$\theta(x) = \frac{T}{\lambda_T G I_t} \left\{ \lambda_T x + \frac{\sinh[\lambda_T (L - x)]}{\cosh(\lambda_T L)} - \tanh(\lambda_T L) \right\} \tag{8.52}$$

At the free end the value of rotation is: $\theta(L) = \frac{T}{\lambda_T G I_t} [\lambda_T L - \tanh(\lambda_T L)]$

Pure torsional moment is expressed as:

$$T_t = T \cdot \left\{ 1 - \frac{\cosh[\lambda_T (L - x)]}{\cosh(\lambda_T L)} \right\} \tag{8.53}$$

Warping moment is expressed as:

$$T_\omega = T \cdot \frac{\cosh[\lambda_T (L - x)]}{\cosh(\lambda_T L)} \tag{8.54}$$

Bimoment is expressed as:

$$B = -\frac{T}{\lambda_T} \cdot \frac{\sinh[\lambda_T(L-x)]}{\cosh(\lambda_T L)} \tag{8.55}$$

At the restraint location ($x = 0$) where rotation and warping are prevented, only the warping moment acts ($T = T_\omega$). At the free end, the applied external torsional moment is balanced by the sole pure torsion ($T = T_t$) where the warping moment is zero. Bimoment B at this location assumes the maximum value; that is: $B(x=0) = -\frac{T}{\lambda_T} \cdot \tanh(\lambda_T L)$.

In order to appraise the distribution of the two types of torsional contributions balancing external torque T applied to the free end of the cantilever beam, the ratios $T_t(x)/T$ and $T_\omega(x)/T$) are plotted versus x/L in Figure 8.12. The rotation is considered and the ratio $\theta(x)/\theta(L)$ is plotted in the figure too.

Furthermore, the case of a beam of length L, simply supported at its ends for torsional moments and loaded at midspan ($x = L/2$) by a torque moment T could also be interesting for design purposes (Figure 8.13).

The general solution expressed by Eq. (8.47) admits as particular solution, θ_p;

$$\theta_p = \frac{T}{2GI_t} x \tag{8.56}$$

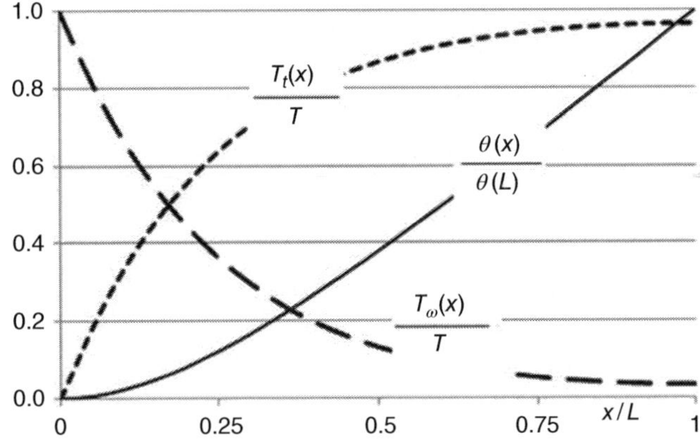

Figure 8.12 Distribution of pure and warping torsion along the cantilever beam ($x = 0$ and $x = L$ indicate fixed end and free end, respectively) and ratio between the rotation at generic cross-section x and the one at the free end.

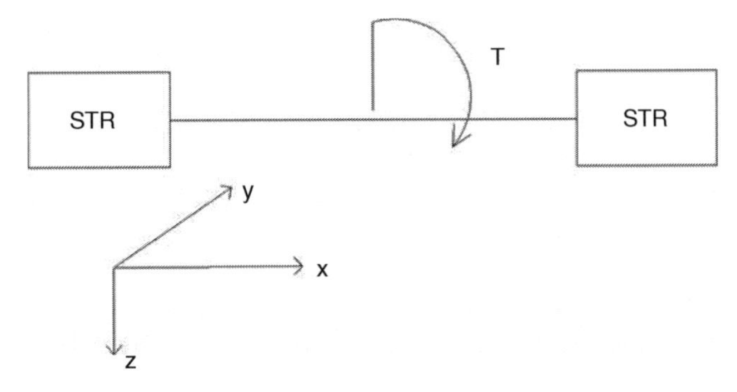

Figure 8.13 Beam restrained at its ends by torsional supports and loaded at the midspan ($x = L/2$) by a torque.

Boundary conditions for this case are:

at the beam ends (i.e. $x = 0$ and $x = L$) rotation is fully restrained ($\theta = 0$) but warping is free and, as a consequence the bimoment is zero $\left(\frac{d^2\theta}{dx^2} = 0\right)$. With reference to the cross-section at the location $x = 0$, we can obtain:

$$a + c = 0 \tag{8.57a}$$

$$c = 0 \tag{8.57b}$$

From these conditions, the result is $a = 0$.

In the loaded cross-section ($x = L/2$) warping is assumed to be completely prevented $\left(\frac{d\theta}{dx} = 0\right)$ and, as a consequence, it results in:

$$b = -\frac{T}{2\lambda_T G I_t}\left[\frac{1}{\cosh(\lambda_T L/2)}\right] \tag{8.58}$$

Rotation $\theta(x)$ is expressed as:

$$\theta(x) = \frac{T}{2\lambda_T G I_t}\left[\lambda x - \frac{\sinh(\lambda_T x)}{\cosh(\lambda_T L/2)}\right] \tag{8.59a}$$

In correspondence of the loaded cross-section, the rotation assumes the maximum value:

$$\theta_{\max}\left(x = \frac{L}{2}\right) = \frac{T}{2\lambda_T G I_t}\left[\lambda_T \frac{L}{2} - \tanh\left(\frac{\lambda_T L}{2}\right)\right] \tag{8.59b}$$

The pure torsional moment is expressed as:

$$T_t = \frac{T}{2}\left\{1 - \frac{\cosh(\lambda_T x)}{\cosh(\lambda_T L/2)}\right\} \tag{8.60a}$$

At the beam ends, the maximum value of the pure torsional moment is:

$$T_t(x = 0) = T_t(x = L) = G I_t \frac{d\theta}{dx} = \frac{T}{2}\left[1 - \frac{1}{\cosh(\lambda_T L/2)}\right] \tag{8.60b}$$

The bimoment is expressed as:

$$B = -E I_\omega \cdot \frac{d^2\theta}{dx^2} = -\frac{T}{2\lambda_T} \cdot \frac{\sinh(\lambda_T x)}{\cosh(\lambda_T L/2)} \tag{8.61a}$$

The maximum value of the bimoment is achieved at the loaded cross-section $\left(x = \frac{L}{2}\right)$:

$$B\left(x = \frac{L}{2}\right) = -E I_\omega \cdot \frac{d^2\theta}{dx^2} = -\frac{T}{2\lambda_T} \cdot \tanh(\lambda_T L/2) \tag{8.61b}$$

The warping moment is expressed as:

$$T_\omega = -EI_\omega \cdot \frac{d^3\theta}{dx^3} = \frac{T}{2} \cdot \frac{\cosh(\lambda_T x)}{\cosh(\lambda_T L/2)} \tag{8.62a}$$

The maximum value of the non-uniform bending moment is at the loaded cross-section:

$$T_\omega\left(\frac{L}{2}\right) = -EI_\omega \cdot \frac{d^3\theta}{dx^3} = \frac{T}{2} \tag{8.62b}$$

The same approach adopted for the beams in Figures 8.11 and 8.13, can be used for other cases of practical interest for routine design. For different load cases and types of restraints Tables 8.5 and 8.6 report key data for torsional design that are related to concentrated and distributed torsional loads, respectively.

8.4 Design in Accordance with the European Procedure

Part 1-1 in EC3 gives limited guidance for the design of torsion members. While both elastic and plastic analyses are generally discussed, only very approximate methods of elastic analysis are specifically discussed for torsion members. Furthermore, while both first yield and plastic design resistances are referred to with regards to bending, only the first yield design resistance is specifically discussed for torsion members: there is no guidance on section classification for torsion members, or on how to account for the effects of local buckling on design resistance.

Table 8.5 Key data for the torsional design in the case of a concentrated torsional load.

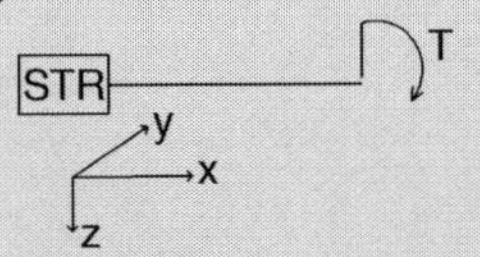	$\theta_{max}(L) = \frac{TL}{I_\omega E\lambda_T^2}$ $\theta'_{max}(L) = \frac{-T}{I_\omega E\lambda_T^2}$ $\theta''_{max}(L) = 0$ $\theta'''_{max}(0) = 0$
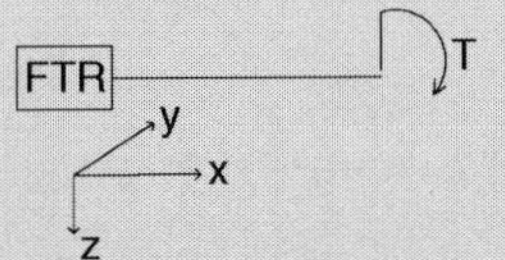	$\theta_{max}(L) = \frac{T}{I_\omega E\lambda_T^3}(\lambda_T L - \tanh\lambda_T L)$ $\theta'_{max}(L) = \frac{-T}{I_\omega E\lambda_T^2}\left(1 - \frac{1}{\cosh\lambda_T L}\right)$ $\theta''_{max}(0) = \frac{T}{I_\omega E\lambda_T}\tanh\lambda_T L$ $\theta'''_{max}(0) = \frac{T}{I_\omega E}$
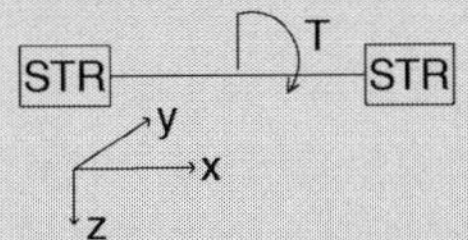	$\theta_{max}\left(\frac{L}{2}\right) = \frac{T}{2I_\omega E\lambda_T^3}\left(\frac{\lambda_T L}{2} - \tanh\frac{\lambda_T L}{2}\right)$ $\theta'_{max}\left(\frac{0}{L}\right) = \frac{\pm T}{2I_\omega E\lambda_T^2}\left[1 - \frac{1}{\cosh(\lambda_T L/2)}\right]$ $\theta''_{max}\left(\frac{L}{2}\right) = \frac{-T}{2I_\omega E\lambda_T}\tanh\frac{\lambda_T L}{2}$ $\theta'''_{max}\left(\frac{L}{2}\right) = \frac{T}{2I_\omega E}$
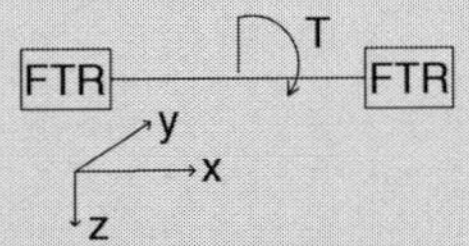	$\theta_{max}\left(\frac{L}{2}\right) = \frac{T}{I_\omega E\lambda_T^3}\left(\frac{\lambda_T L}{4} - \tanh\frac{\lambda_T L}{4}\right)$ $\theta'_{max}\left(\frac{L}{4}\right) = \frac{T}{2I_\omega E\lambda_T^2}\left[1 - \frac{1}{\cosh(\lambda_T L/4)}\right]$ $\theta''_{max}\begin{pmatrix}0\\ L\\ \frac{L}{2}\\ L\end{pmatrix} = \begin{pmatrix}+\\ -\\ +\end{pmatrix}\frac{T}{2I_\omega E\lambda_T}\tanh\frac{\lambda_T L}{4}$ $\theta'''_{max} = \frac{-T}{2I_\omega E}$

Table 8.6 Key data for the torsional design in the case of a uniform torsional load.

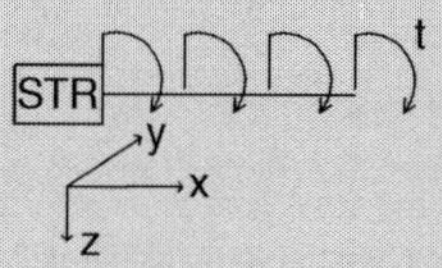

Restraint (load t; axes x, y, z)	Key data
STR	$\theta_{max}(L)=\frac{tL^2}{2I_\omega E\lambda_T^2}$ $\quad\theta'_{max}(0)=\frac{t}{I_\omega E\lambda_T^3}\left(\lambda_T L-\tanh\frac{\lambda_T L}{2}\right)$ $\theta''_{max}\left(\frac{L}{2}\right)=\frac{-t}{I_\omega E\lambda_T^2}\left(1-\frac{1}{\cosh\frac{\lambda_T L}{2}}\right)$ $\quad\theta'''_{max}\left(\frac{0}{L}\right)=\left(\frac{+}{-}\right)\frac{t}{2I_\omega E\lambda_T}\tanh\lambda_T L$
FTR	$\theta_{max}(L)=\frac{t}{I_\omega E\lambda_T^4}\left(1+\frac{\lambda_T^2L^2}{2}-\frac{1+\lambda_T L\sinh\lambda_T L}{\cosh\lambda_T L}\right)$ $\theta''_{max}(0)=\frac{t}{I_\omega E\lambda_T^2}\left(\frac{1+\lambda_T L\sinh\lambda_T L}{\cosh\lambda_T L}-1\right)$ $\quad\theta'''_{max}(0)=\frac{tL}{I_\omega E}$
STR – STR	$\theta_{max}\left(\frac{L}{2}\right)=\frac{t}{I_\omega E\lambda_T^4}\left[\frac{\lambda_T^2L^2}{8}+\frac{1}{\cosh(\lambda_T L/2)}-1\right]$ $\theta'_{max}\left(\frac{0}{L}\right)=\left(\frac{+}{-}\right)\frac{t}{I_\omega E\lambda_T^3}\left(\frac{\lambda_T L}{2}-\tanh\frac{\lambda_T L}{2}\right)$ $\theta''_{max}\left(\frac{L}{2}\right)=\frac{t}{I_\omega E\lambda_T^2}\left[\frac{1}{\cosh(\lambda_T L/2)}-1\right]$ $\quad\theta'''_{max}\left(\frac{0}{L}\right)=\left(\frac{-}{+}\right)\frac{t}{I_\omega E\lambda_T}\tanh\frac{\lambda_T L}{2}$
FTR – FTR	$\theta_{max}(^L/_2)=\frac{tL}{2I_\omega E\lambda_T^3}\left(\frac{\lambda_T L}{4}-\tanh\frac{\lambda_T L}{4}\right)$ $+\theta''_{max}\left(\frac{0}{L}\right)=\frac{t}{I_\omega E\lambda_T^2}\left[\frac{\lambda_T L}{2\tanh(\lambda_T L/2)}-1\right]$ $-\theta''_{max}(^L/_2)=\frac{-t}{I_\omega E\lambda_T^2}\left[1-\frac{\lambda_T L}{2\sinh(\lambda_T L/2)}\right]$ $\quad\theta'''_{max}\left(\frac{0}{L}\right)=\left(\frac{-}{+}\right)\frac{tL}{2I_\omega E}$

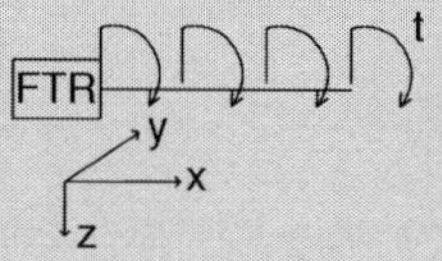

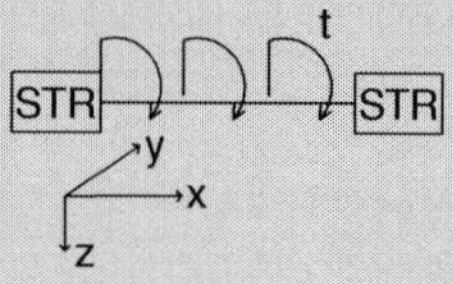

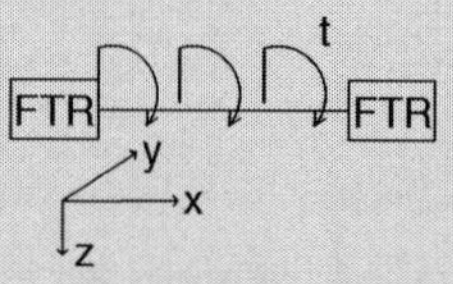

For members subjected to torsion, if distortional deformations may be disregarded, the design value of the torsional moment T_{Ed} at each cross-section has to satisfy the condition:

$$T_{Ed} \leq T_{Rd} \tag{8.63}$$

where T_{Rd} is the design torsional resistance at the cross-section.

In particular, as already discussed, the total torsional moment T_{Ed} at any cross-section should be considered as the sum of two internal effects:

$$T_{Ed} = T_{t,Ed} + T_{w,Ed} \tag{8.64}$$

where $T_{t,Ed}$ is the internal St Venant torsion and $T_{w,Ed}$ is the internal warping torsion.

The values of $T_{t,Ed}$ and $T_{w,Ed}$ at any cross-section may be determined from T_{Ed} by an elastic analysis taking into account the section properties of the member, the conditions of restraint at the supports and the distribution of the actions along the member.

It is required to take into account the following stresses due to torsion:

- shear stresses $\tau_{t,Ed}$ due to St Venant torsion $T_{t,Ed}$;
- normal stresses $\sigma_{w,Ed}$ due to the bimoment B_{Ed} and shear stresses $\tau_{w,Ed}$ due to warping torsion $T_{w,Ed}$.

As a simplification, in cases of a member with a closed hollow cross-section, such as a structural hollow section, EC3 allows the assumption that the effects of torsional warping can be neglected.

In the case of a member with an open cross-section, such as I- or H-shaped profiles, according to EC3 it may be assumed that the effects of St Venant torsion can be neglected.

Furthermore, as already mentioned in Section 7.2.2, for combined shear force and torsional moment the plastic shear resistance accounting for torsional effects should be reduced from $V_{pl,Rd}$ to $V_{pl,T,Rd}$ and the design shear force V_{Ed} must satisfy the condition:

$$V_{Ed} \leq V_{pl,T,Rd} \tag{8.65}$$

The Eq. (7.25) are herein re-proposed for the sake of clarity. For the most common cases $V_{pl,T,Rd}$ is defined as:

- for an I- or H-shaped section:

$$V_{pl,T,Rd} = V_{pl,Rd} \cdot \sqrt{1 - \frac{\tau_{t,Ed}}{\frac{1,25}{\gamma_{M0}} \cdot \left(\frac{f_y}{\sqrt{3}}\right)}} \tag{8.66}$$

- for channel sections:

$$V_{pl,T,Rd} = V_{pl,Rd} \cdot \left[\sqrt{1 - \frac{\tau_{t,Ed}}{\frac{1.25}{\gamma_{M0}} \cdot \left(\frac{f_y}{\sqrt{3}}\right)}} - \frac{\tau_{w,Ed}}{\frac{1}{\gamma_{M0}} \cdot \left(\frac{f_y}{\sqrt{3}}\right)}\right] \tag{8.67}$$

- for structural hollow sections:

$$V_{pl,T,Rd} = V_{pl,Rd} \cdot \left[1 - \frac{\tau_{t,Ed}}{\frac{1}{\gamma_{M0}} \cdot \left(\frac{f_y}{\sqrt{3}}\right)}\right] \tag{8.68}$$

8.5 Design in Accordance with the AISC Procedure

Torsion is addressed in AISC 360-10, Chapter H3. An important help for computing torsion for open shapes comes from AISC Design Guide 9 'Torsional Analysis of Structural Steel Members'.

Chapter H3 deals mainly with torsion for hollow structural sections (HSS). These types of cross-sections can be often subjected to torsional moments. For HSS sections, according to AISC, stresses due to restrained warping can be disregarded and it can be assumed that all the torsional moment is resisted by pure (St Venant) torsional stresses. So for HSS it is possible to define torsional strength in terms of a *resisting torsional moment* and not in terms of *torsional stresses*. This allows for verifying HSS members subjected to compression, bending moment, shear and torsional moments in terms of actions and not stresses. On the contrary, for open cross-section members in which warping is not completely unrestrained, torsional resistance is exhibited as the sum of that due to pure (St Venant) torsion and that due to restrained warping. The contribution of each of them depends on the angle of rotation θ and its derivatives; so it depends, as already discussed, on section properties, type of loads and type of restraints, and it must be determined for each design case. Such open cross-section members are, in most cases, subjected not only to torsion but also to bending moments, shear and axial loads and some details of their design are discussed in Chapter 10. Pure torsion is very unlikely actually and torsion in open cross-section should be avoided with proper design strategies or at least reduced to a secondary action. AISC Specifications prescribe then for open sections to compute stresses due to any kind of generalized forces (axial, bending, shear and torsion) and compare them with defined allowable stresses.

8.5.1 Round and Rectangular HSS

LRFD approach	ASD approach
Using load and resistance factor design (LRFD), *design torsional strength* T_c, is defined as:	Using allowable strength design (ASD), *allowable torsional strength* T_c, is defined as:
$T_c = \phi_T T_n$ (8.69)	$T_c = T_n/\Omega_T$ (8.70)
where $\phi_T = 0.90$ and T_n is the nominal torsional strength	where $\Omega_T = 1.67$ and T_n is the nominal torsional strength

Nominal torsional strength, T_n, is computed with the formula:

$$T_n = F_{cr} C \tag{8.71}$$

where C is the HSS torsional constant and F_{cr} is the critical stress that takes account of local buckling and initial imperfections.

(a) For a round HSS:

$$F_{cr} = \max\left\{\frac{1.23E}{\sqrt{\frac{L}{D}}\cdot\left(\frac{D}{t}\right)^{5/4}};\frac{0.60E}{\left(\frac{D}{t}\right)^{3/2}}\right\} \le 0.60F_y \tag{8.72}$$

where L is the length of the member and D is the outside diameter.

(b) For a rectangular HSS:

$$F_{cr} = 0.60F_y \quad \text{if} \quad h/t \le 2.45\sqrt{\frac{E}{F_y}} \tag{8.73a}$$

$$F_{cr} = \frac{0.6F_y\left(2.45\sqrt{E/F_y}\right)}{\left(\frac{h}{t}\right)} \quad \text{if} \quad 2.45\sqrt{\frac{E}{F_y}} < h/t \le 3.07\sqrt{\frac{E}{F_y}} \tag{8.73b}$$

$$F_{cr} = \frac{0.458\pi^2 E}{\left(\frac{h}{t}\right)^2} \quad \text{if} \quad 3.07\sqrt{\frac{E}{F_y}} < h/t \leq 260 \tag{8.73c}$$

where h is the flat width of longer HSS side and t the design wall thickness.

The torsional constant C should be taken as:

(a) For a round HSS:

$$C = \frac{\pi\left(D^4 - D_i^4\right)}{32D/2} \approx \frac{\pi(D-t)^2 t}{2} \tag{8.74}$$

where D_i is the inside diameter.

(b) For a rectangular HSS:

$$C = 2A_0 t \tag{8.75a}$$

where A_0 is the area bounded by the midline of the section.

Assuming an outside corner radius of $2t$ conservatively, the midline radius is $1.5t$ and the torsional constant C consequently becomes:

$$C = 2(B-t)(H-t)t - 4.5(4-\pi)t^3 \tag{8.75b}$$

If a round or rectangular HSS member is subjected to a torsional moment T_t, computed with LRFD or ASD load combinations and torsion is the only internal action, then verification is:

$$T_t \leq T_c \tag{8.76}$$

where T_c is the design torsional strength (LRFD) or allowable torsional strength (ASD).

In the case of axial load, bending moments and shear, then the verification is performed according to Section 10.3.

8.5.2 Non-HSS Members (Open Sections Such as W, T, Channels, etc.)

For such members, torsion is sustained as pure torsion and restrained warping torsion.

Considering the common case of I- or H-shaped profiles:

(a) pure (St Venant) torsion generates shear stress τ_t in any part of the section (flanges and web);
(b) restrained warping torsion generates normal stress σ_w and shear stress τ_w in the flanges.

As previously discussed, the distribution of torsional moment between pure and restrained warping torsion depends on rotation angle and its derivatives; that is on type of load, restraints and cross-section geometry. In evaluating stresses due to torsion, a helpful tool is the AISC Design Guide 9 and its Appendix B. Verifications for pure torsion are generally meaningless because torsion for this kind of section is very often associated with stresses due to axial load, bending moments and shear. So AISC 360-10 proposes a verification for combined actions that are discussed in Section 10.3.

CHAPTER 9

Members Subjected to Flexure and Axial Force

9.1 Introduction

Members subjected to flexure and axial forces are commonly identified as beam-columns. They are frequently encountered in routine design when:

- the axial force is eccentric with reference to the cross-section centroid;
- the compressed element is also subjected to transverse load inducing flexure (typically, beams in simple frames loaded by gravity loads but also interested by axial forces due to the effects of horizontal forces);
- the vertical elements, which belong to a rigid or to a semi-continuous frame, are loaded at their ends by bending moments transferred by beams;
- thin-walled elements are subjected to axial load on the centroid of the gross section, which does not coincide with the one of the effective cross-section (Figure 4.8a).

When the centre of pressure lies on one of the two main planes of inertia, the cross-section is interested by compression and in-plane bending, while the more general case is related to compression and bi-axial bending. For beam-columns the absence of instability phenomena is very rare and for this reason the more severe design checks are generally the ones related to overall member stability.

Deformability: When the deflection v_{BC} of a beam-column has to be evaluated, a simplified approach can be adopted, which consists of suitably amplifying the deflection v_v due to the loads normal to the beam, to take into account the presence of axial load. In detail, the beam-column displacement v_{BC} can be estimated as:

$$v_{BC} = \frac{1}{1 - \dfrac{N}{N_{cr}}} \cdot v_v \tag{9.1}$$

where N represents the design axial load and N_{cr} the critical buckling load with reference to the bending plane.

Resistance: Resulting from the basis of the theory of structures, in case of members subjected to axial load (N) and bending moment (M), which are suitably constrained against instability, checks must be made on the most stressed cross-section. In order to allow for a general appraisal of the safety of the beam-column, reference can be made to the well-known

Structural Steel Design to Eurocode 3 and AISC Specifications, First Edition. Claudio Bernuzzi and Benedetto Cordova.

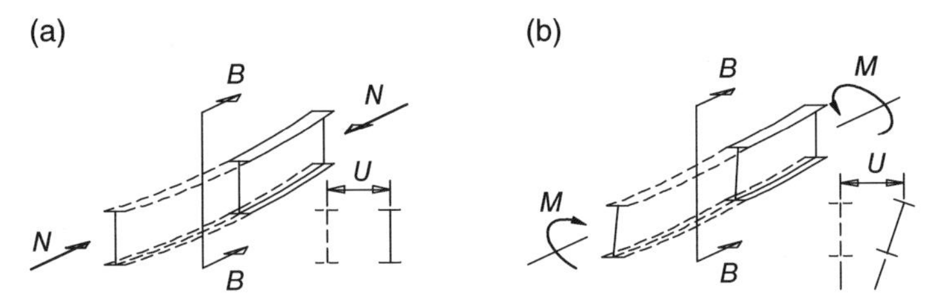

Figure 9.1 Typical deformed shapes for flexural buckling (a) flexural torsional buckling (b).

St Venant's theory and the maximum stress σ, resulting from the linear combination of axial load and bending moment, is expressed as:

$$\sigma = \frac{N}{A} + \frac{M}{W} \tag{9.2}$$

where A and W are the area and the section modulus of the cross-section, respectively.

On the other hand, if verifications refer the performance of the whole cross-section, according to the limit state design philosophy (already introduced with reference to the shear and bending moment interaction), the design bending resistance must be suitably reduced for the presence of axial load.

Stability: A beam-column can be affected by the buckling phenomena already introduced for beams (see Chapter 7) and columns (see Chapter 6), on the basis of the influence of several factors, which can be associated, for example with cross-section geometry, the presence of end and/or intermediate restraints, the load condition and so on. When an unrestrained beam-column is bent about its major axis, it may buckle by deflecting laterally and twisting in correspondence of a load that could be significantly lower than the maximum load predicted by an in-plane analysis. If the shear centre is coincident with the cross-section centroid, two typical kinds of instability (Figure 9.1) can be observed:

flexural buckling, if the member restraints efficiently hamper the sole buckling of the compression flange by means of deflection of the member in the plane that contains the eccentricity of the load;
lateral-torsional (flexural-torsional) buckling, when the instability is associated with the typical deflection due to the buckling of members in bending.

If the shear centre does not coincide with the centroid, design should be governed by flexural-torsional buckling, as well as in the case of a predominant axial load on the bending moment. When stability has to be accounted for into design, the buckling conditions are defined by the interaction between critical axial load (N_{cr}) and bending moment (M_{cr}). As an example, Figure 9.2 presents typical N_{cr}–M_{cr} curves for a simply supported column under uniform (ψ = 1) or gradient ($\psi \neq 1$) end moment distributions.

An efficient and quite simple way to account for the buckling interaction between the axial load and the bending moment is to approximate the critical buckling moment $M_{cr,}(N)$ of the beam-column reducing the one of the beam (M_{cr}) in the presence of the axial load N. In particular, two cases can be distinguished, depending on the symmetry of the cross-section:

- Mono-symmetrical cross-section:

$$M_{cr}(N) = C_1 \frac{\pi^2 EI_z}{(k_z L)^2} \cdot \left\{ \left[\sqrt{\left[\left(\frac{k_z}{k_W} \right)^2 \frac{I_W}{I_z} + \frac{(kL)^2 GI_t}{\pi^2 EI_z} \right] f_M(N) + \left(C_2 z_g - C_3 z_j \right)^2} \right] - \left(C_2 z_g - C_3 z_j \right) \right\} \tag{9.3a}$$

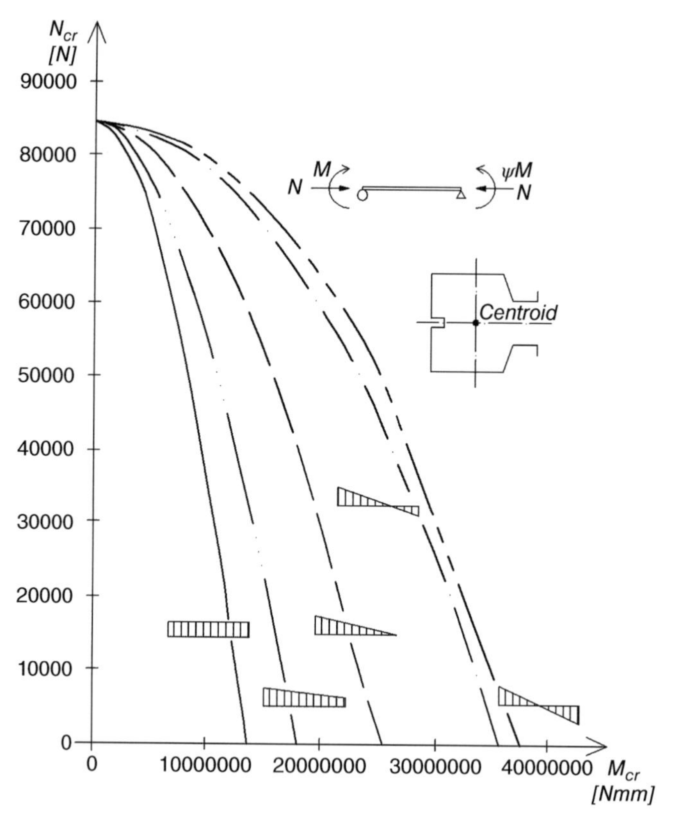

Figure 9.2 Typical axial force-bending moment buckling domains for different values of the end ratio moment (ψ).

where function $f_M(N)$ is defined as:

$$f_N(N) = \left(1 - \frac{N}{N_{cr,y}}\right)\left(1 - \frac{N}{N_{cr,FT(1)}}\right)\left(1 - \frac{N}{N_{cr,FT(2)}}\right) \tag{9.3b}$$

with:

$$N_{cr,FT(1,2)} = \frac{1}{2}\frac{N_{cr,y}}{1-\left(\dfrac{y_0}{i_0}\right)^2}\left[1 + \frac{N_{cr,T}}{N_{cr,y}} \pm \sqrt{\left(1 + \frac{N_{cr,T}}{N_{cr,y}}\right)^2 - 4\left(\frac{y_0}{i_0}\right)^2 \frac{N_{cr,T}}{N_{cr,y}}}\right] \tag{9.3c}$$

All the terms in these equations have already been discussed with reference to the axial buckling of columns (Chapter 6) and to the lateral buckling of beams (Chapter 7).

- Bi-symmetrical cross-section:

$$M_{cr}(N) = M_{cr} \cdot \sqrt{f_B(N)} \tag{9.4a}$$

where M_{cr} is given by Eq. (7.42) for the beam (i.e. element under pure flexure) and $f_B(N)$ is defined as:

$$f_B(N) = \left(1 - \frac{N}{N_{cr,y}}\right)\left(1 - \frac{N}{N_{cr,z}}\right)\left(1 - \frac{N}{N_{cr,T}}\right) \tag{9.4b}$$

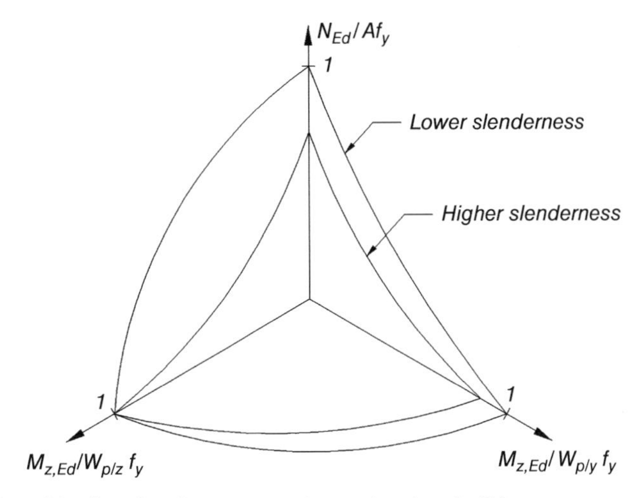

Figure 9.3 Typical axial load (N)-bending moments interaction domain (M).

The flexural ($N_{cr,y}$ and $N_{cr,z}$) and the torsional ($N_{cr,T}$) critical loads have been already defined in Chapter 6.

The influence of buckling phenomena on the response of industrial beam-columns can be evaluated by defining the interaction domains between axial forces and bending moments in the two principal planes of cross-section. With reference to the more general case of compression and biaxial bending in which flexural buckling governs design, typical interaction domains are proposed in Figure 9.3 in a non-dimensional form. In case of compact cross-section profiles, design axial load N_{Ed} is divided by the squash load (N_y) and the design bending moments related to y–y and z–z axis ($M_{ed,y}$ and $M_{ed,z}$, respectively) are divided by the corresponding plastic moment ($M_{Rd,y}$ and $M_{Rd,z}$).
As to the design practice, with reference to contents of the most recent Codes of practice, instead of complex formulations that allow local definition of the interaction domains, simplified criteria suitable for design purposes are proposed that guarantee a safe design.

9.2 Design According to the European Approach

European provisions deal with the most common cases in design practice and in particular provide design rules for members with bi-symmetrical cross-sections. Rules to evaluate both strength and stability of some of the most common cases typical of routine design are discussed in the general part of EC3 (EN 1993-1-1).

9.2.1 The Resistance Checks

It should be noted that, in general, y- and z-axis identify the strong and weak cross-section axes. In cases of profiles with a cross-section in classes 1 and 2, a requirement is that the acting bending moment M_{Ed} does not exceed the design moment resistance $M_{N,Rd}$, which is the plastic moment reduced in the presence of the axial load N_{Ed} on the considered cross-section, that is:

$$M_{Ed} \leq M_{N,Rd} \tag{9.5}$$

For a rectangular solid section without bolt fastener holes, reduced bending resistance $M_{N,Rd}$ can be evaluated as:

$$M_{N,Rd} = M_{pl,Rd} \cdot \left[1 - \left(\frac{N_{Ed}}{N_{pl,Rd}}\right)^2\right] \tag{9.6}$$

where $M_{pl,Rd}$ and $N_{pl,Rd}$ represent the design plastic resistance of the gross cross-section to bending moments (see Chapter 7) and to normal forces (see Chapter 6), respectively.
In case of doubly symmetrical I- and H-shaped sections or other similar sections (typically built-up welded profiles), allowance does not need to be made for the effect of the axial force on the plastic resistance moment about the y–y axis (parallel to the flanges) when both the following conditions are satisfied:

$$N_{Ed} \leq 0.25 \cdot N_{pl,Rd} \tag{9.7a}$$

$$N_{Ed} \leq \frac{0.5 \cdot h_w \cdot t_w \cdot f_y}{\gamma_{M0}} \tag{9.7b}$$

For doubly-symmetrical I- and H-shaped sections, allowance does not need to be made for the effects of the axial force on the plastic resistance moment about the z–z axis (parallel to the web) when:

$$N_{Ed} \leq \frac{h_w \cdot t_w \cdot f_y}{\gamma_{M0}} \tag{9.8}$$

where h_w and t_w are the height and the thickness of the web, respectively.
For cross-sections where bolt fastener holes do not have to be considered, the following approximations may be used for standard rolled I- or H-sections and for welded I- or H-shaped sections with equal flanges:

- bending resistance about the y–y axis:

$$M_{N,y,Rd} = M_{pl,y,Rd} \cdot \frac{1-n}{1-0.5 \cdot a} \tag{9.9a}$$

with the limitation:

$$M_{N,y,Rd} \leq M_{pl,y,Rd} \tag{9.9b}$$

- bending resistance about the z–z axis:
  if $n \leq a$:

$$M_{N,z,Rd} \leq M_{pl,z,Rd} \tag{9.10a}$$

if $n > a$;

$$M_{N,z,Rd} = M_{pl,z,Rd} \cdot \left[1 - \left(\frac{n-a}{1-a}\right)^2\right] \tag{9.10b}$$

where n and a are defined, respectively, as:

$$n = \frac{N_{Ed}}{N_{pl,Rd}} \tag{9.11}$$

$$a = \frac{A - (2 \cdot b \cdot t_f)}{A} \leq 0.5 \tag{9.12}$$

where A is the cross-section area and b and t_f are the width and the thickness of the flange, respectively.

In case of rectangular structural hollow sections of uniform thickness and for welded box sections with equal flanges and equal webs, the following approximations are used if the effect of bolt fastener holes can be neglected:

$$M_{N,y,Rd} = M_{pl,y,Rd} \cdot \frac{(1-n)}{(1-0.5a_w)} \leq M_{pl,y,Rd} \tag{9.13a}$$

$$M_{N,z,Rd} = M_{pl,z,Rd} \cdot \frac{(1-n)}{(1-0.5a_f)} \leq M_{pl,z,Rd} \tag{9.13b}$$

where terms a_w and a_f depend on the type of cross-section.
In particular, the following cases are directly considered by the Code:

hollow profiles (b and h are the width and the height of the cross-section, respectively with a thickness t):

$$a_w = \frac{(A - 2 \cdot b \cdot t)}{A} \leq 0.5 \tag{9.14a}$$

$$a_f = \frac{(A - 2 \cdot h \cdot t)}{A} \leq 0.5 \tag{9.14b}$$

welded box sections (b and h are the width of the flanges of thickness t_f and the height of web of thickness t_w, respectively):

$$a_w = \frac{(A - 2 \cdot h \cdot t_f)}{A} \leq 0.5 \tag{9.15a}$$

$$a_f = \frac{(A - 2 \cdot b \cdot t_w)}{A} \leq 0.5 \tag{9.15b}$$

For bi-axial bending, the member verification can be based on the following criterion:

$$\left(\frac{M_{y,Ed}}{M_{N,y,Rd}}\right)^{\alpha} + \left(\frac{M_{z,Ed}}{M_{N,z,Rd}}\right)^{\beta} \leq 1 \tag{9.16}$$

in which α and β coefficients may conservatively be taken as unity or otherwise can be deduced from Table 9.1 on the basis of the value of n defined by Eq. (9.11).

Table 9.1 Values of α and β coefficients for bi-axial bending verification.

Cross-section type	α	β
I- and H-sections	$\alpha = 2$	$\beta = 5 \cdot n$ with $\beta \geq 1$
Circular hollow sections	$\alpha = 2$	$\beta = 2$
Rectangular hollow sections	$\alpha = \frac{1.66}{1-1.13 \cdot n^2}$ with $\alpha \leq 6$	$\beta = \frac{1.66}{1-1.13 \cdot n^2}$ with $\beta \leq 6$

For class 3 and class 4 cross-sections, when shear force is absent or its effect is negligible, the maximum longitudinal stress $\sigma_{x,Ed}$ due to moment and axial force taking into account the bolt fastener holes if relevant, must fulfil the condition:

$$\sigma_{x,Ed} \leq \frac{f_y}{\gamma_{M0}} \tag{9.17}$$

For a class 4 cross-section, the following additional condition has to be fulfilled:

$$\frac{N_{Ed}}{\frac{A_{eff} \cdot f_y}{\gamma_{M0}}} + \frac{M_{y,Ed} + N_{Ed} \cdot e_{Ny}}{\frac{W_{eff,y} \cdot f_y}{\gamma_{M0}}} + \frac{M_{z,Ed} + N_{Ed} \cdot e_{Nz}}{\frac{W_{eff,z} \cdot f_y}{\gamma_{M0}}} \leq 1 \tag{9.18}$$

where e_{Ny} and e_{Nz} represent the shift of the relevant centroidal axis when the cross-section is subjected to compression only, along the y–y and z–z axis, respectively, A_{eff} is the effective area of the cross-section when subjected to uniform compression and $W_{eff,min}$ is the effective section modulus (corresponding to the fibre with the maximum elastic stress).

9.2.2 The Stability Checks

EC3 proposes a criterion for verification of members under bi-axial bending, which can be applied for the sole case of uniform members with double symmetric cross-sections not susceptible to distortional deformations (Figure 4.1). In particular, two cases are considered:

- members that are not susceptible to torsional deformations, for example circular hollow sections or sections with suitable torsional restraints and hence no lateral-torsional buckling is expected;
- members that are susceptible to torsional deformations, for example members with open cross-sections not restrained against torsion.

The proposed interaction formulas are based on the modelling of simply supported single span members with end fork conditions and with or without continuous lateral restraints, which are subjected to compression forces, end moments and/or transverse loads. Second order effects of the sway system have to be taken into account, either by the end moments of the member or by means of appropriate buckling lengths, respectively.
Members that are subjected to combined bending moments along the y–y and z–z axis, $M_{y,Ed}$ and $M_{z,Ed}$, respectively, and axial compression N_{Ed} must satisfy the conditions:

$$\frac{N_{Ed}}{\frac{\chi_y \cdot N_{Rk}}{\gamma_{M1}}} + k_{yy} \frac{M_{y,Ed} + \Delta M_{y,Ed}}{\chi_{LT} \cdot \frac{M_{y,Rk}}{\gamma_{M1}}} + k_{yz} \frac{M_{z,Ed} + \Delta M_{z,Ed}}{\frac{M_{z,Rk}}{\gamma_{M1}}} \leq 1 \tag{9.19a}$$

Table 9.2 Values for N_{Rk}, M_{Rk} and $\Delta M_{i,Ed}$.

$N_{Rk}=f_y \cdot A_i$; $M_{Rk}=f_y \cdot W_i$

Class	1	2	3	4
A_i	A	A	A	A_{eff}
W_y	$W_{pl,y}$	$W_{pl,y}$	$W_{el,y}$	$W_{eff,y}$
W_z	$W_{pl,z}$	$W_{pl,z}$	$W_{el,z}$	$W_{eff,z}$
$\Delta M_{y,Ed}$	0	0	0	$e_{N,y} \cdot N_{ed}$ [a]
$\Delta M_{z,Ed}$	0	0	0	$e_{N,z} \cdot N_{ed}$ [a]

[a] Terms $e_{N,y}$ and $e_{N,z}$ represent the eccentricity between the gross and the effective cross-section.

$$\frac{N_{Ed}}{\frac{\chi_z \cdot N_{Rk}}{\gamma_{M1}}}+k_{zy}\frac{M_{y,Ed}+\Delta M_{y,Ed}}{\chi_{LT} \cdot \frac{M_{y,Rk}}{\gamma_{M1}}}+k_{zz}\frac{M_{z,Ed}+\Delta M_{z,Ed}}{\frac{M_{z,Rk}}{\gamma_{M1}}}\leq 1 \tag{9.19b}$$

where χ_y and χ_z are the reduction factors due to flexural buckling, χ_{LT} is the reduction factor due to lateral buckling, subscript $_{Rk}$ identifies the characteristic value for resistance to be evaluated in accordance with Table 9.2, additional moment ΔM is due to the shift between the gross and the effective centroid of the cross-section for class 4 members and k_{yy}, k_{yz}, k_{zy} and k_{zz} are the interaction factors. It should be noted that, for members not susceptible to torsional deformation, it is assumed $\chi_{LT}=1.0$; that is no reductions of the bending performance due to lateral-torsional buckling.

The interaction factors k_{yy}, k_{yz}, k_{zy} and k_{zz} depend on the approach, which has to be selected from two alternatives: alternative method 1 (*AM1*) and alternative method 2 (*AM2*), which are considered respectively in Annex A and Annex B of EN 1993-1-1. As to the use of these methods, it should be noted that the AM2 formulation, proposed by Austrian and German researchers, is generally less complex, quicker and simpler than the AM1 developed by a team of French and Belgian researchers. Therefore, AM2 can be regarded as a simplified approach whereas AM1 represents a more exact and general approach.

Finally, it should be noted that the National Annexes of the European countries should give a choice from AM1 or AM2. Furthermore, it is worth mentioning that in the present edition of EN 1993-1-1 no rules are given for the stability verification checks of beam-columns with one axis of symmetry.

9.2.2.1 Alternative Method 1 (AM1)

According to method 1, a member is not susceptible to torsional deformation if the torsional constant, I_T, is not lower than the second moment of area about y-axis, I_y. That is if the following condition is fulfilled:

$$I_T > I_y \tag{9.20}$$

Furthermore, when $I_T \leq I_y$ the following can occur:

if $\bar{\lambda}_o \leq \bar{\lambda}_{o,\lim}$ there is no risk of lateral flexural buckling;
if $\bar{\lambda}_o > \bar{\lambda}_{o,\lim}$ lateral flexural buckling can occur.

Term $\bar{\lambda}_o$ represents the relative slenderness for lateral buckling under constant moment (critical moment $M_{cr,0}$), already defined in Eq. (7.33), here re-proposed for simplicity:

$$\bar{\lambda}_o=\sqrt{\frac{W_{pl,y}f_y}{M_{cr,0}}} \tag{9.21}$$

Term $\bar{\lambda}_{o,\lim}$ is defined, for a doubly symmetrical cross-section as:

$$\bar{\lambda}_{o,\lim} = 0.2\sqrt{C_1}\cdot\sqrt[4]{\left(1-\frac{N_{Ed}}{N_{cr,z}}\right)\left(1-\frac{N_{Ed}}{N_{cr,T}}\right)} \quad (9.22)$$

where axial critical load $N_{cr,z}$ and $N_{cr,T}$ are related to the buckling along z–z axis and the torsional buckling, respectively, and the equivalent uniform moment coefficient C_1 has been already introduced with reference to lateral buckling (Tables 7.6 and 7.7).

In accordance with this method, coefficients k_{yy}, k_{yz}, k_{zy}, k_{zz} can be deduced from Table 9.3 for members not susceptible to torsional deformations and in Table 9.4 for members susceptible to torsional deformations.

Additional terms reported in Tables 9.3 and 9.4 are:

$$\mu_y = \frac{1-\dfrac{N_{Ed}}{N_{cr,y}}}{1-\chi_y\dfrac{N_{Ed}}{N_{cr,y}}} \text{ and } \mu_z = \frac{1-\dfrac{N_{Ed}}{N_{cr,z}}}{1-\chi_z\dfrac{N_{Ed}}{N_{cr,z}}} \quad (9.23a)$$

Table 9.3 Coefficients k_{ij} for members not susceptible to torsional deformations.

Interaction factors	Plastic cross-sectional properties class 1, class 2	Elastic cross-sectional properties class 3, class 4
k_{yy}	$C_{my}\dfrac{\mu_y}{1-\dfrac{N_{Ed}}{N_{cr,y}}}\dfrac{1}{C_{yy}}$	$C_{my}\dfrac{\mu_y}{1-\dfrac{N_{Ed}}{N_{cr,y}}}$
k_{yz}	$C_{mz}\dfrac{\mu_y}{1-\dfrac{N_{Ed}}{N_{cr,z}}}\dfrac{1}{C_{yz}}0.6\sqrt{\dfrac{w_z}{w_y}}$	$C_{mz}\dfrac{\mu_y}{1-\dfrac{N_{Ed}}{N_{cr,z}}}$
k_{zy}	$C_{my}\dfrac{\mu_z}{1-\dfrac{N_{Ed}}{N_{cr,y}}}\dfrac{1}{C_{zy}}0.6\sqrt{\dfrac{w_y}{w_z}}$	$C_{my}\dfrac{\mu_z}{1-\dfrac{N_{Ed}}{N_{cr,y}}}$
k_{zz}	$C_{mz}\dfrac{\mu_z}{1-\dfrac{N_{Ed}}{N_{cr,z}}}\dfrac{1}{C_{zz}}$	$C_{mz}\dfrac{\mu_z}{1-\dfrac{N_{Ed}}{N_{cr,z}}}$

Table 9.4 Coefficients k_{ij} for members susceptible to torsional deformations.

Interaction factors	Plastic cross-sectional properties class 1, class 2	Elastic cross-sectional properties class 3, class 4
k_{yy}	$C_{my}C_{mLT}\dfrac{\mu_y}{1-\dfrac{N_{Ed}}{N_{cr,y}}}\dfrac{1}{C_{yy}}$	$C_{my}C_{mLT}\dfrac{\mu_y}{1-\dfrac{N_{Ed}}{N_{cr,y}}}$
k_{yz}	$C_{mz}\dfrac{\mu_y}{1-\dfrac{N_{Ed}}{N_{cr,z}}}\dfrac{1}{C_{yz}}0.6\sqrt{\dfrac{w_z}{w_y}}$	$C_{mz}\dfrac{\mu_y}{1-\dfrac{N_{Ed}}{N_{cr,z}}}$
k_{zy}	$C_{my}C_{mLT}\dfrac{\mu_z}{1-\dfrac{N_{Ed}}{N_{cr,y}}}\dfrac{1}{C_{zy}}0.6\sqrt{\dfrac{w_y}{w_z}}$	$C_{my}C_{mLT}\dfrac{\mu_z}{1-\dfrac{N_{Ed}}{N_{cr,y}}}$
k_{zz}	$C_{mz}\dfrac{\mu_z}{1-\dfrac{N_{Ed}}{N_{cr,z}}}\dfrac{1}{C_{zz}}$	$C_{mz}\dfrac{\mu_z}{1-\dfrac{N_{Ed}}{N_{cr,z}}}$

$$w_y = \frac{W_{pl,y}}{W_{el,y}} \le 1.5 \text{ and } w_z = \frac{W_{pl,z}}{W_{el,z}} \le 1.5 \tag{9.23b}$$

In absence of lateral-flexural buckling it results in $C_{my} = C_{my,0}$, $C_{mz} = C_{mz,0}$ and $C_{mLT} = 1.0$. In case of lateral-flexural buckling:

$$C_{my} = C_{my,0} + \left(1 - C_{my,0}\right)\frac{\sqrt{\varepsilon_y}a_{LT}}{1 + \sqrt{\varepsilon_y}a_{LT}} \tag{9.24a}$$

$$C_{mz} = C_{mz,0} \tag{9.24b}$$

$$C_{mLT} = C_{my}^2 \frac{a_{LT}}{\sqrt{\left(1 - \frac{N_{Ed}}{N_{cr,z}}\right)\left(1 - \frac{N_{Ed}}{N_{cr,T}}\right)}} \ge 1 \text{ with } a_{LT} = 1 - \frac{I_T}{I_y} \ge 0 \tag{9.24c}$$

- for members of classes 1, 2 and 3:

$$\varepsilon_y = \frac{M_{y,Ed}}{N_{Ed}} \cdot \frac{A}{W_{el,y}} \tag{9.25a}$$

- for members of class 4:

$$\varepsilon_y = \frac{M_{y,Ed}}{N_{Ed}} \cdot \frac{A_{eff}}{W_{eff,y}} \tag{9.25b}$$

Coefficients $C_{my,0}$ and $C_{mz,0}$, accounting for the moment distribution along the overall members, are reported in Table 9.5. Interaction coefficients C_{yy}, C_{yz}, C_{zy} and C_{zz}, which account for plasticity phenomena, are defined as:

$$C_{yy} = 1 + \left(w_y - 1\right)\left[\left(2 - \frac{1.6}{w_y}C_{my}^2\left(\bar{\lambda}_{\max} + \bar{\lambda}_{\max}^2\right)\right) \cdot \frac{\gamma_{M1}N_{Ed}}{f_y A_i} - b_{LT}\right] \ge \frac{W_{el,y}}{W_{pl,y}} \tag{9.26a}$$

$$C_{yz} = 1 + \left(w_z - 1\right)\left[\left(2 - \frac{14}{w_z^5}C_{mz}^2\bar{\lambda}_{\max}^2\right) \cdot \frac{\gamma_{M1}N_{Ed}}{f_y A_i} - c_{LT}\right] \ge 0.6\sqrt{\frac{w_z}{w_y}} \cdot \frac{W_{el,z}}{W_{pl,z}} \tag{9.26b}$$

$$C_{zy} = 1 + \left(w_y - 1\right)\left[\left(2 - \frac{14}{w_y^5}C_{my}^2\bar{\lambda}_{\max}^2\right) \cdot \frac{\gamma_{M1}N_{Ed}}{f_y A_i} - d_{LT}\right] \ge 0.6\sqrt{\frac{w_y}{w_z}}\frac{W_{el,y}}{W_{pl,y}} \tag{9.26c}$$

$$C_{zz} = 1 + \left(w_z - 1\right)\left[\left(2 - \frac{1.6}{w_z}C_{mz}^2\left(\bar{\lambda}_{\max} + \bar{\lambda}_{\max}^2\right)\right) - e_{LT}\right] \cdot \frac{\gamma_{M1}N_{Ed}}{f_y A_i} \ge \frac{W_{el,z}}{W_{pl,z}} \tag{9.26d}$$

Auxiliary terms of the previous equations are:

$$b_{LT} = 0.5 \cdot a_{LT} \cdot \bar{\lambda}_0^2 \cdot \frac{\gamma_{M0}M_{y,Ed}}{\chi_{LT} f_y W_{pl,y}} \cdot \frac{\gamma_{M0}M_{z,Ed}}{f_y W_{pl,z}} \tag{9.27a}$$

$$c_{LT} = 10 \cdot a_{LT} \frac{\bar{\lambda}_0^2}{5 + \bar{\lambda}_z^4} \cdot \frac{\gamma_{M0}M_{y,Ed}}{C_{my}\chi_{LT} f_y W_{pl,y}} \tag{9.27b}$$

Table 9.5 Coefficient $C_{mi,0}$.

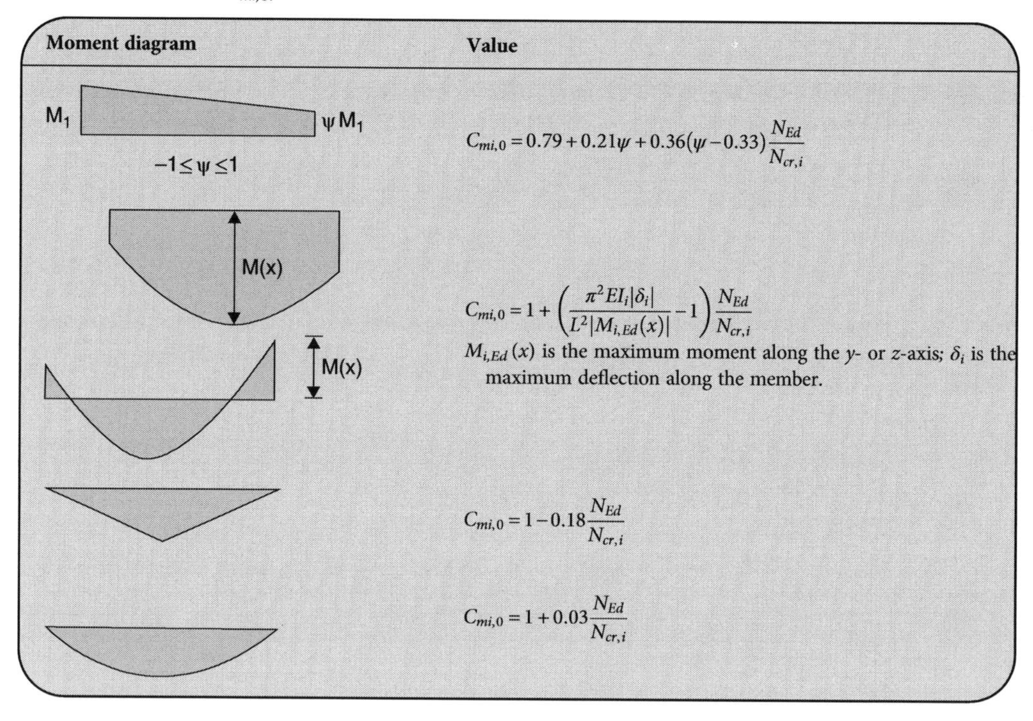

Moment diagram	Value
M_1, ψM_1, $-1 \leq \psi \leq 1$	$C_{mi,0} = 0.79 + 0.21\psi + 0.36(\psi - 0.33)\frac{N_{Ed}}{N_{cr,i}}$
M(x)	$C_{mi,0} = 1 + \left(\frac{\pi^2 EI_i\lvert\delta_i\rvert}{L^2\lvert M_{i,Ed}(x)\rvert} - 1\right)\frac{N_{Ed}}{N_{cr,i}}$ $M_{i,Ed}(x)$ is the maximum moment along the y- or z-axis; δ_i is the maximum deflection along the member.
M(x)	
	$C_{mi,0} = 1 - 0.18\frac{N_{Ed}}{N_{cr,i}}$
	$C_{mi,0} = 1 + 0.03\frac{N_{Ed}}{N_{cr,i}}$

$$d_{LT} = 2 \cdot a_{LT} \frac{\bar{\lambda}_0}{0.1 + \bar{\lambda}_z^4} \cdot \frac{\gamma_{M0} M_{y,Ed}}{C_{my} \chi_{LT} f_y W_{pl,y}} \cdot \frac{\gamma_{M0} M_{z,Ed}}{C_{mz} f_y W_{pl,z}} \tag{9.27c}$$

$$e_{LT} = 1.7 \cdot a_{LT} \frac{\bar{\lambda}_0}{0.1 + \bar{\lambda}_z^4} \cdot \frac{\gamma_{M0} M_{y,Ed}}{C_{my} \chi_{LT} f_y W_{pl,y}} \tag{9.27d}$$

where $\bar{\lambda}_{max} = \max\{\bar{\lambda}_y; \bar{\lambda}_z\}$ with $\bar{\lambda}_y = \sqrt{\frac{A_i f_y}{N_{cr,y}}}$ and $\bar{\lambda}_z = \sqrt{\frac{A_i f_y}{N_{cr,z}}}$.

9.2.2.2 Alternative Method 2 (AM2)

In accordance with the AM2, lateral buckling can be ignored, that is the members can be considered not susceptible to torsional deformations. It happens when:

- members have circular hollow cross-sections.
- members have rectangular hollow sections with $h/b \leq 10\sqrt{\bar{\lambda}_z}$, where $\bar{\lambda}_z$ is the relative slenderness relative to the weak axis (z-axis) and h and b represent the depth and width of the cross-section, respectively.
- members with open cross-sections, such as I- or H-shaped cross-sections, are torsionally and laterally restrained at the compression level;

The values of the interaction coefficient k_{yy}, k_{yz}, k_{zy}, k_{zz} are reported in Tables 9.6 and 9.7 for the cases of member non-susceptible to torsional deformation and member susceptible to torsional deformation, respectively. Table 9.8 presents the values of equivalent uniform moment factor, C_m.

Table 9.6 Interaction factors k_{ij} for members not susceptible to torsional deformations.

Interaction Factors	Type of sections	Design assumptions: Elastic cross-sectional properties class 3, class 4	Design assumptions: Plastic cross-sectional properties class 1, class 2
k_{yy}	I, RHS	$C_{my}\left[1+0.6\cdot\bar{\lambda}_y\frac{N_{Ed}}{\chi_y\cdot N_{Rk}/\gamma_{M1}}\right] \le C_{my}\left(1+0.6\frac{N_{Ed}}{\chi_y\cdot N_{Rk}/\gamma_{M1}}\right)$	$C_{my}\left[1+(\bar{\lambda}_y-0.2)\cdot\frac{N_{Ed}}{\chi_y\cdot N_{Rk}/\gamma_{M1}}\right] \le C_{my}\left(1+0.8\frac{N_{Ed}}{\chi_y\cdot N_{Rk}/\gamma_{M1}}\right)$
k_{yz}	I, RHS	k_{zz}	$0.6\cdot k_{zz}$
k_{zy}	I, RHS	$0.8\cdot k_{yy}$	$0.6\cdot k_{yy}$
k_{zz}	I	$C_{mz}\left[1+0.6\cdot\bar{\lambda}_z\frac{N_{Ed}}{\chi_z\cdot N_{Rk}/\gamma_{M1}}\right] \le C_{mz}\left(1+0.6\frac{N_{Ed}}{\chi_z\cdot N_{Rk}/\gamma_{M1}}\right)$	$C_{mz}\left[1+(2\cdot\bar{\lambda}_z-0.6)\frac{N_{Ed}}{\chi_z\cdot N_{Rk}/\gamma_{M1}}\right] \le C_{mz}\left(1+1.4\frac{N_{Ed}}{\chi_z\cdot N_{Rk}/\gamma_{M1}}\right)$
	RHS		$C_{mz}\left[1+(\bar{\lambda}_z-0.2)\frac{N_{Ed}}{\chi_z\cdot N_{Rk}/\gamma_{M1}}\right] \le C_{mz}\left(1+0.8\frac{N_{Ed}}{\chi_z\cdot N_{Rk}/\gamma_{M1}}\right)$

For I- and H-sections and RHS sections under axial compression and uniaxial bending $M_{y,Ed}$ may be $k_{zy}=0$.

Table 9.7 Interaction factors k_{ij} for members susceptible to torsional deformations.

Interaction factors	Design assumptions: Elastic cross-sectional properties class 3, class 4	Design assumptions: Plastic cross-sectional properties class 1, class 2
k_{yy}	k_{yy} from Table 9.6	k_{yy} from Table 9.3
k_{yz}	k_{yz} from Table 9.6	k_{yz} from Table 9.3
k_{zy}	$\left[1-\frac{0.05\cdot\bar{\lambda}_z}{(C_{mLT}-0.25)}\cdot\frac{N_{Ed}}{\chi_z\cdot N_{Rk}/\gamma_{M1}}\right] \ge \left[1-\frac{0.05}{(C_{mLT}-0.25)}\cdot\frac{N_{Ed}}{\chi_z\cdot N_{Rk}/\gamma_{M1}}\right]$	$\left[1-\frac{0.1\cdot\bar{\lambda}_z}{(C_{mLT}-0.25)}\cdot\frac{N_{Ed}}{\chi_z\cdot N_{Rk}/\gamma_{M1}}\right] \ge \left[1-\frac{0.1}{(C_{mLT}-0.25)}\cdot\frac{N_{Ed}}{\chi_z\cdot N_{Rk}/\gamma_{M1}}\right]$ for $\bar{\lambda}_z \le 0.4$ $k_{zy}=0.6+\bar{\lambda}_z \le 1-\frac{0.1\cdot\bar{\lambda}_z}{(C_{mLT}-0.25)}\cdot\frac{N_{Ed}}{\chi_z\cdot N_{Rk}/\gamma_{M1}}$
k_{zz}	k_{zz} from Table 9.6	k_{zz} from Table 9.6

Table 9.8 Equivalent uniform moment factors C_m in Tables 9.6 and 9.7.

Moment diagram	Range		C_{my} and C_{mz} and C_{mLT} Uniform loading	Concentrated load
M ... ψM	$-1 \le \psi \le 1$		$0.6 + 0.4\,\psi \ge 0.4$	
M_h, M_s, ψM_h; $\alpha_s = M_s/M_h$	$0 \le \alpha_s \le 1$	$-1 \le \psi \le 1$	$0.2 + 0.8\,\alpha_s \ge 0.4$	$0.2 + 0.8\,\alpha_s \ge 0.4$
	$-1 \le \alpha_s \le 0$	$0 \le \psi \le 1$	$0.1 - 0.8\,\alpha_s \ge 0.4$	$-0.8\,\alpha_s \ge 0.4$
		$-1 \le \psi < 0$	$0.1\,(1-\psi) - 0.8\,\alpha_s \ge 0.4$	$0.2\,(-\psi) - 0.8\,\alpha_s \ge 0.4$
M_h, M_s, ψM_h; $\alpha_h = M_h/M_s$	$0 \le \alpha_h \le 1$	$-1 \le \psi \le 1$	$0.95 + 0.05\,\alpha_h$	$0.90 + 0.10\,\alpha_h$
	$-1 \le \alpha_h < 0$	$0 \le \psi \le 1$	$0.95 + 0.05\,\alpha_h$	$0.90 + 0.10\,\alpha_h$
		$-1 \le \psi < 0$	$0.95 + 0.05\,\alpha_h(1 + 2\psi)$	$0.90 + 0.10\,\alpha_h(1 + 2\psi)$

For members with sway buckling mode the equivalent uniform moment factor are $C_{my} = 0.9$ and $C_{mz} = 0.9$

C_{my}, C_{mz} and C_{mLT} are obtained according to the bending moment diagram between the relevant braced points as follows:

Moment factor	Bending axis	Points braced in direction
C_{my}	y–y	z–z
C_{mz}	z–z	y–y
C_{mLT}	y–y	y–y

9.2.3 The General Method

EC3 proposes a quite new and very promising method, the so-called *General Method*, for the stability design of structural components having some geometrical, loading or supporting irregularities. In particular, this method, which generally requires the use of finite element analysis, allows us to assess the lateral and lateral-torsional buckling resistance of steel components that are subject to compression and/or mono-axial bending in the plane, such as single members, built-up or not, uniform or not members, those with complex support conditions and plane frames or sub-frames composed of such members.

The National Annex specifies the field and limits of application of this method, for which it is explicitly required that members do not contain plastic hinges. An elastic structural analysis allows for the evaluation of the internal forces and moments associated with the considered load conditions, accounting for the effects due to in plane geometrical deformation and the global as well as local imperfections. Overall resistance to out-of-plane buckling for any structural component is verified when:

$$\frac{\chi_{op}\alpha_{ult,k}}{\gamma_{M1}} \ge 1.0 \tag{9.28}$$

where $\alpha_{ult,k}$ is the minimum load multiplier with reference to the resistance of the most critical cross-section considering its in-plane behaviour without taking lateral or lateral torsional buckling into account and χ_{op} is the reduction factor for lateral and lateral torsional buckling.

Usually, term $\alpha_{ult,k}$ can be determined via a cross-section resistance check as:

$$\frac{1}{\alpha_{ult,k}} = \frac{N_{Ed}}{N_{Rk}} + \frac{M_{y,Ed}}{M_{y,Rk}} \tag{9.29}$$

The term χ_{op} is evaluated on the basis of the value of the global non-dimensional slenderness $\bar{\lambda}_{op}$ defined as:

$$\overline{\lambda_{op}} = \sqrt{\frac{\alpha_{ult,k}}{\alpha_{cr,op}}} \tag{9.30}$$

where $\alpha_{cr,op}$ is the minimum multiplier for the in plane design loads to reach the elastic critical resistance of the structural component with regards to lateral or lateral torsional buckling without accounting for in plane flexural buckling.
The reduction factor χ_{op} may be determined from either of the following methods:

- the minimum value of χ for lateral buckling according to the equation for compressed elements (see Chapter 6) and χ_{LT} for lateral torsional buckling according to equation for members under flexure (see Chapter 7). Both the χ and χ_{LT} reduction factors have to be evaluated with reference to the global non-dimensional slenderness $\overline{\lambda_{op}}$. It should be noted that if term $\alpha_{ult,K}$ is determined by the cross-sectional check (Eq. 9.29), this method leads to:

$$\frac{N_{Ed}}{N_{Rk}/\gamma_{M1}}+\frac{M_{y,Ed}}{M_{y,Rk}/\gamma_{M1}}\leq\chi_{op} \tag{9.31}$$

- a value interpolated between the values χ and χ_{LT} (as determined in previous point) using the formula for $\alpha_{ult,K}$ corresponding to the critical cross-section. Alternatively, if $\alpha_{ult,k}$ is determined by the cross-section check, this method leads to:

$$\frac{N_{Ed}}{\left(\chi\cdot N_{Rk}/\gamma_{M1}\right)}+\frac{M_{y,Ed}}{\left(\chi_{LT}\cdot M_{y,Rk}/\gamma_{M1}\right)}\leq 1 \tag{9.32}$$

9.3 Design According to the US Approach

AISC 360-10 specifications address design rules for members subjected to flexure and axial force in its Chapter H. This chapter actually contains provisions for 'Design of members for Combined Forces and Torsion', therefore, its scope is more general. AISC 360-10 provides specific rules for the following cases:

(a) Doubly and singly symmetrical members subjected to flexure and compression;
(b) Doubly and singly symmetrical members subjected to flexure and tension;
(c) Doubly symmetrical rolled compact members subjected to single axis flexure and compression;
(d) Unsymmetrical and other members subjected to flexure and axial force.

(a) Doubly and singly symmetrical members subjected to flexure and compression. Provisions of this section (H1.1 in the code) apply typically to:

rolled wide-flange shapes;
welded H sections;
channels;
tee-shapes;
round, square and rectangular HSS;
solid rounds, squares, rectangles and diamonds.

The Code requires that:

(1) When $P_r/P_c \geq 0.2$:

$$\frac{P_r}{P_c}+\frac{8}{9}\left(\frac{M_{rx}}{M_{cx}}+\frac{M_{ry}}{M_{cy}}\right)\leq 1.0 \tag{9.33}$$

(2) When $P_r/P_c < 0.2$:

$$\frac{P_r}{2P_c} + \left(\frac{M_{rx}}{M_{cx}} + \frac{M_{ry}}{M_{cy}}\right) \leq 1.0 \tag{9.34}$$

where P_r is the *required axial strength*, using LFRD or ASD load combinations, P_c is the *available axial strength* and M_{rx}, M_{ry} are the *required flexural strength*, about the x- and y-axes, respectively, using LFRD or ASD load combinations.

LFRD approach	**ASD approach**
Term P_c (*design axial strength*) is defined as: $P_c = \phi_c P_n$ (Section 6.3.3); Terms M_{cx}, M_{cy} are the *available flexural strength* (*design flexural strength*) about the x- and y-axes, respectively, defined (Section 7.3.3) as: $M_{cx} = \phi_b M_{nx}$ $M_{cy} = \phi_b M_{ny}$	Term P_c (*allowable axial strength*) is defined as: $P_c = P_n/\Omega_c$ (Section 6.3.3); Terms M_{cx}, M_{cy} are the *available flexural strength* (*available flexural strength*) about the x- and y-axes, respectively, defined (Section 7.3.3) as: $M_{cx} = M_{nx}/\Omega_b$ $M_{cy} = M_{ny}/\Omega_b$

The method is valid if the compression flange satisfies the following condition:

$$0.1 \leq \frac{I_{yc}}{I_y} \leq 0.9 \tag{9.35}$$

where I_{yc} is the moment of inertia of the compression flange about y-axis and I_c is the moment of inertia of the whole section about y-axis.

This limitation is fulfilled for all the I- and H-shaped hot-rolled profiles (for W, M and HP shapes I_{yc}/I_y ranges from 0.49 to 0.51, S-shapes from 0.57 to 0.62, for C and MC channels from 0.20 to 0.35). For welded sections, obviously it must be checked case by case.

(b) Doubly and singly symmetrical members subjected to flexure and tension (H1.2 in the code). The verification for this case is exactly the same as for (a) case: the conditions (9.33) and (9.34) have to be applied. Taking into account that axial tension increases the bending stiffness of members, and therefore is beneficial for lateral-torsional buckling, AISC Code allows for an increase in the C_b factor using a C'_b factor (modified lateral torsional buckling modification factor), according to the following expression:

$$C'_b = C_b\sqrt{1 + \frac{\alpha P_r}{P_{ey}}}; \quad \text{with}: P_{ey} = \frac{\pi^2 E I_y}{L_b^2} \tag{9.36}$$

where C_b is the lateral torsional buckling modification factor computed as in Section 7.3.3, and α is 1.0 for LFRD and equals 1.6 for ASD.

(c) *Doubly symmetrical rolled compact members subjected to single axis flexure and compression* (H1.3 in the code). For doubly symmetrical hot-rolled compact sections with $(KL)_z \le (KL)_y$ and subjected to bending about the x-axis only ($M_{ry}/M_{cy} < 0.05$), AISC Code gives an optional method for verification. The following conditions must be fulfilled:

when $P_r/P_{cx} \ge 0.2$:

$$\frac{P_r}{P_{cx}} + \frac{8}{9}\left(\frac{M_{rx}}{M_{cx}}\right) \le 1.0 \tag{9.37}$$

when $P_r/P_{cx} < 0.2$:

$$\frac{P_r}{2P_{cx}} + \left(\frac{M_{rx}}{M_{cx}}\right) \le 1.0 \tag{9.38}$$

$$\frac{P_r}{P_{cy}}\left(1.5 - 0.5\frac{P_r}{P_{cy}}\right) + \left(\frac{M_{rx}}{C_b M'_{cx}}\right)^2 \le 1.0 \tag{9.39}$$

where P_{cx} is the *available axial strength*, determined in the plane of bending and M'_{cx} is the *available lateral torsional strength* for x-axis bending, determined using $C_b = 1$.

Equations (9.37) and (9.38) take into account the limit state of in-plane instability, while Eq. (9.39) considers the limit state of out-of-plane buckling and lateral-torsional buckling. For cases where the axial limit state design is in accordance with the out-of-plane buckling and the flexural limit state is the lateral-torsional buckling, case (a) should result quite conservative with respect to the case (c).

(d) *Unsymmetrical and other members subjected to flexure and axial force* (H2 in the code). For generic members, not covered in cases (a), (b) or (c), AISC Code allows for use of the following equation:

$$\left|\frac{f_{ra}}{F_{ca}} + \frac{f_{rbw}}{F_{cbw}} + \frac{f_{rbz}}{F_{cbz}}\right| \le 1.0 \tag{9.40}$$

where, f_{ra} is the *required axial stress* at the point of consideration, F_{ca} is the *available axial stress* at the point of consideration, f_{rbw}, f_{rbz} are the *required flexural stresses* at the point of consideration and F_{cbw}, F_{cbz} are the *available flexural stresses* at the point of consideration;

Furthermore, it should be noted that the subscripts w and z indicate the major and the minor principal axis of the cross-section.

LFRD approach	ASD approach
$F_{ca} = \phi_c F_{cr}$	$F_{ca} = \frac{F_{cr}}{\Omega_c}$
$F_{cbw,cbz} = \frac{\phi_c M_{nw,nz}}{S_{w,z}}$	$F_{cbw,cbz} = \frac{M_{nw,nz}}{\Omega_b S_{w,z}}$

where M_{nw}, M_{nz} are the *nominal flexural strengths* in the w- and z-directions (see Section 8.3.3).

Equation (9.40) allows us to then verify the section using stress values and not strengths and it could also be used for members covered by the design case (a).

9.4 Worked Examples

Example E9.1 Beam-Column Design According to the EU Approach

Verify whether a S275 HEA 260 column, belonging to a braced frame, subjected to combined compression force and bending in both axes, is able to support the design axial force and moments listed here (second order acting along both cross-sectional effects are included). The column is pinned at its end and subjected to a uniformly distributed load along both the axes. The unbraced length is 4 m (13.1 ft) in both axes. The column is not braced along its height (lateral-torsional buckling is not prevented).

Geometrical properties:

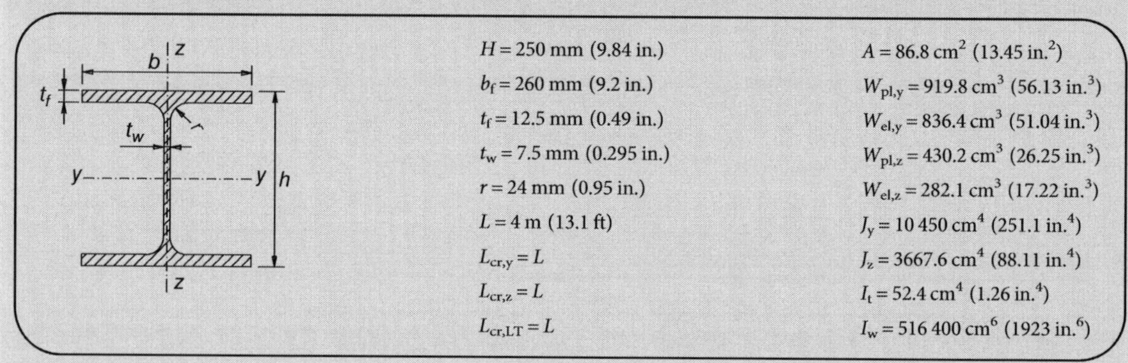

Material properties:

$$\text{Steel}: \text{S275} \quad f_y = 275 \text{ MPa} (34 \text{ ksi}) \quad f_u = 430 \text{ MPa} (62 \text{ ksi})$$

Axial load and maximum moments at the member midspan:

$$N_{Ed} = 400 \text{ kN} (89.9 \text{ kips})$$
$$M_{y,Ed} = 71 \text{ kNm} (52.4 \text{ kips-ft})$$
$$M_{z,Ed} = 30 \text{ kNm} (22.1 \text{ kips-ft})$$

AM1 Check According to EC3 Alternative Method 1.

(a) Section Classification

- Compression:

$$\text{Flange}: \quad (c/t_f) = [260 - 7.5 - (2\times 24)]/(2\times 12.5) = 8.2 \le 8.3 \quad \text{Class 1}$$
$$\text{Web}: \quad (d/t_w) = [250 - (2\times 12.5) - (2\times 24)]/7.5 = 23.6 \le 30.3 \quad \text{Class 1}$$
$$\text{Section}: \quad \text{Class 1}$$

- Bending:

$$\text{Flange}: \quad (c/t_f) = [260 - 7.5 - (2\times 24)]/(2\times 12.5) = 8.2 \le 8.3 \quad \text{Class 1}$$
$$\text{Web}: \quad (d/t_w) = [250 - (2\times 12.5) - (2\times 24)]/7.5 = 23.6 \le 66 \quad \text{Class 1}$$
$$\text{Section}: \quad \text{Class 1}$$

For combined axial load and bending, the section is therefore classified as Class 1.

(b) Compression

$$\text{Compression: unbraced length}: L_{cr,y} = \beta_y L = 1.0 \times 4.0 = 4.0 = 4\ \text{m}(13.1\ \text{ft});$$

$$L_{cr,z} = \beta_z L = 1.0 \times 4.0 = 4.0 = 4\ \text{m}(13.1\ \text{ft})$$

Evaluation of the elastic critical buckling load about both principal axes:

$$N_{cr,y} = \pi^2 \frac{EJ_y}{L_{cr,y}^2} = \pi^2 \times \frac{21000 \times 10450}{400^2} = 13536.8\,\text{kN}\,(3043\,\text{kips})$$

$$N_{cr,z} = \pi^2 \frac{EJ_z}{L_{cr,z}^2} = \pi^2 \times \frac{21000 \times 3667.6}{400^2} = 4750.9\,\text{kN}\,(1068\,\text{kips})$$

Evaluation of the relative slenderness and associated reduction factor χ about both axes:

$$\bar{\lambda}_y = \sqrt{\frac{A \cdot f_y}{N_{cr,y}}} = \sqrt{\frac{86.8 \times 27.50}{13536.8}} = 0.420; \bar{\lambda}_z = \sqrt{\frac{A \cdot f_y}{N_{cr,z}}} = \sqrt{\frac{86.8 \times 27.50}{4750.9}} = 0.709$$

$$\Phi_y = 0.5\left[1 + \alpha_y(\bar{\lambda}_y - 0.2) + \bar{\lambda}_y^2\right] = 0.5 \times \left[1 + 0.34 \times (0.420 - 0.2) + 0.420^2\right] = 0.626$$

$$\chi_y = \frac{1}{\Phi_y + \sqrt{\Phi_y^2 - \bar{\lambda}_y^2}} = \frac{1}{0.626 + \sqrt{0.626^2 - 0.420^2}} = 0.918$$

$$\Phi_z = 0.5\left[1 + \alpha_z(\bar{\lambda}_z - 0.2) + \bar{\lambda}_z^2\right] = 0.5 \times \left[1 + 0.49 \times (0.709 - 0.2) + 0.709^2\right] = 0.876$$

$$\chi_z = \frac{1}{\Phi_z + \sqrt{\Phi_z^2 - \bar{\lambda}_z^2}} = \frac{1}{0.876 + \sqrt{0.876^2 - 0.709^2}} = 0.719$$

Available column strength in axial compression about the y–y and z–z axes are:

$$N_{by,Rd} = \frac{\chi_y A_i f_y}{\gamma_{M1}} = \frac{0.918 \times 86.8 \times 27.50}{1.00} = 2191.9\,\text{kN}\,(492.8\,\text{kips})$$

$$N_{bz,Rd} = \frac{\chi_z A_i f_y}{\gamma_{M1}} = \frac{0.719 \times 86.8 \times 27.50}{1.00} = 1717.0\,\text{kN}\,(386\,\text{kips})$$

(c) Lateral-Torsional Buckling (EC3 Section 6.3.2.3 Method)

Consider $k = k_w = 1$ (rotation around the vertical axis and warping not prevented at both ends), $z_g = 0$ (load applied at centroid); and $C_1 = 1.127$; $C_2 = 0.454$ (uniformly distributed load).

Critical moment:

$$M_{cr} = C_1 \frac{\pi^2 EJ_z}{(kL_{cr,LT})^2} \left\{ \sqrt{\left(\frac{k}{k_w}\right)^2 \frac{I_w}{J_z} + \frac{(kL_{cr,LT})^2 GI_t}{\pi^2 EJ_z} + (C_2 z_g)^2} - C_2 z_g \right\}$$

$$= 1.127 \times \frac{\pi^2 \times 21000 \times 3667.6}{(1 \times 400)^2} \sqrt{\left(\frac{1}{1}\right)^2 \times \frac{516400}{3667.6} + \frac{(1 \times 400)^2 \times 8077 \times 52.4}{\pi^2 \times 21000 \times 3667.6}} \cdot 10^{-2}$$

$$= 811.7\,\text{kNm}\ (598.7\,\text{kips-ft})$$

Terms containing z_g have been disregarded because $z_g = 0$.

$$\bar{\lambda}_{LT} = \sqrt{\frac{W_y f_y}{M_{cr}}} = \sqrt{\frac{919.4 \times 27.50}{811.7 \cdot 10^2}} = 0.558; \quad \bar{\lambda}_{LT,0} = 0.2; \quad \beta = 1;$$

$$\alpha_{LT} = 0, \ k_c = 0.94;$$

$$f = 1 - 0.5(1 - k_c)\left[1 - 2.0(\bar{\lambda}_{LT} - 0.8)^2\right] = 1 - 0.5 \times (1 - 0.94) \times \left[1 - 2.0 \times (0.558 - 0.8)^2\right] = 0.974$$

$$\Phi_{LT} = 0.5\left[1 + \alpha_{LT}(\bar{\lambda}_{LT} - \bar{\lambda}_{LT,0}) + \beta \bar{\lambda}_{LT}^2\right] = 0.5 \times \left[1 + 0.34 \times (0.558 - 0.2) + 1 \times 0.558^2\right] = 0.717$$

$$\chi_{LT} = \frac{1}{\Phi_{LT} + \sqrt{\Phi_{LT}^2 - \bar{\lambda}_{LT}^2}} = \frac{1}{0.717 + \sqrt{0.717^2 - 0.558^2}} = 0.857 \leq 1$$

$$\chi_{LT,\text{mod}} = \chi_{LT}/f = 0.857/0.974 = 0.880$$

Calculate flexural strength for bending along the y–y and z–z axes:

$$M_{b,Rd} = \chi_{LT,\text{mod}} \frac{W_y f_y}{\gamma_{M1}} = 0.880 \times \frac{919.8 \times 27.50}{1.00} \cdot 10^{-2} = 222.8\,\text{kNm}\ (164.2\,\text{kips-ft})$$

$$M_{z,Rd} = \frac{W_y f_y}{\gamma_{M1}} = \frac{430.2 \times 27.50}{1.00} \cdot 10^{-2} = 118.3\,\text{kNm}\ (87.3\,\text{kips-ft})$$

(d) Combined compression and bending

$$\mu_y = \frac{1 - \dfrac{N_{Ed}}{N_{cr,y}}}{1 - \chi_y \dfrac{N_{Ed}}{N_{cr,y}}} = \frac{1 - \dfrac{400}{13536.8}}{1 - 0.918 \times \dfrac{400}{13536.8}} = 0.998;$$

$$w_y = \frac{W_{pl,y}}{W_{el,y}} = \frac{919.8}{836.4} = 1.100$$

$$\mu_z = \frac{1-\dfrac{N_{Ed}}{N_{cr,z}}}{1-\chi_z\dfrac{N_{Ed}}{N_{cr,z}}} = \frac{1-\dfrac{400}{4750.9}}{1-0.719\times\dfrac{400}{4750.9}} = 0.975;$$

$$w_z = \frac{W_{pl,z}}{W_{el,z}} = \frac{430.2}{282.1} = 1.525 > 1.500 \quad \text{hence}: w_z = 1.500$$

Calculate $C_{my,0}$ and $C_{mz,0}$ parameters accounting for a bending moment diagram shape.

$$C_{my,0} = 1+0.03\frac{N_{Ed}}{N_{cr,y}} = 1+0.03\times\frac{400}{13536.8} = 1.001$$

$$C_{mz,0} = 1+0.03\frac{N_{Ed}}{N_{cr,z}} = 1+0.03\times\frac{400}{4750.9} = 1.003$$

Check if the section is subjected to flexural-torsional buckling.

$$I_t = 52.4\ \text{cm}^4 < I_y = 10450\ \text{cm}^4$$

In accordance with the criteria proposed in Eq. (9.8) the result is that the cross-section should be susceptible of torsional deformations. Then calculate:

$$M_{cr,0} = \frac{\pi^2 EI_z}{L^2_{cr,LT}}\sqrt{\frac{I_w}{I_z}+\frac{L^2_{cr,LT}GI_t}{\pi^2 EI_z}}$$

$$= \frac{\pi^2\times 21000\times 3667.6}{400^2}\times\sqrt{\frac{516400}{3667.6}+\frac{400^2\times 8077\times 52.4}{\pi^2\times 21000\times 3667.6}}\cdot 10^{-2}$$

$$= 720.3\,\text{kNm}\,(531.3\,\text{kips}-\text{ft})$$

$$N_{cr,TF} = N_{cr,T} = \frac{A}{I_y+I_z}\left(GI_t+\frac{\pi^2 EI_w}{L^2_{cr,T}}\right)$$

$$= \frac{86.8}{10450+3667.6}\times\left(8077\times 52.4+\frac{\pi^2\times 21000\times 516400}{400^2}\right) = 6715.1\,\text{kN}\,(1510\,\text{kips})$$

$$\bar{\lambda}_{0\,\text{lim}} = 0.2\sqrt{C_1}\sqrt[4]{\left(1-\frac{N_{Ed}}{N_{cr,z}}\right)\left(1-\frac{N_{Ed}}{N_{cr,TF}}\right)} = 0.2\times\sqrt{1.127}\times,\sqrt[4]{\left(1-\frac{400}{4750.9}\right)\times\left(1-\frac{400}{6715.1}\right)} = 0.205$$

$$\bar{\lambda}_0 = \sqrt{\frac{W_{pl,y}f_y}{M_{cr,0}}} = \sqrt{\frac{919.8\times 27.50}{720.3\cdot 10^2}} = 0.593 > \bar{\lambda}_{0\,\text{lim}} = 0.205;$$

Being $\bar{\lambda}_0 > \bar{\lambda}_{0\,\text{lim}}$ the section can be subjected to flexural-torsional buckling.

Compute the C_{my}, C_{mz} and C_{mLT} parameters.

$$a_{LT} = 1-\frac{I_T}{I_y} = 1-\frac{52.4}{10450} = 0.995 \geq 0;\ \varepsilon_y = \frac{M_{y,Ed}}{N_{Ed}}\cdot\frac{A}{W_{el,y}} = \frac{71\times 100\times 86.8}{400\times 836.4} = 1.842;$$

$$C_{my} = C_{my,0} + (1 - C_{my,0})\frac{\sqrt{\varepsilon_y}a_{LT}}{1+\sqrt{\varepsilon_y}a_{LT}} = 1.001 + (1-1.001)\times\frac{\sqrt{1.842}\times 0.995}{1+\sqrt{1.842}\times 0.995} = 1.00;\; C_{mz} = C_{mz,0} = 1.003$$

$$C_{mLT} = C_{my}^2\frac{a_{LT}}{\sqrt{\left(1-\frac{N_{Ed}}{N_{cr,z}}\right)\left(1-\frac{N_{Ed}}{N_{cr,T}}\right)}} = 1.00^2\times\frac{0.995}{\sqrt{\left(1-\frac{400}{4750.9}\right)\times\left(1-\frac{400}{6715.1}\right)}} = 1.073 \geq 1$$

Compute the interaction parameters C_{yy}, C_{yz}, C_{zy}, C_{zz}:

$$\bar{\lambda}_{\max} = \max\{\bar{\lambda}_y;\bar{\lambda}_z\} = \max\{0.427;\; 0.709\} = 0.709;$$

$$b_{LT} = 0.5a_{LT}\bar{\lambda}_0^2\frac{\gamma_{M0}M_{y,Ed}}{\chi_{LT}f_yW_{pl,y}}\cdot\frac{\gamma_{M0}M_{z,Ed}}{f_yW_{pl,z}} = 0.5\times 0.995\times 0.593^2\times\frac{1.00\times 71\cdot 10^2}{0.857\times 27.50\times 919.8}\times\frac{1.00\times 30\cdot 10^2}{27.50\times 430.2} = 0.014$$

$$c_{LT} = 10a_{LT}\frac{\bar{\lambda}_0^2}{5+\bar{\lambda}_z^4}\cdot\frac{\gamma_{M0}M_{y,Ed}}{C_{my}\chi_{LT}f_yW_{pl,y}}$$

$$= 10\times 0.995\times\frac{0.593^2}{5+0.709^4}\times\frac{1.00\times 71\cdot 10^2}{1.00\times 0.880\times 27.50\times 919.8} = 0.212$$

$$d_{LT} = 2a_{LT}\frac{\bar{\lambda}_0}{0.1+\bar{\lambda}_z^4}\cdot\frac{\gamma_{M0}M_{y,Ed}}{C_{my}\chi_{LT}f_yW_{pl,y}}\cdot\frac{\gamma_{M0}M_{z,Ed}}{C_{mz}f_yW_{pl,z}}$$

$$= 2\times 0.995\times\frac{0.593}{0.1+0.709^4}\times\frac{1.00\times 71\cdot 10^2}{1.00\times 0.880\times 27.50\times 919.8}\times\frac{1.00\times 30\cdot 10^2}{1.003\times 27.50\times 430.2} = 0.269$$

$$e_{LT} = 1.7a_{LT}\frac{\bar{\lambda}_0}{0.1+\bar{\lambda}_z^4}\cdot\frac{\gamma_{M0}M_{y,Ed}}{C_{my}\chi_{LT}f_yW_{pl,y}}$$

$$= 1.7\times 0.995\times\frac{0.593}{0.1+0.709^4}\times\frac{1.00\times 71\cdot 10^2}{1.00\times 0.880\times 27.50\times 919.8} = 0.906$$

$$C_{yy} = 1 + (w_y - 1)\left[\left(2 - \frac{1.6}{w_y}C_{my}^2\left(\bar{\lambda}_{\max} + \bar{\lambda}_{\max}^2\right)\right)\cdot\frac{\gamma_{M1}N_{Ed}}{f_yA_i} - b_{LT}\right]$$

$$= 1 + (1.1-1)\times\left[\left(2-\frac{1.6}{1.1}1.0^2\times(0.709+0.709^2)\right)\times\frac{1.00\times 400}{27.50\times 86.8} - 0.014\right] = 1.003$$

$$\geq \frac{W_{el,y}}{W_{pl,y}} = \frac{836.4}{919.8} = 0.909$$

$$C_{yz} = 1 + (w_z - 1)\left[\left(2 - \frac{14}{w_z^5}C_{mz}^2\bar{\lambda}_{\max}^2\right)\cdot\frac{\gamma_{M1}N_{Ed}}{f_yA_i} - c_{LT}\right]$$

$$= 1 + (1.5-1)\times\left[\left(2-\frac{14}{1.5^5}1.003^2\times 0.709^2\right)\times\frac{1.00\times 400}{27.50\times 86.8} - 0.212\right] = 0.984$$

$$\geq 0.6\sqrt{\frac{w_z}{w_y}}\frac{W_{el,z}}{W_{pl,z}} = 0.6\times\sqrt{\frac{1.5}{1.1}}\times\frac{282.1}{430.2} = 0.459$$

$$C_{zy} = 1 + (w_y - 1)\left[\left(2 - \frac{14}{w_y^5}C_{my}^2\bar{\lambda}_{\max}^2\right)\cdot\frac{\gamma_{M1}N_{Ed}}{f_yA_i} - d_{LT}\right]$$

$$= 1 + (1.1 - 1)\times\left[\left(2 - \frac{14}{1.1^5}1.00^2\times 0.709^2\right)\times\frac{1.00\times 400}{27.50\times 86.8} - 0.269\right] = 0.933$$

$$\geq 0.6\sqrt{\frac{w_y}{w_z}\frac{W_{el,y}}{W_{pl,y}}} = 0.6\times\sqrt{\frac{1.1}{1.5}}\times\frac{836.4}{919.8} = 0.467$$

$$C_{zz} = 1 + (w_z - 1)\left[\left(2 - \frac{1.6}{w_z}C_{mz}^2\left(\bar{\lambda}_{\max} + \bar{\lambda}_{\max}^2\right)\right) - e_{LT}\right]\cdot\frac{\gamma_{M1}N_{Ed}}{f_yA_i}$$

$$= 1 + (1.5 - 1)\times\left[\left(2 - \frac{1.6}{1.5}1.003^2\times\left(0.709 + 0.709^2\right)\right) - 0.906\right]\times\frac{1.00\times 400}{27.50\times 86.8} = 0.983$$

$$\geq \frac{W_{el,z}}{W_{pl,z}} = \frac{282.1}{430.2} = 0.656$$

Compute k_{ij}:

$$k_{yy} = C_{my}C_{mLT}\frac{\mu_y}{1 - \frac{N_{Ed}}{N_{cr,y}}}\frac{1}{C_{yy}} = 1.0\times 1.073\times\frac{0.998}{\left(1 - \frac{400}{13536.8}\right)\times 1.003} = 1.100$$

$$k_{yz} = C_{mz}\frac{\mu_y}{1 - \frac{N_{Ed}}{N_{cr,z}}}\frac{1}{C_{yz}}0.6\sqrt{\frac{w_z}{w_y}} = 1.003\times\frac{0.998}{\left(1 - \frac{400}{4750.9}\right)\times 0.984}\times 0.6\times\sqrt{\frac{1.5}{1.1}} = 0.778$$

$$k_{zy} = C_{my}C_{mLT}\frac{\mu_z}{1 - \frac{N_{Ed}}{N_{cr,y}}}\frac{1}{C_{zy}}0.6\sqrt{\frac{w_y}{w_z}} = 1.0\times 1.073\times\frac{0.975}{\left(1 - \frac{400}{13536.8}\right)\times 0.933}\times 0.6\times\sqrt{\frac{1.1}{1.5}} = 0.593$$

$$k_{zz} = C_{mz}\frac{\mu_z}{1 - \frac{N_{Ed}}{N_{cr,z}}}\frac{1}{C_{zz}} = 1.003\times\frac{0.975}{\left(1 - \frac{400}{4750.9}\right)\times 0.983} = 1.086$$

And finally compute Eqs. (9.19a) and (9.19b):

$$\frac{N_{Ed}}{N_{by,Rd}} + k_{yy}\frac{M_{y,Ed} + e_{N,y}N_{Ed}}{M_{b,Rd}} + k_{yz}\frac{M_{z,Ed} + e_{N,z}N_{Ed}}{M_{z,Rd}} = \frac{400}{2191.9} + 1.10\times\frac{71}{222.8} + 0.778\times\frac{30}{118.3}$$
$$= 0.18 + 0.35 + 0.20 = 0.73 \leq 1$$

$$\frac{N_{Ed}}{N_{bz,Rd}} + k_{zy}\frac{M_{y,Ed} + e_{N,y}N_{Ed}}{M_{b,Rd}} + k_{zz}\frac{M_{z,Ed} + e_{N,z}N_{Ed}}{M_{z,Rd}} = \frac{400}{1717.0} + 0.593\times\frac{71}{222.8} + 1.086\times\frac{30}{118.3}$$
$$= 0.23 + 0.19 + 0.28 = 0.70 \leq 1$$

AM2 Check according to EC3 Alternative Method 2
Compute parameters C_{my}, C_{mz} and C_{mLT}, using Table B.3 of EN 1993-1-1 Annex B.

$$\alpha_h = M_h/M_s = 0.0/71.0 = 0; \quad \psi = 0$$

(same value for the moment about the z–z axis)

$$C_{my} = C_{mz} = C_{mLT} = 0.95 + 0.05\alpha_h = 0.95 + 0.05 \times 0 = 0.95$$

Compute k_{ij}.

$$k_{yy} = C_{my}\left[1 + (\bar{\lambda}_y - 0.2)\frac{\gamma_{M1} N_{Ed}}{\chi_y f_y A_i}\right] = 0.95 \times \left[1 + (0.420 - 0.2) \times \frac{1.00 \times 400}{0.918 \times 27.50 \times 86.8}\right] = 0.936$$

$$\leq C_{my}\left(1 + 0.8\frac{\gamma_{M1} N_{Ed}}{\chi_y f_y A_i}\right) = 0.95 \times \left(1 + 0.8 \times \frac{1.00 \times 400}{0.918 \times 27.50 \times 86.8}\right) = 1.089$$

$$k_{zz} = C_{mz}\left[1 + (2\bar{\lambda}_z - 0.6)\frac{\gamma_{M1} N_{Ed}}{\chi_z f_y A_i}\right] = 0.95 \times \left[1 + (2 \times 0.709 - 0.6) \times \frac{1.00 \times 400}{0.719 \times 27.50 \times 86.8}\right] = 1.071$$

$$\leq C_{mz}\left(1 + 1.4\frac{\gamma_{M1} N_{Ed}}{\chi_z f_y A_i}\right) = 0.95 \times \left(1 + 1.4 \times \frac{1.00 \times 400}{0.719 \times 27.50 \times 86.8}\right) = 1.260$$

$k_{yz} = 0.6 k_{zz} = 0.6 \times 1.071 = 0.643$; $\bar{\lambda}_z = 0.709 > 0.4$; hence:

$$k_{zy} = \left[1 - \frac{0.1\bar{\lambda}_z}{(C_{mLT} - 0.25)}\frac{\gamma_{M1} N_{Ed}}{\chi_z f_y A_i}\right] = 1 - \frac{0.1 \times 0.709}{(0.95 - 0.25)} \times \frac{1.00 \times 400}{0.719 \times 27.50 \times 86.8} = 0.975$$

$$\geq \left[1 - \frac{0.1}{(C_{mLT} - 0.25)}\frac{\gamma_{M1} N_{Ed}}{\chi_z f_y A_i}\right] = 1 - \frac{0.1}{(0.95 - 0.25)} \times \frac{1.00 \times 400}{0.719 \times 27.50 \times 86.8} = 0.967$$

And finally compute Eqs. (9.19a) and (9.19b):

$$\frac{N_{Ed}}{N_{by,Rd}} + k_{yy}\frac{M_{y,Ed} + e_{N,y} N_{Ed}}{M_{b,Rd}} + k_{yz}\frac{M_{z,Ed} + e_{N,z} N_{Ed}}{M_{z,Rd}} = \frac{400}{2191.9} + 0.936 \times \frac{71}{222.8} + 0.643 \times \frac{30}{118.3}$$

$$= 0.18 + 0.30 + 0.16 = 0.64 \leq 1$$

$$\frac{N_{Ed}}{N_{bz,Rd}} + k_{zy}\frac{M_{y,Ed} + e_{N,y} N_{Ed}}{M_{b,Rd}} + k_{zz}\frac{M_{z,Ed} + e_{N,z} N_{Ed}}{M_{z,Rd}} = \frac{400}{1717.0} + 0.967 \times \frac{71}{222.8} + 1.071 \times \frac{30}{118.3}$$

$$= 0.23 + 0.32 + 0.27 = 0.82 \leq 1$$

Summarizing main results obtained with both methods, safety indexes are directly compared in the following table:

Method	Safety index
AM1-Method 1 – EN 1993-1-1 Annex A	0.73
AM2-Method 2 – EN 1993-1-1 Annex B	0.82

Example E9.2 Beam-Column Design According to the US Approach H1.1

Verify whether an ASTM A99 W10 × 49 column, belonging to a braced frame, subjected to combined compression and bending along in both axes, is able to support the axial forces and moments listed here (second order effects included) The column is pinned at its ends and subjected to a uniformly distributed load along the cross-section axes. The unbraced length is 13.5 ft (4.1 m) along both axes. The column is not braced along its height (lateral-torsional buckling not prevented).

Geometrical properties:

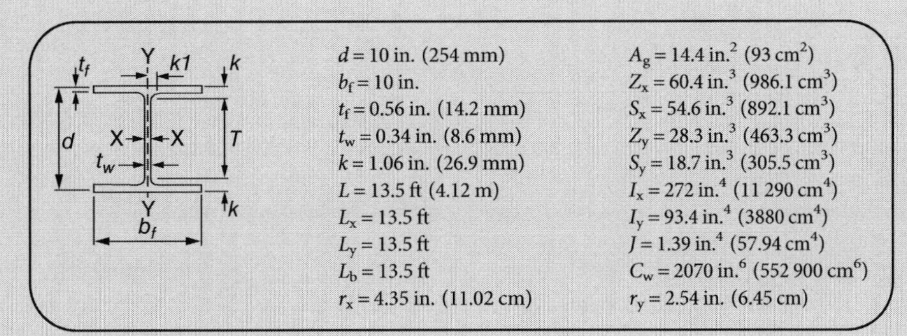

Material properties:

$$\text{Steel: ASTM A992} \quad F_y = 50 \text{ ksi}(345 \text{ MPa}) \quad F_u = 65 \text{ ksi}(448 \text{ MPa})$$

Axial load and maximum moments at the midspan:

LFRD	ASD
$P_u = 100$ kips (445 kN)	$P_a = 65$ kips (289 kN)
$M_{ux} = 53$ kips-ft (71.9 kNm)	$M_{ax} = 35$ kips-ft (47.5 kNm)
$M_{uy} = 22$ kips-ft (29.8 kNm)	$M_{ay} = 14.7$ kips-ft (19.9 kNm)

Verification according to AISC 360-10 H1.1: ‘Doubly and Singly Symmetric Members Subjected to Flexure and Compression’.

(1) Axial Strength

Section classification for local buckling.

Flange:

$$b/t = (0.5 \times 10)/0.56 = 8.92 < 0.56\sqrt{E/F_y} = 0.56 \times \sqrt{29000/50} = 13.49 \rightarrow \text{non-slender}$$

Web:

$$h/t_w = (10 - 2 \times 1.06)/0.34 = 23.17 < 1.49\sqrt{E/F_y} = 1.49 \times \sqrt{29000/50} = 35.88 \rightarrow \text{non-slender}$$

ASTM A36 steel W10 × 49, subjected to axial load, is a non-slender section ($Q = 1$).Check slenderness ratio about both axes (assuming $K = 1.0$)

$$\frac{KL_x}{r_x} = \frac{1 \times (13.5 \cdot 12)}{4.35} = 37.2$$

$$\frac{KL_y}{r_y} = \frac{1 \times (13.5 \cdot 12)}{2.54} = 63.8 \text{ governs}$$

Calculate the elastic critical buckling stress (F_e).

$$F_e = \frac{\pi^2 E}{\left(\frac{KL_y}{r_y}\right)^2} = \frac{\pi^2 \times 29000}{(63.8)^2} = 70.4\,\text{ksi}\,(485\,\text{MPa})$$

Calculate flexural buckling stress (F_{cr}).

$$\text{Check limit}: 4.71\sqrt{\frac{E}{QF_y}} = 4.71 \times \sqrt{\frac{29000}{1 \times 50}} = 113.4 > 63.8$$

Because $\frac{KL_y}{r_y} \leq 4.71\sqrt{\frac{E}{(QF_y)}}$ then:

$$F_{cr} = \left[0.658^{\frac{(QF_y)}{F_e}}\right](QF_y) = \left[0.658^{\frac{1\times 50}{70.4}}\right] \times (1 \times 50) = 37.14\,\text{ksi}\,(256\,\text{MPa})$$

Compute the nominal compressive strength (P_n).

$$P_n = F_{cr} A_g = 37.14 \times 14.4 = 534.8\,\text{kips}\,(2379\,\text{kN})$$

Compute the available strength.

LFRD: $\phi_c P_n = 0.90 \times 534.8 = 481.3\,\text{kips}\,(2141\,\text{kN})$
ASD: $P_n/\Omega_c = 534.8/1.67 = 320.2\,\text{kips}\,(1424\,\text{kN})$

(2) Flexural Strength
Section classification for local buckling.

Flange:

$$b/t = (0.5 \times 10)/0.56 = 8.92 < 0.38\sqrt{E/F_y} = 0.38 \times \sqrt{29000/50} = 9.15 \rightarrow \text{compact}$$

Web:

$$h/t_w = (10 - 2 \times 1.06)/0.34 = 23.17 < 3.76\sqrt{E/F_y} = 3.76 \times \sqrt{29000/50} = 90.55 \rightarrow \text{compact}$$

ASTM A36 steel W10 × 49, subjected to flexure, is a compact section.
Plastic moment:

$$M_p = F_y \times Z_x = 50 \times 60.4/12 = 251.7\ \text{kips-ft}(341.3\ \text{kNm})$$

Flexural strength corresponding to lateral torsional buckling limit state (M_{LT}):
$C_b = 1.136$ (uniformly distributed load)

$$L_p = 1.76 r_y \sqrt{\frac{E}{F_y}} = 1.76 \times 2.54 \times \sqrt{\frac{29000}{50}} = 107.7\,\text{in.}/12 = 8.98\,\text{ft}\,(2.74\,\text{m})$$

$$r_{ts} = \sqrt{\frac{\sqrt{I_y C_w}}{S_x}} = \sqrt{\frac{\sqrt{93.4 \times 2070}}{54.6}} = 2.838\,\text{in.}\,(7.21\,\text{cm})$$

$c = 1$ (doubly-symmetric I-shape)

$$h_o = d - t_f = 10 - 0.56 = 9.44 \text{ in.}$$

$$L_r = 1.95 r_{ts} \frac{E}{0.7F_y} \sqrt{\frac{J \cdot c}{S_x h_0} + \sqrt{\left(\frac{J \cdot c}{S_x h_0}\right)^2 + 6.76\left(\frac{0.7F_y}{E}\right)^2}}$$

$$= 1.95 \times 2.838 \times \frac{29000}{0.7 \times 50} \sqrt{\frac{1.39 \times 1}{54.6 \times 9.44} + \sqrt{\left(\frac{1.39 \times 1}{54.6 \times 9.44}\right)^2 + 6.76 \times \left(\frac{0.7 \times 50}{29000}\right)^2}}$$

$$= 379 \text{ in.} / 12 = 31.6 \text{ ft} \, (9.63 \text{ m})$$

$L_p = 8.98$ ft $< L_b = 13.5$ ft $\leq L_r = 31.6$ ft; Hence:

$$M_{LT} = C_b \left[M_p - \left(M_p - 0.7F_y S_x\right)\left(\frac{L_b - L_p}{L_r - L_p}\right) \right]$$

$$= 1.136 \times \left[251.7 - (251.7 - 0.7 \times 50 \times 54.6/12)\left(\frac{13.5 - 8.98}{31.6 - 8.98}\right) \right] = 264.9 \text{ kips} - \text{ft} \, (359.2 \text{ kNm.})$$

Compute the nominal flexural strength (M_n).

$$M_n = \min\left(M_p; M_{LT}\right) = \min(251.7; 264.9) = 251.7 \text{ kips-ft}(341.3 \text{ kNm})$$

Compute the available strength.

LFRD: $\phi_b M_n = 0.90 \times 251.7 = 226.5$ kips (307 kN)
ASD: $M_n / \Omega_b = 251.7/1.67 = 150.7$ kips (204.3 kN)

(3) Verification for Axial Load and Bending Moments

LFRD:

$$P_r = P_u = 100 \text{ kips}(445 \text{ kN})$$

$$M_{rx} = M_{ux} = 53 \text{ kips-ft}(71.9 \text{ kNm})$$

$$M_{ry} = M_{uy} = 22 \text{ kips-ft}(29.8 \text{ kNm})$$

$$P_c = \phi_c P_n = 481.3 \text{ kips} \, (2141 \text{ kN})$$

$$M_{cx} = \phi_b M_n = 226.5 \text{ kips} \, (307 \text{ kN})$$

$$F_y Z_y = 50 \times 28.3/12 = 117.9 \text{ kips-ft} \leq 1.6 F_y S_y = 1.6 \times 50 \times 18.7/12 = 124.7 \text{ kips-ft}$$

Hence:

$$M_{cy} = \phi_b F_y Z_y = 0.90 \times 117.9 = 1061 \text{ kips} - \text{ft} \, (143.9 \text{ kNm})$$

$$\frac{P_r}{P_c} = \frac{100}{481.3} = 0.21 \geq 0.20$$

$$\frac{P_r}{P_c} + \frac{8}{9}\left(\frac{M_{rx}}{M_{cx}} + \frac{M_{ry}}{M_{cy}}\right) = \frac{100}{481.3} + \frac{8}{9}\left(\frac{53}{226.5} + \frac{22}{106.1}\right) = 0.21 + \frac{8}{9}(0.23 + 0.21) = 0.60 < 1.0$$

ASD:

$$P_r = P_a = 65 \text{ kips}(289 \text{ kN})$$

$$M_{rx} = M_{ax} = 35 \text{ kips-ft}(47.5 \text{ kNm})$$

$$M_{ry} = M_{ay} = 14.7 \text{ kips-ft}(19.9 \text{ kNm})$$

$$P_c = P_n/\Omega_c = 320.2 \text{ kips } (1424 \text{ kN})$$

$$M_{cx} = M_n/\Omega_b = 150.7 \text{ kips } (204.3 \text{ kN})$$

$$F_y Z_y = 50 \times 28.3/12 = 117.9 \text{ kips-ft} \leq 1.6 F_y S_y = 1.6 \times 50 \times 18.7/12 = 124.7 \text{ kips-ft}$$

Hence:

$$M_{cy} = F_y Z_y/\Omega_b = 117.9/1.67 = 70.6 \text{ kips-ft } (95.7 \text{ kNm})$$

$$\frac{P_r}{P_c} = \frac{65}{320.2} = 0.21 \geq 0.20$$

$$\frac{P_r}{P_c} + \frac{8}{9}\left(\frac{M_{rx}}{M_{cx}} + \frac{M_{ry}}{M_{cy}}\right) = \frac{65}{320.2} + \frac{8}{9}\left(\frac{35}{150.7} + \frac{14.7}{70.6}\right) = 0.21 + \frac{8}{9}(0.23 + 0.21) = 0.60 < 1.0$$

Example E9.2 Beam-Column Design According to the US Approaches H1.1, H1.3 and H2

Verify whether an ASTM A992 W40 × 264 member, subjected to combined compression and bending about the x-axis only, is able to support the axial forces and bending moment listed here (second order effects included), using AISC 360-10 H1.1, H1.3 and H2 methods. The member is pinned and subjected to a uniformly distributed load along the x-axis. The unbraced length is 45 ft along both axes. The member is not braced along its height (lateral-torsional buckling not prevented).

Geometrical properties:

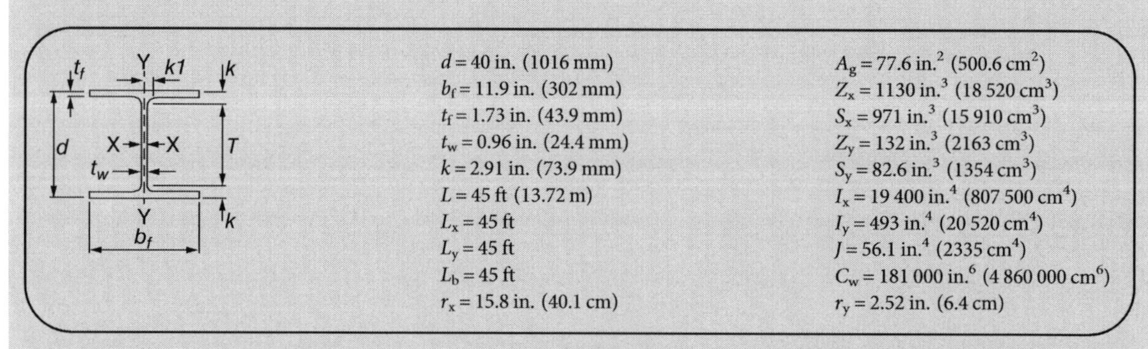

$d = 40$ in. (1016 mm)
$b_f = 11.9$ in. (302 mm)
$t_f = 1.73$ in. (43.9 mm)
$t_w = 0.96$ in. (24.4 mm)
$k = 2.91$ in. (73.9 mm)
$L = 45$ ft (13.72 m)
$L_x = 45$ ft
$L_y = 45$ ft
$L_b = 45$ ft
$r_x = 15.8$ in. (40.1 cm)

$A_g = 77.6$ in.^{2} (500.6 cm^2)
$Z_x = 1130$ in.^{3} (18 520 cm^3)
$S_x = 971$ in.^{3} (15 910 cm^3)
$Z_y = 132$ in.^{3} (2163 cm^3)
$S_y = 82.6$ in.^{3} (1354 cm^3)
$I_x = 19\,400$ in.^{4} (807 500 cm^4)
$I_y = 493$ in.^{4} (20 520 cm^4)
$J = 56.1$ in.^{4} (2335 cm^4)
$C_w = 181\,000$ in.^{6} (4 860 000 cm^6)
$r_y = 2.52$ in. (6.4 cm)

Material properties:

$$\text{Steel: ASTM A992} \quad F_y = 50 \text{ ksi}(345 \text{ MPa}) \quad F_u = 65 \text{ ksi}(448 \text{ MPa})$$

Axial load and maximum moments acting at the midspan:

LFRD	ASD
$P_u = 500$ kips (2224 kN)	$P_a = 325$ kips (1446 kN)
$M_{ux} = 1000$ kips-ft (1356 kNm)	$M_{ax} = 660$ kips-ft (895 kNm)
$M_{uy} = 0$	$M_{ay} = 0$

(a) Verification According to AISC 360-10 H1.1: 'Doubly and Singly-Symmetric Members Subjected to Flexure and Compression'.
(a1) Axial Strength
Section classification for local buckling.

Flange:

$$b/t = (0.5 \times 11.9)/1.73 = 3.44 < 0.56\sqrt{E/F_y} = 0.56 \times \sqrt{29000/50} = 13.49 \rightarrow \text{non-slender}$$

Web:

$$h/t_w = (40 - 2 \times 2.91)/0.96 = 35.6 < 1.49\sqrt{E/F_y} = 1.49 \times \sqrt{29000/50} = 35.88 \rightarrow \text{non-slender}$$

ASTM A36 steel W40 × 264, subjected to axial load, is a non-slender section ($Q = 1$).
Check slenderness ratio about both axes (assuming $K = 1.0$)

$$\frac{KL_x}{r_x} = \frac{1 \times (45 \cdot 12)}{15.8} = 34.2$$

$$\frac{KL_y}{r_y} = \frac{1 \times (45 \cdot 12)}{2.52} = 107.1 \text{ governs}$$

Calculate the elastic critical buckling stress (F_e).

$$F_e = \frac{\pi^2 E}{\left(\frac{KL_y}{r_y}\right)^2} = \frac{\pi^2 \times 29000}{(107.1)^2} = 24.9 \text{ ksi (172 MPa)}$$

Calculate flexural buckling stress (F_{cr}).

Check limit: $4.71\sqrt{\frac{E}{QF_y}} = 4.71 \times \sqrt{\frac{29000}{1 \times 50}} = 113.4 > 107.1$

Because $\frac{KL_y}{r_y} \leq 4.71\sqrt{\frac{E}{(QF_y)}}$ then:

$$F_{cr} = \left[0.658^{\frac{(QF_y)}{F_e}}\right](QF_y) = \left[0.658^{\frac{1 \times 50}{24.9}}\right] \times (1 \times 50) = 21.6 \text{ ksi (149 MPa)}$$

Compute the nominal compressive strength (P_n).

$$P_n = F_{cr} A_g = 21.6 \times 77.6 = 1676\,\text{kips}\,(7455\,\text{kN})$$

Compute the available strength.

LFRD: $\phi_c P_n = 0.90 \times 1676 = 1508.4\,\text{kips}\,(6710\,\text{kN})$
ASD: $P_n/\Omega_c = 1676/1.67 = 1003.7\,\text{kips}\,(4465\,\text{kN})$

(a2) Flexural Strength
Section classification for local buckling.

Flange:

$$b/t = (0.5 \times 11.9)/1.73 = 3.44 < 0.38\sqrt{E/F_y} = 0.38 \times \sqrt{29000/50} = 9.15 \rightarrow \text{compact}$$

Web:

$$h/t_w = (10 - 2 \times 2.91)/0.96 = 35.6 < 3.76\sqrt{E/F_y} = 3.76 \times \sqrt{29000/50} = 90.6 \rightarrow \text{compact}$$

ASTM A36 steel W40 × 264, subjected to flexure, is a compact section.
Plastic moment:

$$M_p = F_y \times Z_x = 50 \times 1130/12 = 4708\ \text{kips-ft}(6383\ \text{kNm})$$

Flexural strength corresponding to lateral torsional buckling limit state M_{LT}:
$C_b = 1.136$ (uniformly distributed load)

$$L_p = 1.76 r_y \sqrt{\frac{E}{F_y}} = 1.76 \times 2.52 \times \sqrt{\frac{29000}{50}} = 106.8\,\text{in.}/12 = 8.9\,\text{ft}\,(2.71\,\text{m})$$

$$r_{ts} = \sqrt{\frac{\sqrt{I_y C_w}}{S_x}} = \sqrt{\frac{\sqrt{493 \times 181000}}{971}} = 3.12\,\text{in.}\,(7.92\,\text{cm})$$

$c = 1$ (doubly symmetrical I-shape)

$$h_o = d - t_f = 40 - 1.73 = 38.27\ \text{in.}$$

$$L_r = 1.95 r_{ts} \frac{E}{0.7F_y} \sqrt{\frac{J \cdot c}{S_x h_0} + \sqrt{\left(\frac{J \cdot c}{S_x h_0}\right)^2 + 6.76\left(\frac{0.7F_y}{E}\right)^2}}$$

$$= 1.95 \times 3.12 \times \frac{29000}{0.7 \times 50} \sqrt{\frac{56.1 \times 1}{971 \times 38.27} + \sqrt{\left(\frac{56.1 \times 1}{971 \times 38.27}\right)^2 + 6.76 \times \left(\frac{0.7 \times 50}{29000}\right)^2}}$$

$$= 356.1\,\text{in.}/12 = 29.7\,\text{ft}\,(9.04\,\text{m})$$

$L_b = 45$ ft $> L_r = 29.7$ ft; Hence:

$$F_{cr} = \frac{C_b \pi^2 E}{\left(\frac{L_b}{r_{ts}}\right)^2} \sqrt{1 + 0.078 \frac{J \cdot c}{S_x h_0} \left(\frac{L_b}{r_{ts}}\right)^2}$$

$$= \frac{1.136 \times \pi^2 \times 29000}{\left(\frac{45/12}{3.12}\right)^2} \sqrt{1 + 0.078 \frac{56.1 \times 1}{971 \times 38.27} \left(\frac{45/12}{3.12}\right)^2} = 23.09 \, \text{ksi} \, (159.2 \, \text{MPa})$$

$$M_{LT} = F_{cr} S_x = 23.1 \times 971/12 = 1868 \, \text{kips-ft} \, (2533 \, \text{kNm})$$

Compute the nominal flexural strength (M_n).

$$M_n = \min\left(M_p; M_{LT}\right) = \min(4708; 1868) = 1868 \text{ kips-ft}(2533 \text{ kNm})$$

Compute the available strength.

LFRD: $\phi_b M_n = 0.90 \times 1868 = 1682 \, \text{kips-ft} \, (2280 \, \text{kNm})$
ASD: $M_n/\Omega_b = 1868/1.67 = 1119 \, \text{kips-ft} \, (1517 \, \text{kNm})$

(a3) Verification for Axial Load and Bending Moments

LFRD:

$$P_r = P_u = 500 \text{ kips}(2224 \text{ kN})$$

$$M_{rx} = M_{ux} = 1000 \text{ kips-ft}(1356 \text{ kNm})$$

$$M_{ry} = M_{uy} = 0$$

$$P_c = \phi_c P_n = 1508 \, \text{kips} \, (6710 \, \text{kN})$$

$$M_{cx} = \phi_b M_n = 1682 \, \text{kips} \, (2280 \, \text{kN})$$

$$\frac{P_r}{P_c} = \frac{500}{1508} = 0.33 \geq 0.20$$

$$\frac{P_r}{P_c} + \frac{8}{9}\left(\frac{M_{rx}}{M_{cx}}\right) = \frac{500}{1508} + \frac{8}{9}\left(\frac{1000}{1682}\right) = 0.33 + \frac{8}{9}(0.59) = 0.86 < 1.0$$

ASD:

$$P_r = P_a = 325 \text{ kips}(1446 \text{ kN})$$

$$M_{rx} = M_{ax} = 660 \text{ kips-ft}(895 \text{ kNm})$$

$$M_{ry} = M_{ay} = 0$$

$$P_c = P_n/\Omega_c = 1004 \, \text{kips} \, (4465 \, \text{kN})$$

$$M_{cx} = M_n/\Omega_b = 1119 \, \text{kips-ft} \, (1517 \, \text{kNm})$$

$$\frac{P_r}{P_c} = \frac{325}{1004} = 0.32 \geq 0.20$$

$$\frac{P_r}{P_c} + \frac{8}{9}\left(\frac{M_{rx}}{M_{cx}} + \frac{M_{ry}}{M_{cy}}\right) = \frac{325}{1004} + \frac{8}{9}\left(\frac{660}{1119}\right) = 0.32 + \frac{8}{9}(0.59) = 0.85 < 1.0$$

(b) Verification According to AISC 360-10 H1.3: 'Doubly-Symmetric Rolled Compact Members Subjected to Single Flexure and Compression'.
(b1) Axial Strength for In-Plane Instability
Check slenderness ratio about the x-axis only (assuming $K = 1.0$)

$$\frac{KL_x}{r_x} = \frac{1 \times (45 \cdot 12)}{15.8} = 34.2$$

Calculate the elastic critical buckling stress (F_e).

$$F_e = \frac{\pi^2 E}{\left(\frac{KL_x}{r_x}\right)^2} = \frac{\pi^2 \times 29000}{(34.2)^2} = 245\,\text{ksi}\,(1689\,\text{MPa})$$

Calculate the flexural buckling stress (F_{cr}).

Check limit: $4.71\sqrt{\frac{E}{QF_y}} = 4.71 \times \sqrt{\frac{29000}{1 \times 50}} = 113.4 > 34.2$

Because $\frac{KL_y}{r_y} \leq 4.71\sqrt{\frac{E}{(QF_y)}}$ then:

$$F_{cr} = \left[0.658^{\frac{(QF_y)}{F_e}}\right](QF_y) = \left[0.658^{\frac{1 \times 50}{245}}\right] \times (1 \times 50) = 45.9\,\text{ksi}\,(317\,\text{MPa})$$

Compute the nominal compressive strength (P_n).

$$P_n = F_{cr} A_g = 45.9 \times 77.6 = 3562\,\text{kips}\,(15850\,\text{kN})$$

Compute the available strength.

LFRD: $\phi_c P_n = 0.90 \times 3562 = 3206\,\text{kips}\,(14260\,\text{kN})$
ASD: $P_n/\Omega_c = 3562/1.67 = 2133\,\text{kips}\,(9489\,\text{kN})$

(b2) Axial Strength for Out-of-Plane Instability
Check slenderness ratio about the y-axis only (assuming $K = 1.0$)

$$\frac{KL_y}{r_y} = \frac{1 \times (45 \cdot 12)}{2.52} = 107.1$$

Calculate the elastic critical buckling stress (F_e).

$$F_e = \frac{\pi^2 E}{\left(\frac{KL_y}{r_y}\right)^2} = \frac{\pi^2 \times 29000}{(107.1)^2} = 24.9\,\text{ksi}\,(172\,\text{MPa})$$

Calculate flexural buckling stress (F_{cr}).

Check limit: $4.71\sqrt{\frac{E}{QF_y}} = 4.71 \times \sqrt{\frac{29000}{1 \times 50}} = 113.4 > 107.1$

Because $\frac{KL_y}{r_y} \leq 4.71\sqrt{\frac{E}{(QF_y)}}$ then:

$$F_{cr} = \left[0.658^{\frac{(QF_y)}{F_e}}\right](QF_y) = \left[0.658^{\frac{1\times 50}{24.9}}\right] \times (1 \times 50) = 21.6\,\text{ksi}\,(149\,\text{MPa})$$

Compute the nominal compressive strength (P_n).

$$P_n = F_{cr}A_g = 21.6 \times 77.6 = 1676\,\text{kips}\,(7456\,\text{kN})$$

Compute the available strength.

LFRD: $\phi_c P_n = 0.90 \times 1676 = 1509\,\text{kips}\,(6710\,\text{kN})$

ASD: $P_n/\Omega_c = 1676/1.67 = 1004\,\text{kips}\,(4465\,\text{kN})$

(b3) Flexural Strength in the Plane of Bending

Plastic moment:

$$M_n = M_\text{p} = F_\text{y} \times Z_\text{x} = 50 \times 1130/12 = 4708 \text{ kips-ft}(6383 \text{ kNm})$$

Compute the available strength.

LFRD: $\phi_b M_n = 0.90 \times 4708 = 4237\,\text{kips}-\text{ft}\,(5745\,\text{kNm})$

ASD: $M_n/\Omega_b = 4708/1.67 = 2819\,\text{kips}-\text{ft}\,(3923\,\text{kNm})$

(b4) Flexural Strength Out of the Plane of Bending

Plastic moment:

$$M_\text{p} = F_\text{y} \times Z_\text{x} = 50 \times 1130/12 = 4708 \text{ kips-ft}\,(6383 \text{ kNm})$$

Flexural strength corresponding to lateral torsional buckling limit state M_{LT} determined using $C_b = 1$:

$$L_p = 8.9\,\text{ft}\,(2.71\,\text{m})$$

$$L_r = 29.7\,\text{ft}\,(9.04\,\text{m})$$

$L_b = 45$ ft $> L_r = 29.7$ ft; Hence:

$$F_{cr} = \frac{C_b \pi^2 E}{\left(\frac{L_b}{r_{ts}}\right)^2} \sqrt{1 + 0.078 \frac{J \cdot c}{S_x h_0} \left(\frac{L_b}{r_{ts}}\right)^2}$$

$$= \frac{1.0 \times \pi^2 \times 29000}{\left(\frac{45/12}{3.12}\right)^2} \sqrt{1 + 0.078 \frac{56.1 \times 1}{971 \times 38.27} \left(\frac{45/12}{3.12}\right)^2} = 20.3\,\text{ksi}\,(140.1\,\text{MPa})$$

Compute the nominal flexural strength (M_n).

$$M_n = M_{LT} = F_{cr}\,S_x = 20.3 \times 971/12 = 1643 \text{ kips-ft } (2230 \text{ kNm})$$

Compute the available strength.

LFRD: $\phi_b M_n = 0.90 \times 1643 = 1480\,\text{kips} - \text{ft}\,(2007\,\text{kNm})$
ASD: $M_n/\Omega_b = 1643/1.67 = 985\,\text{kips} - \text{ft}\,(1335\,\text{kNm})$

(b5) Verification for In-Plane Instability

LFRD:

$$P_r = P_u = 500 \text{ kips}(2224 \text{ kN})$$

$$M_{rx} = M_{ux} = 1000 \text{ kips-ft}(1356 \text{ kNm})$$

$$P_c = \phi_c P_n = 3206\,\text{kips}\,(14260\,\text{kN})$$

$$M_{cx} = \phi_b M_n = 4237\,\text{kips} - \text{ft}\,(5745\,\text{kNm})$$

$$\frac{P_r}{P_c} = \frac{500}{3206} = 0.16 < 0.20$$

$$\frac{P_r}{2P_c} + \left(\frac{M_{rx}}{M_{cx}}\right) = \frac{500}{2 \times 3206} + \left(\frac{1000}{4237}\right) = 0.08 + (0.24) = 0.32 < 1.0$$

ASD:

$$P_r = P_a = 325 \text{ kips}(1446 \text{ kN})$$

$$M_{rx} = M_{ax} = 660 \text{ kips-ft } (895 \text{ kNm})$$

$$P_c = P_n/\Omega_c = 2133\,\text{kips}\,(9489\,\text{kN})$$

$$M_{cx} = M_n/\Omega_b = 2819\,\text{kips} - \text{ft}\,(3923\,\text{kNm})$$

$$\frac{P_r}{P_c} = \frac{325}{2133} = 0.15 < 0.20$$

$$\frac{P_r}{2P_c} + \left(\frac{M_{rx}}{M_{cx}}\right) = \frac{325}{2 \times 2133} + \left(\frac{660}{2819}\right) = 0.08 + (0.23) = 0.31 < 1.0$$

(b6) Verification for Out-of-Plane Buckling and Lateral-Torsional Buckling

LFRD:

$$P_r = P_u = 500 \text{ kips}(2224 \text{ kN})$$
$$M_{rx} = M_{ux} = 1000 \text{ kips-ft } (1356 \text{ kNm})$$
$$P_{cy} = \phi_c P_n = 1509 \text{ kips } (6710 \text{ kN})$$
$$M'_{cx} = \phi_b M_n = 1480 \text{ kips} - \text{ft } (2007 \text{ kNm})$$

$$\frac{P_r}{P_{cy}}\left(1.5 - 0.5\frac{P_r}{P_{cy}}\right) + \left(\frac{M_{rx}}{C_b M'_{cx}}\right) = \frac{500}{1509}\left(1.5 - 0.5\frac{500}{1509}\right) + \left(\frac{1000}{1.136 \times 1480}\right)^2 = 0.80 < 1.0$$

ASD:

$$P_r = P_a = 325 \text{ kips}(1446 \text{ kN})$$
$$M_{rx} = M_{ax} = 660 \text{ kips-ft}(895 \text{ kNm})$$
$$P_{cy} = P_n/\Omega_c = 1004 \text{ kips } (4465 \text{ kN})$$
$$M'_{cx} = M_n/\Omega_b = 985 \text{ kips} - \text{ft } (1335 \text{ kNm})$$

$$\frac{P_r}{P_{cy}}\left(1.5 - 0.5\frac{P_r}{P_{cy}}\right) + \left(\frac{M_{rx}}{C_b M'_{cx}}\right) = \frac{325}{1004}\left(1.5 - 0.5\frac{325}{1004}\right) + \left(\frac{660}{1.136 \times 985}\right)^2 = 0.78 < 1.0$$

(c) Verification According to AISC 360-10 H2: 'Unsymmetric and Other Members Subjected to Flexure and Axial Force'.

LFRD:

$$f_{ra} = \frac{P_u}{A_g} = \frac{500}{77.6} = 6.44 \text{ ksi } (44.4 \text{ MPa})$$

$$f_{rbw} = \frac{M_{ux}}{S_x} = \frac{1000 \cdot 12}{971} = 12.4 \text{ ksi } (85.2 \text{ MPa})$$

$$F_{ca} = \phi_c F_{cr} = 0.90 \times 21.6 = 19.4 \text{ ksi } (134 \text{ MPa})$$

$$F_{cbw} = \frac{\phi_b M_n}{S_x} = \frac{0.90 \times 1869 \cdot 12}{971} = 20.8 \text{ ksi } (143 \text{ MPa})$$

$$\left|\frac{f_{ra}}{F_{ca}} + \frac{f_{rbw}}{F_{cbw}}\right| = \left|\frac{6.44}{19.4} + \frac{12.4}{20.8}\right| = |0.33 + 0.60| = 0.93 < 1.0$$

ASD:

$$f_{ra} = \frac{P_a}{A_g} = \frac{325}{77.6} = 4.19 \text{ ksi } (28.9 \text{ MPa})$$

$$f_{rbw} = \frac{M_{ux}}{S_x} = \frac{660 \cdot 12}{971} = 8.16 \text{ ksi } (56.2 \text{ MPa})$$

$$F_{ca} = \frac{F_{cr}}{\Omega_c} = \frac{21.6}{1.67} = 12.9\,\text{ksi}\,(89.2\,\text{MPa})$$

$$F_{cbw} = \frac{M_n}{\Omega_b S_x} = \frac{1869{\cdot}12}{1.67 \times 971} = 13.8\,\text{ksi}\,(95.3\,\text{MPa})$$

$$\left|\frac{f_{ra}}{F_{ca}} + \frac{f_{rbw}}{F_{cbw}}\right| = \left|\frac{4.19}{12.9} + \frac{8.16}{13.8}\right| = |0.32 + 0.59| = 0.91 < 1.0$$

Verification results are summarized in the following table in terms of values of the safety index (SI):

Verification according to:	Safety index	
	LFRD	ASD
(a) H1.1	0.86	0.85
(b) H1.3	0.80	0.78
(c) H2	0.93	0.91

In this case, verification according to paragraph H1.1 of AISC 360-10 produces slightly more conservative results than the one according to paragraph H1.3. Verifications according to paragraph H2 are much more conservative. Curves representing the bilinear interaction equations of paragraph H1.1 and the parabolic interaction equation of paragraph H1.3, for the LFRD verification are shown in Figure E9.3.1.

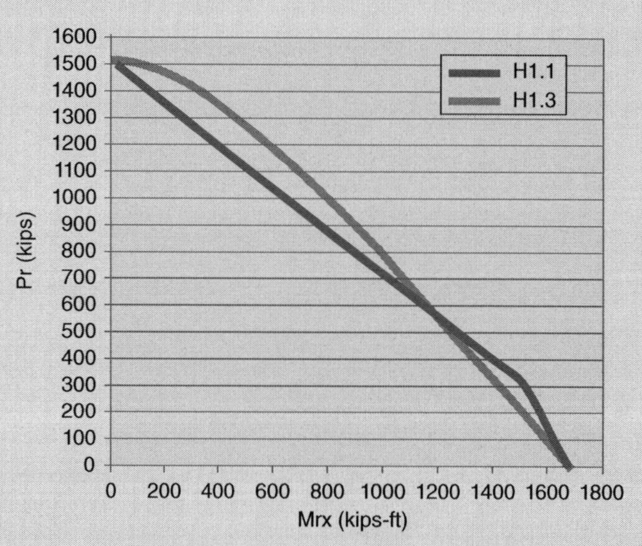

Figure E9.3.1 *P-M* interaction curves; Example E9.3, LFRD verification.

CHAPTER 10

Design for Combination of Compression, Flexure, Shear and Torsion

10.1 Introduction

In routine design, structural analysis is usually carried out by means of commercial finite element (FE) analysis packages. For all the mono-dimensional elements (columns, beams, beam-columns, diagonals, etc.) used to model the skeleton frame, a set of internal forces and moments is proposed from these very refined software tools for each load condition. As a consequence, the designer has to refer, for verification checks, to the values of axial load, shear forces and bending moments about the principal cross-section axes and torsional moment, which act at the time.

In many cases, the member cross-sections present two axes of symmetry and hence the shear centre (*C*) is coincident with the centroid (*O*). Structural analysis is usually carried out via FE analysis programs with libraries offering 6 degrees of freedom (DOFs) beam formulations: for each node, three displacements (u, v and w) and three rotations (φ_x, φ_y and φ_z) are employed (Figure 10.1) to evaluate displacements, internal forces and moments, and hence to obtain the output data necessary to develop the required design verification checks.

Furthermore, if open mono-symmetrical cross-sections are used, the shear centre does not coincide with the centroid and the warping of the cross-section remarkably influences the member response. Suitable 7 DOF beam formulations accounting for warping effects have already been proposed in literature and are now implemented in few FE general purpose commercial analysis packages. As shown in Figure 10.2, in the case of FE formulation for mono-symmetrical cross-sections, the eccentricity between the shear centre (*C*) and the centroid (*O*) has to be taken into account. Usually, reference is made to the shear centre for the definition of all the internal displacements except for the axial displacement u_O. Shear forces (F_y and F_z), uniform torsional moment (M_t) and bimoment (B) are defined on the shear centre (*C*) (Figure 10.2) while the axial force (N) and bending moments (M_y and M_z) refer to the centroid (*O*). Cross-section warping θ, that is the 7 DOF necessary to model non-bi-symmetric cross-section, is defined as:

$$\theta = \theta(x) = -\frac{d\varphi_x}{dx} \tag{10.1}$$

As to the FE 7 DOF formulation details, warping terms can be found in the torsional coefficients of the local elastic stiffness matrix $[\mathrm{K}]_j^{\mathrm{E}}$ of the beam element. From a practical point

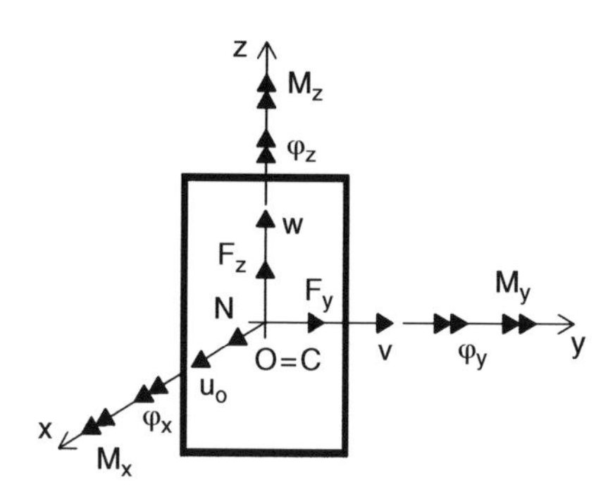

Figure 10.1 Displacements, internal forces and moments for bi-symmetrical cross-section members.

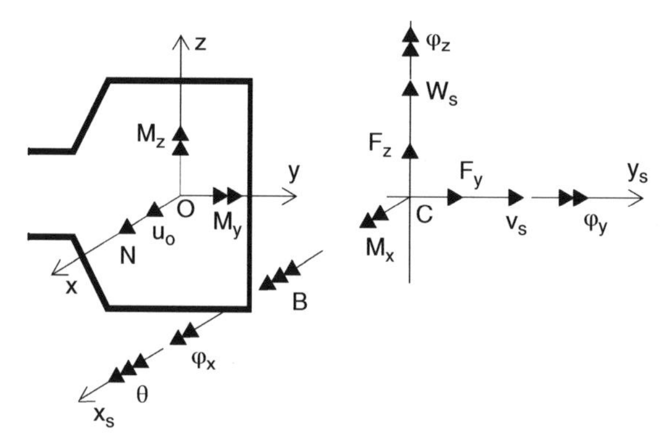

Figure 10.2 Displacements, internal forces and moments for mono-symmetrical cross-section members.

of view, in addition to the presence of bimoment (B) and a different value of the torsional moment (M_t), a relevant influence of warping is expected also in the value of the bending moments and consequently, in the shear forces. It is worth mentioning that the complex mutual interactions of the transferred end member forces, governed by the traditional equilibrium and compatibility principles, should lead to significant differences of the set of displacements, internal forces and moments with respect to the ones associated with a 6 DOF beam formulation.

For members with a mono-symmetrical cross-section, FE beam formulations are significantly different if compared with the ones adopted for bi-symmetrical cross-sections. Let us denote j and k as the two nodes of the generic beam, the governing matrix displacement equations of the FE element can be written in a general form, valid with reference to both elastic $[\mathrm{K}]^{\mathrm{E}}$ and geometric $[\mathrm{K}]^{\mathrm{G}}$ stiffness matrices, such as:

$$\begin{bmatrix} [K]_{jj} & [K]_{jk} \\ [K]_{kj} & [K]_{kk} \end{bmatrix} \begin{bmatrix} \{u\}_j \\ \{u\}_k \end{bmatrix} = \begin{bmatrix} \{f\}_j \\ \{f\}_k \end{bmatrix} \tag{10.2}$$

With reference to the more general case of 7 DOFs beam formulation, the nodal displacement, $\{u\}_j$ and $\{u\}_k$ and the associated force vectors, $\{f\}_j$ and $\{f\}_k$, can be defined (Figure 10.2), respectively, as:

$$\{u\}_j = \begin{bmatrix} u_o \\ v_s \\ w_s \\ \varphi_x \\ \varphi_y \\ \varphi_z \\ (\theta) \end{bmatrix} \tag{10.3a}$$

$$\{f\}_j = \begin{bmatrix} N \\ F_y \\ F_z \\ M_t \\ M_y \\ M_z \\ (B) \end{bmatrix} \tag{10.3b}$$

These FE formulations are very complex, especially for that which concerns the definition of the geometric stiffness matrix $[K]^G$, otherwise these are quite simple with reference to the elastic matrix $[K]^E$. If reference is made to a beam element of length L_b by considering its area (A), moments of inertia (I_z and I_y) along the principal axes, uniform and non uniform torsional constants (I_t and I_w, respectively) and assuming E and G representing the Young's and tangential material modulus, respectively, the stiffness elastic sub-matrices $[K]_{jj}^E$ (or equivalently $[K]_{kk}^E$) and $[K]_{jk}^E$ (or $[K]_{kj}^E$) can be written as:

$$[K]_{jj}^E = \left[\begin{array}{cccccc:c}
\frac{EA}{L_b} & 0 & 0 & 0 & 0 & 0 & 0 \\
 & \frac{12EI_z}{L_b^3} & 0 & 0 & 0 & \frac{6EI_z}{L_b^2} & 0 \\
 & & \frac{12EI_y}{L_b^3} & 0 & -\frac{6EI_y}{L_b^2} & 0 & 0 \\
 & & & \frac{GI_t}{L_b} + \left(\frac{12EI_w}{L_b^3} + \frac{1}{5}\frac{GI_t}{L_b}\right) & 0 & 0 & -\left(\frac{6EI_w}{L_b^2} + \frac{3}{30}GI_t\right) \\
 & & \mathit{Symmetric} & & \frac{4EI_y}{L_b} & 0 & 0 \\
 & & & & & \frac{4EI_z}{L_b} & 0 \\
\hdashline
 & & & & & & \left(\frac{4EI_w}{L_b} + \frac{4}{30}GI_tL_b\right)
\end{array}\right] \tag{10.4a}$$

$$[K]_{jk}^E = \left[\begin{array}{cccccc:c}
-\frac{EA}{L_b} & 0 & 0 & 0 & 0 & 0 & 0 \\
 & -\frac{12EI_z}{L_b^3} & 0 & 0 & 0 & \frac{6EI_z}{L_b^2} & 0 \\
 & & -\frac{12EI_y}{L_b^3} & 0 & -\frac{6EI_y}{L_b^2} & 0 & 0 \\
 & & & -\frac{GI_t}{L_b} + \left(\frac{12EI_w}{L_b^3} + \frac{1}{5}\frac{GI_t}{L_b}\right) & 0 & 0 & -\left(\frac{6EI_w}{L_b^2} + \frac{3}{30}GI_t\right) \\
 & & \frac{6EI_y}{L_b^2} & & \frac{2EI_y}{L_b} & 0 & 0 \\
 & -\frac{6EI_z}{L_b^2} & & & & \frac{2EI_z}{L_b} & 0 \\
\hdashline
 & & & \left(+\frac{6EI_w}{L_b^2} + \frac{3}{30}GI_t\right) & & & \left(\frac{2EI_w}{L_b} - \frac{1}{30}GI_tL_b\right)
\end{array}\right] \tag{10.4b}$$

Terms between brackets relate to the sole formulations including the 7 DOF (warping), which directly influences also the terms associated with uniform torsion; that is term (4,4). It should be noted that classical 6 DOF beam formulations are characterized, for that concerning the elastic stiffness matrix $[K]^E$ and, in particular, the uniform torsion contribution, by the presence of term $\frac{GI_t}{L_b}$, while in the 7 DOF formulation the contribution $\left(\frac{12EI_w}{L_b^3}+\frac{1}{5}\frac{GI_t}{L_b}\right)$ has to be directly added (for $[K]_{jj}^E$) in sub-matrix Eq. (10.4a) or subtracted (for $[K]_{jj}^E$) in sub-matrix Eq. (10.4b) to $\frac{GI_t}{L_b}$.

Furthermore, with reference to the geometric stiffness matrix, $[K]^G$, the traditional 6 DOF beam formulations allow us to satisfactorily approximate the geometric non-linearities based on the sole value of the internal axial load *N*. Otherwise, in the case of beam formulations including warping, bending moments (M_y and M_z), torsional moment (M_t), bi-moment (B_w) and shear actions (F_y and F_z) also contribute significantly to the geometric stiffness, $[K]^G$, containing terms also strictly dependent on the distance between the load application point and shear centre.

Furthermore, these formulations that take into account the coupling between flexure and torsion are the only ones also capable of directly capturing the overall flexural-torsional buckling of the frame as well as of isolated columns, beams and beam-columns.

Current design practice neglects warping effects for both analysis and verification checks and this could lead, in a few cases, to a very non-conservative design. More adequate resistance check criteria are required for mono-symmetrical profiles also including the contribution due to bi-moment (B_{Ed}) acting on the cross-section.

Accounting for warping torsion, reference should be made to the equation:

$$\frac{N_{Ed}}{N_{Rd}}+\frac{M_{y,Ed}}{M_{y,Rd}}+\frac{M_{z,Ed}}{M_{z,Rd}}+\frac{B_{Ed}}{B_{Rd}}\leq 1 \tag{10.5a}$$

where B_{Rd} is the bimoment section capacity defined as:

$$B_{Rd}=\frac{I_w}{\omega_{\max}}\cdot f_y \tag{10.5b}$$

where I_w is the warping constant and $\omega_{\max}$ is the maximum value of the static moment of the sectorial area.

Furthermore, it is worth mentioning that in case of axial force, shear, bending and torsion an accurate check of the state of stresses is required in each cross-section of interest. As discussed in Chapter 8, normal $\sigma_{w,Ed}(y,z)$ and shear $\tau_{w,Ed}(y,z)$ stresses due to bimoment B_{Ed} in a general point *P* of coordinate (*y*,*z*) defined with reference to the cross-section centroid (Figure 10.2), can be expressed as:

$$\sigma_{w,Ed}(y,z)=\frac{B_{Ed}}{I_w}\cdot\omega(y,z) \tag{10.6a}$$

$$\tau_{w,Ed}(y,z)=\frac{T_w}{I_w}\cdot\frac{S_\omega(y,z)}{t} \tag{10.6b}$$

where T_w represents the non-uniform torsional moment and *t* is the thickness of the cross-section.

The use of Eq. (10.5a) in resistance checks could lead to a slightly conservative design, owing to the fact that the maximum of the sectorial area ($\omega_{\max}$), and its first moment of area ($S_{\omega,max}$), are generally not at the same location where stresses due to bending moments reach the maximum values. As a consequence, it should be more appropriate, in order to guarantee an optimal use of the material, to evaluate the local distribution of the normal stresses summing the values of the stresses occurring at the same point of the cross-section. As an example, the distributions of the sectorial area $\omega(y,z)$ and of its first moment $S_{\omega,max}(y,z)$ are presented in Figure 10.3 for a typical mono-symmetrical cross-section used for the upright for vertical members in industrial storage systems.

With reference to the sole case of axial load, bending moments and the bimoment acting on the cross-section, the influence of warping effects on the location of the maximum normal stress can be appraised via Figure 10.4. Assuming the sign conventions of Figure 10.2, maximum normal stress is in point D′ if the sole axial load and positive bending moments are considered. Otherwise, if bimoment B_{Ed} acts on cross-section, maximum stress is in correspondence of point F′ $(B_{Ed} > 0)$ or point B′ if $(B_{Ed} < 0)$.

More generally, this figure also indicates the point where the normal stress is maximum when the axial load is negative (compression) and moments are positive or negative. It appears from Table 10.1 that if the normal stresses due to warping are neglected, that is $\sigma = \sigma(N, M_y, M_z)$, or considered, that is $\sigma = \sigma(N, M_y, M_z, B_w)$, the point with the maximum stress coincides only if all the moments are negative, otherwise, as already mentioned, a moderate member oversizing is possible when Eq. (10.5a,b) is used.

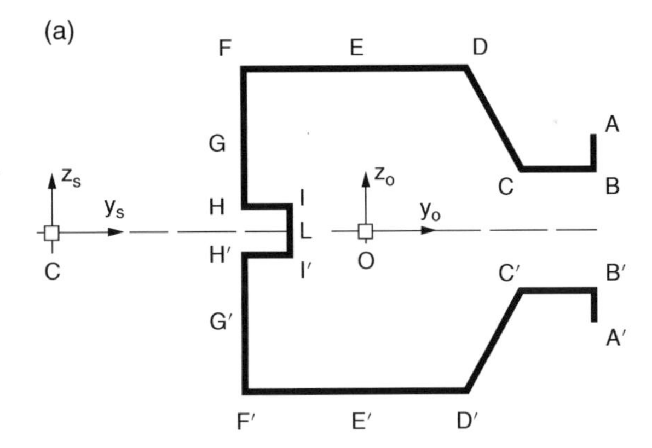

(b)

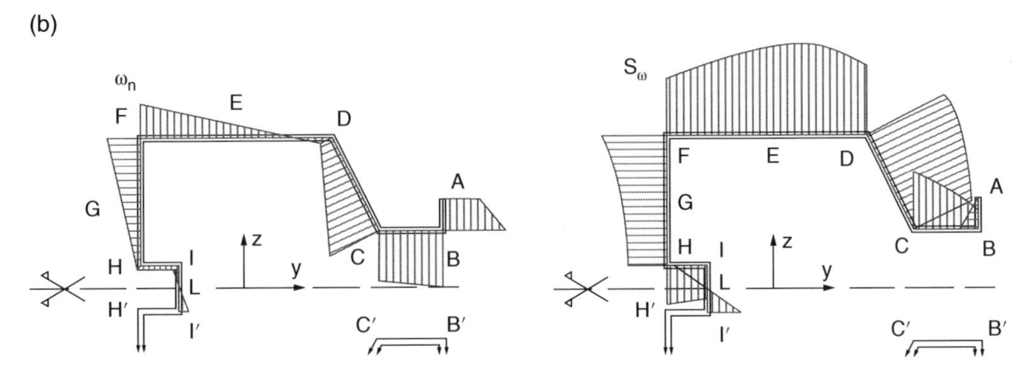

Figure 10.3 Example of a mono-symmetrical cross-section and distribution of the sectorial area ω_n (a) and the static moment of the sectorial area S_ω (b).

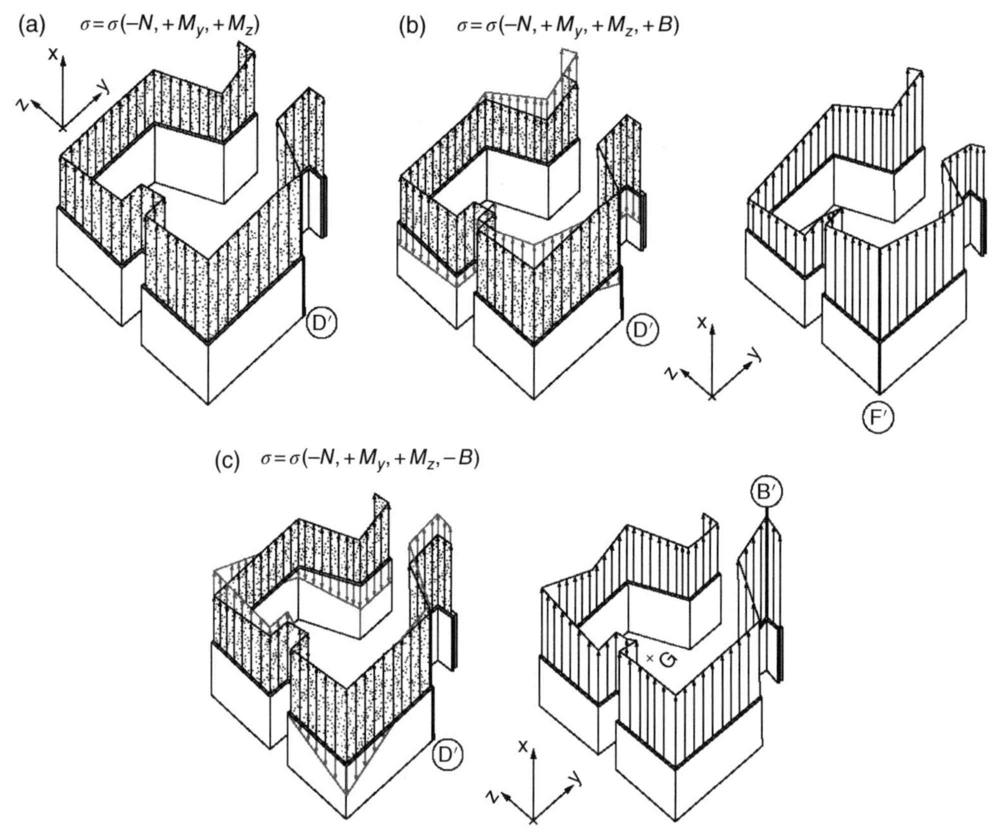

Figure 10.4 Example of the influence of the bimoment B on the location of the maximum normal stress in the cross-section (a–c).

Table 10.1 Influence of warping on the location of the more stressed cross-section point.

F E D A B C G H I L Z_o Y_o O H′ I′ C′ B′ G′ A′ F′ E′ D′

N	M_z	M_y	$\sigma = \sigma$ (N, M_y, M_z) point	B	$\sigma = \sigma$ (N, M_y, M_z, B) point
−	+	+	D′	+	F′
				−	B′
−	+	−	D	+	B
				−	F
−	−	+	F′	+	F′
				−	B′
−	−	−	F	+	B
				−	F

10.2 Design in Accordance with the European Approach

In Part 1-1 of EC3, which regards the general rules and the rules for building, the non-coincidence between the shear centre and the centroid of the cross-section is ignored and the verification checks of beam-columns are referred mainly to bi-symmetrical I-shaped and hollow cross-sections. Several research activities are currently in progress in Europe to improve these rules in order to also include the case of I-shaped (with unequal flanges) cross-section members but no adequate attention seems to have been paid, up to now, to the more complex case of mono-symmetrical cross-sections. As to a resistance check, a very general yield criterion is proposed in the European codes for elastic verification: with reference to the critical point of the cross-section, the following condition has to be fulfilled:

$$\left(\frac{\sigma_{x,Ed}}{f_y}\right)^2 + \left(\frac{\sigma_{z,Ed}}{f_y}\right)^2 - \left(\frac{\sigma_{x,Ed}}{f_y}\right)\cdot\left(\frac{\sigma_{z,Ed}}{f_y}\right) + 3\cdot\left(\frac{\tau_{Ed}}{f_y}\right)^2 \leq 1 \tag{10.6}$$

where $\sigma_{x,Ed}$ and $\sigma_{z,Ed}$ are the design values of the local longitudinal and transverse stress, respectively, τ_{Ed} is the design value of the local shear stress and f_y represents the design yielding stress (i.e. the value of the yielding stress divided by the material safety factor associated with the considered code).

It should be noted that it is clearly recommended in EC3 part 1-1 to account for the stresses due to torsion in Eq. (10.6) and, in particular:

- the shear stress τ_{Ed} has to include the contribution $\tau_{t,Ed}$ due to St Venant torsion $T_{t,Ed}$ and $\tau_{w,Ed}$ due to the warping torsion $T_{w,Ed}$;
- the normal stress $\sigma_{x,Ed}$ has to include $\sigma_{w,Ed}$ due to the bimoment B_{Ed}.

No practical indications are provided to designers for the correct evaluation of stresses $\tau_{w,Ed}$ and $\sigma_{w,Ed}$, which usually could require very complex computations due to the mono-symmetry of the cross-section.

As to cold-formed members, which are considered in Part 1-3 of EC3, it should be noted that very general statements are provided with regard to the possible influence of torsional moments. The direct stresses ($\sigma_{N,Ed}$) due to the axial force N_{Ed} and the ones associated with bending moments $M_{y,Ed}$ ($\sigma_{My,Ed}$) and $M_{z,Ed}$ ($\sigma_{Mz,Ed}$), respectively, should be based on the relative effective cross-sections. Properties of the gross cross-section have to be considered to evaluate the shear stresses τ due to transverse shear forces, $\tau_{Fy,Ed}$ and $\tau_{Fz,Ed}$, the shear stresses due to uniform torsion, $\tau_{t,Ed}$, and both the normal, $\sigma_{w,Ed}$, and shear stresses, $\tau_{w,Ed}$, due to warping.

The total direct stress $\sigma_{tot,Ed}$ and the total shear stress $\tau_{tot,Ed}$ must be, respectively, obtained as:

$$\sigma_{tot,Ed} = \sigma_{N,Ed} + \sigma_{My,Ed} + \sigma_{Mz,Ed} + \sigma_{w,Ed} \tag{10.6a}$$

$$\tau_{tot,Ed} = \tau_{Fy,Ed} + \tau_{Fz,Ed} + \tau_{t,Ed} + \tau_{w,Ed} \tag{10.6b}$$

In cross-sections subjected to torsion, it is a requirement that the following conditions have to be satisfied:

$$\sigma_{tot,Ed} \leq f_{ya} \tag{10.7a}$$

$$\tau_{tot,Ed} \leq f_{ya}/\sqrt{3} \tag{10.7b}$$

$$\sqrt{\sigma_{tot,Ed}^2 + 3\cdot\tau_{tot,Ed}^2} \leq 1.1\cdot f_{ya} \tag{10.7c}$$

where f_{ya} is the increased average yield strength due to the forming process, already defined by Eq. (1.4).

10.3 Design in Accordance with the US Approach

Elements subjected to combined stresses due to axial load, bending moment, shear and torsion, are addressed in AISC 360-10 at Chapter H3. Two cases are treated: (i) round and rectangular HSS sections and (ii) all other cases; that is, open sections.

10.3.1 Round and Rectangular HSS

If an HSS member is subjected to an axial load P_r, a bending moment M_r a shear V_r and a torsional moment T_r, all computed with LRFD (load resistance factor design) or ASD (allowable stress design) loading combinations, then the verification is:

$$\left(\frac{P_r}{P_c}+\frac{M_r}{M_c}\right)+\left(\frac{V_r}{V_c}+\frac{T_r}{T_c}\right)^2 \leq 1.0 \tag{10.8}$$

where P_c is the design (LRFD) or allowable (ASD) tensile or compressive strength, (see Chapters 5 and 6), M_c is the design (LRFD) or allowable (ASD) flexural strength, (see Chapter 7), V_c is the design (LRFD) or allowable (ASD) shear strength (see Chapter 7) and T_c is the design (LRFD) or allowable (ASD) torsional strength, (see Chapter 8).

Torsional effects can be neglected if $T_r \leq 0.20\ T_c$.

10.3.2 Non-HSS Members (Open Sections Such as W, T, Channels, etc.)

For such members, verification for combined stresses has to be performed comparing the design stresses with limit stress for the limit state considered. The procedure according to AISC 360-10 is the following:

(a) choose a member cross-section (midspan, support, etc.);
(b) compute the normal stresses σ in any point of the selected cross-section due to axial load and bending moments (according to LRFD or ASD loading combinations);
(c) compute the shear stresses τ in any point of the selected cross-section due to shear (according to LRFD or ASD loading combinations);
(d) compute shear stress τ_t in any point of the selected section due to pure (St Venant) torsion (according to LRFD or ASD loading combinations) (see Chapter 8);
(e) compute normal stress σ_w and shear stress τ_w in any point of the selected section due to restrained warping torsion (according to LRFD or ASD loading combinations) (see Chapter 8);
(f) sum in any point the normal stresses $(\sigma+\sigma_w)$ and the shear stresses $(\tau+\tau_t+\tau_w)$, and find the maximum values σ_{max} and τ_{max};
(g) for limit state of yielding under normal stress verify that:

LRFD approach	ASD approach
$\sigma_{max} \leq \varphi_T F_y$	$\sigma_{max} \leq F_y/\Omega_T$

(h) for limit state of yielding under shear stress verify that:

LRFD approach	ASD approach
$\tau_{max} \leq \varphi_T(0.6F_y)$	$\tau_{max} \leq (0.6F_y)/\Omega_T$

where $\varphi_T = 0.90$; $\Omega_T = 1.67$.

(i) the F_{cr} tension associated with the buckling limit states (usually lateral-torsional buckling and local buckling) has to be evaluated and stresses shall be compared with F_{cr} value, if $F_{cr} < F_y$.

CHAPTER 11

Web Resistance to Transverse Forces

11.1 Introduction

In many cases, beams are attached to other members at their ends via connections (i.e. web cleats, header plates by other components introduced in Chapter 15) with large forces applied at these locations. Similarly, concentrated loads at intermediate locations may be applied by other beams connected to the web of the main beam. Furthermore, in case of beams subjected to heavy concentrated loads applied directly through their flanges, the associated local effects are very relevant (Figure 11.1) and appropriate verification checks are required. In particular, forces applied through one flange may be resisted by shear forces in the web or transferred through the web directly to the other beam flange.

Three different failure modes have to be considered when transverse forces are applied on web (Figure 11.2):

- *web crushing* (*a*): high compressive stresses developed in relatively thin webs cause crushing failure directly adjacent to the flange. The load is spread from the stiff bearing length over an appropriate length to the beam web;
- *web crippling* (*b*): a localized buckling phenomenon is associated with the crushing of the web close to the flange that is directly loaded, accompanied by plastic flange deformations;
- *web buckling* (*c*): web failure is due to a compression load as a result of web buckling as a vertical strut. Effective buckling length depends on the combinations of rotational and out-of-plane displacement restraints provided by flanges. This failure mode occurs when the forces are directly transferred through the web to a reaction at the other flange.

When a web has non-adequate bearing capacity, it may be strengthened by adding one or more pairs of load-bearing stiffeners. These stiffeners increase the yield and buckling resistances significantly improving the performance of the original members.

Independent of the values of the acting force, stiffeners are strongly recommended in correspondence of each cross-section where concentrated loads are applied as well as in correspondence of the member restraints.

Structural Steel Design to Eurocode 3 and AISC Specifications, First Edition. Claudio Bernuzzi and Benedetto Cordova.

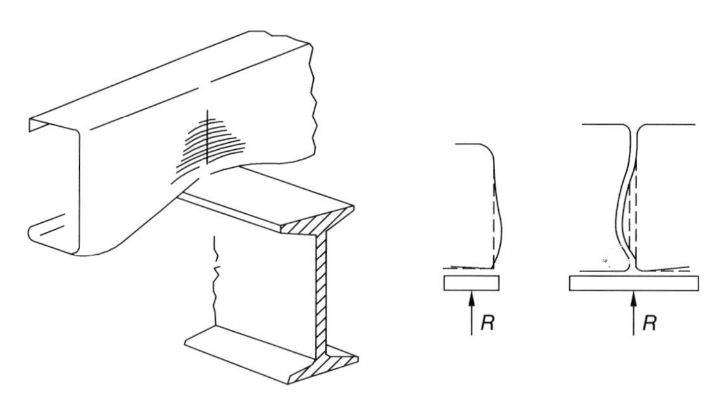

Figure 11.1 Example of failure due to large transverse forces on the beam web.

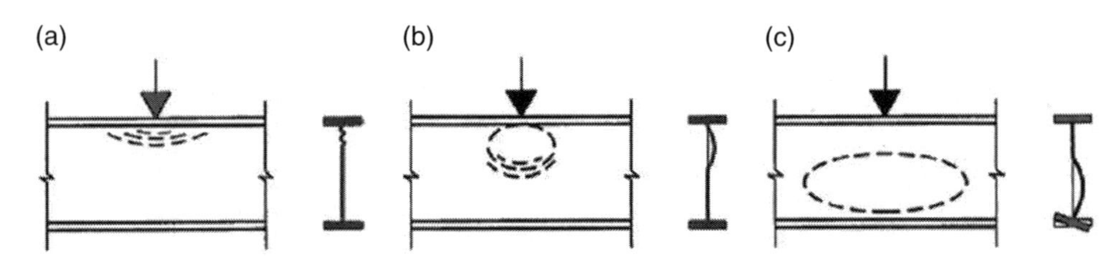

Figure 11.2 Different types of patch loading and buckling k_F coefficients: web crushing (a), web crippling (b) and web buckling (c).

11.2 Design Procedure in Accordance with European Standards

Practical indications to evaluate the design resistance of the webs of rolled beams and welded girders are given in EN 1993-1-5 (Plated structural elements). In previous ENV 1993-1-1 three different equations were proposed to check separately the web with reference to the previously introduced failure modes (Figure 11.2), that is resistance, web crippling and web buckling. Now a unified approach is proposed, which is based on a combined method. An essential prerequisite is to have the compression flange adequately restrained in the lateral direction and the following cases are directly considered (Figure 11.3):

- load applied through the flange and resisted by shear forces in the web (case a);
- load applied through one flange and transferred through the web directly (case b);
- load applied through one flange adjacent to an unstiffened end (case c).

For unstiffened or stiffened webs, loaded by a web transverse force F_{Ed}, it is required that:

$$F_{Ed} \leq F_{Rd} \tag{11.1}$$

where F_{Rd} is the design resistance to local buckling defined as:

$$F_{Rd} = \frac{f_{yw} \cdot L_{\text{eff}} \cdot t_w}{\gamma_{M1}} \tag{11.2}$$

where f_{yw} and t_w are the yielding strength and the thickness of the web, respectively, L_{eff} is the effective length and γ_{M1} is the partial safety coefficient.

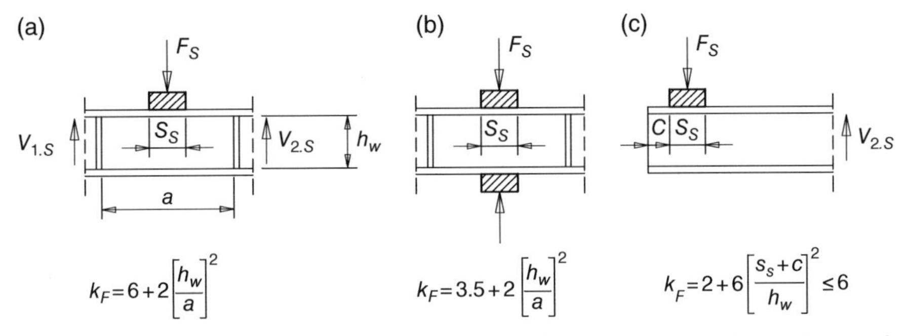

Figure 11.3 Different types of patch loading and buckling k_F coefficients: loads applied to the flange and resisted by shear forces in the web (a), transferred through the web directly (b) and adjacent to an unstiffened end (c).

Effective length for resistance to transverse forces depends on the effective loaded length, l_y, and on a reduction factor due to local buckling (χ_F):

$$L_{\text{eff}} = \chi_F \cdot l_y \tag{11.3}$$

The reduction factor for effective length for resistance (χ_F) is defined as:

$$\chi_F = \frac{0.5}{\bar{\lambda}_F} \leq 1.0 \tag{11.4}$$

Relative slenderness $\bar{\lambda}_F$ is defined as:

$$\bar{\lambda}_F = \sqrt{\frac{l_y \cdot t_w \cdot f_{yw}}{F_{cr}}} \tag{11.5}$$

where F_{cr} is the critical force, approximated as:

$$F_{cr} = 0.9 \cdot k_F \cdot \frac{E \cdot t_w^3}{h_w} \tag{11.6}$$

where E is the Young's modulus and the coefficient k_F depends on the type of loading and the geometry of the loaded zones.

The proposed buckling coefficients were derived assuming that the rotation of the flange is prevented at the load application point, as generally occurs.

For webs with transversal stiffeners, the recommended values of k_F are presented in Figure 11.3. In the case of longitudinal stiffeners, in absence of more specific indications directly provided by the National Annex, term k_F is defined as:

$$k_F = 6 + 2\left[\frac{h_w}{a}\right]^2 + \left[5.44\frac{b_1}{a} - 0.21\right] \cdot \sqrt{\gamma_s} \tag{11.7}$$

where b_1 is the depth of the loaded sub-panel (clear distance between the loaded flange and the stiffener) and the coefficient γ_s is:

$$\gamma_s = 10.9\frac{I_{sl,1}}{h_w \cdot t_w^3} \leq 13 \cdot \left[\frac{a}{h_w}\right]^3 + 210 \cdot \left[0.3 - \frac{b_1}{a}\right] \tag{11.8}$$

where $I_{sl,1}$ is the moment of inertia of the stiffener closest to the loaded flange, including contributing parts of the web.

It should be noted that Eq. (11.7) is valid for the case (a) in Figure 11.3 and if:

$$0.05 \leq \frac{b_1}{a} \leq 0.3 \tag{11.9a}$$

$$\frac{b_1}{h_w} \leq 0.3 \tag{11.9b}$$

A very important parameter is the length l_y over which the web is supposed to yield. The first step is to determine the loaded length on top of the flange (S_s), which has to be done in accordance with the Figure 11.4 assuming a load spread angle of 45°.

For the case of loading from two rollers the model requires two checks:

- check for the combined influence of the two loads with s_s as the distance between the loads;
- check for the loads considered individually with $s_s = 0$.

For types (a) and (b) in Figure 11.4 length l_y is defined, with the limitation of not exceeding the distance between adjacent transverse stiffeners (i.e. $l_y < a$) as:

$$l_y = s_s + 2t_f(1 + \sqrt{m_1 + m_2}) \tag{11.10}$$

Coefficients m_1 and m_2 are defined, respectively, as:

$$m_1 = \frac{f_{yf} \cdot b_f}{f_{yw} \cdot t_w} \tag{11.11a}$$

$$m_2 = 0.02 \cdot \left(\frac{h_w}{t_f}\right)^2 \text{ if } \bar{\lambda}_F > 0.5 \tag{11.11b}$$

$$m_2 = 0 \text{ if } \bar{\lambda}_F \leq 0.5 \tag{11.11c}$$

For type (c) the effective loaded length l_y has to be assumed as the smallest value obtained from the following three equations:

$$l_y = l_e + t_f \sqrt{\frac{m_1}{2} + \left(\frac{l_e}{t_f}\right)^2 + m_2} \tag{11.12a}$$

$$l_y = l_e + t_f \sqrt{m_1 + m_2} \tag{11.12b}$$

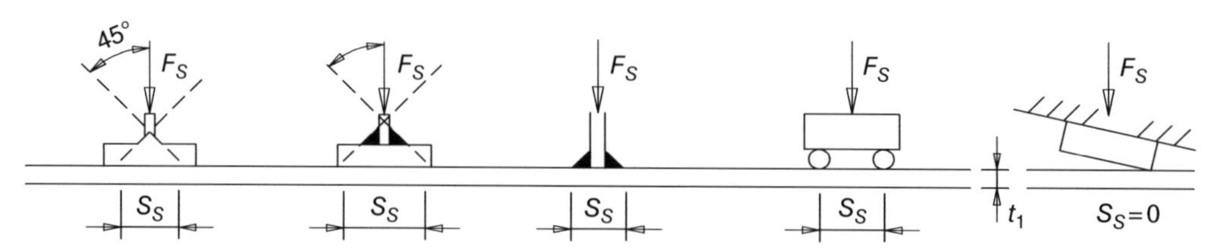

Figure 11.4 Definition of stiff loaded length.

$$l_e = \frac{k_F \cdot E \cdot t_w^2}{2f_{yw} \cdot h_w} \le s_s + c \tag{11.12c}$$

A summary of the procedure is proposed next:

- use Eq. (11.11a): $m_1 = \frac{f_{yf} \cdot b_f}{f_{yw} \cdot t_w}$
- use Eq. (11.11b): $m_2 = 0.02\left(\frac{h_w}{t_f}\right)^2$ if $\bar{\lambda}_F > 0.5$
- use Eq. (11.11c): $m_2 = 0$ if $\bar{\lambda}_F \le 0.5$
- use Eq. (11.10): $l_y = s_s + 2t_f\left(1 + \sqrt{m_1 + m_2}\right)$
- use Eq. (11.6): $F_{cr} = 0.9 \cdot k_F \cdot \frac{E \cdot t_w^3}{h_w}$
- use Eq. (11.5): $\bar{\lambda}_F = \sqrt{\frac{l_y \cdot t_w \cdot f_{yw}}{F_{cr}}}$
- use Eq. (11.4): $\chi_F = \frac{0.5}{\bar{\lambda}_F}$
- use Eq. (11.3): $L_{\text{eff}} = \chi_F \cdot l_y$
- use Eq. (11.3): $F_{Rd} = \frac{f_{yw} \cdot L_{\text{eff}} \cdot t_w}{\gamma_{M1}}$

When checking the buckling resistance in the presence of stiffeners, the resisting cross-section may be taken as the gross area comprising the stiffeners plus a plate width equal to $15\varepsilon \cdot t_w$ on either side of the stiffeners, avoiding any overlap of contributing parts to adjacent stiffeners (Figure 11.5).

The effective length of the compression member is taken as the stiffener length h_w, or as $0.75h_w$ if flange restraints act to reduce the stiffener end rotations during buckling, and reference has to be made to stability curve *c*, already introduced for compression members (see Chapter 6).

Intermediate transverse stiffeners that act as a rigid support at the boundary of inner web panels have to be checked for strength and stiffness. Minimum stiffness for an intermediate transverse stiffener to be considered rigid is expressed in terms of minimum moment of inertia, I_{st}, as:

$$I_{st} \ge 1.5 \frac{h_w^3 \cdot t_w^3}{a^2} \text{ if } \frac{h}{a_w} < \sqrt{2} \tag{11.13a}$$

$$I_{st} \ge 0.75 \cdot h_w \cdot t_w^3 \text{ if } \frac{h}{a_w} \ge \sqrt{2} \tag{11.13b}$$

If the relevant requirements are not met, transverse stiffeners are considered flexible. Requirements provided in Eqs. (11.13a) and (11.13b) assure that at the ultimate shear resistance the lateral

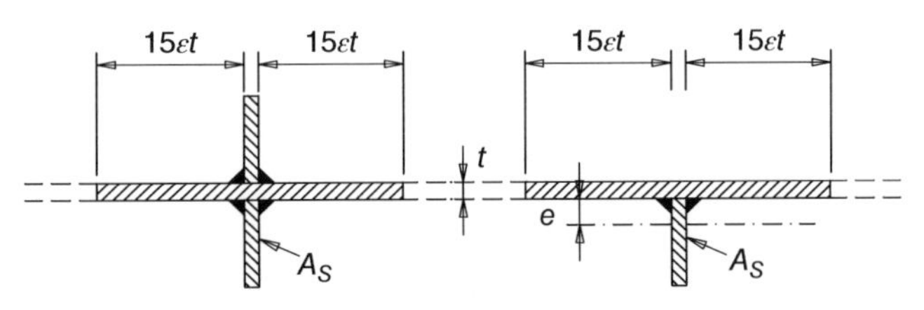

Figure 11.5 Effective cross-section of stiffener.

deflection of intermediate stiffeners remains small if compared with the one of the web. These conditions were derived from linear elastic buckling theory but the minimum stiffness was increased from 3 (for long panels) to 10 times (for short panels) to take the post-buckling behaviour into account. These requirements are relatively easy to meet and do not require very strong stiffeners.

11.3 Design Procedure in Accordance with US Standards

At paragraph J10, AISC 360-10 addresses the problem of webs with concentrated forces.

A different formula is presented for each web failure mode.

(1) *Web local yielding*: This failure mode is what has been previously identified as *web crushing* (Figure 11.2a). The force that causes web yielding can be compression or tension.
The *available strength* is defined as:

$$\phi R_n \ \text{for LFRD}\ (\phi = 1.00);$$
$$R_n/\Omega \ \text{for ASD}\ (\Omega = 1.50)$$

R_n is the nominal strength and shall be determined as follows:

(a) when the concentrated force to be resisted is applied at a distance from the member end greater or equal than member depth d:

$$R_n = F_{yw} t_w (5k + l_b) \tag{11.14a}$$

(b) when the concentrated force to be resisted is applied at a distance from the member end less then member depth:

$$R_n = F_{yw} t_w (2.5k + l_b) \tag{11.14b}$$

where F_{yw} is the minimum yield stress of web material, k is the distance from outer face of the flange to the web toe of the fillet, l_b, is the length of bearing and t_w is web thickness.

(2) *Web local crippling*: The force that causes web crippling (Figure 11.2b) can be compression only.
The *available strength* is defined as:

$$\phi R_n \ \text{for LFRD}\ (\phi = 0.75);$$
$$R_n/\Omega \ \text{for ASD}\ (\Omega = 2.00)$$

R_n is the nominal strength and shall be determined as follows:

(a) when the concentrated force to be resisted is applied at a distance from the member end $\geq 0.5d$:

$$R_n = 0.80 t_w^2 \left[1 + 3\left(\frac{l_b}{d}\right)\left(\frac{t_w}{t_f}\right)^{1.5} \right] \sqrt{\frac{E F_{yw} t_f}{t_w}} \tag{11.15a}$$

(b) when the concentrated force to be resisted is applied at a distance from the member end $<0.5d$ and $l_b/d \leq 0.2$:

$$R_n = 0.40 t_w^2 \left[1 + 3\left(\frac{l_b}{d}\right)\left(\frac{t_w}{t_f}\right)^{1.5} \right] \sqrt{\frac{E F_{yw} t_f}{t_w}} \tag{11.15b}$$

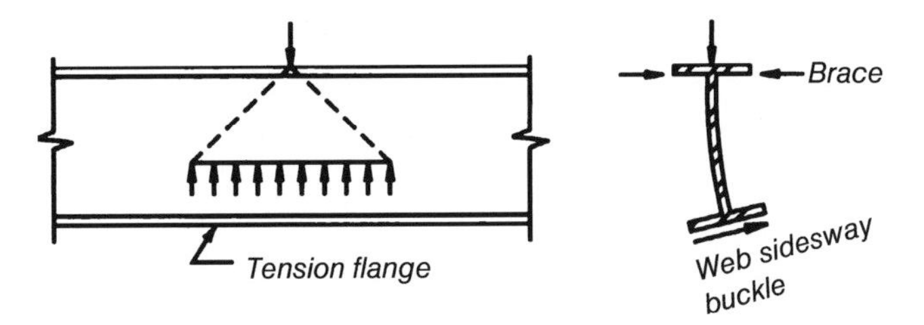

Figure 11.6 Web sidesway buckling.

(c) when the concentrated force to be resisted is applied at a distance from the member end $<0.5d$ and $l_b/d > 0.2$:

$$R_n = 0.40t_w^2\left[1 + \left(\frac{4l_b}{d} - 0.2\right)\left(\frac{t_w}{t_f}\right)^{1.5}\right]\sqrt{\frac{EF_{yw}t_f}{t_w}} \quad (11.15c)$$

where d is the full nominal depth of the section.

(3) *Web sidesway buckling*: This failure mode happens when a concentrated load, acting downward, is applied on the top flange of a beam supported at ends. The upper flange can be braced or not. The web is then compressed and buckles sidesway (Figure 11.6).

The *available strength* is defined as:

$$\phi R_n \text{ for LFRD } (\phi = 0.85);$$

$$R_n/\Omega \text{ for ASD } (\Omega = 1.76)$$

R_n is the nominal strength and shall be determined as follows:

(a) when compression flange is restrained against rotation and $(h/t_w)/(L_b/b_f) \leq 2.3$:

$$R_n = \frac{C_r t_w^3 t_f}{h^2}\left[1 + 0.4\left(\frac{h/t_w}{L_b/b_f}\right)^3\right] \quad (11.16a)$$

If $(h/t_w)/(L_b/b_f) > 2.3$, the limit state of web sidesway buckling does not apply.

(b) when compression flange is not restrained against rotation and $(h/t_w)/(L_b/b_f) \leq 1.7$:

$$R_n = \frac{C_r t_w^3 t_f}{h^2}\left[0.4\left(\frac{h/t_w}{L_b/b_f}\right)^3\right] \quad (11.16b)$$

If $(h/t_w)/(L_b/b_f) > 1.7$, the limit state of web sidesway buckling does not apply.

The constant C_r has to be assumed equal to:

- 960 000 ksi ($6.62 \cdot 10^6$ MPa) when $M_u < M_y$ (LRFD (load and resistance factor design)) or $1.5M_a < M_y$ (ASD (allowable strength design)), where M_u is the required flexural strength using LRFD combinations, M_a is the required flexural strength using ASD combinations and M_y is the moment at the location of the force;
- 480 000 ksi ($3.31 \cdot 10^6$ MPa) when $M_u \geq M_y$ (LRFD) or $1.5M_a \geq M_y$ (ASD).

(4) *Web compression buckling.* This failure mode happens when two concentrated compression forces are applied at both flanges, causing web buckling (Figure 11.3b).
The *available strength* is:

$$\phi R_n \text{ for LFRD } (\phi = 0.90);$$

$$R_n/\Omega \text{ for ASD } (\Omega = 1.67)$$

R_n is the nominal strength and shall be determined as follows:

$$R_n = \frac{24t_w^3\sqrt{EF_{yw}}}{h} \tag{11.17}$$

If the two forces are applied at a distance from the member end $<d/2$, R_n will be reduced by 50%.

When the web is not able to bear the stresses computed here, stiffeners have to be added.

Regarding dimensioning of transverse full depth bearing stiffeners, the member properties have to be determined using an effective length of $0.75h$ and a cross section composed of two stiffeners, and a strip of web having a width of $25t_w$ at interior stiffeners and $12t_w$ at the ends of members (a direct comparison with Figure 11.6 allows us to directly appraise the differences from European procedures).

CHAPTER 12

Design Approaches for Frame Analysis

12.1 Introduction

The routine design of a steel structure is usually carried out following two different steps:

(1) structural analysis of the overall frame, which has to be executed selecting the method of analysis and evaluating, on the basis of the most severe load combinations, internal forces and moments for each member;
(2) safety checks of each member and joint, which have to be carried out on the basis of the criteria discussed in the previous chapters.

The structural analysis of the overall frame also allows estimation of the most relevant displacement associated with the considered load conditions, which have to be defined for services as well as ultimate loading conditions (see Chapter 2).

As to the frame horizontal limit displacement, in absence of direct indications, European practice should to be referred to the limits presented in the previous version of EC3 (ENV 1993-1-1) and proposed in Table 12.1 in terms of maximum frame displacement ratio Δ/H and inter-storey drift ratio δ/h, with H and h representing the overall and the inter-storey height, respectively.

Also, the AISC Specifications prescribe the evaluation of total frame drift ratio Δ/H and inter-storey drift ratio $\delta/\ h$ under service loads in order to guarantee the serviceability of the structure (integrity of interior partitions and external cladding, mainly). Drift under ultimate load combinations has to be evaluated in order to avoid collision with adjacent structures. No specific maximum values are listed directly in Specifications. In the Commentary, some typical drift limit ratios are suggested (see Table 12.2) to designers.

Structural analysis according to EC3 and AISC code is discussed in the following paragraphs. It was decided to move this chapter to follow the ones related to isolated member verifications for the sake of clarity, especially with regard mainly to the proposed comparative example.

12.2 The European Approach

As already discussed with reference to the methods of analysis, when the conditions expressed by Eqs. (3.4a) and (3.4b) are not fulfilled, according to EC3 code it is necessary to perform a second order analysis taking into account initial imperfections (out-of-straightness imperfections of

Structural Steel Design to Eurocode 3 and AISC Specifications, First Edition. Claudio Bernuzzi and Benedetto Cordova.

Table 12.1 Deflection limit ratios for structures under horizontal load according to ENV 1993-1-1.

Type of framed system	Recommended limiting values for horizontal deflections according to EC3	
	δ/h	Δ/H
Portal frames without gantry cranes	1/150	
Other portal frames	1/300	
Multi-storey frames	1/300	1/500

Table 12.2 Deflection limit ratios for structures under horizontal load according to AISC.

Every type of framed system	Recommended limiting values for horizontal deflections	
	δ/h	Δ/H
Range of common values	1/200	1/100
	1/600	1/600
Most widely used values	1/400	1/400
	1/500	1/500

single members and lack-of-verticality imperfections of the whole structural system) and second order effects.

EC3 in Section 5.2.2 (clauses (3) and (7)) proposes the following methods for performing such analysis, identified (and named) by authors as:

- EC3-1 rigorous second order analysis with global and local imperfections;
- EC3-2a rigorous second order analysis with global imperfections;
- EC3-2b approximated second order analysis with global imperfections;
- EC3-3 first order analysis.

A summary of main features of EC3 four methods for frame analysis and verifications is proposed in Table 12.3.

12.2.1 The EC3-1 Approach

The *rigorous second order analysis with global and local imperfections* takes into account:

- all initial imperfections (out-of-straightness imperfections of single members and lack-of-verticality imperfections of the whole structural system);
- second order effects.

Both lack-of-verticality imperfections and out-of-straightness imperfections of single members can be taken into account (a) by a direct modelling or (b) by means of equivalent horizontal distributed loads, according to Section 3.5.1. If such analysis has been performed, no further member stability verifications are required and hence designers have to execute only resistance checks.

Table 12.3 Summary of the key features of the EC3 methods of analysis for frames.

Feature	Methods according to EC3			
	1)	2a)	2b)	3)
Type of analysis	Second order analysis		First order analysis	
Lack-of-verticality imperfections	Yes (direct modelling or notional nodal loads)			No
Out-of-straightness imperfections	Yes (direct modelling or notional nodal loads)	No		
Global second order effects	Yes by direct analysis		Yes amplifying lateral loads by $\frac{1}{1-(1/\alpha_{cr})}$	No
Member (local) second order effects	Yes by direct analysis	No		
Member stability checks	No	Yes according to EC3, Section 6.3		
Buckling length	—	System (geometrical) lengths		Effective lengths (buckling analysis or determined using graphs of ENV code)

12.2.2 The EC3-2a Approach

The *rigorous second order analysis with global imperfections* takes into account the lack-of-verticality imperfections but neglects the member out-of-straightness imperfections. However, for frames sensitive to second order effects, out-of-straightness imperfections should be taken into account, provided that the member is not pinned at both ends and:

$$N_{Ed} > 0.25N_{cr} \tag{12.1}$$

in which N_{Ed} is the design axial force acting on the element and N_{cr} is the critical elastic buckling load for the member.

After this type of second order analysis, member stability has to be checked according to relevant criteria of Section 6.3 of EC3. In such verifications, the code declares that, *structure shall be considered a no-sway frame and buckling lengths shall be made equal to system (geometrical) lengths*. EC3 formulas for stability verifications of beam-columns actually take into account both the member second order effects and the member (out-of-straightness) imperfections, disregarded in global second order analysis as said before.

12.2.3 The EC3-2b Approach

The *approximated second order analysis with global imperfections* neglects the out-of-straightness imperfections of single members but considers all structural system imperfections. This method is applicable if:

$$10 > \alpha_{cr} > 3$$

Table 12.4 Second order amplification factor as a function of α_{cr}.

α_{cr}	$\frac{1}{1-\frac{1}{\alpha_{cr}}}$	α_{cr}	$\frac{1}{1-\frac{1}{\alpha_{cr}}}$
3	1.50	7	1.17
4	1.33	8	1.14
5	1.25	9	1.13
6	1.20	10	1.11

Frames must have a regular distribution of horizontal and vertical loads, as well as member stiffnesses at the various stories. As the method is based on a first order analysis, second order effects have to be considered in an approximate way, amplifying the moments by means of factor β, already introduced in Eq. (3.34), depending on the critical load multiplier of the frame (α_{cr}), defined as:

$$\beta = \frac{1}{1-\frac{1}{\alpha_{cr}}} \tag{12.2}$$

As alternative to the evaluation of α_{cr} via a finite element (FE) buckling analysis, the critical load multiplier should be calculated in regular framed systems by means of the following approximated formula, based on Horne's method:

$$\alpha_{cr} = \frac{H_{Ed}}{V_{Ed}} \cdot \frac{h}{\delta_{H,Ed}} \tag{12.3}$$

where H_{Ed} is the design value of horizontal reaction at the bottom of the storey to the applied horizontal loads and to the fictitious horizontal loads, simulating frame imperfections, V_{Ed} is the total vertical load on the structure at the bottom of the storey, h is the height of the storey and $\delta_{H,Ed}$ is the drift of the storey.

Values of the second order amplification factor defined by Eq. (12.2) are listed in Table 12.4 as a function of α_{cr}. It can be noted that such a factor varies from 1.11 to 1.50 when this type of analysis is admitted.

After the structural analysis, member stability has to be checked according to relevant criteria of Section 6.3 of EC3. In such verifications, the code declares that a *structure shall be considered a no-sway frame and buckling lengths shall be put equal to system (geometrical) lengths* as in method EC3-2a. It is worth mentioning that method EC3-2b is actually similar to the EC3-2a one, with the difference that the direct second order analysis is substituted by first order analysis with a simplified evaluation of second order effects.

12.2.4 The EC3-3 Approach

The *first order analysis* neglects the imperfections and second order effects and member stability is checked according to relevant criteria of Section 6.3 of EC3. The code declares that *buckling lengths shall not be the system (geometrical) lengths but effective lengths evaluated on the basis of global buckling mode of the frame, considered a sway frame.*

EC3 code does not provide any further detail about the more convenient and reliable approach to determine members buckling lengths.

12.3 AISC Approach

Frame analysis and design is addressed mainly in Chapter C 'Design for Stability' of AISC 360-10, and also in Appendix 6–8 of the same specifications. The aim of Chapter C is providing methods for assuring:

- the stability of the whole frame;
- the stability of each element (beam, column, bracing) of the frame.

In order to assure stability, the following effects have to be taken into account:

(a) flexural, shear and axial member deformations;
(b) second order effects: global effects of loads acting on displaced structure (P-Δ), local effects of loads acting on the deflected shape of a member between joints (P-δ);
(c) geometric imperfections;
(d) stiffness reduction due to partial yielding and residual stresses;
(e) uncertainty in stiffness and strength.

AISC 360-10, although allows use of 'any rational method of design for stability that considers all of the listed effects', it actually suggests three methods for design:

(1) the Direct Analysis Method;
(2) the Effective Length Method;
(3) the First Order Analysis method.

12.3.1 The Direct Analysis Method (DAM)

This is the main suggested method and it is addressed in Sections C2 and C3 of the specifications. It can be applied in every case without limitations. The designer has to follow these steps:

(1) to compute the required strength of each member of the frame;
(2) to define the member available strengths;
(3) to verify that they are greater or at least equal to the required strength values.

For computing the required strength, direct analysis method (DAM) approach requires a *second order analysis*, considering both P-δ and P-Δ effects, together with flexural, shear and axial member deformations.

All the P-δ effects can be neglected in the analysis of the structure, but they must be taken into account in the strength evaluation for individual members subjected to compression and flexure (beam-columns), if the following conditions are satisfied:

(a) the structure supports gravitational loads through vertical elements (columns and/or walls and/or frames);
(b) no more than one-third of total vertical loads is supported by columns belonging to the moment-resisting frame acting in the direction being considered;
(c) in all stories:

$$\alpha \cdot \Delta_{2nd-order,\max} / \Delta_{1st-order,\max} \leq 1.7$$

where $\Delta_{2nd-order,\max}$ is the maximum second order drift, $\Delta_{1st-order,\max}$ is the maximum first order drift, and $\alpha = 1$ for load and resistance factor design (LRFD) and $= 1.6$ for allowable strength design (ASD).

It should be stated that:

(1) in order to evaluate second order effects, structural analysis has to be performed with limit loads and not with allowable loads that are lower. So, even if the designer uses the ASD method, analysis has to be done with limit loads. The mean ratio between limit state (LRFD) and allowable (ASD) loads is 1.6 from Code; then $\alpha = 1.6$ is the coefficient to use for switching from LRFD to ASD verifications;

(2) a second order analysis by hand calculations cannot be performed, except in few very simple cases, but it is necessary in most cases to carry out a structural analysis by means of FE method computer programs and several commercial FE analysis packages are nowadays adequate to perform such analysis considering both *P-Δ* and *P-δ* effects. Furthermore, AISC 360-10 provides two simple benchmark problem cases for verifying the software capability to take into account one or both the second order effect (i.e. *P-δ* and *P-Δ* effects). Such test cases are summarized in Figures 12.1 and 12.2;

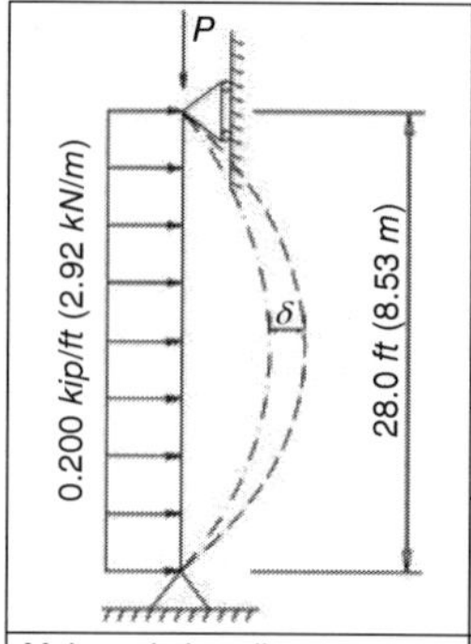

Axial force, *P* [kips]	0	150	300	450
M_{mid} [kips-in.]	235 [235]	270 [269]	316 [313]	380 [375]
δ_{mid} [in.]	0.202 [0.197]	0.230 [0.224]	0.269 [0.261]	0.322 [0.311]
Axial force, *P* [kN]	0	667	1334	2001
M_{mid} [kNm]	26.6 [26.6]	30.5 [30.4]	35.7 [35.4]	43.0 [42.4]
δ_{mid} [mm]	5.13 [5.02]	5.86 [5.71]	6.84 [6.63]	8.21 [7.91]

Major axis bending
W14×48 (W360×72)
E=29,000 ksi (200 GPa)

Analyses include axial, flexural and shear deformation
[Values in brackets] neglecting shear deformations

Figure 12.1 AISC benchmark problem Case 1 (*P-δ* effects). From Figure C-C2.2 of AISC 360-10.

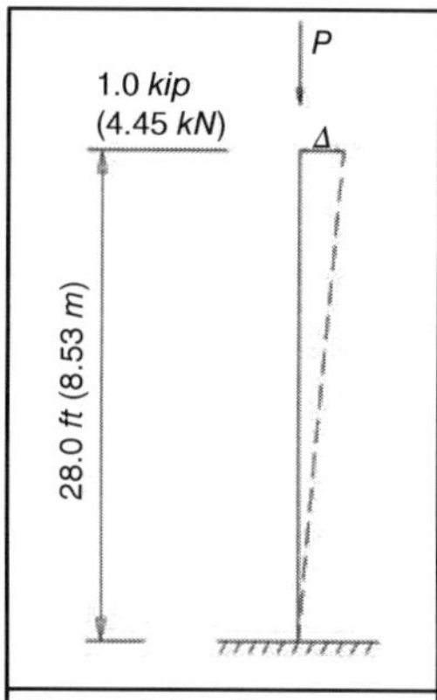

Axial force, *P* [kips]	0	100	150	200
M_{base} [kips-in.]	336 [336]	470 [469]	601 [598]	856 [848]
Δ_{tid} [in.]	0.907 [0.901]	1.34 [1.33]	1.77 [1.75]	2.60 [2.56]
Axial force, *P* [kN]	0	445	667	890
M_{base} [kNm]	38.0 [38.0]	53.2 [53.1]	68.1 [67.7]	97.2 [96.2]
Δ_{tip} [mm]	23.1 [22.9]	34.2 [33.9]	45.1 [44.6]	66.6 [65.4]

Major axis bending
W14×48 (W360×72)
E=29,000 ksi (200 GPa)

Analyses include axial, flexural and shear deformation
[Values in brackets] neglecting shear deformations

Figure 12.2 AISC benchmark problem Case 2 (*P-Δ* effects). From Figure C-C2.3 of AISC 360-10.

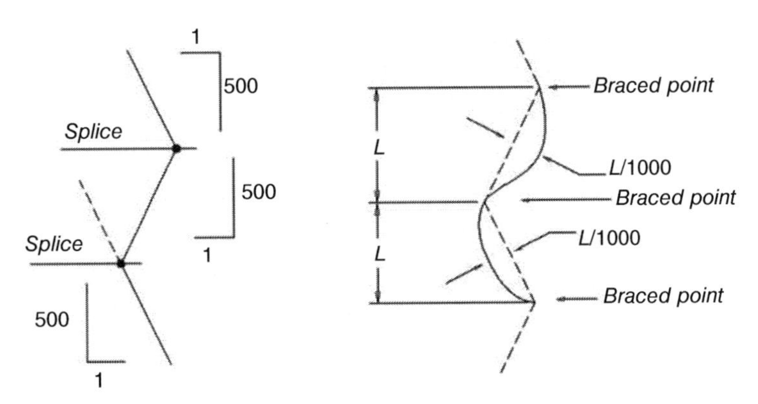

Figure 12.3 AISC 303-10 column out-of-plumbness tolerances.

(3) AISC specifications allow, in lieu of a rigorous second order analysis, to use the approximate method of the second order analysis approach provided in Appendix 8 of the Specification (see Section 12.3.4);

(4) As a principle, *P-δ* effects influence *P-Δ* effects, and then they should be taken into account in structural analysis. It should be noted that, when second order effects are not very high ($\alpha{\cdot}\Delta_{2nd-order,\max}/\Delta_{1st-order,\max} \leq 1.7$), it is permitted to consider them separately: *P-Δ* effects in structural analysis and *P-δ* effects in verifications of isolated members. In such case, *P-δ* effects can be evaluated in a simplified way using the method defined in Appendix 8 of the Specification (see Section 12.3.4), hence allowing a more usual second order analysis (with *P-Δ* effects only).

The second order structural analysis model also has to take into account the *initial imperfections* of the structure, which mainly consists of the out-of-plumbness of columns. Their maximum acceptable values are addressed in AISC 303-10 'Code of Standard Practice for Steel Building and Bridges' and reported in Figure 12.3. The standard tolerance is $H/500$, where H is the storey height, which is reduced to $L/1000$ when L is the column length between brace or framing point.

Such an initial imperfection must be taken into account in one of the two following modes:

(1) by direct modelling;
(2) by notional loads.

If the designer chooses to incorporate out-of-plumbness values in the structural model, their entity has to be not smaller than the recommended values previously introduced. Otherwise, if the designer prefers to use the (easier) method of notional loads, (i.e. reference is made to a perfect frame), at every floor level lateral loads are applied to frame, whose magnitude is:

$$N_i = 0.002\alpha Y_i \tag{12.4}$$

where $\alpha = 1$ for LRFD and $= 1.6$ for ASD, N_i is the notional load applied at level i and Y_i is the gravity load applied at level i, from LRFD or ASD load combination.

The value of notional loads is based, as mentioned before, on the tolerance of $H/500$ on out-of-plumbness of columns. The value of horizontal notional force N_i generating the same additional moment when a vertical load Y_i acts, can be easily derived from the equivalence (Figure 12.4):

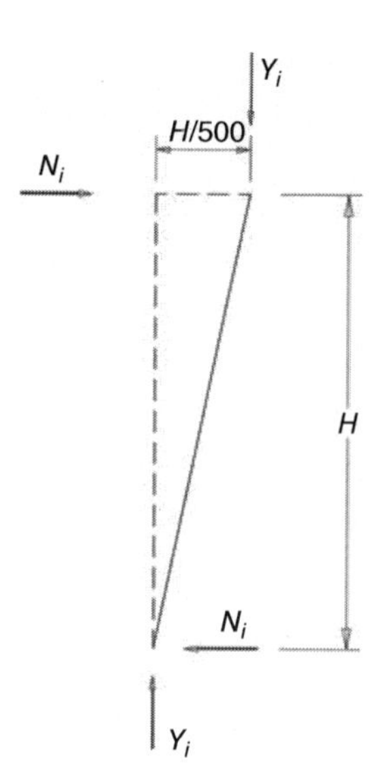

Figure 12.4 AISC notional loads N_i.

$$N_i \cdot H = Y_i \cdot \frac{H}{500} \tag{12.5}$$

Hence:

$$N_i = 0.002 \cdot Y_i \tag{12.6}$$

Notional loads have to be applied in all the load combinations but then they can be applied in load combinations containing gravitational loads only, when:

$$\alpha \cdot \Delta_{2nd-order,\max} / \Delta_{1st-order,\max} \leq 1.7$$

A further very important aspect that has to be included in the structural analysis model is the *reduction in stiffness*. Due to the *partial yielding* occurring in structural elements because of high levels of stress, accentuated by the presence of *residual stresses*, the structure suffers a reduction in stiffness that in turn decreases stability. To take this into account, AISC prescribes to *reduce to* 80% all the stiffnesses of the members that contribute to the stability of the whole structure. In addition, all the flexural stiffnesses of members that are considered to contribute to the stability of the structure have to be reduced by a factor τ_b computed as:

if $\frac{\alpha P_r}{P_y} \leq 0.5$:

$$\tau_b = 1.0 \tag{12.7}$$

if $\frac{\alpha P_r}{P_y} > 0.5$:

$$\tau_b = 4 \frac{\alpha P_r}{P_y}\left(1 - \frac{\alpha P_r}{P_y}\right) \tag{12.8}$$

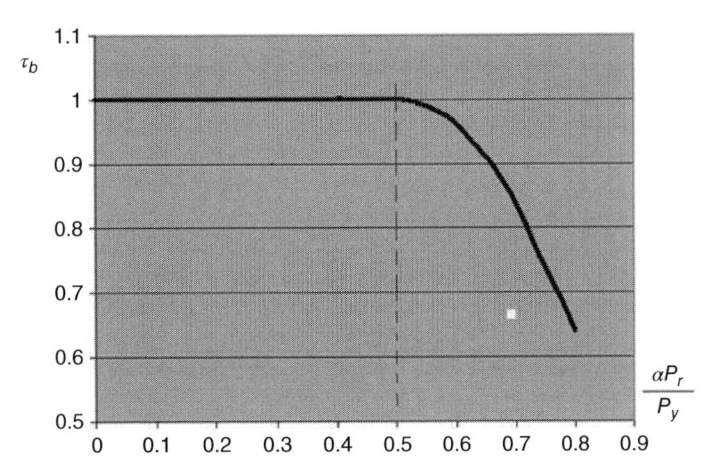

Figure 12.5 τ_b coefficient.

where $\alpha = 1$ for LRFD and $=1.6$ for ASD, P_r is the required axial compression strength computed using LRFD or ASD load combinations and $P_y = F_y A_g$ is the axial yield strength.

Figure 12.5 shows the variation of τ_b coefficient versus the ratio $\alpha P_r/P_y$.

It is possible to apply notional loads at all floor levels that are added to those of Eq. (12.6), in lieu of using τ_b lower than unity ($\tau_b < 1.0$). These additive notional loads are defined as

$$N_i = 0.001\alpha Y_i \tag{12.9}$$

If τ_b is used, from a practical point of view the designer has to change the axial stiffness EA to $0.8EA$, and the flexural stiffness EI to $0.8\tau_b EI$. When $\tau_b = 1$, every stiffness has to be multiplied by 0.8. In this case the easiest way to reduce stiffness is to change directly the elastic modulus E to $0.8E$.

As an alternative to a second order analysis, an *approximate* (easier) second order analysis can be performed, according to the method outlined in Appendix 8 of the Specification (see Section 12.3.4).

Once the designer has computed the required strength of each member of the frame by means of the second order analysis (rigorous or approximate), the available member strengths have to be defined. This step has to be performed in accordance with the relevant provisions for elements in compression, bending and shear as discussed in Chapters 6 and 7. In such verifications the code declares that *the effective length factor K shall be taken to be unity*. As said before, if a rigorous second order analysis has been performed but *P*-δ effects have not been incorporated, they have to be considered in the beam-column verifications, amplifying the moments by factor B_1, according to Appendix 8 (see Section 12.3.4). A conceptual scheme of DAM is presented in Table 12.5.

12.3.2 The Effective Length Method (ELM)

This method is an alternative to the DAM, and it is addressed in the Appendix 7. The effective length method (ELM) can be applied if the following conditions are fulfilled:

(a) the structure supports gravitational loads through vertical elements (columns and/or walls and/or frames);

(b) in all stories: $\alpha \cdot \Delta_{2nd-order,\max}/\Delta_{1st-order,\max} \leq 1.5$ or: $B_2 \leq 1.5$

Table 12.5 Summary of the direct analysis method.

<table>
<tr><td colspan="7">Applicability: always</td></tr>
<tr><td colspan="5">Second order analysis</td><td colspan="2">Member verifications</td></tr>
<tr><td colspan="2">$\frac{\alpha \cdot \Delta_{2nd-order,max}}{\Delta_{1st-order,max}} > 1.7?$</td><td colspan="2">Initial imperfections</td><td>Adjustments to stiffness</td><td>P-δ and P-Δ effects in second order analysis?</td><td>P-Δ effects only in second order analysis?</td></tr>
<tr><td>YES</td><td>NO</td><td rowspan="2">Direct modelling</td><td rowspan="2">Notional loads (0.2% of vertical loads)</td><td rowspan="2">0.8EA
$0.8\tau_b EI$ or 0.8EA + notional loads (0.1% of vertical loads)</td><td rowspan="3">Internal actions from second order analysis and K = 1</td><td rowspan="3">Internal actions from second order analysis, but moments amplified by B_1 for beam-column (Appendix 8) and K = 1</td></tr>
<tr><td>P-δ and P-Δ effects</td><td>P-Δ effects only (P-δ effects in member verifications!)</td></tr>
<tr><td colspan="5">Alternative to second order analysis:
Approximate second order analysis (Appendix 8)</td></tr>
</table>

The definition of term B_2 is proposed in Section 12.3.4.

According to this method, the designer also has to compute the required strength of each member of the frame; then designer has to define the available member strengths and verify that they are greater or at least equal to the required strength values.

For computing the required strength, ELM prescribes to perform a *second order analysis*, as in DAM but with the following differences:

(a) stiffness reduction has not to be applied;
(b) imperfections has to be taken into account by notional loads, applied in gravity-only load combinations.

Being the ratio $\alpha \cdot \Delta_{2nd-order,max}/\Delta_{1st-order,max} \leq 1.5$, and therefore <1.7, it is permitted to consider *P-Δ* effects in the structural analysis and *P-δ* effects in the verifications of isolated members.

Once the designer has computed the required strength of each member, he has to define the available member strengths. This has to be performed with the relevant provisions for elements in compression, bending and shear (see Chapters 6 and 7). In such verifications the code declares that *the effective length factor K shall be taken as follows*:

(1) In *braced frame systems*, or shear wall systems, that is when lateral loads are not sustained by flexural resistance of beams and columns:

$$K = 1$$

(2) In *moment frame systems*, that is when lateral loads are sustained by flexural resistance of beams and columns:

(a) If $\alpha \cdot \Delta_{2nd-order,max}/\Delta_{1st-order,max} \leq 1.1$:

$$K = 1$$

Table 12.6 Summary of the effective length method.

<table>
<tr><td colspan="5">Applicability: $\alpha \cdot \Delta_{2nd-order,max} / \Delta_{1st-order,max} \leq 1.5$ or: $B_2 \leq 1.5$</td></tr>
<tr><td colspan="3">Second order analysis</td><td colspan="2">Member verifications</td></tr>
<tr><td rowspan="2">P-Δ effects only (P-δ effects in member verifications!)</td><td>Initial imperfections</td><td>Adjustments to stiffness</td><td></td><td></td></tr>
<tr><td>Notional loads (0.2% of vertical loads)</td><td>NO</td><td colspan="2">Internal actions from second order analysis, but moments amplified by B_1 for beam-columns (Appendix 8)</td></tr>
<tr><td colspan="3">Alternative to second order analysis:
Approximate second order analysis (Appendix 8)</td><td>$K = 1$</td><td>K from alignment charts</td></tr>
</table>

(b) If $\alpha \cdot \Delta_{2nd-order,max} / \Delta_{1st-order,max} > 1.1$:$K$ or F_e (elastic critical buckling stress) determined from a *sidesway buckling analysis*, or for isolated columns with simple conditions of end restraints, using the values in Table 6.8.

A conceptual scheme of the Effective Length Method is presented in Table 12.6.

12.3.3 The First Order Analysis Method (FOM)

Another method alternative to the DAM is *the First Order Analysis Method*, also addressed in Appendix 7. It can be applied if the following conditions are fulfilled:

(a) the structure supports gravitational loads through vertical elements (columns and/or walls and/or frames);
(b) in all stories: $\alpha \cdot \Delta_{2nd-order,max} / \Delta_{1st-order,max} \leq 1.5$ or $B_2 \leq 1.5$;
(c) the required axial compressive strengths of all members whose flexural stiffness contributes to lateral stability of the structure satisfy the limitation:

$$\frac{\alpha P_r}{P_y} \leq 0.5$$

where $\alpha = 1$ for LRFD and = 1.6 for ASD, P_r is the requited axial compression strength, computed using LRFD or ASD load combinations and $P_y = F_y A_g$ is the axial yield strength.

According to this method the designer has to compute the required strength of each member of the frame; then he has to define the member available strengths and verify that they are greater or, at least, equal to the required strength values. For computing the required strength, FOM, unlike DAM and ELM, prescribes performance of a *first order analysis*, with the following additional requirement:

additional lateral loads have to be applied to every load combination, together with relevant loads, and their values are defined as:

$$N_i = 2.1\alpha \left(\frac{\Delta}{L}\right)_{max} \cdot Y_i \geq 0.0042 Y_i \qquad (12.10)$$

where $\alpha = 1$ for LRFD and = 1.6 for ASD N_i is the additional lateral load applied at level i, Y_i is the gravity load applied at level i, from LRFD or ASD load combination, Δ is the first order inter-storey drift and L is the height of the storey.

Table 12.7 Summary of the first order analysis method.

Applicability: $\alpha \cdot \Delta_{2nd-order,max}/\Delta_{1st-order,max} \le 1.5$ or: $B_2 \le 1.5$ $\alpha P_r/P_y \le 0.5$				
First order analysis			**Member verifications**	
P-Δ effects by means of additional lateral loads (P-δ effects in member verifications!)	Initial imperfections and P-Δ effects	Adjustments to stiffness	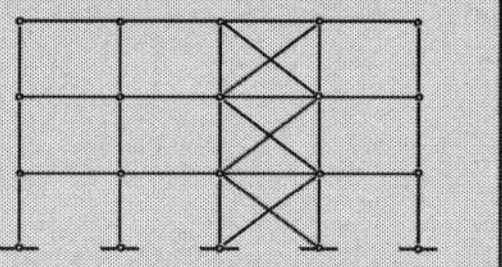	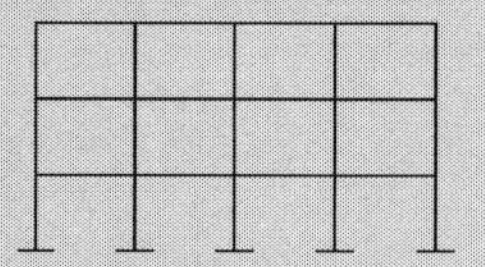
	Additional lateral loads ($2.1(\Delta/L)_{max}$ % of vertical loads, but not less than 0.42%)	NO	Internal actions from second order analysis, but moments amplified by B_1 for beam-columns (Appendix 8)	
			$K = 1$	

The minimum value $0.0042Y_i$ is based on the assumption of a minimum first order drift ratio due to any effect of $\Delta/L = 1/500$.

First order analysis is permitted without any reduction in stiffnesses.

The additional lateral loads take into account P-Δ effects in structural analysis. For P-δ effects in verifications of single members, designer has to apply to beam-column moments the B_1 amplifier defined in Appendix 8 of the Specification (see Section 12.3.4). Members will be verified assuming $K = 1$.

A conceptual scheme of First Order Analysis Method is presented in Table 12.7.

12.3.4 Method for Approximate Second Order Analysis

As already mentioned, AISC specifications allow use of an alternative procedure for performing a second order analysis required by both the DAM and the Effective Length Method. Such alternative procedure is called the *Approximate Second Order Analysis* and is outlined in Appendix 8, which deals with first order analysis, simulating second order effects, both P-δ and P-Δ, by amplifications of internal stresses in members.

The procedure can be applied if the structure supports gravity loads mainly through vertical columns, walls or frames. The method consists of the following steps:

(1) create a structural FE model;
(2) restrain the model against sidesway by means of additional (fictitious) restraints;
(3) for each load combination (LRFD or ASD) compute first order moments M_{nt} and axial loads P_{nt}; compute lateral reactions at additional restraints;
(4) remove additional restraints, load the model with the previously computed lateral reactions, compute first order moments M_{lt} and axial loads P_{lt};
(5) compute, for each member subjected to axial load and bending moments and for each direction of bending, the B_1 multiplier that takes into account for P-δ effects;
(6) compute, for each storey and each direction of lateral translation, the B_2 multiplier that takes into account P-Δ effects;
(7) compute required second order flexural strength M_r and axial strength P_r for all members with the following formulas:

$$M_r = B_1 M_{nt} + B_2 M_{lt} \tag{12.11}$$

$$P_r = P_{nt} + B_2 P_{lt} \tag{12.12}$$

For structures where gravity loads cause negligible lateral displacements, this procedure can be simplified as follows:

(1) create a structural model;
(2) load the structure with *gravity loads only* of each load combination (LRFD or ASD) and compute first order moments M_{nt} and axial loads P_{nt};
(3) load the structure with *lateral loads only* of each load combination (LRFD or ASD) and compute first order moments M_{lt} and axial loads P_{lt};
(4) compute, for each member subjected to axial load and bending moments and each direction of bending, the B_1 multiplier that takes into account P-δ effects;
(5) compute, for each storey and each direction of lateral translation, the B_2 multiplier that takes into account P-Δ effects;
(6) compute required second order flexural strength M_{r} and axial strength P_{r} for all members as for step (7) in the previous list.

The multiplier B_1, as previously mentioned, takes into account P-δ effects. It has to be calculated for each member subjected to axial load and bending, and for each direction of bending, with the following formula:

$$B_1 = \frac{C_m}{1 - \frac{\alpha P_r}{P_{e1}}} \geq 1 \tag{12.13}$$

where $\alpha = 1$ for LRFD and $= 1.6$ for ASD, P_r is the requited axial compression strength, computed using LRFD or ASD load combinations and P_{el} is the elastic critical buckling strength of the member in the plane of bending, which is defined as:

$$P_{e1} = \pi^2 \frac{(EI)^*}{L^2} \tag{12.14}$$

where $(EI)^*$ is the flexural stiffness depending on the method of analysis used and C_m is a coefficient that takes into account the shape of bending moment diagram, and has to be computed for each plane of bending.

It should be noted that $(EI)^*$ has to be assumed equal to $0.8\tau_b EI$ for DAM or to EI for effective length and the first order analysis method.

For beam-columns not directly loaded between supports in the plane of bending, term C_m is given by:

$$C_m = 0.6 - 0.4 \frac{M_1}{M_2} \tag{12.15}$$

where M_1 is the smaller moment and M_2 the larger one and the ratio M_1/M_2 is positive if the member is bent in reverse curvature, negative if it is bent in simple curvature.

For beam-columns subjected to transverse loads between supports:

$$C_m = 1 \tag{12.16}$$

It must be noted that B_1 is computed *for each bending direction*, so actually the designer has to apply distinct $B_{1\mathrm{x}}$ and $B_{1\mathrm{y}}$ multipliers for the two member axes.

The multiplier B_2, on the other hand, takes into account P-Δ effects. It has to be calculated for each storey and for each direction of bending, with the following formula:

$$B_2 = \frac{1}{1 - \dfrac{\alpha P_{story}}{P_{e,story}}} \geq 1 \tag{12.17}$$

where $\alpha = 1$ for LRFD and $= 1.6$ for ASD, P_{story} is the total vertical load supported by the storey (according to LRFD or ASD load combinations), including columns that are not part of the lateral force resisting system and $P_{e,story}$ is the elastic critical buckling strength for the storey in the direction of translation considered, determined by sidesway buckling analysis as:

$$P_{e,story} = R_M \frac{H \cdot L}{\Delta_H} \tag{12.18}$$

where $R_M = 1 - 0.15 \dfrac{P_{mf}}{P_{story}}$ accounts for the influence of P-δ effects on P-Δ, L is the height of the storey, H is the storey shear, in the direction of translation considered produced by the lateral forces used to compute Δ_H (first order inter-storey drift due to lateral forces acting in the direction of translation considered and computed using the required stiffness) and P_{mf} is the total vertical load in columns that are part of the lateral force resisting system.

If Δ_H varies over the plan area, an average drift value (or the maximum value, conservatively) shall be used.

The R_M factor varies from 0.85, when all the columns of the storey belong to a moment resisting frame, to 1.0, when there are no moment frames in the storey.

12.4 Comparison between the EC3 and AISC Analysis Approaches

As can be observed from the previous paragraphs, there are some similarities and differences between EC3 and AISC methods for frame stability, as stated directly in the AISC 360-10 Commentary, C2, about the DAM:

> *While the precise formulation of this method is unique to AISC Specification, some of its features have similarities to other major design specifications around the world, including the Eurocodes, the Australian standard, the Canadian standard and ACI 318 (ACI, 2008).*

Some of the differences between the two codes are:

(1) AISC 360-10 appears to be easier to use. The three methods, the Direct Analysis Method, the Effective Length Method and the First Order Analysis Method, are clearly explained and the procedures to apply them are well illustrated. EC3 methods are not so clearly defined: the names of methods (EC3-1, EC3-2a, EC3-2b and EC3-3) we reported in Table 12.3 are not present in the Code and have been introduced by authors here for the sake of clarity.

(2) AISC DAM and EC3-1 are quite similar: both prescribe a second order analysis that takes into any initial imperfections. There are some relevant differences anyway:

- **(a)** AISC method prescribes a reduction in stiffness, not addressed in EC3: remarkable differences are hence expected for analysis results.
- **(b)** Out-of-straightness member imperfections can be taken into account by means of notional member loads according to EC3, and this is necessary if axial load in columns is greater than 25% of critical load. AISC on the contrary states that such imperfections do not need to be considered in the analysis because they are already accounted for in the compression member design rules.

(c) If second order effects and initial imperfections have been taken into account in the analysis, EC3 states that no further stability member verification is needed. AISC, on the contrary, prescribes performing member verifications with $K = 1$.

(d) According to both codes, if the structure is not very sensitive to second order effects, the analysis can consider *P-Δ* effects only and disregard *P-δ* effects, but includes them in member verifications. This is possible, according to EC3, if:

$$\alpha_{cr} \geq 3 \tag{12.19}$$

according to AISC, if:

$$\alpha \cdot \Delta_{2nd-order,\max} / \Delta_{1st-order,\max} \leq 1.7 \tag{12.20}$$

These two limitations are conceptually similar. The ratio of second order drift to first order drift in a storey may be taken, actually, as the B_2 multiplier (Eq. (12.17)) defined in AISC Appendix 8. Taking into account the Eq. (12.18), we can write:

$$B_2 = \frac{1}{1 - \dfrac{\alpha P_{story}}{P_{e,story}}} = \frac{1}{1 - \dfrac{1}{R_M\left(\dfrac{H \cdot L}{\alpha P_{story} \cdot \Delta_H}\right)}} \leq 1.7 \tag{12.21}$$

The quantity $HL/\alpha P_{story}\Delta_H$ of AISC is equivalent to the term α_{cr} contained in EC3, Eq. (12.3). Considering also that R_M, if all the columns are part of the moment-resisting frame, is equal to 0.85, Eq. (12.21) becomes:

$$B_2 = \frac{1}{1 - \dfrac{\alpha P_{story}}{P_{e,story}}} = \frac{1}{1 - \dfrac{1}{0.85 \cdot \alpha_{cr}}} \leq 1.7 \tag{12.22}$$

For values of B_2 less or equal to 1.7, this means:

$$\alpha_{cr} \geq 2.85 \tag{12.23}$$

As can be seen, Eq. (12.23), derived from AISC, is very close to Eq. (12.19) belonging to EC3.

(3) EC3 method-2a is then equivalent to AISC DAM if *P-δ* effects are not taken into account.

(4) If *P-δ* effects can be disregarded in analysis, they must be taken into account in member verifications. Using AISC Specification, this is done applying the B_1 multiplier defined in Appendix 8 because AISC formulas for beam-columns do not take into account *P-δ* effects. If EC3 code is used this amplification is not necessary because EC3 formulas take into account directly *P-δ* effects.

(5) EC3 method-2b is quite similar to AISC DAM, when *P-δ* effects are not taken into account and a rigorous second order analysis is substituted by an approximate second order analysis according to Appendix 8. Both EC3 and AISC methods prescribe first order analysis with amplification of the effects due to lateral loads. It should be noted that B_2 multiplier, defined in AISC code, takes into account the interaction between *P-Δ* and *P-δ* effects by means of R_M coefficient, while EC3 multiplier Eq. (12.2) does not.

(6) EC3 method-3 can be considered similar to AISC effective length method. Graphs to be employed for computing effective buckling lengths are different between the two codes.

(7) AISC first order method has no correspondence in EC3 methods. It uses first order analysis like EC3 method-3, but the notional loads that take into account the out-of-straightness imperfections are incremented to simulate *P-Δ* effects, and buckling lengths are computed using $K = 1$.

12.5 Worked Example

Example 12.1 Structural Analysis According to the EC3 and US Codes

Verify an ASTM A992 W10 × 60 cantilever column (Figure E12.1.1). Its height L is 15 ft (4.57 m).The nominal loads are a vertical concentrated load P of 200 kips (889.6 kN) and a horizontal concentrated load H of 9.7 kips (43.1 kN). In correspondence with the external flange of the column, a metal sheeting is efficiently connected so that the column can be considered fully braced for out-of-plane buckling as well as for lateral-torsional buckling.

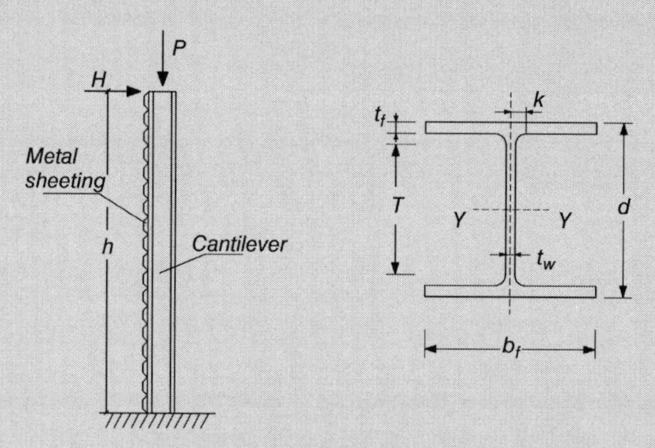

Figure E12.1.1

Geometrical properties of cantilever:

$$\begin{aligned} &d = 10.2 \text{ in. } (259 \text{ mm}) && A_g = 17.6 \text{ in.}^2 \,(113{,}5 \text{ cm}^2) \\ &b_f = 10.1 \text{ in. } (257 \text{ mm}) && Z_y(W_{pl}) = 74.6 \text{ in.}^3 \,(1222 \text{ cm}^3) \\ &t_f = 0.68 \text{ in. } (17.3 \text{ mm}) && S_y(W_{el}) = 66.7 \text{ in.}^3 \,(1093 \text{ cm}^3) \\ &t_w = 0.42 \text{ in. } (10.7 \text{ mm}) && I_y = 341 \text{ in.}^4 \,(14190 \text{ cm}^4) \\ &k = 1.18 \text{ in. } (30.0 \text{ mm}) && r_y = 4.39 \text{ in. } (11.15 \text{ cm}) \\ &L = 15 \text{ ft } (4.57 \text{ m}) && \end{aligned}$$

Material properties:

$$\text{Steel: ASTM A992} \quad F_y = 50 \text{ ksi}(345 \text{ MPa}) \quad F_u = 65 \text{ ksi}(448 \text{ MPa})$$

European Approaches

EC3-1: in accordance with this approach, a second order analysis is required accounting for all the imperfections (both global and local). Verification is developed by considering only the resistance check of the more stressed member cross-section members.

Global imperfections (out-of-plumbness angle Φ_0) are taken into account via a notional horizontal load (N_i) evaluated in accordance with Eq. (3.19):

$$\Phi_0 = 1/200; \text{m} = 1; \alpha_m = \sqrt{0.5\left(1+\frac{1}{m}\right)} = 1; \alpha_h = \frac{2}{\sqrt{L}} = \frac{2}{\sqrt{4.57}} = 0.936$$

$$H_i = 0.00468P = 0.00468 \times 889.6 = 4.16 \text{ kN}(0.935 \text{ kips})$$

Local (bow) imperfections are taken into account by assuming an out-of-straightness in according to the European procedure (Chapter 3). With reference at the stability curve b in plastic analysis (Table 3.1):

$$e_0 = \frac{1}{200};\quad k \cdot e_0 = 0.5 \cdot \frac{4570}{200} = 11.42\,mm$$

Term k ($k = 0.5$) has been introduced because of the later-torsional buckling is prevented. Owing to the absence of clear indication in EC3 about the design procedure, it was decided to simulate bow imperfection via a uniformly distributed notional load.

A suitable value of the uniformly distributed horizontal load q_i applied along the member (Figure E12.1.2) has been evaluated, with second order analysis, in order to have the same top displacement.

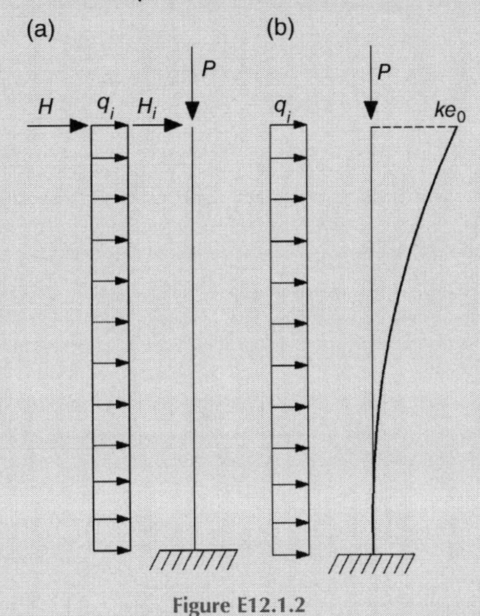

Figure E12.1.2

An iterative procedure based on a second order analysis has been adopted to evaluate the q_i value corresponding to the top displacement of 11.42 mm. From FE analysis, this results in $q_i = 4.50$ N/m

Key data associated with the load condition in Figure E12.1.2, assuming E = 210 000 MPa (30 460 ksi):

$$\begin{array}{ll} \text{First order displacement:} & \Delta_{H,1st} = 58.46\text{ mm (2.30 in.)} \\ \text{Second order displacement:} & \Delta_{H,2nd} = 81.33\text{ mm (3.21 in.)} \\ \text{First order moment:} & M_I = 262.98\text{ kNm (2327 kip-in.)} \\ \text{Second order moment at the base:} & M_{II} = 335.55\text{ kNm (3146 kip-in.)} \end{array}$$

Then a resistance check of the column base cross-section is performed via the criterion described in Section 9.2.1.

At first it is necessary to check if the axial load influences the bending resistance: being

$$n = \frac{N_{Ed}}{N_{pl,Rd}} = 0.227 < 0.25 \text{ but } N_{Ed} > \frac{0.5 \cdot h_w \cdot t_w \cdot f_y}{\gamma_{M0}} \equiv\triangleright 889600 > 367304$$

Reference has to be made to the reduced bending resistance for verification. In accordance with Eq. (9.9a), term a has to be computed:

$$a = \frac{A - (2 \cdot b \cdot t_f)}{A} \le 0.5$$

It results in $a = 0.216$ and hence:

$$M_{N,y,Rd} = M_{pl,y,Rd} \cdot \frac{1-n}{1-0.5 \cdot a} = 421.59 \cdot 10^6 \cdot 0.867 = 365.51 \cdot 10^6 \text{Nmm}$$

Evaluation of the safety index SI^{EC3-1}:

$$SI^{EC3-1} = \frac{M_{Ed}}{M_{N,Rd}} = \frac{335.55}{365.51} = 0.92$$

EC3-2a: in accordance with this approach a second order analysis is only required for accounting the global imperfection via the additional horizontal force $F_i = 4.16$ kN (0.935 kips), the computation of which has been already presented with reference to the EC3 method 1; load condition is presented in Figure E12.1.3. A stability check is permitted with reference to the system length. A second order analysis has been executed on the structures presented in Figure E12.1.3 and the main results are:

Figure E12.1.3

First order displacement:	$\Delta_{H,1st} = 50.5$ mm (1.98 in.)
Second order displacement:	$\Delta_{H,2nd} = 72.61$ mm (2.86 in.)
First order moment:	$M_I = 215.99$ kNm (1915 kip-in.)
Second order moment at the base:	$M_{II} = 281.81$ kNm (2494 kip-in.)

Method 1:
From the original approach expressed by Eq. (9.19) it has been assumed that reduction coefficient χ_z and χ_{LT} are equal to unity, being flexural buckling about a weak axis and flexural torsional buckling embedded by the metal sheeting. Hence, reference has to be made to the following equation for verification checks:

$$\frac{N_{Ed}}{\frac{\chi_y \cdot N_{Rk}}{\gamma_{M1}}} + k_{yy} \frac{M_{y,Ed}}{\frac{M_{y,Rk}}{\gamma_{M1}}} \leq 1$$

Evaluation of term χ_y:

$$N_{cr,y} = \frac{\pi^2 E \cdot I_y}{L_{0,y}^2} = \frac{\pi^2 \cdot 210000 \cdot (14190 \cdot 10^4)}{4570^2} = 14082 \cdot 10^3 N$$

$$\bar{\lambda}_y = \sqrt{\frac{A \cdot f_y}{N_{cr}}} = \sqrt{\frac{11350 \cdot 345}{14082 \cdot 10^3}} = 0.527$$

$$\varphi = 0.5 \cdot \left[1 + \alpha(\bar{\lambda} - 0.2) + \bar{\lambda}^2\right] = 0.5 \cdot \left[1 + 0.34 \cdot (0.527 - 0.2) + 0.527^2\right] = 0.695$$

$$\chi_y = \frac{1}{\phi + \sqrt{\phi^2 - \bar{\lambda}_y^2}} = \frac{1}{0.695 + \sqrt{0.695^2 - 0.527^2}} = 0.871$$

Coefficients k_{ij} are evaluated with reference to both the approaches recommended in Appendices A and B of EC3.

Alternative Method 1 in Appendix A of EC3 (AM1)

$$\mu_y = \frac{1 - \frac{N_{Ed}}{N_{cr,y}}}{1 - \chi_y \frac{N_{Ed}}{N_{cr,y}}} = 0.992$$

It has been assumed:

$$C_{my} = C_{mi,0} = 0.79 + 0.21\psi + 0.36(\psi - 0.33)\frac{N_{Ed}}{N_{cr,i}} = 0.782$$

It results in:

$$C_{yy} = 1 + (w_y - 1)\left[\left(2 - \frac{1.6}{w_y} C_{my}^2 \left(\bar{\lambda}_{max} + \bar{\lambda}_{max}^2\right)\right) \cdot \frac{\gamma_{M1} N_{Ed}}{f_y A_i} - b_{LT}\right] \geq \frac{W_{el,y}}{W_{pl,y}}$$

where

$$\bar{\lambda}_{max} = \bar{\lambda}_y . \; w_y = \frac{W_{pl,y}}{W_{el,y}} \leq 1.5 \quad w_y = \frac{1222}{1093} = 1.118$$

$$b_{LT} = 0.5 \cdot a_{LT} \cdot \bar{\lambda}_0^2 \cdot \frac{\gamma_{M0} M_{y,Ed}}{\chi_{LT} f_y W_{pl,y}} \cdot \frac{\gamma_{M0} M_{z,Ed}}{f_y W_{pl,z}} = 0$$

$$C_{yy} = 1 + (w_y - 1)\left[\left(2 - \frac{1.6}{w_y} C_{my}^2 \left(\bar{\lambda}_{max} + \bar{\lambda}_{max}^2\right)\right) \cdot \frac{\gamma_{M1} N_{Ed}}{f_y A_i} - b_{LT}\right] \geq \frac{W_{el,y}}{W_{pl,y}} = 1.034$$

$$k_{yy} = C_{my} \frac{\mu_y}{1 - \frac{N_{Ed}}{N_{cr,y}}} \frac{1}{C_{yy}} = 0.782 \frac{0.992}{1 - \frac{889.6}{14082}} \cdot \frac{1}{1.034} = 0.799$$

Evaluation of the safety index $SI_{AM1}^{EC3-2a)}$:

$$SI_{AM1}^{EC3-2a)} = \frac{889.6 \cdot 10^3}{\frac{0.872 \cdot (11350 \cdot 345)}{1}} + 0.799 \frac{281.81 \cdot 10^6}{\frac{1222 \cdot 10^3 \cdot 345}{1}} = 0.26 + 0.54 = 0.80$$

Alternative Method 2 in Appendix B of EC3 (AM2)

$C_{my} = 0.6 + 0.4\psi \geq 0.4 \psi = 0$ due to the gradient moment distribution with a top moment equal to zero.

$$k_{yy} = C_{my}\left[1 + (\bar{\lambda}_y - 0.2)\cdot\frac{N_{Ed}}{\chi_y \cdot N_{Rk}/\gamma_{M1}}\right]$$
$$\le C_{my}\left(1 + 0.8\frac{N_{Ed}}{\chi_y \cdot N_{Rk}/\gamma_{M1}}\right)$$

It results in $k_{yy} = 0.651$

Evaluation of the safety index $SI_{AM2}^{EC3-2a)}$:

$$SI_{AM2}^{EC3-2a)} = \frac{889.6\cdot 10^3}{\frac{0.872\cdot(11350\cdot 345)}{1}} + 0.651\frac{281.81\cdot 10^6}{\frac{1222\cdot 10^3\cdot 345}{1}} = 0.26 + 0.44 = 0.70$$

EC3-2b: in accordance with this approach, the effects of global imperfection are accounted for in simplified way, increasing the first order moment via the amplification coefficient β.

There are two different but equivalent ways to estimate the amplification coefficient β:

(1) The first one with the use of the elastic critical load multiplier obtained by a FE buckling analysis:

$$\alpha_{cr,FEM} = 3.95 \qquad \beta_{FEM} = \frac{1}{1 - \frac{1}{\alpha_{cr}}} = 1.34$$

(2) The second one with the use of the elastic critical load multiplier obtained by Horne's method:

$$\alpha_{cr,H} = \frac{H}{P}\frac{h}{\Delta_{H,1st}} = \frac{(43.1 + 4.16)\cdot 4570}{889.6\cdot 52.98} = 4.583. \ \beta_H = \frac{1}{1 - \frac{1}{\alpha_{cr}}} = 1.279$$

Taking into account that method 2b should be used when second order FE analysis packages are not available, it has been decided to use the $\alpha_{cr,H}$ value estimated via Horne's approach, which corresponds to $\beta = 1.279$.

First order displacement: $\Delta_{H,1st} = 50.5$ mm(1.98 in.)
First order moment: $M_I = 215.99$ kNm(1912 kip-in.)
Second order moment at the base: $M_{II} = 276.03$ kNm(2443 kip-in.)

Method 1:

Terms χ_y and k_{yy} have been already evaluated with reference to the method 2a. Alternative Method 1 in Appendix A of EC3

Evaluation of the safety index $SI_{AM1}^{EC3-2b)}$:

$$SI_{AM1}^{EC3-2b)} = \frac{889.6\cdot 10^3}{\frac{0.872\cdot(11350\cdot 345)}{1}} + 0.799\frac{276.03\cdot 10^6}{\frac{1222\cdot 10^3\cdot 345}{1}} = 0.26 + 0.52 = 0.78$$

Alternative Method 2 in Appendix B of EC3

Evaluation of the safety index $SI_{AM2}^{EC3-2b)}$:

$$SI_{AM2}^{EC3-2b)} = \frac{889.6\cdot 10^3}{\frac{0.872\cdot(11350\cdot 345)}{1}} + 0.651\frac{276.03\cdot 10^6}{\frac{1222\cdot 10^3\cdot 345}{1}} = 0.26 + 0.43 = 0.69$$

EC3-3: in accordance with this approach, a first order analysis is required without the effects of imperfection.

$$M_I = 41.3 \times 4.57 = 196.97 \text{ kNm}(1743 \text{ kip-in.})$$

The column check will be performed with $P = 889.6$ kN and $M = 196.97$ kNm and using as effective length $L_{eff} = 2 \times 4.57 = 9.14$ m (359.8 in.).

As to the beam-column verification check, reference is made to the moment distribution presented in Figure E12.1.4.

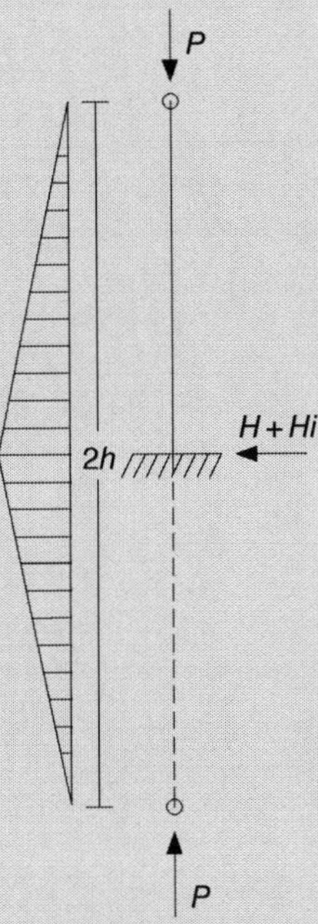

Figure E12.1.4

Evaluation of term χ_y:

$$N_{cr,y} = \frac{\pi^2 E \cdot I_y}{L_{0,y}^2} = \frac{\pi^2 \cdot 210000 \cdot (14190 \cdot 10^4)}{(2 \cdot 4570)^2} = 3520.5 \cdot 10^3 N$$

$$\bar{\lambda}_y = \sqrt{\frac{A \cdot f_y}{N_{cr}}} = \sqrt{\frac{11350 \cdot 345}{3520.5 \cdot 10^3}} = 1.055$$

$$\varphi = 0.5 \cdot \left[1 + \alpha(\bar{\lambda} - 0.2) + \bar{\lambda}^2\right] = 0.5 \cdot \left[1 + 0.34 \cdot (1.055 - 0.2) + 1.055^2\right] = 1.202$$

$$\chi_y = \frac{1}{\phi + \sqrt{\phi^2 - \bar{\lambda}_y^2}} = \frac{1}{1.202 + \sqrt{1.202^2 - 1.055^2}} = 0.563$$

Coefficients k_{ij} are evaluated with reference to both the approaches recommended in the appendices A and B of EC3.

Alternative Method 1 in Appendix A of EC3

$$\mu_y = \frac{1 - \dfrac{N_{Ed}}{N_{cr,y}}}{1 - \chi_y \dfrac{N_{Ed}}{N_{cr,y}}} = 0.871$$

Following the indication of the Code, for the case in Figure E12.1.4, it has been assumed:

$$C_{my,0} = 1 - 0.18 \frac{N_{Ed}}{N_{cr,y}} = 0.955$$

$$\bar{\lambda}_{0,\lim} = 0.2 \cdot \sqrt{C_1} \sqrt[4]{\left(1 - \frac{N_{Ed}}{N_{cr,z}}\right)\left(1 - \frac{N_{Ed}}{N_{cr,T}}\right)} = 0.232$$

It was obtained by considering the column that was not subjected to a torsional and out-of-plane buckling mode ($N_{cr,z}$ and $N_{cr,T}$ tends to infinity).

$$\bar{\lambda}_0 = \sqrt{\frac{M_{y,Rk}}{M_{cr,0}}} = 0.701 > 0.232 = \bar{\lambda}_{0,\lim}$$

$$C_{my} = C_{my,0} + \left(1 - C_{my,0}\right)\frac{\sqrt{\varepsilon_y} a_{LT}}{1 + \sqrt{\varepsilon_y} a_{LT}} = 0.955$$

$a_{LT} = 1 - \dfrac{I_t}{I_y} \geq 0$ I_t is greater than I_y, therefore, $a_{LT} = 0$

It results in:

$$C_{yy} = 1 + \left(w_y - 1\right)\left[\left(2 - \frac{1.6}{w_y} C_{my}^2 \left(\bar{\lambda}_{\max} + \bar{\lambda}_{\max}^2\right)\right) \cdot \frac{\gamma_{M1} N_{Ed}}{f_y A_i} - b_{LT}\right] \geq \frac{W_{el,y}}{W_{pl,y}}$$

$$w_y = \frac{W_{pl,y}}{W_{el,y}} \leq 1.5 \quad w_y = \frac{1222}{1093} = 1.118$$

$$b_{LT} = 0.5 \cdot a_{LT} \cdot \bar{\lambda}_0^2 \cdot \frac{\gamma_{M0} M_{y,Ed}}{\chi_{LT} f_y W_{pl,y}} \cdot \frac{\gamma_{M0} M_{z,Ed}}{f_y W_{pl,z}} = 0$$

$$C_{yy} = 1 + \left(w_y - 1\right)\left[\left(2 - \frac{1.6}{w_y} C_{my}^2 \left(\bar{\lambda}_{\max} + \bar{\lambda}_{\max}^2\right)\right) \cdot \frac{\gamma_{M1} N_{Ed}}{f_y A_i} - b_{LT}\right] \geq \frac{W_{el,y}}{W_{pl,y}} = 0.978$$

$$\mathrm{k}_{yy} = C_{my} \frac{\mu_y}{1 - \dfrac{N_{Ed}}{N_{cr,y}}} \frac{1}{C_{yy}} = 0.955 \frac{0.871}{1 - \dfrac{889.6}{3520.5}} \cdot \frac{1}{0.978} = 1.138$$

Evaluation of the safety index $SI_{AM1}^{EC3-3)}$:

$$SI_{AM1}^{EC3-3)} = \frac{889.6 \cdot 10^3}{\dfrac{0.563 \cdot (11350 \cdot 345)}{1}} + 1.138 \frac{196.97 \cdot 10^6}{\dfrac{1222 \cdot 10^3 \cdot 345}{1}} = 0.40 + 0.53 = 0.93$$

Alternative Method 2 in Appendix B of EC3

$C_{my} = 0.90 + 0.10\alpha_h$ $\alpha_h = 0$ due to the moment distribution in Figure E12.1.4.

$$k_{yy} = C_{my}\left[1 + \left(\bar{\lambda}_y - 0.2\right) \cdot \frac{N_{Ed}}{\chi_y \cdot N_{Rk}/\gamma_{M1}}\right]$$

$$\leq C_{my}\left(1 + 0.8 \frac{N_{Ed}}{\chi_y \cdot N_{Rk}/\gamma_{M1}}\right)$$

It is assumed $k_{yy} = 1.191$.

Evaluation of the safety index $SI_{AM2}^{EC3-3)}$:

$$SI_{AM2}^{EC3-3)} = \frac{889.6 \cdot 10^3}{\frac{0.563 \cdot (11350 \cdot 345)}{1}} + 1.191 \frac{196.97 \cdot 10^6}{\frac{1222 \cdot 10^3 \cdot 345}{1}} = 0.40 + 0.56 = 0.96$$

US-DAMta: Direct Analysis Method with true second order analysis
Reduction of the yield strength for compression:

$$F_{el} = \frac{\pi^2 E}{(KL/r_y)^2} = \frac{286218.53}{(15 \cdot 12/4.39)^2} = 170.2\,\text{ksi}\ (1173.8\,\text{MPa})$$

$$F_{cr} = \left[0.658^{\frac{F_y}{F_{el}}}\right] F_y = \left[0.658^{\frac{50}{170.24}}\right] 50 = 44.22\,\text{ksi}\ (304.9\,\text{MPa})$$

Notional horizontal load to take into account out-of-plumbness imperfections:

$$N_i = 0.002P = 0.002 \times 200 = 0.4 \text{ kips } (1.78 \text{ kN})$$

Total horizontal load:

$$H_{tot} = 9.7 + 0.4 = 10.1 \text{ kips } (44.9 \text{ kN})$$

Displacement at the top computed by taking into account second order effects and stiffness reduction (E = 0.8 × 29 000 ksi = 23 200 ksi (160 000 MPa)):

$$\Delta_{H,2nd} = 3.79 \text{ in. } (96.21 \text{ mm})$$

Displacement of the first order at the top:

$$\Delta_{H,1st} = 2.48 \text{ in. } (62.92 \text{ mm})$$

Second order moment at the base:

$$M_{II} = 2574 \text{ kip-in. } (290.97 \text{ kNm})$$

First order moment is:

$$M_I = 10.1 \times 15 \cdot 12 = 1818 \text{ kip-in. } (205.4 \text{ kNm})$$

The increment of the moment considering second order effects is now lower than in the previous exercise:

$$2574/1818 = 1.42$$

Evaluation of the safety index $SI^{US-DAMta}$ (with $P = 200$ kips and $M = 2574$ kip-in.) will be:

$$SI^{US-DAMta} = \frac{P}{\phi_c P_n} + \frac{8}{9}\left(\frac{M}{\phi_b M_n}\right) = \frac{200}{0.9 \cdot 778.21} + \frac{8}{9}\left(\frac{2574}{0.9 \cdot 3730}\right) = 0.28 + 0.68 = 0.96$$

The check has been done with $K = 1$.

US-DAMapp: Direct Analysis Method with approximate second order analysis
Compute:

$$R_M = 1 - 0.15\frac{P_{mf}}{P_{story}} = 1 - 0.15\frac{200}{200} = 0.85$$

$$B_2 = \frac{1}{1 - \frac{\alpha P_{story}}{P_{e,story}}} = \frac{1}{1 - 1R_M\frac{H \cdot L}{P \cdot \Delta_{H,1st}}} = \frac{1}{1 - \frac{1}{0.85\frac{10.1 \times 15 \cdot 12}{200 \times 2.48}}} = 1.473$$

Second order moment at the base:

$$M_{II} = B_2 \cdot M_I = 1.473 \times 1818 = 2677 \text{ kip-in.} (302.50 \text{ kNm})$$

Evaluation of the safety index $SI^{US-DAMapp}$ (with $P = 200$ kips and $M = 2677$ kip-in.) will be:

$$SI^{US-DAMapp} = \frac{P}{\varphi_c P_n} + \frac{8}{9}\left(\frac{M}{\varphi_b M_n}\right) = \frac{200}{0.9 \cdot 778.21} + \frac{8}{9}\left(\frac{2677}{0.9 \cdot 3730}\right) = 0.28 + 0.71 = 0.99$$

Now the approximate method for second order analysis gives a result that is also very close to that obtained with a true second order analysis.

It must be outlined that B_2, by means of the R_M coefficient, accounts for the influence of P-δ effects on P-Δ.

AISC-ELM:
The method is applicable if:

$$\frac{\Delta_{2nd-order}}{\Delta_{1st-order}} \leq 1.5 \text{ or } B_2 \leq 1.5$$

In our case:

$$\frac{\Delta_{2nd-order}}{\Delta_{1st-order}} = \frac{3.69}{2.48} = 1.49 < 1.5 \text{ and } B_2 = 1.473 < 1.5$$

So the method is applicable.

It has to be noted that the previous limitation has been evaluated with displacements obtained with direct analysis; that is, with reduced stiffness. By applying the effective length method we can use the nominal (non-reduced) stiffness. Then we compute the structure performing a second order analysis with nominal stiffness (E = 29 000 ksi (199 900 MPa)) and with P = 200 kips (889.6 kN) and H = 10.1 kips (44.9 kN). We get:

$$\Delta_{H,2nd} = 2.69 \text{ in.} (68.31 \text{ mm})$$

Displacement of the first order at the top:

$$\Delta_{H,1st} = 1.98 \text{ in.} (50.41 \text{ mm})$$

Second order moment at the base:

$$M_{II} = 2368 \text{ kip-in.} (267.64 \text{ kNm})$$

First order moment is:

$$M_{\text{I}} = 10.1 \times 15{\cdot}12 = 1818 \text{ kip-in. } (205.4 \text{ kNm})$$

The increment of the moment considering second order effects is now:

$$2368/1818 = 1.30$$

The column check shall be performed with the effective length of a cantilever ($K = 2$) and with $P = 200$ kips and $M = 2368$ kip-in. The result will be:

$$F_{el} = \frac{\pi^2 E}{(KL/r_y)^2} = \frac{286218.53}{(2{\cdot}15{\cdot}12/4.39)^2} = 42.56\,\text{ksi}\,(293.451\,\text{MPa})$$

$$F_{cr} = \left[0.658^{\frac{F_y}{F_{el}}}\right] F_y = \left[0.658^{\frac{50}{42,56}}\right] 50 = 30.59\,\text{ksi}\,(210.916\,\text{MPa})$$

Evaluation of the safety index SI^{US-ELM} (with $P = 200$ kips and $M = 2368$ kip-in.) will be:

$$SI^{US-ELM} = \frac{P}{\phi_c P_n} + \frac{8}{9}\left(\frac{M}{\phi_b M_n}\right) = \frac{220}{0.9{\cdot}538.384} + \frac{8}{9}\left(\frac{2368}{0.9{\cdot}3730}\right) = 0.41 + 0.63 = 1.04$$

AISC-FOM:

The method is applicable because $\frac{\Delta_{2nd-order}}{\Delta_{1st-order}} \leq 1.5$ and $B_2 \leq 1.5$.

With this method we perform a first order analysis but we compute horizontal notional loads as follows:

$$N_i = 2.1 \frac{\Delta_{H,1st}}{L} P = 2.1 \frac{1.98}{15{\cdot}12} 200 = 4.62\,\text{kips}\,(20.60\,\text{kN})$$

Total horizontal load:

$$H_{\text{tot}} = 9.7 + 0.4 + 4.62 = 14.72 \text{ kips } (65.47 \text{ kN})$$

First order moment:

$$M_{\text{I}} = 14.72 \times 15{\cdot}12 = 2650 \text{ kip-in. } (299.45 \text{ kNm})$$

The first order moment must be amplified by B_1 to take into account P-δ effects:

$$P_{\text{e1}} = \pi^2 \frac{EI}{(KL)^2} = \pi^2 \frac{29000 \times 341}{(1 \times 15{\cdot}12)^2} = 3012\,\text{kips}\,(13400\,\text{kN});\ C_{\text{m}} = 0.6$$

$$B_1 = \frac{C_m}{1 - \frac{P_r}{P_{\text{e1}}}} = \frac{0.6}{1 - \frac{200}{3012}} = 0.64 < 1 \text{ then } B_1 = 1$$

The safety index SI^{US-FOM} shall be performed with the nominal length ($K=1$) and with $P=200$ kips and $M=2650$ kip-in. The result will be:

$$SI^{US-FOM}=\frac{P}{\phi_c P_n}+\frac{8}{9}\left(\frac{M}{\phi_b M_n}\right)=\frac{200}{0.9\cdot 778.21}+\frac{8}{9}\left(\frac{2650}{0.9\cdot 3730}\right)=0.28+0.72=1.00$$

In Table E12.1.1 a summary of verification results is reported.

In addition to the value of the Safety Index, the design value of the bending moment (M_{Ed}) and the shear (V_{Ed}) at the base restraint are also reported together with the lateral displacement at the top end (δ_{top}).

Table E12.1.1 Summary of verification results.

Method		N_{Ed} (kN) (P_r (kips))	M_{Ed} (kNm) (M_r (kip-in.))	Safety index	V_{Ed} (kN) (V_r (kips))	δ_{top} (mm) (δ_{top} (in.))
EC3-1		889.60 (200)	335.55 (3146)	0.92	67.78 (15.25)	81.33 (3.21)
EC3-2a	AM1	889.60 (200)	281.81 (2494)	0.80	47.26 (10.63)	72.61 (2.86)
	AM2			0.70		
EC3-2b	AM1	889.60 (200)	276.03 (2443)	0.78	63.32 (14.24)	67.67 (2.66)
	AM2			0.69		
EC3-3	AM1	889.60 (200)	196.97 (1746)	0.93	43.10 (9.69)	44.09 (1.74)
	AM2			0.96		
US-DAMta		889.60 (200)	290.97 (2574)	0.96	44.90 (10.10)	96.21 (3.79)
US-DAMapp		889.60 (200)	302.50 (2677)	0.99	66.14 (14.87)	92.68 (3.65)
US-ELM		889.60 (200)	267.64 (2368)	1.04	44.90 (10.10)	68.31 (2.69)
US-FOM		889.60 (200)	299.45 (2650)	1.00	65.47 (14.73)	69.89 (2.75)

Remarks

From Table E12.1.1 it can be observed that European approaches lead to Safety Index values characterized by a great dispersion, ranging from 0.69 to 0.96. The US approaches are characterized by Safety Index values with a more limited dispersion, ranging from 0.96 to 1.04.

Finally it is worth to mention that, in general, US approaches are more severe than the European ones, leading to greater Safety Index value.

EC3-1 rigorous second order analysis with all imperfections;
EC3-2a rigorous second order analysis with global imperfections;
EC3-2b approximated second order analysis with global imperfections;
EC3-3 first order analysis;
US-DAMta AISC direct analysis method with true second order analysis;
US-DAMapp AISC direct analysis method with approximate second order analysis;
US-ELM AISC effective length method;
US FOM AISC first order analysis method.

CHAPTER 13
The Mechanical Fasteners

13.1 Introduction

Mechanical fasteners are generally realized by means of bolts, pins and rivets, which make possible the erection of the skeleton frame in a much reduced time frame, especially when compared with the one required when site welds are employed. The most common mechanical fasteners are the ones using bolts. They are generally composed of (Figure 13.1):

- a bolt, that is a metal pin with a head (usually hexagonal) and a partially or totally threaded shank (Figure 13.1a). Bolt diameter for structural applications ranges between 12 and 36 mm in accordance with European practice and between ½ and 1-½ in. in accordance with US practice;
- a nut (usually hexagonal, Figure 13.1b);
- one or more washers (usually round, Figure 13.1c), when necessary.

Where vibrations could occur and the nut might loosen, lock nuts or spring washers can be used efficiently.

As mentioned in Chapter 1, various steel components are available with different grades and steel grades for bolts; nuts and washers have to be selected in accordance with the requirements of specifications.

Basic concepts regarding the design approaches for bolted connections are presented here, leaving the discussion about the requirements associated with EU and US standards to the last sub-sections.

13.2 Resistance of the Bolted Connections

Design strength of bolted connections is usually evaluated by conventional approaches that, through suitable formulas, allow an interpretation of the actual behaviour of connections and stress distributions. In many cases, in fact, it is impossible to determine the effective distribution of stresses in the connection, due to the great variability of geometrical as well as mechanical parameters influencing connection response and hence a realistic assumption of internal forces, in equilibrium with the external forces on the connection, appears adequate in many cases for design purposes. Despite the fact that nowadays the refined capabilities of finite element analysis packages allow development

Structural Steel Design to Eurocode 3 and AISC Specifications, First Edition. Claudio Bernuzzi and Benedetto Cordova.

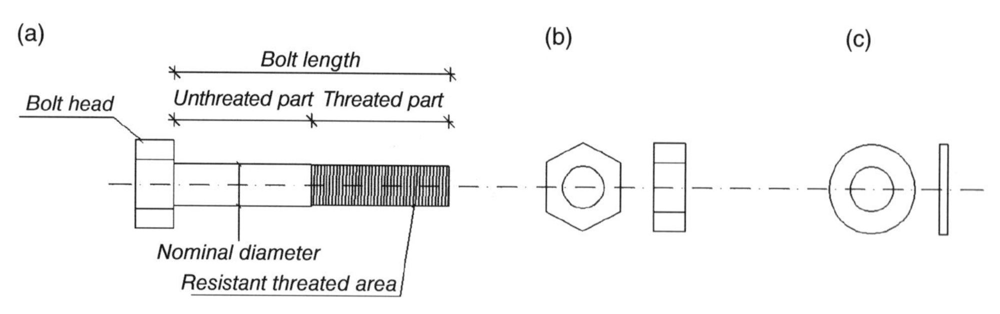

Figure 13.1 Examples of bolt (a), nut (b) and washer (c).

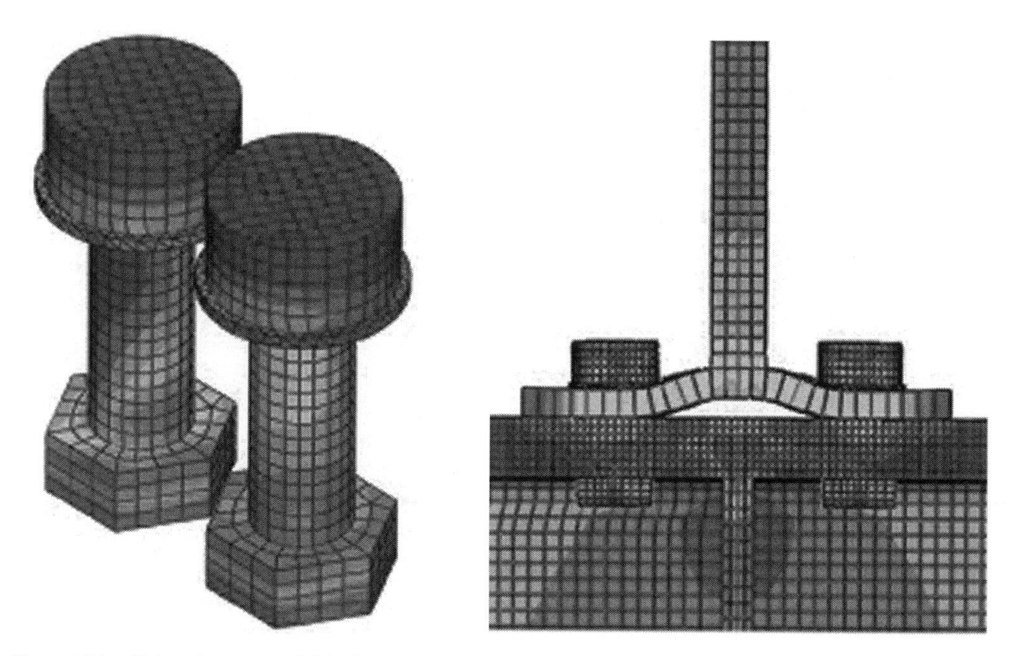

Figure 13.2 Finite element model for beam-to-column joints.

of advanced connection models to simulate response of main connection components (Figure 13.2), design of connections is generally based on simplified models that require in many cases only hand written calculations. These models are based on the first elementary principle of the limit analysis theory (the so called *static theorem* or *theorem on the lower edge of the limit loads*):

> *...any distribution of forces, where all internal forces (in this case, bolt forces) are in equilibrium with the external forces in such a way that nowhere the internal load-carrying resistance (the design resistance of the bolts) is exceeded, gives a lower bound to the design resistance of the connection.*

To use this principle on the safe side, brittle and buckling phenomena have to be avoided and the geometry of the connection must fulfil special requirements for the newly introduced approaches to simple unions and based on simplifying assumptions (e.g. 'static equal commitment of each bolt').

The distribution of forces in the connection may, hence, be arbitrarily determined in whatever rational way is best, provided that:

- the assumed internal forces are balanced with the applied design forces and moments;
- each part of the connection is able to resist the applied forces and moments;
- the deformations imposed by the chosen distribution are within the deformation capacity of the fasteners, welds and the other key parts of the connection.

Each bolt it usually modelled as point mass, making reference to its centroid. A uniform stress distribution is taken along the holes and both the deformation of the plates as well as the stress concentration in correspondence of the holes, due to pluri-axial state of stresses, are usually neglected.

Connections can be classified on the basis of the acting loads as follows:

- connections in shear;
- connections in tension;
- connections simultaneously in tension and shear.

13.2.1 Connections in Shear

A connection is affected by shear when the plates connected via bolts are loaded by forces parallel to the contact planes. The basic case presented in Figure 13.3 is related to a connection subjected to an external force F_v, which is applied to one plate and is transmitted to the other two plates through one bolt connecting all three plates together. The bolt can be considered to be a simply supported beam loaded at its midspan. External plates are considered as support restraints while the central one loads the structure. Two shear planes can be distinguished, each of them associated with the common surface of two contiguous plates.

Different responses are expected, depending on two different modes to transfer the shear load, which make possible the distinction between bearing connections and slip-resistant connections.

13.2.1.1 Bearing Connections

It is required that the plates must be connected to each other achieving a firm contact and no tightening of the bolt is required. With reference to Figure 13.3, it should be noted that when the load increases, the bolt shank comes into contact with the surface of the hole plates, causing the spread of plasticity in the contact zone due to the hole diameter greater than the one of the bolt. When increasing the load, the extension of the part of the plate in contact with the bolt shank in plastic range increases too. From the design point of view, these effects are neglected, plasticity being located in a very limited portion of the connection, without any influence on the overall connection performances.

In the pre-sizing phase, or when using the allowable stress design approach, the effects of the forces transferred between bolts and plates can be appraised directly with reference to the tangential stress (τ) acting on each shear plane, which is evaluated on the basis of the effective bolt area of the shear plane. The acting force can be transferred through the unthreaded area (A) or the threaded area (A_{res}) and hence τ can be expressed as:

$$\tau = \frac{V}{n \cdot A_{res}} \tag{13.1a}$$

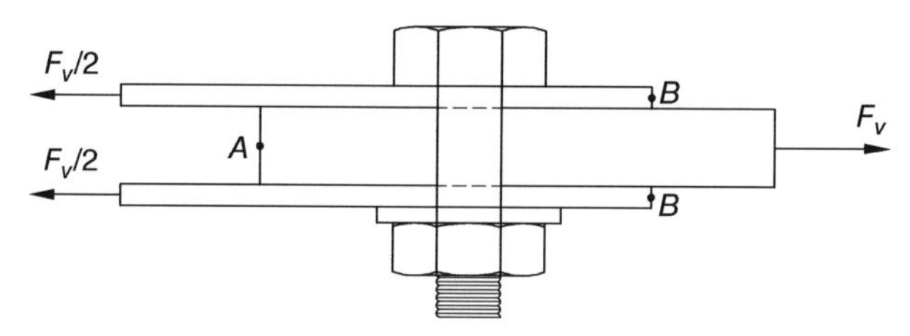

Figure 13.3 Typical connection in shear.

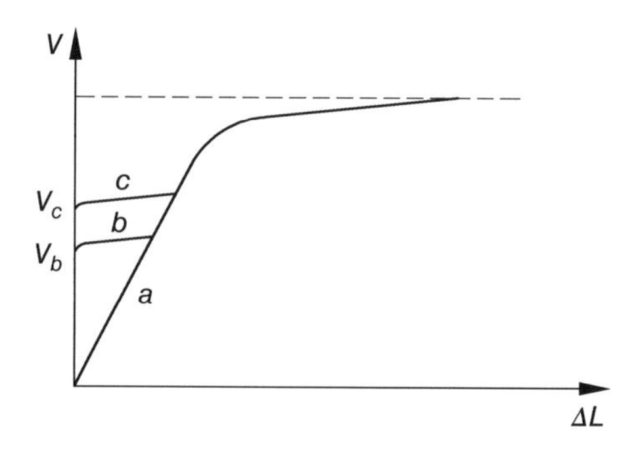

Figure 13.4 Influence of degree of tightening on the behaviour of bolted joints: relationship between the applied load and the relative displacement of plates in Figure 13.3.

$$\tau = \frac{V}{n \cdot A} \tag{13.1b}$$

where V is the total shear force on the bolt and n is the number of shear planes.

Curve (a) in Figure 13.4 presents the connection response in terms of the relationship between the applied shear force, V, versus the relative displacement, ΔL, between points A and B - (Figure 13.3) in two adjacent connected plates. The response is linear in the first branch up to when the yielding of components is achieved. Failure of the shear connection can be due to one of the following mechanisms:

- bolt failure (Figure 13.5a)
- plate bearing (Figure 13.5b)
- tension failure of the plate (Figure 13.5c)
- shear failure of the plate (Figure 13.5d).

In Figure 13.6 typical bearing failure is presented. Due to the difficulty of correctly evaluating the actual bearing pressure distribution, a conventional value is assumed. In particular, bearing pressure between bolt and plate can be approximated with reference to the mean value of the bearing stress, σ_{bear}:

$$\sigma_{\text{bear}} = \frac{V}{t \cdot d} \tag{13.2}$$

where V is the acting shear force per shear plane, t is the minimum thickness of connected plates and d is the bolt diameter.

As to the conventional bearing resistance, a design value is considered, which is based on strength of the plate suitably increased to account for the benefits associated with the complex spatial stress distribution along the hole.

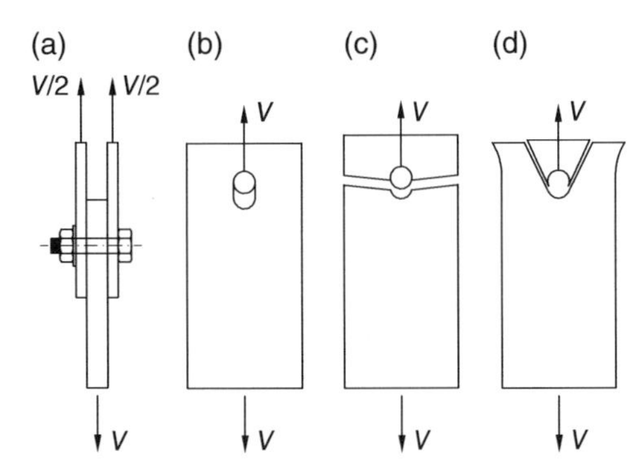

Figure 13.5 Typical failure modes for a shear connection: bolt failure (a), plate bearing (b), tension failure of the plate (c) and shear failure of the plate (d).

Figure 13.6 Typical deformation holes due to a bearing.

13.2.1.2 Slip Resistant Connection or Connection with Pre-Loaded Joints

Pre-loading of bolts can be explicitly required for slip resistance, seismic connections, fatigue resistance, execution purposes or as a quality measure (e.g. for durability). Tightening involves the application of a twisting moment to the bolt before external forces are applied. This produces connected plate shortening and bolt shank elongation. Only a portion of the twisting moment applied to tighten the bolt is absorbed by friction between plate and bolt on one side, and plate and nut on the other side of the mechanical fastener. The remaining twisting moment is carried by the bolt shank. Thus, once the bolt is tightened, the joint is loaded by self-balanced stresses associated with the bolt in tension and the compression in the plates and with the torsion of the bolt and plate/bolt friction. Tightening increases joint performance, mainly with reference to serviceability limit states. Furthermore, it should be noted that:

- in shear joints, tightening prevents plate slippage and, therefore, inelastic settlements in the structure;

- in tension joints, tightening prevents plate separation (reducing corrosion dangers) and significantly improves fatigue resistance. However, tightening must not exceed a certain limit, to avoid attaining joint ultimate capacity.

As to curves (b) and (c) in Figure 13.4, four different branches can be identified:

(1) The load increases from zero but no relative displacement is observed; force transmission is due to friction between the plates until friction limit of the joint is reached, which depends on the degree of preload. Curve (c) is related to a connection with a pre-load degree greater than the one of case (b);
(2) For a level of shear equal to the friction limit, slippage occurs suddenly due to the bolt-hole clearance and it stops when the shank bolt is in contact with the plate holes. During this phase, the applied load is practically constant, coincident with the friction limit load;
(3) After the contact, the V-ΔL relationship coincides with the one of the bearing connection (curve (a)). The response is practically linear until the elastic limit of the connection is reached either in the connected plates or in the bolt;
(4) Finally, in plastic range, a significant deformation takes place for moderate load increments until the connection failure load is achieved.

Slip-resistant joints are required when inelastic settlements of connections must be avoided in order to reduce the deformability of the structure or to fulfil some functional requirements. The value of the force at which slippage occurs depends upon bolt tightening, surface treatment and number of surfaces in contact (n_f). With reference to the case of a connection with a single bolt, the maximum value of the force transferred by friction, F_{Lim}, can be estimated as:

$$F_{\mathrm{Lim}} = n_f \cdot \mu \cdot N_s \tag{13.3}$$

where μ is the friction coefficient.

As far as the tightening of the bolts is concerned, the following methods are commonly used:

- *torque method*: bolts are tightened using a torque wrench offering a suitable operating range. Hand or power operated wrenches may be used. Impact wrenches may be used for the first step of tightening for each bolt. The tightening torque has to be applied continuously and smoothly;
- *combined method*: bolts are tightened using the torque method until a significant degree of preload is reached and then a specified part turn is applied to the turned part of the assembly.
- *HRC tightening method*: this method is used with special bolts, named HRC bolts (Figure 13.7). They are tightened via specific shear wrench equipped with two co-axial sockets, which react by torque one against the other: the outer socket, which engages the nut, rotates clockwise while the inner socket, which engages the spline end of the bolt, rotates counter-clockwise.
- *direct tension indicator (DTI)*: this method requires the use of special compressible washers (Figure 13.8), such as DTIs, which indicate that the required minimum preload has been achieved, monitoring the force in the bolt.

When several bolts are placed in a row, as indicated in Figure 13.9a, and assuming elastic behaviour, an uneven distribution of forces occurs. This distribution can easily be found when two extreme situations are considered. Assuming infinitely stiff bolts and weak plates, all the bolts remain undeformed and parallel to each other. Each piece of plate between two contiguous bolts therefore has the same length, the same strain and, consequently, also the same stress. It can be noted from Figure 13.9b that the forces in the plates between bolt 1 and bolt 2 are: 0.5 F, 1.0 F and

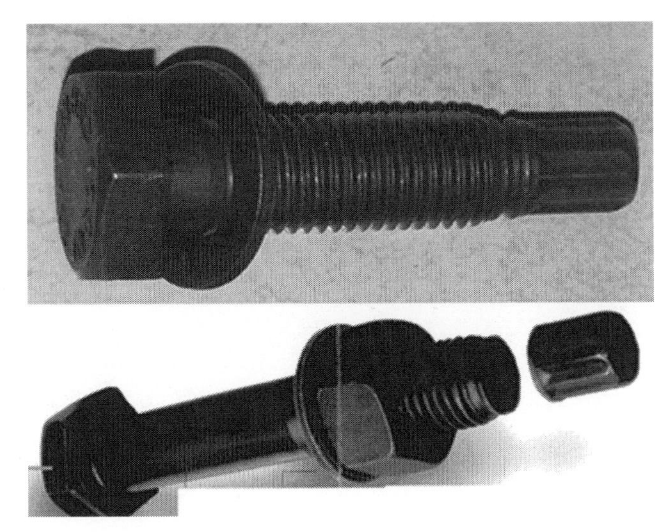

Figure 13.7 Example of an HRC bolt.

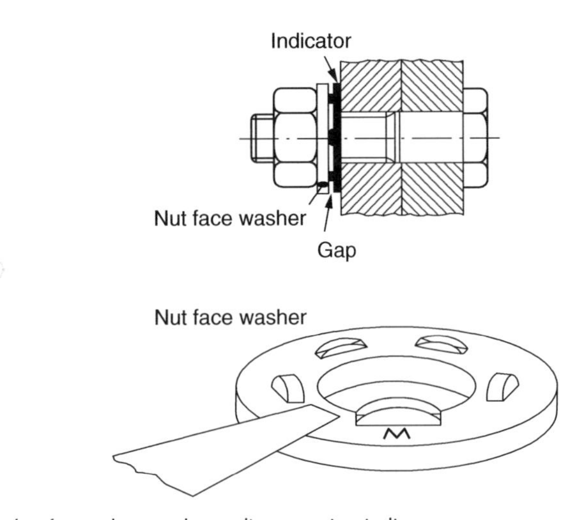

Figure 13.8 Example of a washer used as a direct tension indicator.

0.5 F. But this also applies to the plates between bolts 2 and 3 and between bolts 3 and 4. As a conclusion based on the equilibrium condition, bolts 1 and 4 transmit the full load F while the other bolts are unloaded, Otherwise, by considering infinitely stiff plates and weak bolts, plates between the bolts do not deform. Every bolt has the same deformation and therefore is loaded to the same extent. As appears to be the case from Figure 13.9c, every bolt carries 0.5 F, that is 0.25 F per shear area. The effective distribution of forces in routine design cases is between these two extremes.

The difference between the forces in the outer bolts and the inner bolts is greater when the stiffness of the plates is low. This situation occurs generally when the connection is longer (several bolts) and the plate thickness is quite small compared to the bolt diameter.

The part of the connection between the outer bolts must be designed to be as short and stiff as possible in order to minimize the differences between the values of the force in each bolt. In practice, however, it is normally admitted to assume an even distribution of forces, owing to the plastic

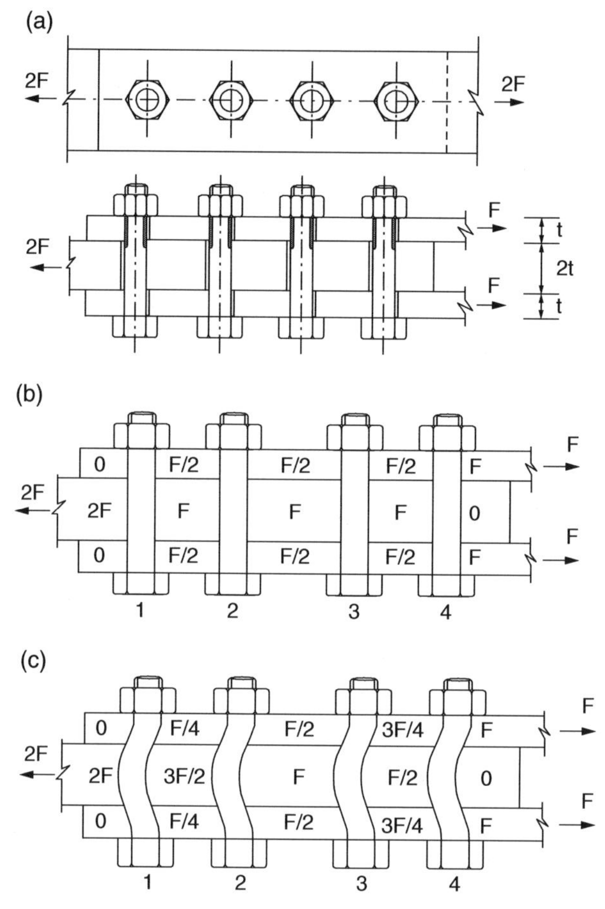

Figure 13.9 Bolted connection with four bolts (a): bolt forces per shear area in the case of stiff bolts and weak plates and (b) in the cases of weak bolts and stiff plates (c).

deformation capacity of the bolts and plates and to the force redistribution occurring between bolts.

As the deformation capacity of plates is generally much higher than the deformation capacity of the bolts, it is strongly recommended to design the connection such that yielding of the plates in bearing occurs before yielding of the bolts in shear, in order to guarantee a ductile failure rather than a brittle failure.

With reference to the resistance of the plate, as previously mentioned, stress distribution in the correspondence of the hole is non-uniform, with higher stress values in correspondence of the hole. Furthermore, plastic redistribution at failure occurs with a uniform stress distribution and this justifies the use in design of a mean value of stress, assumed for sake of simplicity constant in elastic range (Figure 13.10) and conventionally considered equal to:

$$\sigma = \frac{V}{A_n} \tag{13.4}$$

where V is the shear force and A_n the net area of the cross-section of the plate, that is the gross area reduced for the presence of the hole.

In connections with more than one bolt, a correct evaluation of the resistant area for the plates could become complex, depending on the ultimate load for tension and shear as a function of the possible failure path (Figure 13.11). Following an empirical rule, from the safe side, the resistant area can be considered to be the one corresponding to the shortest path passing through one or more holes (Figure 13.11): the main rules for estimating an appropriate value of the reduced area have already been introduced for tension member verification (Chapter 6).

To minimize the weakness of cross-section for the presence of holes, it is possible to increase the number of the holes from the end to the centre of the connection, as shown in Figure 13.12. It is worth noting that this causes an increase in the dimension of the joint.

As it happens in some practical cases discussed in Chapter 15 dealing with joints, the design load F_v can be eccentric with reference to the centroid of the fasteners, the result of this is the connection is subject to shear and torsion (Figure 13.13). If e identifies the value of the force eccentricity, the connection is subjected to a torsional moment $T = F_v \cdot e$. Using the superimposition principle, bolt design can be conventionally based on the evaluation of the shear force acting on the bolt associated with the shear (F_v) force and the torsion moment (T), identified as V and $V_{T,I}$, respectively.

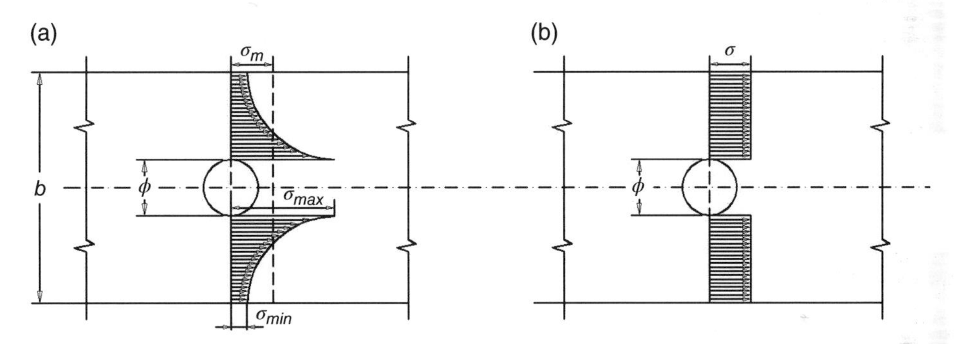

Figure 13.10 Distribution of the stress in the plate of a bearing connection in elastic (a) and plastic (b) range.

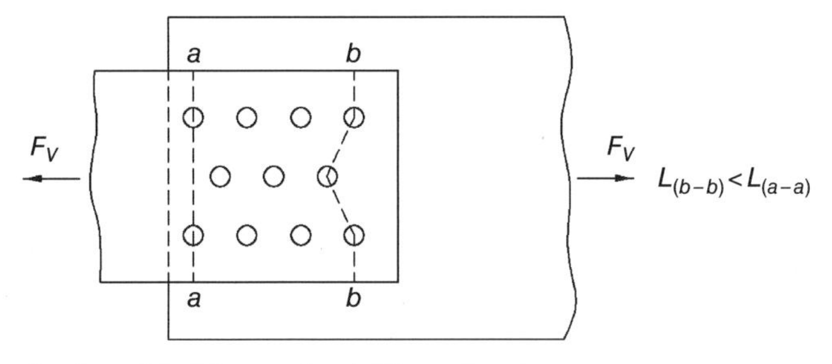

Figure 13.11 Example of possible failure paths of different lengths.

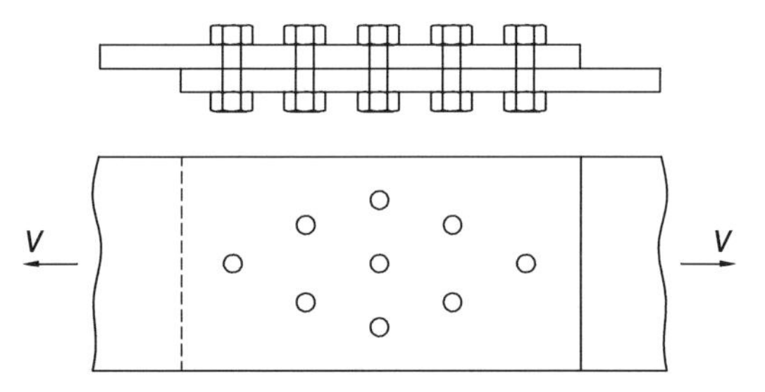

Figure 13.12 Bearing connection with a different number of bolts per line.

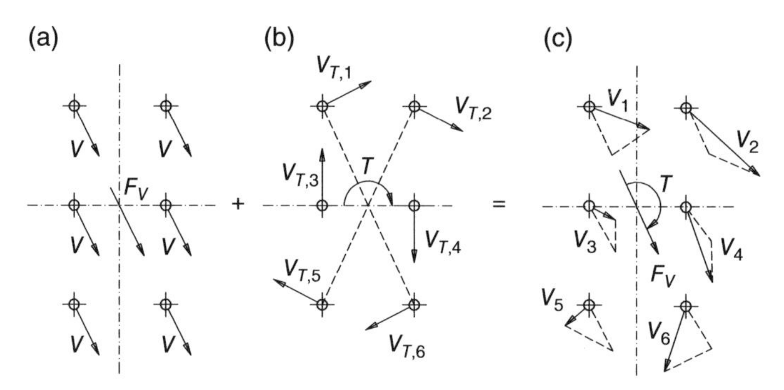

Figure 13.13 Typical force distribution due to a bearing connection with eccentric shear.

The actual response of this connection is quite difficult to be predicted. The position of the instantaneous centre of rotation of the joint is not constant. As the applied loads increase, the irregular distribution of friction forces, the material elastic behaviour and the hole/bolt clearance modify its position. Furthermore, considering the force redistribution from the most to the less loaded bolt due to local plasticity around the holes, it is preferable to use a simplified approach, which assumes, on the safe side, infinitely stiff plates and perfectly elastic bolts.

Shear force, F_v, is equally divided between the bolts in the same direction (Figure 13.13a) and the shear force V per each shear plane of each bolt can be evaluated as:

$$V = \frac{F_V}{n_f \cdot n} \tag{13.5}$$

where n_f is the number of shear resisting plane per bolt and n is the number of the bolts.

The torsional moment is balanced by shear forces acting on the bolts normal to the line joining the bolt to the centre of gravity and proportionally to the distance from the bolt centroid to the centre of gravity of all the bolts (Figure 13.13b). With reference to the generic i-bolt, the shear force $V_{T,i}$ per shear plane can be evaluated as:

$$V_{T,i} = \frac{(F_V \cdot e) \cdot a_i}{n_f \cdot \sum_{i=1}^{n} a_i^2} \tag{13.6}$$

where a_i is the distance between the centroid of all the bolts and that of the single i-bolt.

The value of the resulting force on each bolt can be obtained via a vectorial sum of the contributes V and $V_{T,i}$ (Figure 13.13c) and, for design purposes, reference has to be made to the maximum force value. In case of only one bolt row, contribution V is perpendicular to $V_{T,i}$ and the resulting force V_i is obtained as:

$$V_i = \sqrt{V^2 + V_{T,i}^2} \tag{13.7}$$

13.2.2 Connections in Tension

Tension occurs when the plates connected via bolts are loaded by a force normal to the contact plane; that is parallel to the bolt axis. As in case of bearing connection, the response of a connection in tension is quite difficult to predict. In order to analyse this problem from a qualitative point of view, a brief discussion is proposed with reference to the very simple tension connection in Figure 13.14. If the flange is sufficiently stiff, its deformation can be disregarded and the bolts can be assumed to be in pure tension (case *a*) and the failure of the joint is expected to be due to failure of the bolts. Otherwise, if the flange is more flexible, the presence of prying forces, Q, depending on the stiffness of both flanges and bolts as well as on the applied load, increases the value of the axial load transferred via bolts. Connection failure may be due to bolts, flange or to both components.

In order to better appraise the tightening effects, reference can be made to the response of the tension connection presented in Figure 13.15a, which is realized by one bolt. Figure 13.15b presents two typical relationships between the applied external load N to the connection and bolt elongation ΔL, which are related to the case of non-tightened bolt (curve *a*) and tightened bolt (curve *b*). In addition, the applied external load N is plotted versus the axial force acting in the bolt shank N_b. in Figure 13.15c. It can be noted that:

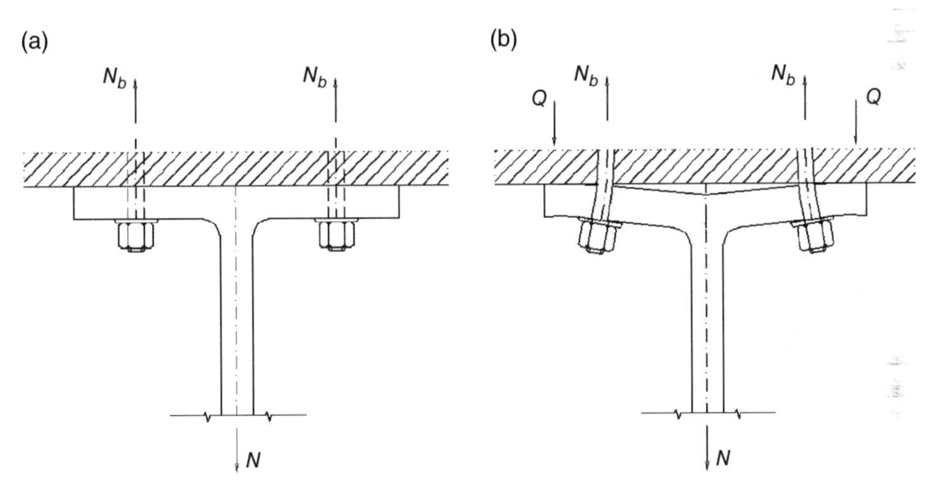

Figure 13.14 Influence of the stiffness of plate and bolts of the force transfer mechanism: stiff plate and flexible bolts (a) or flexible plate and stiff bolts (b).

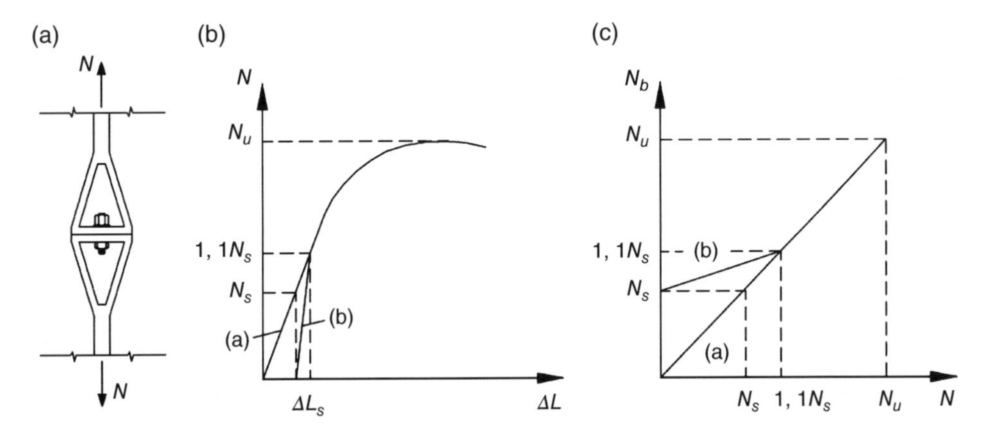

Figure 13.15 Connection in tension (a): relationship (b) between the applied tension force (N) and the bolt shank elongation (ΔL) and relationship (c) between the applied tension force (N) and the axial load on the bolt shank (N_b).

- in case of non-tightened bolt (curve *a*), as N increases, N_b increases too by an equal amount (Figure 13.15c) and once the elastic limit has been reached, the response is in the plastic range until failure is achieved at a load level equal to N_u.
- in case of tightened bolt (curve *b*), in absence of external force, the tightened tensile force N_s causes a shank elongation ΔL_s. As the external load N is applied and increases, tension force N_b in the shank increases very slowly (Figure 13.15b), due to the fact that N is mainly transferred by decompression of plates. For a value of N slightly greater than N_S (approximately, 1.1 N_S), the separation of the plates occurs, load is transferred via the bolt and the ultimate load N_u is independent on tightening effects.

In case of tension force applied on the centroid of the bolts, it is assumed that the design load is balanced by forces equal on each bolt. Otherwise, if a bending moment also acts, the evaluation of the bolt forces is usually based on the assumption of stiff plate.

As an example, the connection in Figure 13.16 can be considered, which is composed of two equal leg angles bolted to the web of a stocky cantilever and to the column flange of H-shaped profiles. Angle legs on the plane *a* are subjected to shear force and torsion moment and, as a consequence, the shear force on the bolts can be evaluated via the approach already discussed for the case of shear force eccentric with respect to the centroid of the bolts (Figure 13.13). Angle legs on plane *b* are subjected to shear force and bending moment, and the bolts are subjected to shear and tension force. The behaviour of this cross-section may be considered similar to that of a typical reinforced concrete cross-section: in this case, tension is absorbed by the bolts and compression by the contact pressure between the angle legs and the column flange. Assumptions similar to the one related to the allowable stress design approach for concrete structures can be adopted in this case and, in particular, linear elastic behaviour of the materials, bolts not resisting compression, the plate not resisting tension and full planarity of the cross-sections.

To evaluate x, which identifies the distance between the neutral axis and the leg bottom (i.e. the zone of the leg where the maximum stress, $\sigma_{\max}$, acts), the equilibrium translation condition can be imposed. The resultant force of the compression stresses on the plate is balanced by the resultant force of the tension bolt forces. Considering the previously mentioned hypotheses, it can be deduced:

$$\frac{1}{2}\cdot\left[(2\cdot B)\cdot x^2\right]=\sum_{i=1}^{n}A_{bi}\cdot(y_i-x) \tag{13.8a}$$

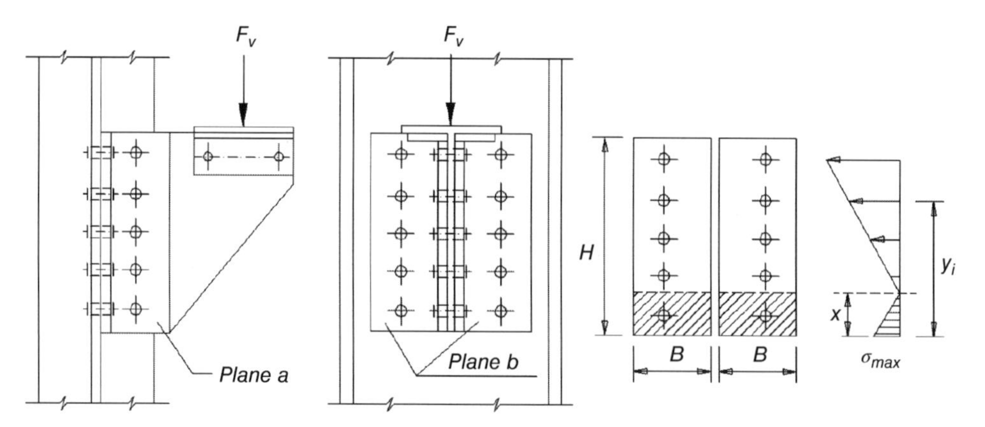

Figure 13.16 Example of connection under shear and torsion (plane *a*) and shear and bending (plane *b*).

where y_i is the distance between the centroid of the bolt (having area A_{bi}) and the bottom of the leg and B is the width of the single leg.

Variable x can be obtained by solving the correspondent quadratic equation:

$$B \cdot x^2 + \sum_{i=1}^{n} (A_{bi} \cdot x) - \sum_{i=1}^{n} (A_{bi} \cdot y_i) = 0 \tag{13.8b}$$

The solution of practical interest is:

$$x = \frac{1}{2 \cdot B} \cdot \left[-\sum_{i=1}^{n} A_{bi} + \sqrt{\left(\sum_{i=1}^{n} A_{bi} \right)^2 + (4 \cdot B) \sum_{i=1}^{n} (A_{bi} \cdot y_i)} \right] \tag{13.9}$$

If bolts are located in the leg zone under compression (for the i-bolt, it occurs if $x > y_i$) the position of the neutral axis has to be re-evaluated by excluding the contribution of the bolts in the compression zone. On the basis of the computed value of x, internal forces and maximum stresses can be directly evaluated. At first, it is necessary to evaluate the moment of inertia of the effective resisting cross-section, which is given by:

$$J = \frac{(2 \cdot B) \cdot x^3}{3} + \sum_{i=k+1}^{n} A_{bi} \cdot (y_i - x)^2 \tag{13.10}$$

Maximum stress in the compression zone is:

$$\sigma_{max} = \frac{M \cdot x}{J} \tag{13.11a}$$

Maximum tension force on the bolts is given by:

$$N_{max} = \frac{M \cdot A_i \cdot (y_{max} - x)}{J} \tag{13.11b}$$

As clearly already stated when discussing the limit state approaches to design connections, there are different possibilities to evaluate internal forces and stresses. Other criteria can hence be followed to evaluate an equilibrated force distribution. As an example, all the bolts should be considered to be in tension and hence the neutral axis coincides with the bottom of the leg ($x = 0$). As a consequence, the maximum tension bolt force, obtained from Eq. (13.11b) considering $x = 0$, is:

$$N_{max} = \frac{M \cdot A_{bi} \cdot y_{max}}{\sum_{j=k+1}^{n} A_{bj} \cdot y_j^2} \tag{13.12}$$

In a similar way, it should be possible to avoid the direct calculation of the position of neutral axis by putting arbitrarily $x = H/6$, where H is the leg depth. Maximum compressive stress at the bottom of the leg is:

$$\sigma_{max} = \frac{M \cdot \frac{H}{6}}{I} = \frac{M \cdot \frac{H}{6}}{\frac{B \cdot H^3}{324} + \sum_{i=k+1}^{n} A_{bi} \cdot \left(y_i - \frac{H}{6}\right)^2} \tag{13.13a}$$

Maximum tensile force on the bolt is:

$$N_{max} = \frac{M \cdot A_{bi} \cdot \left(y_{max} - \frac{H}{6}\right)}{\frac{B \cdot H^3}{324} + \sum_{j=k+1}^{n} A_{bj}\left(y_j - \frac{H}{6}\right)^2} \tag{13.13b}$$

13.2.3 Connection in Shear and Tension

The approaches previously introduced for the case of sole shear force and sole tension force on the connection can be combined each other in order to be used for the more general case of shear and tension. In many practical cases, tension and shear act simultaneously on bolts, as it occurs also on the bolts of the plane b in Figure 13.16. According to the various limit states, different interaction formulas have been defined that can be applied to assess bolt strength.

In case of pre-loaded bolts, slippage load is reduced by the presence of axial load. As to the ultimate resistance, a simplified domain is used for the design under combined axial tension and shear on the shank. More details about the requirements for verification are presented in the following parts, in accordance with European and United States design provisions.

13.3 Design in Accordance with European Practice

The main criteria associated with assembly techniques are presented first. A short description of the structural verification criteria in accordance with European practice is given afterwards.

13.3.1 European Practice for Fastener Assemblages

Particular care should be paid to assembly techniques of site bolted connections. To this aim, it should be noted that Chapter 8 (*Mechanical fastenings*) of the EN 1090-2 (*Execution of steel structures and aluminium structures – Part 2: Technical requirements for steel structures*) deals with mechanical fasteners, giving important and useful information, some of them herein summarized.

13.3.1.1 Bolts

As mentioned in the Introduction in Section 13.1, the minimum nominal fastener diameter used for structural bolting is 12 mm (M12 bolt), except for thin gauge components and sheeting. Bolt length has to be chosen such that after tightening appropriate requirements are met for bolt end protrusion beyond the nut face and the thread length. Furthermore, length of protrusion is required to be at least the length of one thread pitch measured from the outer face of the nut to the end of the bolt.

- For non-preloaded bolts, at least one full thread (in addition to the thread run out) are required to remain clear between the bearing surface of the nut and the unthreaded part of the shank.
- For preloaded bolts according to EN 14399-3 and EN 14399-7, at least four full threads (in addition to the thread run out) have to remain clear between the bearing surface of the nut

and the unthreaded part of the shank. For preloaded bolts according to EN 14399-4 and EN 14399-8, clamp lengths have to be in accordance with those specified in Table A.1 of EN 14399-4.

13.3.1.2 Nuts

It is required that nuts run freely on their partnering bolt, which is easily checked during hand assembly. Any nut and bolt assembly where the nut does not run freely has to be discarded. If a power tool is used, either of the following two checks may be used:

- for each new batch of nuts or bolts, their compatibility may be checked by hand assembly before installation;
- for mounted bolt assemblies but prior to tightening, sample nuts may be checked for free-running by hand after initial loosening.

Nuts have to be assembled so that their designation markings are visible for inspection afterwards.

13.3.1.3 Washers

Washers are not required when non-preloaded bolts are used in normal round holes, but recommended anyway to avoid damage to steel painting. If used, it must be specified as to whether washers must be placed under the nut or the bolt head (whichever is rotated) or both. For single lap connections with only one bolt row, washers are required under both bolt head and the nut. Washers used under heads of preloaded bolts have to be chamfered according to EN 14399-6 and positioned with the chamfer towards the bolt head. Washers according to EN 14399-5 have to be used only under nuts. Plain washers (or, if necessary, hardened taper washers) have to be used for preloaded bolts as follows:

- for 8.8 bolts, a washer has to be used under the bolt head or the nut, whichever is to be rotated;
- for 10.9 bolts, washers have to be used under both the bolt head and the nut.

Plate washers have to be used for connections with long slotted and oversized holes. One additional plate washer or up to three washers with a maximum combined thickness of 12 mm may be used in order to adjust the grip length of bolt assemblies, to be placed on the side that is not turned. Dimensions and steel grades of plate washers have to be clearly specified (never thinner than 4 mm).

The EN 1090-2 standard also contains detailed guidance on tightening systems for high strength bolts, which is briefly summarized in the following. Prior to assembly, for any mechanical fasteners it is strongly recommended to free the contact surfaces from all contaminants, such as oil, dirt or paint. Burrs have to be removed preventing solid seating of the connected parts. Furthermore, it is important to guarantee that uncoated surfaces have to be freed from all films of rust and other loose material. Particular care is required in order not to damage or smooth the roughened surface. Areas around the perimeter of the tightened connection have to be left untreated until any inspection of the connection has been completed.

The connected components have to be drawn together in order to achieve firm contact, eventually using shims to adjust the fit. Each bolt assembly has to be brought at least to a snug-tight condition, which can generally be taken as the one achievable by the effort of one man using a normal sized spanner without an extension arm and can be set as the point at which a percussion wrench starts hammering.

The tightening process has to be carried out from bolt to bolt of the group, starting from the most rigid part of the connection and moving progressively towards the least rigid part. In order to achieve a uniform snug-tight condition, more than one cycle of tightening may be necessary. As an example, it is worth mentioning that the most rigid part of a cover plate connection of an I-shaped section is commonly in the middle of the connection bolt group and the most rigid parts of end plate connections of I-shaped cross-sections are usually beside the flanges.

The bolt has to protrude from the face of the nut after tightening not less than one full thread pitch.

Torque wrenches used in all steps of the torque method have to guarantee accuracy no greater than ±4% according to EN ISO 6789 (*Assembly tools for screws and nuts. Hand torque tools – Requirements and test methods for design conformance testing, quality conformance testing and recalibration procedure*). Each wrench has to be checked for accuracy at least weekly and, in case of pneumatic wrenches, every time the hose length is changed. For torque wrenches used in the first step of the combined method these requirements are modified to ±10% for the accuracy and yearly for the periodicity.

The following methods are considered for tightening the bolts:

(1) *Torque method*: The bolts have to be tightened using a torque wrench offering a suitable operating range. Hand or power operated wrenches may be used as well as impact wrenches for the first step of tightening of each bolt. The tightening torque has to be applied continuously and smoothly. Tightening by the torque method comprises the two following steps at least:
   (a) the wrench has to be set to a torque value of about 0.75 of the torque reference values. This first step has to be completed for all bolts in one connection prior to commencement of the second step;
   (b) the wrench has to be set to a torque value of 1.10 of the torque reference values.

(2) *Combined method*: Tightening by the combined method is applied in two subsequent steps:
   (a) a torque wrench offering a suitable operating range has to be used. The wrench has to be set to a torque value of about 0.75 torque reference values. This first step has to be completed for all bolts of one connection before the commencement of the second step.
   (b) a specified part turn is applied to the turned part of the assembly. The position of the nut relative to the bolt threads has to be marked after the first step, using a marking crayon or marking paint, so that the final rotation of the nut relative to the thread in this second step can be easily determined. The second step has to be executed in accordance with the values indicated in Table 13.1.

In Tables 13.2a and 13.2b the minimum free space for tightening hexagonal screws, bolts and nuts are reported for single-head wrench and slugging wrench, respectively. Symbols used in the tables are presented in Figure 13.17.

Table 13.1 Additional rotation for the combined method (8.8 and 10.9 bolts).

	Rotation value to apply for the second step (II) of tightening	
	Degrees	**Part turn**
$t < 2d$	60	1/6
$2d \leq t < 6d$	90	1/4
$6d \leq t \leq 10d$	120	1/3

t = total nominal thickness of the parts to be connected, including all packs and washers and d = diameter of the bolt.

Table 13.2a Minimum free space for tightening hexagonal screws, bolts and nuts (mm) for engineer's wrench (single-head) and box wrench (single-head).

		Engineer's wrench (single-head)		Box wrench (single-head)	
Bolt diameter	*S*	*f*	*g*	*h*	*K*
M12	22	23.5	45	18.25	35
M14	24	25	48	19.75	38
M16	27	28	55	21.75	42
M18	30	30	60	23.75	46
M20	32	31.5	62.5	25.25	49
M22	36	37	73	28.25	55
M24	41	41.5	82.5	32.25	63
M27	46	45	90	36.25	71
M30	50	47	96.5	39.25	77
M36	60	51.5	109.5	48	93

All values are in millimetres.

Table 13.2b Minimum free space for tightening hexagonal screws, bolts and nuts (mm) for slugging wrench (open end) and slugging wrench (box).

		Slugging wrench (open end)		Slugging wrench (box)	
Bolts diameter	*S*	*f*	*g*	*h*	*K*
M16	27	32	58	24.5	47
M18	30	32	60	27	52
M20	32	34	64	28	54
M22	36	37	70	31	60
M24	41	41	80	34	66
M27	46	46	89	38.5	75
M30	50	51	98	41	80
M36	60	60	116	48	94

All values are in millimetres.

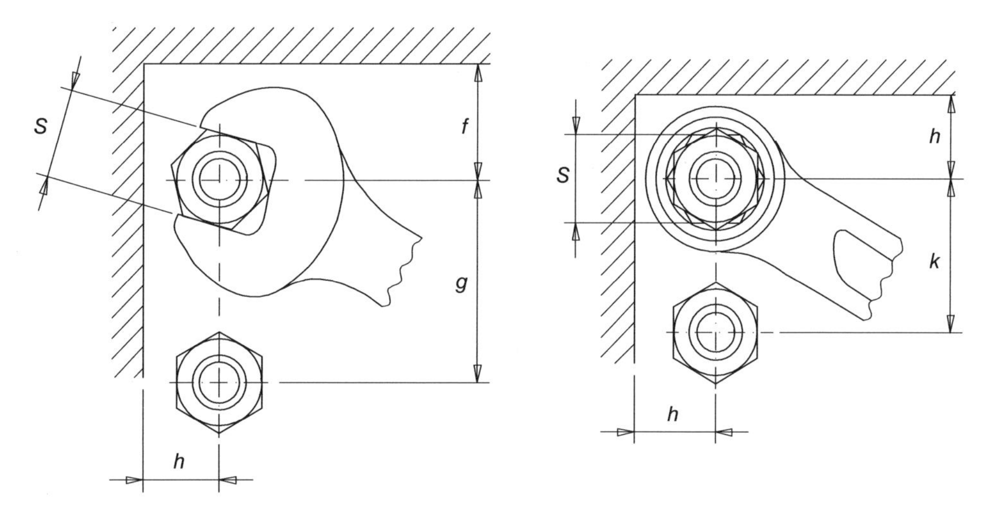

Figure 13.17 Symbols used to define the minimum free space.

13.3.1.4 Clearances for Bolts and Pins

The definition of the nominal hole diameter combined with the nominal diameter of the bolt to be used in the hole determines whether the hole is 'normal' or 'oversize'. The terms 'short' and 'long' applied to slotted holes refer to the two types of holes used for the structural design of preloaded bolts. These terms may be used also to designate clearances for non-preloaded bolts. Special dimensions should be specified for movement joints. No indications are given in EN 1993-1-8 on the nominal clearance for bolts and pins, which are reported in Table 13.3 and are derived directly from EN 1090-2, where this topic is dealt with.

As to the positioning of the holes for fasteners, minimum and maximum spacing and end and edge distances for bolts and rivets are given in Table 13.4, which refers to the symbols presented in

Table 13.3 Nominal clearances for bolts and pins (values in millimetres).

	Nominal bolt or pin diameter d (mm)							
Type of holes	**12**	**14**	**16**	**18**	**20**	**22**	**24**	**≥27**
Normal round holes	1		2					3
Oversize round holes	3		4				6	8
Short slotted holes (on the length)	4		6				8	10
Long slotted holes (on the length)	1.5 d							

Table 13.4 Minimum and maximum spacing, end and edge distances (using millimetres).

		Maximum		
		Structures made from steels conforming to EN 10025 except steels conforming to EN 10025-5		**Structures made from steels conforming to EN 10025-5**
Distances and spacings (Figure 13.12)	**Minimum**	**Steel exposed to the weather or other corrosive influences**	**Steel not exposed to the weather or other corrosive influences**	**Steel used unprotected**
End distance e_1	$1.2d_0$	$4t + 40$ mm	—	The larger of $8t$ or 125 mm
End distance e_2	$1.2d_0$	$4t + 40$ mm	—	The larger of $8t$ or 125 mm
Distance e_3 in slotted holes	$1.5d_0$	—	—	—
Distance e_4 in slotted holes	$1.5d_0$	—	—	—
Spacing p_1	$2.2d_0$	The smaller of $14t$ or 200 mm	The smaller of $14t$ or 200 mm	The smaller of $14t_{min}$ or 175 mm
Spacing $p_{1,0}$	—	The smaller of $14t$ or 200 mm	—	—
Spacing $p_{1,i}$	—	The smaller of $28t$ or 200 mm	—	—
Spacing p_2	$2.4d_0$	The smaller of $14t$ or 200 mm	The smaller of $14t$ or 200 mm	The smaller of $14t_{min}$ or 175 mm

t is the thickness of the thinner outer connected part.

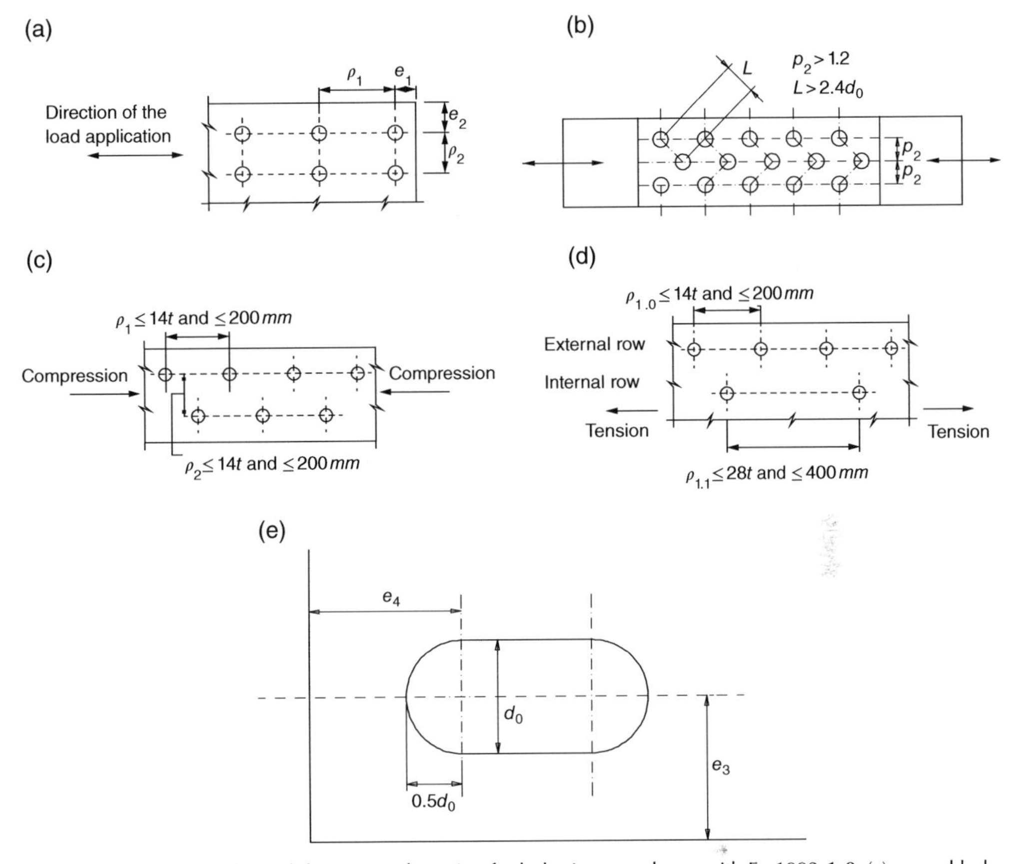

Figure 13.18 Symbols for end distance and spacing for holes in accordance with En 1993-1-8: (a) normal holes, (b) staggered holes, (c) staggered holes for compression members, (d) tension member and (e) slotted holes.

Figure 13.18. In case of end and edge distances for connections in structures subjected to fatigue, reference has to be made to the requirements given in EN 1993-1-8.

Maximum values for spacing, edge and end distance are unlimited, except for exposed tension members to prevent corrosion and for compression members, to avoid local buckling and to prevent corrosion in exposed members.

13.3.2 EU Structural Verifications

Part 1–8 of Eurocode 3, that is EN 1993-1-8 (*Design of Steel Structures – Part 1–8: Design of Joints*) deals with connections that are divided in two groups depending on the type of loading: shear and tension connections.

As to shear connections, bolted connections loaded by shear forces should be designed according to one of the following categories:

- *Category A – Bearing type*: bolts from class 4.6 up to and including class 10.9 should be used. No preloading and special provisions for contact surfaces are required. The design ultimate shear load should not exceed the design shear resistance.

- *Category B – Slip resistant at serviceability limit states*: high resistant preloaded bolts are used and slip does not occur at the serviceability limit state. The design serviceability shear load should not exceed the design slip resistance. The design ultimate shear load should not exceed either the design shear resistance or the design bearing resistance.
- *Category C – Slip resistant at ultimate limit states*: preloaded high resistant bolts are used and slip does not occur at the ultimate limit state. The design ultimate shear load should not exceed either the design slip resistance or the design bearing resistance. In addition to connections in tension, the design plastic resistance of the net cross-section in correspondence with the bolt holes has to be checked at the ultimate limit state.

As to tension connections, the design has to be developed with reference to one of the following categories:

- *Category D – tension connection non-preloaded*: bolts from class 4.6 up to and including class 10.9 are used and no preloading is required. This category must not be used where the connections are frequently subjected to variations of the tensile force. However, they may be used in connections designed to resist normal wind loads.
- *Category E – tension connection preloaded*: preloaded 8.8 and 10.9 bolts with controlled tightening are used. This category shall be used in connections designed to resist to seismic loads.

Table 13.5 summarizes the verifications required for the five different categories of connection.

Table 13.5 Categories of connections in accordance with EN 1993-1-8.

Category	Criteria	Remarks
Shear connections		
A	$F_{v,Ed} \le F_{v,Rd}$	No preloading required
Bearing type	$F_{v,Ed} \le F_{b,Rd}$	Bolt classes from 4.6 to 10.9 are used
B	$F_{v,Ed} \le F_{v,Rd}$	Preloaded 8.8 or 10.9 bolts are used
Slip-resistant at serviceability	$F_{v,Ed} \le F_{b,Rd}$	For slip resistance at serviceability
	$F_{v,Ed,ser} \le F_{s,Rd,ser}$	
C	$F_{v,Ed} \le F_{s,Rd}$	Preloaded 8.8 or 10.9 bolts are used
Slip-resistant at ultimate	$F_{v,Ed} \le F_{b,Rd}$	For slip resistance at ultimate make reference to the net area
	$F_{v,Ed} \le N_{net,Rd}$	
Tension connections		
D	$F_{t,Ed} \le F_{t,Rd}$	No preloading required
Non preloaded	$F_{t,Ed} \le B_{p,Rd}$	Bolt classes from 4.6 to 10.9 are used
E	$F_{t,Ed} \le F_{t,Rd}$	Preloaded 8.8 or 10.9 bolts are used
Preloaded	$F_{t,Ed} \le B_{p,Rd}$	

Where

$F_{v,Ed,ser}$ is the design shear force per bolt for the serviceability limit state;
$F_{v,Ed}$ is the design shear force per bolt for the ultimate limit state;
$F_{v,Rd}$ is the design shear resistance per bolt;
$F_{b,Rd}$ is the design bearing resistance per bolt;
$F_{s,Rd,ser}$ is the design slip resistance per bolt at the serviceability limit state;
$F_{s,Rd}$ is the design slip resistance per bolt at the ultimate limit state;
$F_{t,Ed}$ is the design tensile force per bolt for the ultimate limit state;
$F_{t,Rd}$ is the design tension resistance per bolt;
$B_{p,Rd}$ is the design punching shear resistance of the bolt head and the nut.

13.3.2.1 Tension Resistance

The tension resistance per bolt at ultimate limit states, $F_{t,Rd}$, is defined as:

$$F_{t,Rd} = \frac{k_2 \cdot f_{ub} \cdot A_s}{\gamma_{M2}} \tag{13.14}$$

where k_2 accounts for the type of bolts ($k_2 = 0.63$ for countersunk bolts, otherwise $k_2 = 0.9$), A_s is the threaded area of the bolt, f_{ub} is the ultimate tensile strength of the bolt and γ_{M2} is the safety factor.

Countersunk bolts must have sizes and geometry in accordance with their reference standards, otherwise the tension resistance has to be adjusted accordingly.

Punching shear resistance $B_{p,Rd}$ of the plate is defined as:

$$B_{p,Rd} = \frac{0.6 \cdot \pi \cdot d_m \cdot t_p \cdot f_u}{\gamma_{M2}} \tag{13.15}$$

where f_u and t_p are the ultimate tensile strength and the thickness of the plate, respectively, and d_m is the minimum between the nut diameter and the mean value of the bolt head.

13.3.2.2 Shear Resistance per Shear Plane

Two different cases are distinguished, depending on the portion of bolt subjected to the shear force:

(a) if shear plane passes through the threaded portion of the bolt (A_s is the threaded area of the bolt) the shear resistance, $F_{v,Rd}$, is:

for classes 4.6, 5.6 and 8.8:

$$F_{v,Rd} = \frac{0.6 \cdot f_{ub} \cdot A_s}{\gamma_{M2}} \tag{13.16a}$$

for classes 4.8, 5.8, 6.8 and 10.9:

$$F_{v,Rd} = \frac{0.5 \cdot f_{ub} \cdot A_s}{\gamma_{M2}} \tag{13.16b}$$

(b) if shear plane passes through the unthreaded portion of the bolt (A is the gross cross area of the bolt), the shear resistance, $F_{v,Rd}$, is:

$$F_{v,Rd} = \frac{0.6 \cdot f_{ub} \cdot A}{\gamma_{M2}} \tag{13.16c}$$

13.3.2.3 Combined Shear and Tension Resistance

If the bolt is subjected to combined design shear, $F_{v,Ed}$ and tension, $F_{t,Ed}$, the resistance of the bolt is defined as:

$$\frac{F_{v,Ed}}{F_{v,Rd}} + \frac{F_{t,Ed}}{1.4 \cdot F_{t,Rd}} \leq 1 \tag{13.17}$$

where $F_{v,Rd}$ and $F_{t,Rd}$, are the design shear resistance per bolt (Eq. (13.16a–c)) and the design tension resistance per bolt (Eq. (13.14)), respectively.

13.3.2.4 Bearing Resistance

The bearing resistance per bolt, $F_{b,Rd}$, is:

$$F_{b.Rd} = \frac{k_1 \cdot a_b \cdot f_u \cdot d \cdot t}{\gamma_{M2}} \tag{13.18}$$

where d is the bolt diameter, t and f_u are the thickness and the ultimate strength of the plate, respectively, γ_{M2} is the material safety factor and terms k_1 and α_b depend on the materials and the connection geometry.

In particular, in accordance with the symbols presented in Figure 13.18, these terms are distinguished on the basis of the transfer load direction:

In case of bolts in the direction of load:

for edge bolts:

$$\alpha_b = \min\left\{\frac{e_1}{3 \cdot d_0}; \frac{f_{ub}}{f_u}; 1.0\right\} \tag{13.19a}$$

for internal bolts:

$$\alpha_b = \min\left\{\frac{p_1}{3 \cdot d_0} - \frac{1}{4}; \frac{f_{ub}}{f_u}; 1.0\right\} \tag{13.19b}$$

where f_{ub} and d_0 are the ultimate resistance of the bolt and the diameter of the hole, respectively.

In case of bolts perpendicular to the direction of the load:

for edge bolts:

$$k_1 = \min\left\{\frac{2.8 \cdot e_2}{d_0} - 1.7; 2.5\right\} \tag{13.20a}$$

for internal bolts:

$$k_1 = \min\left\{\frac{1.4 \cdot p_2}{d_0} - 1.7; 2.5\right\} \tag{13.20b}$$

A reduction of the bearing resistance, $F_{b,Rd}$, has to be considered in the following cases:

- bolts in oversized holes, for which a bearing resistance of $0.8F_{b,Rd}$ (reduction of 20% with reference to the case of normal holes) has to be considered;
- bolts in slotted holes, where the longitudinal axis of the slotted hole is perpendicular to the direction of the force transfer, for which a bearing resistance of $0.6F_{b,Rd}$ (reduction of 40% with reference to the case of normal holes) has to be considered.

For a countersunk bolt, the bearing resistance should be based on a plate thickness equal to the thickness of the connected plate minus half the depth of the countersinking.

13.3.2.5 Slip-Resistant Connection

In case of slip-resistant connection, the design pre-loading force for high strength class 8.8 or 10.9 bolt, $F_{p,Cd}$, has to be taken, as recommended in EN 1090-2, as:

$$F_{p,Cd} = \frac{0.7 \cdot f_{ub} \cdot A_s}{\gamma_{M7}} \tag{13.21}$$

The design slip resistance, $F_{S,Rd}$, of a preloaded class 8.8 or 10.9 bolt is:

$$F_{s,Rd} = \frac{k_s \cdot n \cdot \mu}{\gamma_{M3}} \cdot F_{p,C} \tag{13.22a}$$

The design pre-loading force $F_{p,C}$, to use in Eq. (13.22a) is defined as:

$$F_{p,C} = 0.7 \cdot f_{ub} \cdot A_s \tag{13.22b}$$

where μ is the friction coefficient, γ_{M3} and γ_{M7} are safety factors and coefficient k_s accounts for the type of holes and assumes the following values:

- $k_s = 1$ for bolts in normal holes;
- $k_s = 0.85$ for bolts in either oversized holes or short slotted holes with the axis of the slot perpendicular to the direction of load transfer;
- $k_s = 0.7$ for bolts in long slotted holes with the axis of the slot perpendicular to the direction of load transfer;
- $k_s = 0.76$ for bolts in short slotted holes with the axis of the slot parallel to the direction of load transfer;
- $k_s = 0.63$ for bolts in long slotted holes with the axis of the slot parallel to the direction of load transfer.

It should be noted that when the preload is not explicitly used in design calculations for shear resistances (i.e. reference is made to a bearing connection) but it is required for execution purposes or as a quality measure (e.g. for durability), then the level of preloading can be specified in the National Annex.

From EN 1090-2, in absence of experimental data, the surface treatments that may be assumed to provide the minimum slip factor according to the specified class of the friction surface are:

Class A – $\mu = 0.5$	Surfaces blasted with shot or grit with loose rust removed, not pitted;
Class B – $\mu = 0.4$	Surfaces blasted with shot or grit: **(a)** spray-metallized with an aluminium or zinc based product; **(b)** with alkali-zinc silicate paint with a thickness of 50–80 μm
Class C – $\mu = 0.3$	Surfaces cleaned by wire-brushing or flame cleaning, with loose rust removed;
Class D – $\mu = 0.2$	Surfaces as rolled.

No practical indications are given in EN 1993-1-8 about the tightening, which is considered in EN 1090-2 for both cases of slip-resistant connection and connections with pre-loaded bolts.

Further details are reported for high strength class 8.8 and 10.9 bolts in EN 14399 (*High-strength structural bolting assemblies for preloading*), which deals with the delivery conditions. It is required that fasteners have to be supplied to the purchaser either in the original unopened, single sealed container or, alternatively, in separate sealed containers by the manufacturer of the assemblies.

The manufacturer of the assembly must specify the suitable methods for tightening in accordance with EN 1090-2. Assemblies can be supplied in one of the following alternatives (EN 14399-1):

- bolts, nuts and washers supplied by one manufacturer. The elements of an assembly are packed together in one package that is labelled with an assembly lot number and the manufacturer's identification;
- bolts, nuts and washers supplied by one manufacturer. Each element is packed in separate packages that are labelled with the manufacturing lot number of the components and the manufacturer's identification. The elements in an assembly are freely interchangeable within the deliveries of one nominal thread diameter.

All the components used in assemblies for high strength structural bolting, which are suitable for preloading, have to be marked with the identification mark of the manufacturer of the assemblies and with the letter 'H'. Additional letters defining the system (e.g. R for HR or V for HV) have to be added to the H for bolts and nuts. All the components of any assembly have to be marked with the same identification mark.

Furthermore, the manufacturer has to declare the value of the k factor to assess the tightening moment M_s on the basis of preloading force $F_{p,C}$ and of the diameter of the bolt, d, as

$$M_s = k \cdot d \cdot F_{p,C} \tag{13.23}$$

13.3.2.6 Combined Tension and Shear

In case of slip-resistant connections subjected to a design tensile force ($F_{t,Ed}$ or $F_{t,Ed,\text{ser}}$) and a design shear force ($F_{v,Ed}$ or $F_{v,Ed,\text{ser}}$), the design slip resistance per bolt has to be reduced in accordance with the following rules:

- slip-resistant connection at serviceability limit state (category B):

$$F_{s,Rd,ser} = \frac{k_s n \mu \cdot \left(F_{p,C} - 0.8 F_{t,Ed,ser}\right)}{\gamma_{M3,ser}} \tag{13.24a}$$

- slip-resistant connection at ultimate limit state (category C):

$$F_{s,Rd} = \frac{k_s n \mu \cdot \left(F_{p,C} - 0.8 F_{t,Ed}\right)}{\gamma_{M3}} \tag{13.24b}$$

13.3.2.7 Long Joints

Where the distance L_j between the centres of the end fasteners in a joint, measured in the direction of force transfer (Figure 13.19), is more than $15d$, the design shear resistance $F_{v,Rd}$ of all the fasteners has to be reduced by multiplying it by a reduction factor β_{Lf} defined as:

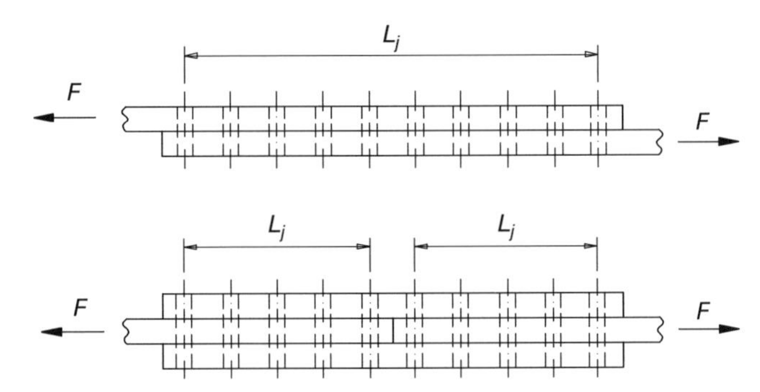

Figure 13.19 Examples of long joints.

$$\beta_{Lf} = 1 - \frac{L_j - 15\,d}{200d} \tag{13.25}$$

with the limitation $\beta_{Lf} \leq 1.0$ and $\beta_{Lf} \geq 0.75$.

13.4 Bolted Connection Design in Accordance with the US Approach

Bolts are treated in AISC 360-10, Section J.3 mainly, and in the *RCSC Specification for Structural Joints Using High-Strength Bolts*, that is referred to by AISC 360-10.

13.4.1 US Practice for Fastener Assemblage

The two main US standard structural bolt types are ASTM A325 and ASTM A490. In particular:

- ASTM A325 bolts are available in diameters from ½ to 1-½ in. (from 16 to 36 mm, ASTM A325M) with a minimum tensile strength of 120 ksi (827 MPa) for diameters of 1 in. (25.4 mm) and less and 105 ksi (724 MPa) for sizes over 1–1-½ in. (38.1 mm). They also come in two types. Type 1 is a medium carbon steel and can be galvanized; Type 3 is a weathering steel that offers atmospheric corrosion resistance. They have a fixed threaded length shorter than the total bolt length. But if nominal length is equal to or shorter than four times the nominal bolt diameter then A325 bolts can be threaded full length.
- ASTM A490 bolts are available in diameters from ½ to 1-½ in. (from 16 to 36 mm, ASTM A490M) with a minimum tensile strength of 150 ksi (1034 MPa) and maximum tensile strength of 170 ksi (1172 MPa) for all diameters, and are offered in two types. Type 1 is alloy steel, and Type 3 is a weathering steel that offers atmospheric corrosion resistance.

ASTM F1852 and ASTM F2280 bolts can be used in lieu, respectively, of ASTM A325 and ASTM A490. They have the same mechanical and chemical characteristics and differ only for a splined end that extends beyond the threaded portion of the bolt. During installation, this splined end is gripped by a specially designed wrench chuck and provides a mean for turning the nut relative to the bolt, as discussed in the following.

Structural bolts are specifically designed for use with heavy hex nuts. The nuts for structural connections have to be conforming to ASTM A563 or ASTM A194 and the washers used for structural connections to ASTM F436 specifications. This specification covers both flat circular and bevelled washers.

ASTM Specifications permit the galvanizing of ASTM A325 bolts, but ASTM A490 bolts should not be galvanized or electroplated. The major problem with hot dip galvanizing and electroplating A490 bolts is the potential hydrogen embrittlement. This scenario may occur when atomic hydrogen is introduced during the pickling process that takes place prior to plating or hot dip galvanizing process. In 2008, ASTM approved the use of zinc/aluminium protective coatings per ASTM F1136 for use on A490 structural bolts.

Both A325 and A490 are heavy hex bolts intended for structural usage and for connecting steel profiles and/or plates, so they are not very long and their diameter is limited as indicated above. If different diameters and/or different length have to be used, reference can be made to:

- ASTM A449 in lieu of ASTM A325 bolts. They range in diameter from ¼ to 3 in. (from 6.4 to 76.2 mm) and are far more flexible in their configuration: they can be a headed bolt, a straight rod with threads or a bend bolt such as a right angle bend foundation bolt. From a material point of view, there are no relevant differences between A449 and A325 bolts.
- ASTM A354 grade BD in lieu of ASTM A490 bolts. A354 grade BD bolts are quenched and tempered alloy steel bolts, and are equal in strength to ASTM A490 bolts. They range in diameter from ¼ to 4 in. (from 6.4 to 102 mm) and can be used for anchor bolts and threaded rods. A354 grade BC has a lower strength then Grade BD, closer to that of ASTM A325 bolts.

AISC 360-10 also allows the usage of ASTM A307 bolts. The ASTM A307 specification covers carbon steel bolts and studs ranging from ¼ to 4 in. diameter (from 6.4 to 102 mm), with a minimum tensile strength of 60 ksi (414 MPa). There are three different steel grades (A, B and C), which denote configuration and application:

- Grade A is for headed bolts, threaded rods and bent bolts intended for general applications.
- Grade B is for heavy hex bolts and studs intended for flanged joints in piping systems with cast iron flange.
- Grade C denotes non-headed anchor bolts, either bent or straight, intended for structural anchorage purpose. Grade C has been now substituted by ASTM F1554 Grade 36, low carbon, 36 ksi (248 MPa) yield steel anchor bolts.

AISC 360-10 lists all bolt specifications into two groups according to their tensile strength:

- Group A (lower strength): ASTM A325, A325M, F1852, A354 Grade BC and A449;
- Group B (higher strength): ASTM A490, A490M, F2280 and A354 Grade BD.

Bolts belonging to the same group have almost the same ultimate tensile strength:

Group A: 120 ksi (827 MPa) minimum up to 1 in. diameter, 105 ksi (724 MPa) for higher diameters (yielding strength: 92 ksi (634 MPa) minimum up to 1 in. diameter, 81 ksi (558 MPa) for higher diameters);
Group B: 150 ksi (1034 MPa) minimum, 170 ksi (1172 MPa) maximum (yielding strength: 130 ksi (896 MPa) minimum).

ASTM A307 bolts are not grouped because their ultimate strength (60 ksi, 414 MPa) is lower than the one of group A. As to the holes for bolts, AISC 360-10 individuates four types of holes:

- normal holes;
- oversized holes;
- short-slotted holes;
- long-slotted holes.

The maximum allowed size for holes are listed in Tables 13.6a and 13.6b.

Oversized holes are permitted in any or all plies of slip-critical connections, but *they cannot be used in bearing-type connections*. Hardened washers have to be installed over oversized holes in an outer ply.

Short-slotted holes are permitted in any or all plies of slip-critical or bearing-type connections. The slots are permitted without regard to direction of loading in slip critical connections, but *the length has to be normal to the direction of the load in bearing-type connections*. Washers have to be installed over short-slotted holes in an outer ply.

Long-slotted holes are permitted in only one of the connected parts of either a slip critical or bearing-type connection at an individual faying surface. Long-slotted holes are permitted without regard to direction of loading in slip-critical connections, but *have to be normal to the direction of load in bearing-type connections*. Where long-slotted holes are used in an outer ply, plate washers or a continuous bar with standard holes, of a size sufficient to completely cover the slot after installation, have to be provided.

Table 13.6a Nominal holes (dimensions in inches) (from Table J3.3 of AISC 360-10).

	Hole dimensions, in. (mm)			
Bolt diameter, in. (mm)	**Standard (diameter)**	**Oversize (diameter)**	**Short-slot (width × length)**	**Long-slot (width × length)**
1/2 (12.7)	9/16 (14.3)	5/8 (7.9)	9/16 × 11/16 (14.3 × 17.5)	9/16 × 1–1/4 (14.3 × 331.8)
5/8 (15.9)	11/16 (17.5)	13/16 (20.6)	11/16 × 7/8 (17.5 × 22.2)	11/16 × 1–9/16 (17.5 × 39.7)
3/4 (19.1)	13/16 (20.6)	15/16 (23.8)	13/16 × 1 (20.6 × 25.4)	13/16 × 1–7/8 (20.6 × 47.6)
7/8 (22.2)	15/16 (23.8)	1–1/16 (27.0)	15/16 × 1–1/8 (23.8 × 28.6)	15/16 × 2–3/16 (23.8 × 55.6)
1 (25.4)	1–1/16 (27.0)	1–1/4 (31.8)	1–1/16 × 1–5/16 (27.0 × 33.3)	1–1/16 × 2–1/2 (27.0 × 63.5)
≥1-1/8 (≥28.6)	$d + 1/16$ $(d + 1.6)$	$d + 5/16$ $(d + 7.9)$	$(d + 1/16) \times (d + 3/8)$ $((d + 1.6) \times (d + 9.5))$	$(d + 1/16) \times (2.5 \times d)$ $((d + 1.6) \times (2.5 \times d))$

Table 13.6b Nominal holes (dimensions in millimetres) (from Table J3.3M of AISC 360-10).

	Hole dimensions, mm (in.)			
Bolt diameter, in. (mm)	**Standard (diameter)**	**Oversize (diameter)**	**Short-slot (width × length)**	**Long-slot (width × length)**
M16 (0.630)	18 (0.709)	20 (0.787)	18 × 22 (0.709 × 0.866)	18 × 40 (0.709 × 1.575)
M20 (0.787)	22 (0.866)	24 (0.945)	22 × 26 (0.866 × 1.024)	22 × 50 (0.866 × 1.969)
M22 (0.866)	24 (0.945)	28 (1.102)	24 × 30 (0.945 × 1.181)	24 × 55 (0.945 × 2.165)
M24 (0.945)	27 (1.063)	30 (1.181)	27 × 32 (1.063 × 1.260)	27 × 60 (1.063 × 2.362)
M27 (1.06)	30 (1.181)	35 (1.378)	30 × 37 (1.181 × 1.457)	30 × 67 (1.181 × 2.638)
M30 (1.18)	33 (1.299)	38 (1.496)	33 × 40 (1.299 × 1.575)	33 × 75 (1.299 × 2.953)
≥ M36 (≥1.42)	$d + 3$ $(d + 0.118)$	$d + 8$ $(d + 0.315)$	$(d + 3) \times (d + 10)$ $((d + 0.118) \times (d + 0.394))$	$(d + 3) \times (2.5 \times d)$ $((d + 0.118) \times (2.5 \times d))$

The distance between centres of standard, oversized or slotted holes should not be less than 2–2/3 times the nominal diameter, d, of the fastener; a distance of $3d$ is preferred. The distance from the centre of a standard hole to an edge of a connected part in any direction should not be less than either the applicable value from Tables 13.7a and 13.7b.

The distance from the centre of an oversized or slotted hole to an edge of a connected part should not be less than the one required for a standard hole to an edge of a connected part plus the applicable increment, C_2, from Tables 13.8a and 13.8b.

The maximum distance from the centre of any bolt to the nearest edge of parts in contact is 12 times the thickness of the connected part under consideration, but cannot exceed 6 in. (150 mm). The longitudinal spacing of fasteners between elements consisting of a plate and a shape or two plates in continuous contact has to be as follows:

- for painted members or unpainted members not subject to corrosion, the spacing is limited to 24 times the thickness of the thinner part or 12 in. (305 mm);
- for unpainted members of weathering steel subject to atmospheric corrosion, the spacing is limited to 14 times the thickness of the thinner part or 7 in. (180 mm).

According to the RCSC Specification and AISC 360-10, for structural applications there are generally three types of connections in which bolts are used: *snug-tightened*, *pre-tensioned* and *slip critical* connections.

Table 13.7a Minimum edge distance (*a*) from the centre of standard hole (*b*) to the edge of connected part (dimensions in inches) (from Table J3.4 of AISC 360-10).

Bolt diameter, in. (mm)	Minimum edge distance, in. (mm)
1/2 (12.7)	3/4 (19.1)
5/8 (15.9)	7/8 (22.2)
3/4 (19.1)	1 (25.4)
7/8 (22.2)	1–1/8 (28.6)
1 (25.4)	1–1/4 (31.75)
1–1/8 (28.6)	1–1/2
1–1/4 (31.75)	1–5/8
Over 1–1/4 (Over 31.75)	1–1/4 × d

Table 13.7b Minimum edge distance (*a*) from the centre of standard hole (*b*) to the edge of connected part (dimensions in millimetres) (from Table J3.4M of AISC 360-10).

Bolt diameter, mm (in.)	Minimum edge distance, mm (in.)
16 (0.630)	22 (0.866)
20 (0.787)	26 (1.024)
22 (0.866)	28 (1.102)
24 (0.945)	30 (1.181)
27 (1.063)	34 (1.339)
30 (1.181)	38 (1.496)
36 (1.417)	46 (1.811)
Over 36	1.25 × d

Table 13.8a Values of edge distance increment C_2 – dimensions in inches (mm) (from Table J3.5 of AISC 360-10).

Nominal diameter of fastener, in. (mm)	Oversized holes, in. (mm)	Slotted holes: Long axis perpendicular to edge – Short slots, in. (mm)	Slotted holes: Long axis perpendicular to edge – Long slots, in. (mm)	Slotted holes: Long axis parallel to edge
≤7/8 (22.2)	1/16 (1.588)	1/8 (3.175)	(3/4)d	0
1 (25.4)	1/8 (3.175)	1/8 (3.175)		
≥1–1/8 (28.6)	1/8 (3.175)	3/16 (4.763)		

Table 13.8b Values of edge distance increment C_2 – dimensions in millimetres (inches) (from Table J3.5M of AISC 360-10).

Nominal diameter of fastener, mm (in.)	Oversized holes, mm (in.)	Slotted holes: Long axis perpendicular to edge – Short slots, mm (in.)	Slotted holes: Long axis perpendicular to edge – Long slots, mm (in.)	Slotted holes: Long axis parallel to edge
≤22 (0.866)	2 (0.0787)	3 (0.118)	0.75d	0
24 (0.945)	3 (0.118)	3 (0.118)		
≥27 (1.063)	3 (0.118)	5 (0.197)		

13.4.1.1 Snug-Tightened Connections

Bolts are permitted to be installed to the snug-tight condition when used in bearing-type connections, with bolts in shear, in tension or combined shear and tension. There are no special requirements for the faying surfaces.

Only Group A bolts in tension or combined shear and tension and Group B bolts in shear, where loosening or fatigue are not the parameters governing the design, are permitted to be installed snug tight. Washers are not required, except for sloping surfaces. The snug-tight condition is defined as the *tightness required to bring the connected plies into firm contact.*

13.4.1.2 Pretensioned Connections

Pretension of bolts is required, but ultimate strength does not depend on slip resistance but on shear/bearing behaviour. In other words, the connection is still a bearing type connection.

RCSC Specification prescribes that bearing type connections have to be pretensioned in the following circumstances:

- joints that are subjected to significant reversal load;
- joints that are subjected to fatigue load with no reversal of the loading direction;
- joints with ASTM A325 or F1852 bolts that are subjected to tensile fatigue;
- joints with ASTM A490 and F2280 bolts that are subjected to tension or combined shear and tension, with or without fatigue.

On the other side, AISC 360-10 prescribes that bearing type connections have to be pretensioned in the following circumstances:

- column splices in buildings with high ratios of height to width;
- connections of members that provide bracing to columns in tall buildings;

- various connections in buildings with cranes over 5-ton capacity;
- connections for supports of running machinery and other sources of impact or stress reversal.

Bolts are to be pretensioned to tension values not less than those given in Tables 13.9a and 13.9b. Such values are equal to 0.70 times the minimum tensile strength of the bolt.

13.4.1.3 Slip-Critical Connections

Slip resistance is required at the faying surfaces subjected to shear or combined shear and tension. Slip resistance has to be checked at either the factored-load level or service-load level, as a choice of the designer. RCSC Specification states that slip-critical joints are only required in the following applications involving shear or combined shear and tension:

- joints that are subjected to fatigue load with reversal of the loading direction;
- joints that use oversized holes;
- joints that use slotted holes, except those with applied load approximately normal (within 80–100°) to the direction of the long dimension of the slot;
- joints in which slip at the faying surfaces would be detrimental to the performance of the structure.

Washers are not required in pretensioned joints and slip-critical joints, with the following exceptions:

- ASTM F436 washers under both the bolt head and nut, when ASTM A490 bolts are pretensioned in the connected material of a specified minimum yield strength less than 40 ksi (276 MPa);

Table 13.9a Minimum bolt pretension (from Table J3.1 of AISC 360-10).

Bolt size, in. (mm)	Group A (e.g. A325 bolts), kips (kN)	Group B (e.g. A490 bolts), kips (kN)
1/2 (12.7)	12 (53.4)	15 (66.7)
5/8 (15.9)	19 (84.5)	24 (106.8)
3/4 (19.1)	28 (124.6)	35 (155.7)
7/8 (22.2)	39 (173.5)	49 (218.0)
1 (25.4)	51 (226.9)	64 (284.7)
1-1/8 (28.6)	56 (249.1)	80 (355.9)
1-1/4 (31.8)	71 (315.8)	102 (453.7)
1-3/8 (34.9)	85 (378.1)	121 (538.2)
1-1/2 (38.1)	103 (458.2)	148 (658.3)

Table 13.9b Minimum bolt pretension (from Table J3.1M of AISC 360-10).

Bolt size, mm (in.)	Group A (e.g. A325 bolts), kN (kips)	Group B (e.g. A490 bolts), kN (kips)
M16 (0.630)	91 (20.5)	114 (25.6)
M20 (0.787)	142 (31.9)	179 (40.2)
M22 (0.866)	176 (39.6)	221 (49.7)
M24 (0.945)	205 (46.1)	257 (57.8)
M27 (1.06)	267 (60.0)	334 (75.1)
M30 (1.18)	326 (73.3)	408 (91.7)
M36 (1.42)	475 (106.8)	595 (133.8)

- an ASTM F436 washer under the turned element, when the calibrated wrench pretensioning method is used;
- an ASTM F436 washer under the nut as part of the fastener assembly, when the twist-off-type tension-control bolt pretensioning method is used;
- an ASTM F436 washer under the side (head or nut) that is turned, when the direct-tension-indicator pretensioning method is used;
- a washer or a continuous bar of sufficient size to completely cover the hole, when an oversized or slotted hole occurs in an outer ply.

Bolts in slip-critical connections are pretensioned at the same values of bolts in pretensioned connections as in Tables 13.9a and 13.9b.

There are four methods of installation procedures admitted by the AISC to achieve the tension required for the pretensioned bearing connections or the slip critical connections:

(1) Turn-of-Nut method;
(2) Twist-Off-Type Tension-Control Bolt Pretensioning;
(3) Direct Tension Indicating method (DTI);
(4) Calibrated Wrench method.

For pretensioned joints and slip-critical joints, a *tension calibrator* has to be used before bolt installation. A tension calibrator is a hydraulic device indicating the pretension that is developed in a bolt that is installed in it. A representative sample of no fewer than three complete fastener assemblies of each combination of diameter, length, grade and lot to be used in the work has to be checked at the site of installation in the tension calibrator to verify that the pretensioning method develops a pretension equal to or greater than 1.05 times than the one specified in Tables 13.9a and 13.9b.

A bolt tension calibrator is essential for:

(1) the pre-installation verification of the suitability of the fastener assembly, including the lubrication that is applied by the manufacturer;
(2) verifying the adequacy and the proper use of the specified pretensioning method;
(3) determining the installation torque for the calibrated wrench pretensioning method. Actually according to AISC 360-10 *torque values determined from tables or from equations that claim to relate torque to pretension without verification shouldn't be used.* For the calibrated wrench pretensioning method, installation procedures have to be calibrated on a daily basis.

(1) *Turn-of-Nut Method*: This method involves tightening the fastener to a low initial 'snug tight' condition and then applying a prescribed amount of turn to develop the required preload. The actual preload depends on how far the nut is turned. Special attention has to be paid when this method is used, in particular, it is important:
  (a) to snug the joint to bring the assembly into firm contact.
  (b) to inspect the joint to verify 'snug tight'.
  (c) to match mark bearing face of the nut and end of the bolt with a single straight line.
  (d) to use a systematic approach that would involve the appropriate bolting pattern, apply the required turns as given in the Table 13.10.
(2) *Twist-Off-Type Tension-Control Bolt Pretensioning*: Tension control bolts use design features that indirectly indicate tension. The most common is the twist-off bolt or tension control

Table 13.10 Nut rotation from the snug-tight condition for turn-of-nut pretensioning.

Bolt length	Rotation
$\leq 4\ d_b$	120°
$4 < d_b \leq 8$	180°
$8 < d_b \leq 12$	240°

d_b = bolt nominal diameter.

(TC) bolt. An assembly tool holds this bolt from the nut end while an inner spindle on the tool grips a spline section connected to the end of the bolt. An outer spindle on the tool turns the nut and tightens the fastener. When the designated torque has been reached, the spline snaps off. This type of torque control system allows for quick inspection, if the spline is gone, in theory, the bolt has been properly tightened.

Twist-off-type tension-control bolt assemblies that meet the requirements of ASTM F1852 and F2280 have to be used.

(3) *Direct Tension Indicating Method (DTI)*: This method requites DTI washers according to ASTM F959. The most common type of washer involves the use of hollow bumps on one side of the washer. These bumps are flattened as the fastener is tightened. A feeler gauge is used to measure the gap developed by the bumps. When the fastener has developed the proper tension, the feeler gauge will no longer fit in the gap. Some washer types fill the void under the bumps with coloured silicone that squirts out once the bumps are compressed, thereby indicating proper tension is reached (Figure 13.8). The pre-installation verification procedure illustrated previously has to be applied to demonstrate that, when the pretension in the bolt reaches 1.05 times the one specified for installation, the gap is no less than prescribed by ASTM F959.

(4) *Calibrated Wrench Method*: Bolts are initially snug-tightened. Subsequently, the installation torque determined in the pre-installation verification of the fastener assembly has to be applied to all bolts in the joint. The scatter in the installed pretension can be significant with this installation method. The relationship between torque and pretension is affected by many factors: the finish and tolerance on the bolt and nut threads, the lubrication, the shop or jobsite conditions that contribute to dust and dirt or corrosion on the threads, the friction between the turned element and the supporting surface, the variability of the air supply parameters on impact wrenches, the condition, lubrication and power supply for the torque wrench. For these reasons RCSC Specification and AISC 360-10 put emphasis on daily wrench calibration activity and, as said before, it is not valid to use published values based on a torque-tension relationship.

13.4.2 US Structural Verifications

Bolt structural verifications are addressed in AISC 360-10, chapter J3. Nominal strength of fasteners has to be according to the values listed in Table 13.11.

The nominal tensile strength values in Table 13.11 are obtained from the equation:

$$F_{nt} = 0.75 F_u \tag{13.26}$$

Table 13.11 Nominal strength of fasteners and threaded parts, ksi (MPa) (from Table J3.2 of AISC 360-10).

Description of fasteners	Nominal tensile strength, F_{nt}, ksi (MPa)	Nominal shear strength in bearing type connections, F_{nv}, ksi (MPa)
A307 bolts	45 (310)	27 (188)
Group A (e.g. A325) bolts, when threads are not excluded from shear planes	90 (620)	54 (372)
Group A (e.g. A325) bolts, when threads are excluded from shear planes	90 (620)	68 (457)
Group B (e.g. A490) bolts, when threads are not excluded from shear planes	113 (780)	68 (457)
Group B (e.g. A490) bolts, when threads are excluded from shear planes	113 (780)	84 (579)
Threaded parts of anchor roads and threaded roads, according to ASTM F1554, when threads are not excluded from shear planes	$0.75F_u$	$0.450F_u$
Threaded parts of anchor roads and threaded roads, according to ASTM F1554, when threads are excluded from shear planes	$0.75F_u$	$0.563F_u$

The factor of 0.75 accounts for the approximate ratio of the effective tension area of the threaded portion of the bolt to the area of the shank of the bolt for common sizes. Thus in verification formulas the gross (unthreaded) area has to be used.

The values of nominal shear strength in Table 13.11 are obtained from the following equations:

(a) When threads are excluded from the shear planes:

$$F_{nv} = 0.563F_u \tag{13.27}$$

The factor 0.563 accounts for the effect of a shear/tension ratio of 0.625 and a 0.90 length reduction factor, with joint lengths up to and including 38 in. (965 mm).

(b) When threads are not excluded from the shear plane

$$F_{nv} = 0.450F_u \tag{13.28}$$

It is worth mentioning that the factor of 0.450 is 80% of 0.563, which accounts for the reduced area of the threaded portion of the fastener when the threads are not excluded from the shear plane.

In tension or compression long joints (i.e. with length longer than approximately 16 in. or 406 mm), the differential strain produces an uneven distribution of load between fasteners, those near the end taking a disproportionate part of the total load, so the maximum strength per fastener is reduced. In this case, AISC 360-10 requires that the initial 0.90 factor should be replaced by 0.75 when determining bolt shear strength for connections longer than 38 in. (965 mm). In lieu of another column of design values, the appropriate values are obtained by multiplying the tabulated values by 0.75/0.90 = 0.833.

13.4.2.1 Tensile or Shear Strength of Bolts

LRFD approach

The *design tensile strength* ϕR_n of a snug-tightened or pretensioned high-strength bolt has to be greater or equal to the required tensile strength T_u:

$$\phi R_n = \phi F_{nt} A_b \geq T_u \quad (13.29)$$

where $\phi = 0.75$, F_{nt} is the nominal tensile stress (Table 13.11), A_b is the nominal unthreaded bolt area and T_u is the required tensile strength.

The *design shear strength* ϕR_n of a snug-tightened or pretensioned high-strength bolt has to be greater or equal to the required shear strength V_u:

$$\phi R_n = \phi F_{nv} A_b \geq V_u \quad (13.31)$$

where $\phi = 0.75$, F_{nv} is the nominal shear stress (Table 13.11), A_b is the nominal unthreaded bolt area and V_u is the required shear strength for LRFD combinations.

ASD approach

The *allowable tensile strength* R_n/Ω of a snug-tightened or pretensioned high-strength bolt has to be greater or equal to the required tensile strength T_a:

$$R_n/\Omega = F_{nt} A_b/\Omega \geq T_a \quad (13.30)$$

where $\Omega = 2.00$, F_{nt} is the nominal tensile stress (Table 13.11), A_b is the nominal unthreaded bolt area and T_a is the required tensile strength.

The *allowable shear strength* R_n/Ω of a snug-tightened or pretensioned high-strength bolt has to be greater or equal to the required shear strength V_a:

$$R_n/\Omega = F_{nv} A_b/\Omega \geq V_a \quad (13.32)$$

where $\Omega = 2.00$, F_{nv} is the nominal shear stress (Table 13.11), A_b is the nominal unthreaded bolt area and V_a is the required shear strength for ASD combinations.

LRFD, load and resistance factor design and ASD, allowable strength design.

13.4.2.2 Combined Tension and Shear in Bearing-Type Connections

LRFD approach

The *design tensile strength* ϕR_n of a snug-tightened or pretensioned high-strength bolt subjected to combined tension and shear, has to be greater or equal to the required tensile strength T_u.

$$\phi R_n = \phi F'_{nt} A_b \geq T_u \quad (13.33)$$

The available shear stress of the bolt ϕF_{nv} has to be equal or to exceed the required shear stress, f_{rv}:

$$\phi F_{nv} \geq f_{rv} \quad (13.35)$$

where $\phi = 0.75$ and F'_{nt} is the nominal tensile stress (Table 13.11), modified to take into account the shear effects as:

$$F'_{nt} = 1.3F_{nt} - \frac{F_{nt}}{\phi F_{nv}} f_{rv} \leq F_{nt}$$

where F_{nt} and F_{nv} are the nominal tensile and the nominal shear stress (Table 13.11), respectively and f_{rv} is the required shear stress defined as:

$$f_{rv} = V_u/A_b$$

where A_b is the nominal unthreaded bolt area and V_u is the required shear strength.

ASD approach

The *allowable tensile strength* R_n/Ω of a snug-tightened or pretensioned high-strength bolt has to be greater or equal to the required tensile strength T_a:

$$R_n/\Omega = F'_{nt} A_b/\Omega \geq T_a \quad (13.34)$$

The available shear stress of the bolt F_{nv}/Ω has to be equal or to exceed the required shear stress, f_{rv}:

$$F_{nv}/\Omega \geq f_{rv} \quad (13.36)$$

where $\Omega = 2.00$ and F'_{nt} is the nominal tensile stress (Table 13.11), modified to take into account the shear effects as:

$$F'_{nt} = 1.3F_{nt} - \frac{\Omega F_{nt}}{F_{nv}} f_{rv} \leq F_{nt}$$

where F_{nt} and F_{nv} are the nominal tensile and the nominal shear stress (Table 13.11), and f_{rv} is the required shear stress defined as:

$$f_{rv} = V_a/A_b$$

where A_b is the nominal unthreaded bolt area and V_a is the required shear strength.

When the required stress, f, in either shear or tension, is less than or equal to 30% of the corresponding available stress, the effects of combined stresses need not to be investigated.

13.4.2.3 Slip-Critical Connections

LRFD	ASD
The *design slip resistance* ϕR_n of a pretensioned high-strength bolt in a slip-critical connection has to be greater or equal to the required shear strength V_u:	The *allowable slip resistance* R_n/Ω of a pretensioned high-strength bolt in a slip-critical connection has to be greater or equal to the required shear strength V_a:
$\phi R_n \geq V_u$ (13.37)	$R_n/\Omega \geq V_a$ (13.38)

As to the safety coefficients, it must be assumed:

$\phi = 1.00$ or $\Omega = 1.50$	for standard size and short-slotted holes perpendicular to the direction of the load;
$\phi = 0.85$ or $\Omega = 1.76$	for oversized and short-slotted holes parallel to the direction of the load;
$\phi = 0.70$ or $\Omega = 2.14$	for long-slotted holes.

The decrease of ϕ (and the increase of Ω) reflects the fact that consequences of exceeding slip limit state become more severe from standard size holes through long-slotted holes.

The *available slip resistance* has to be determined as follows:

$$R_n = \mu D_u h_f T_b n_s k_{sc} \tag{13.39}$$

where T_b is the minimum fastener tension given in Tables 13.9a and 13.9b, $D_u = 1.13$ is the ratio of the mean installed bolt pretension to the specified minimum bolt pretension, n_s is the number of slip planes, μ is the mean slip coefficient (see later) and $h_f = 1$ if elements under stress are connected directly by bolts or by means of one interposed filler (Figure 13.20).

When fillers are more than one, it is assumed that $h_f = 0.85$.

Term k_{sc} has to be assumed = 1 if there is no tension action applied to connection. If there is, then it assumes the following values:

$$k_{sc} = 1 - \frac{T_u}{D_u T_b n_b} \text{(LFRD)} \tag{13.40a}$$

$$k_{sc} = 1 - \frac{1.5 T_a}{D_u T_b n_b} \text{(ASD)} \tag{13.40b}$$

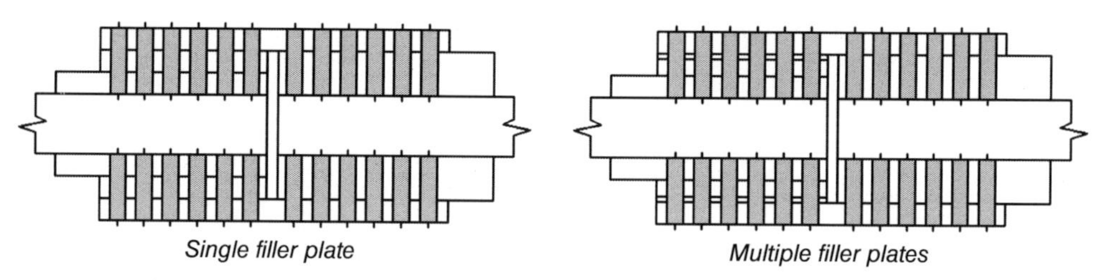

Figure 13.20 Single and multiple filler plates.

where T_a and T_u are the required tension force using ASD and LRFD load combinations respectively, and n_b is the number of bolts carrying the applied tension.

The mean slip coefficient μ is referred by AISC 360-10 and RCSC Specification to two kind of faying surfaces (classes A and B):

- $\mu = 0.30$ for class A faying surfaces: uncoated clean mill scale steel surfaces, surfaces with class A coatings on blast-cleaned steel;
- $\mu = 0.50$ for class B faying surfaces: uncoated blast-cleaned steel surfaces with class B coatings on blast-cleaned steel
- $\mu = 0.35$ for class C surfaces: roughened hot-dip galvanized surfaces.

If faying surfaces have to be protected by a zinc primer coating, the coating has to be tested and qualified according to appendix A of RCSC Specification to determine if it can be classified as Class A or B, in order to assign a slip coefficient equal to 0.30 (if class A) or 0.50 (if class B).

The mean slip coefficient for clean hot-dip galvanized surfaces is in the order of 0.19, but can be significantly improved by treatments such as hand wire brushing to the value of 0.35 as indicated by RCSC Specification.

AISC 360-10 states that 'Slip-critical connections have to be designed to prevent slip and for the limit states of bearing-type connections'. This means that a slip-critical connection has to be designed to prevent slip at service load and AISC 360-10 design procedure guarantees this. So at ultimate loads the connection can slip, then the Specifications requires verification of the connection for bearing and shear at ultimate loads.

In AISC 360-05 it was stated:

> *High-strength bolts in slip-critical connections are permitted to be designed to prevent slip either as a serviceability limit state or at the required strength limit state. The connection must also be checked for shear strength in accordance with Sections J3.6 and J3.7 and bearing strength in accordance with Sections J3.1 and J3.10 […]. Connections with standard holes or slots transverse to the direction of the load have to be designed for slip at serviceability limit state. Connections with oversized holes or slots parallel to the direction of the load have to be designed to prevent slip at the required strength level.*

So the 2005 Code was slightly different: slip-critical connections could have been designed for preventing slip at service loads or at ultimate loads. This was achieved by changing resistance and safety factors:

$\phi = 1.00$; $\Omega = 1.50$	For preventing slip at service loads
$\phi = 0.85$; $\Omega = 1.76$	For preventing slip at ultimate loads

In AISC 360-10 more severe values for resistance and safety factors have been maintained for oversized and short-slotted holes parallel to the direction of the load. So 2010 Code allows for design of slip-critical connections with oversized loads for preventing slip at loads higher than service loads, even if not explicitly stated, because consequences of slip are worse with oversized holes than with regular holes.

13.4.2.4 Bearing Strength at Bolt Holes

LRFD	ASD
The *design bearing strength* ϕR_n at a bolt hole has to be greater or equal to the required shear strength V_u:	The *allowable bearing strength* R_n/Ω at a bolt hole has to be greater or equal to the required shear strength V_a:
$\phi R_n \geq V_u$ (13.41)	$R_n/\Omega \geq V_a$ (13.42)
where $\phi = 0.75$.	where $\Omega = 2.00$.

The nominal bearing strength R_n has to be determined as follows:

(a) for standard, oversized and short-slotted holes, if deformation at service loads *is not* a design consideration:

$$R_n = 1.5 l_c t F_u \leq 3.0 d t F_u \tag{13.43}$$

(b) for standard, oversized and short-slotted holes, if deformation at service loads *is* a design consideration:

$$R_n = 1.2 l_c t F_u \leq 2.4 d t F_u \tag{13.44}$$

(c) for a bolt in a connection with long-slotted holes with the slot perpendicular to the direction of force:

$$R_n = 1.0 l_c t F_u \leq 2.0 d t F_u \tag{13.45}$$

where F_u is the specified minimum tensile strength of the connected material, d is the nominal bolt diameter, l_c is the clear distance in the direction of the force between the edge of the hole and the edge of the adjacent hole or edge of the material and t is the thickness of connected material.

The use of oversized holes and short- and long-slotted holes parallel to the line of force is restricted to slip-critical connections. Bearing-type connections can be used with standard holes only.

Bearing resistance has to be checked for both bearing-type and slip-critical connections.

The strength of a single bolt is the smaller between its shear strength and the bearing strength at the bolt hole. The strength of a connection is the sum of the strengths of its individual bolts.

13.5 Connections with Rivets

Riveting techniques were used extensively until the first decades of the twentieth century (Figure 13.21) and they have now virtually disappeared in construction practice in favour of bolted and welded connection techniques, which are cheaper nowadays.

Riveting is a method of connecting plates: ductile metal pins are inserted into holes and riveted to form a head at each end to prevent the joint from coming apart (Figure 13.22). Rivets were often used in the same way as ordinary structural bolts are currently used in shear and bearing and in tension joints. There is usually less slip in a riveted joint with respect to a bearing bolted joint because of the tendency for the rivet holes to be filled by the rivets when being hot-driven. Shop riveting was cheaper in the past than site riveting and for this reason shop-riveting was often combined with site-bolting. However, now this connection technique is used only in some historical refurbishments. The principal factor that delayed immediate acceptance of the bolts was the high cost of materials, including washers. Since then, because of the higher labour costs together with the modern approach to designing connections requiring fewer bolts than rivets, riveting is used only in some cases for historical buildings and bridges, owing to their non-competitive costs.

Figure 13.21 Example of a riveted joint in an historical bridge.

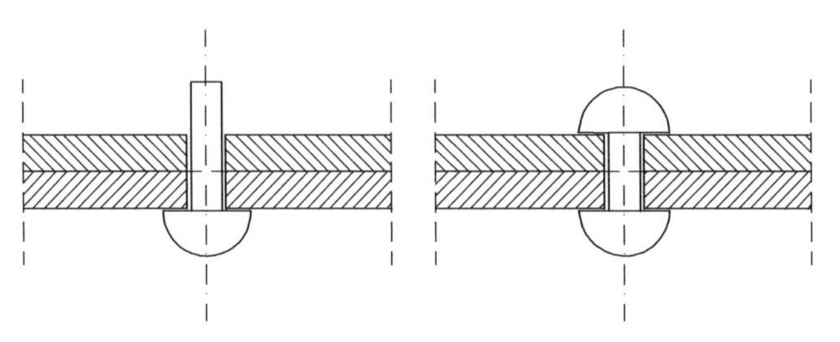

Figure 13.22 Riveting of the pin.

13.5.1 Design in Accordance with EU Practice

Criteria for the rivet verifications are reported in EN 1993-1-8 and are very similar to those proposed for bolted connections. In particular, if A_0 and f_{ur} identify the area of the hole and the ultimate strength of the pin, respectively, the following design resistance can be evaluated.

Shear: Design shear resistance is:

$$F_{v,Rd} = \frac{0.6 \cdot f_{ur} \cdot A_0}{\gamma_{M2}} \tag{13.46}$$

Bearing: The same equations already introduced for bolted connections are proposed.
Tension: Design tension resistance is:

$$F_{t,Rd} = \frac{0.6 \cdot f_{ur} \cdot A_0}{\gamma_{M2}} \tag{13.47}$$

Shear and tension: The same equations for bolted connections are proposed.

13.5.2 Design in Accordance with US Practice

Rivets are not addressed any more either in AISC 360-10 or in RCSC Specification. The second document indicates the 'Guide to Design Criteria for Bolted and Riveted Joints Second Edition' (1987), as a reference document for rivets. Such a guide recognizes three structural rivet steels:

- ASTM A502 grade 1, carbon rivet steel for general purposes;
- ASTM A502 grade 2, carbon-manganese rivet steel suitable for use with high-strength carbon and high strength low-alloy structural steels;
- ASTM A502 grade 3, similar to grade 2 but with enhanced corrosion resistance.

The following design rules are taken from the 'Guide to Design Criteria for Bolted and Riveted Joints, Second Edition'.

Tension: The tensile capacity B_u of a rivet is equal to the product of the rivet cross-sectional area A_b and its tensile strength σ_u. The cross section is generally taken as the undriven cross section area of the rivet:

$$B_u = A_b \sigma_u \tag{13.48}$$

A reasonable lower bound estimate of the rivet tensile capacity σ_u is 60 ksi (414 MPa) for A502 grade 1 rivets and 80 ksi (552 MPa) for A502 grade 2 or grade 3 rivets. Since ASTM specifications do not specify the tensile capacity, these values can be used.

Shear: The ratio of the shear strength τ_u to the tensile strength σ_u of a rivet was found to be independent on the rivet grade, installation procedure, diameter and grip length.

Tests indicate the ratio to be about 0.75. Hence:

$$\tau_u = 0.75 \sigma_u \tag{13.49}$$

The shear resistance of a rivet is directly proportional to the available shear area and the number of critical shear planes. If a total of m critical shear planes pass through the rivet, the maximum shear resistance S_u of the rivet is equal to:

$$S_u = 0.75 m A_b \sigma_u \tag{13.50}$$

where A_b is the cross-section area of undriven rivet.

Shear and tension: The following equation must be verified:

$$\left(\frac{\tau/\tau_u}{0.75}\right)^2 + (\sigma/\sigma_u)^2 \le 1 \tag{13.51}$$

where τ is the shear stress on the rivet shear plane, σ is the tensile stress of the rivet, τ_u is the shear strength of the rivet and σ_u is the tensile strength of the rivet.

13.6 Worked Examples

Example E13.1 Verification of a Bearing Connection According to EC3

Verify, according to EC3, the connection in Figure E13.1.1 (all dimensions are in millimetres). It is a single lap bearing type joint with one shear plane. Ultimate design load, N_{Sd}, is 140 kN (31.5 kips). Bolts have a 16 mm (0.63 in.) diameter, class 8.8, not preloaded and the threaded portion of the shank is located in the bearing length. Holes have 18 mm (0.709 in.) diameter. Plates to be connected by bolts are 150 mm (5.91 in.) wide and 5 mm (0.197 in.) thick. The steel of the plates is S235.

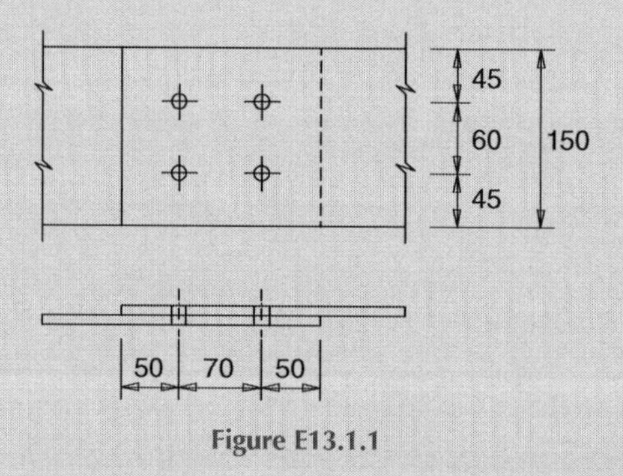

Figure E13.1.1

Procedure

The verification of this bearing type bolted connection goes through the following steps:

- check the positioning of the holes (spacing and end and edge distances);
- evaluation of shear design force for each shear plane of each bolt (V_{Ed});
- evaluation of the design shear resistance for each shear plane ($F_{v.Rd}$);
- evaluation of the design bearing resistance ($F_{b.Rd}$);
- evaluation of design ultimate tensile resistance of the connected plate net cross-section at holes ($N_{u,Rd}$).

Solution

- Check of positioning of the holes: with reference to EC3 prescriptions about spacing and edge distances of holes (Table 13.4) we get:

$p_1 \geq 2.2 \cdot d_0$	$70\text{ mm} \geq (2.2 \cdot 18)$	$= 39.6\text{ mm}$
$p_1 \leq \min(14t_{\min};200\text{ mm})$	$70\text{ mm} \leq \min(14 \cdot 5;200\text{ mm})$	$= 70.0\text{ mm}$
$p_2 \geq 2.4 \cdot d_0$	$60\text{ mm} \geq (2.4 \cdot 18)$	$= 43.2\text{ mm}$
$p_2 \leq \min(14t_{\min};200\text{ mm})$	$60\text{ mm} \leq \min(14 \cdot 5;200\text{ mm})$	$= 70.0\text{ mm}$
$e_1 \geq 1.2 \cdot d_0$	$50\text{ mm} \geq (1.2 \cdot 18)$	$= 21.6\text{ mm}$
$e_1 \leq 40\text{ mm} + 4t_{\min}$	$50\text{ mm} \leq (40 + 4 \cdot 5)$	$= 60.0\text{ mm}$
$e_2 \geq 1.2 \cdot d_0$	$45\text{ mm} \geq (1.2 \cdot 18)$	$= 21.6\text{ mm}$
$e_2 \leq 40\text{ mm} + 4t_{\min}$	$45\text{ mm} \leq (40 + 4 \cdot 5)$	$= 60.0\text{ mm}$

Design shear load for each bolt: $F_{v,Sd} = \frac{N_{Sd}}{4} = \frac{140}{4} = 35\,\text{kN}\,(7.87\,\text{kips})$

- Design shear resistance per bolt (Eq. (13.16a)):

$$F_{v,Rd} = \frac{0.6 \cdot f_{ub} \cdot A_s}{\gamma_{M2}} = \frac{0.6 \times 800 \times 157}{1.25} \cdot 10^{-3} = 60.3\,\text{kN}\,(13.56\text{ kips})$$

Check:

$$F_{V,Sd} = 35\,\text{kN} \leq F_{V,Rd} = 60.3\,\text{kN} \quad \text{OK}$$

- Design bearing resistance (Eqs. (13.18), (13.19a,b) and (13.20a,b))
  edge row bolt:

$$\alpha_b = \min\left\{\frac{e_1}{3 \cdot d_0}; \frac{f_{ub}}{f_u}; 1.0\right\} = \min(0.926; 1.123; 2.222; 1) = 0{,}926$$

$$k_1 = \min\left\{2.8 \cdot \frac{e_2}{d_0} - 1.7; 2.5\right\} = \min\{5.3; 2.5\} = 2.5$$

$$F_{b,Rd,ext} = \frac{k_1 \cdot \alpha_b \cdot f_u \cdot d \cdot t}{\gamma_{M2}} = \frac{2.5 \times 0.926 \times 360 \times 16 \times 5}{1.25} \cdot 10^{-3} = 53.3\,\text{kN}\,(11.98\text{ kips})$$

  internal row bolt:

$$\alpha_b = \min\left\{\frac{p_1}{3 \cdot d_0} - \frac{1}{4}; \frac{f_{ub}}{f_u}; 1{,}0\right\} = \min(1.046; 2.222; 1) = 1.000$$

$$k_1 = \min\left\{1.4 \cdot \frac{p_2}{d_0} - 1.7; 2.5\right\} = \min\{2.769; 2.5\} = 2.5$$

$$F_{b,Rd,\text{int}} = \frac{k_1 \cdot \alpha_b \cdot f_u \cdot d \cdot t}{\gamma_{M2}} = \frac{2.5 \times 1.000 \times 360 \times 16 \times 5}{1.25} \cdot 10^{-3} = 57.6\text{ kN}\,(12.95\text{ kips})$$

Check:

$$F_{v,Sd} = 35\text{kN} \leq \min\left(F_{b,Rd,\text{int}}; F_{b,Rd,ext}\right) = \min(53.3; 57.6) = 53.3\text{kN} = F_{b,Rd,ext}\ \text{OK}$$

- Plate design tensile resistance (Eqs. (5.3a) and (5.3b))
Compute gross area, A

$$A = 5\times150 = 750\,\text{mm}^2\ (1.163\ \text{in.}^2)$$

Compute net area, A_{net}

$$A_{net} = (5\times150) - 2\times(5\times18) = 570\,\text{mm}^2\ (0.884\ \text{in.}^2)$$

$$N_{pl,Rd} = \frac{A\cdot f_y}{\gamma_{M0}} = \frac{750\times235}{1.00}\cdot10^{-3} = 176.3\,\text{kN}\ (39.6\ \text{kips})$$

$$N_{u,Rd} = 0.9\frac{A_{net}\cdot f_u}{\gamma_{M2}} = \frac{0.9\times570\times360}{1.25}\cdot10^{-3} = 147.7\ \text{kN}\ (33.2\ \text{kips})$$

Check:

$$N_{Ed} = 140\ \text{kN} \le \min\{N_{pl,Rd}; N_{u,Rd}\} = \min\{176.3;147.7\} = 147.7\ \text{kN}\quad \text{OK}$$

Example E13.2 Verification of a Bearing Connection According to AISC 360-10

Verify, according to AISC 360-10, the connection in Figure E13.2.1 (all dimensions are in inches). It is a single lap bearing-type joint with one shear plane. The required shear strength for LRFD combinations V_u is 31.5 kips (140.1 kN). The required shear strength for ASD combinations V_a is 21 kips (93.4 kN). Bolts are 5/8 in. (15.9 mm) in diameter, ASTM A325, not preloaded and the threaded portion of the shank is located in the bearing length. Holes have a 11/16 in. (17.5 mm) diameter. Plates to be connected by bolts are 6 in. (152.4 mm) wide and 0.2 in. (5.1 mm) thick. The steel of the plates is ASTM A36: $F_y = 36$ ksi (248 MPa), $F_u = 58$ ksi (400 MPa).

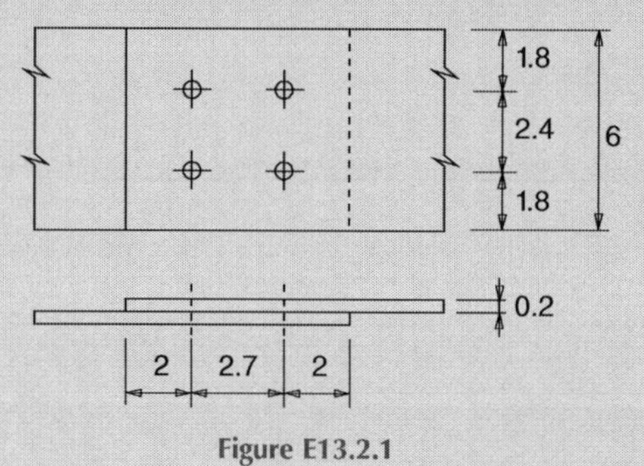

Figure E13.2.1

Check of positioning of the holes
The distance between centres of standard holes should not be less than 2–2/3 times the nominal diameter:

$$d = 5/8\ \text{in.} = 0.625\ \text{in.}\,(15.9\ \text{mm})$$

$$\min\{2.4;2.7\}/d = 2.4/d = 2.4/0.625 = 3.8 > 2-2/3\quad \text{OK}$$

Check of design shear strength
Minimum edge distance from centre of standard hole to every edge of connected part (see Table 13.7a) for 5/8 bolt: 7/8 in. = 0.875 in. (22.2 mm)

$$\min(2;1.8) = 1.8\ \text{in.} > 0.875\ \text{in.}\quad \text{OK}$$

Nominal shear strength of a bolt in bearing-type connection (see Table 13.11):

$$F_{nv} = 54 \text{ ksi}(372 \text{ MPa}).$$

$$\text{Nominal unthreaded bolt area}: A_b = 0.307 \text{ in.}^2 \left(198 \text{ mm}^2\right)$$

LRFD approach:
Design shear strength of a bolt (1 shear plane):

$$\phi R_n = \phi F_{nv} A_b = 0.75 \times 54 \times 0.307 = 12.43 \text{ kips } (55.3 \text{ kN})$$

Check of design bearing strength
Design bearing strength at the bolt hole (deformation at service loads *is not* a design consideration):

edge row bolt:

$$\phi \cdot 1.5 l_c t F_u = 0.75 \times 1.5 \times [2 - 0.5 \times (11/16)] \times 0.2 \times 58 = 21.61 \text{ kips}$$

$$\phi \cdot 3.0 dt F_u = 0.75 \times 3.0 \times (5/8) \times 0.2 \times 58 = 16.31 \text{ kips}$$

$$\phi R_n = \min\{\phi \cdot 1.5 l_c t F_u; \phi \cdot 3.0 dt F_u\} = \min\{21.61; 16.31\} = 16.31 \text{ kips } (72.6 \text{ kN})$$

internal row bolt:

$$\phi \cdot 1.5 l_c t F_u = 0.75 \times 1.5 \times [2.7 - (11/16)] \times 0.2 \times 58 = 26.26 \text{ kips}$$

$$\phi \cdot 3.0 dt F_u = 0.75 \times 3.0 \times (5/8) \times 0.2 \times 58 = 16.31 \text{ kips}$$

$$\phi R_n = \min\{\phi \cdot 1.5 l_c t F_u; \phi \cdot 3.0 dt F_u\} = \min\{26.26; 16.31\} = 16.31 \text{ kips } (72.6 \text{ kN})$$

Design strength of a bolt (minimum between shear and bearing strength):

$$\phi R_n = \min\{12.43; 16.31\} = 12.43 \text{ kips } (55.3 \text{ kN})$$

Design strength of the connection:

$$12.43 \times 4 = 49.7 \text{ kips } (221 \text{ kN}) > V_u = 31.5 \text{ kips}(140.1 \text{ kN}) \quad \text{OK}$$

ASD approach:
Allowable shear strength R_n/Ω of a bolt:

$$R_n/\Omega = F_{nv} A_b/\Omega = 54 \times 0.307/2.00 = 8.29 \text{ kips } (36.9 \text{ kN})$$

Design bearing strength at the bolt hole (deformation at service loads *is not* a design consideration):

edge row bolt:

$$1.5 l_c t F_u/\Omega = 1.5 \times [2 - 0.5 \times (11/16)] \times 0.2 \times 58/2.00 = 14.41 \text{ kips } (64.1 \text{ kN})$$

$$3.0 dt F_u/\Omega = 3.0 \times (5/8) \times 0.2 \times 58/2.00 = 10.88 \text{ kips } (48.4 \text{ kN})$$

$$R_n/\Omega = \min\{1.5 l_c t F_u/\Omega; 3.0 dt F_u/\Omega\} = \min\{14.41; 10.88\} = 10.88 \text{ kips } (48.4 \text{ kN})$$

internal row bolt:

$$1.5l_c tF_u/\Omega = 1.5\times[2.7-(11/16)]\times 0.2\times 58/2.00 = 17.51\,\text{kips}\,(77.9\text{ kN})$$

$$3.0dtF_u/\Omega = 3.0\times(5/8)\times 0.2\times 58/2.00 = 10.88\,\text{kips}\,(48.4\text{ kN})$$

$$R_n/\Omega = \min\{1.5l_c tF_u/\Omega; 3.0dtF_u/\Omega\} = \min\{17.51;10.88\} = 10.88\,\text{kips}\,(48.4\text{ kN})$$

Design strength of a bolt (minimum between shear and bearing strength):

$$\phi R_n = \min\{8.29;10.88\} = 8.29\,\text{kips}\,(36.9\text{ kN})$$

Design strength of the connection:

$$8.29\times 4 = 33.2\text{ kips}\,(147.7\text{ kN}) > V_a = 21.0\text{ kips}(93.4\text{ kN})\quad\text{OK}$$

But the maximum tensile force that can be sustained by the connected elements depends on their tensile resistance.

Connected elements tensile strength

Refer to Chapter 5 for comprehending formulas and symbols.

Tensile yielding in gross section:

$$A_g = 6.0\times 0.2 = 1.20\text{ in.}^2\,(7.74\text{ cm}^2)$$

$$LFRD: \phi A_g F_y = 0.90\times 1.20\times 36 = 38.9\,\text{kips}\,(173\text{ kN})$$

$$ASD: A_g F_y/\Omega = 1.20\times 36/1.67 = 25.9\,\text{kips}\,(115\text{ kN})$$

Tensile rupture in net section:

$$A_e = [6.0-2\times(11/16+1/16)]\times 0.2 = 0.90\text{ in.}^2\,(5.81\text{ cm}^2)$$

$$LFRD: \phi A_e F_u = 0.75\times 0.90\times 58 = 39.2\,\text{kips}\,(174\text{ kN})$$

$$ASD: A_e F_u/\Omega = 0.90\times 58/2.00 = 26.1\text{ kips}\,(116\text{ kN})$$

Tensile strength:

$$\text{LFRD: }\min\{38.9;39.2\} = 38.9\text{ kips}(173\text{ kN}) > V_u = 31.5\text{ kips}(140.1\text{ kN})\quad\text{OK}$$

$$\text{ASD: }\min\{25.9;26.1\} = 25.9\text{ kips}(115\text{ kN}) > V_a = 21.0\text{ kips}(93.4\text{ kN})\quad\text{OK}$$

So the connection is stronger than the connected plates.

Example E13.3 Evaluation of the Resistance of a Slip-Resistant Connection Subjected to Shear Force According to EC3

Evaluate shear design resistance, according to EC3, of connection illustrated in Figure E13.3.1 (all dimensions are in millimetres). The connection is category C slip-resistant at an ultimate limit state (see Table 13.5). Bolts have a 20 mm (0.787 in.) diameter, class 10.9, preloaded. Holes have a 22 mm (0.866 in.) diameter. Each bolt has two friction surfaces. Elements to be connected and cover plates are made of S235 steel. Friction surfaces are in class A (surfaces blasted with shot or grit with loose rust removed, not pitted).

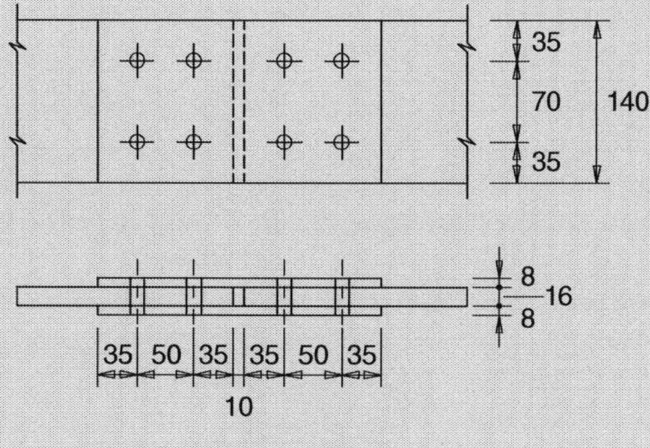

Figure E13.3.1

Procedure

Connection design resistance is computed with the following steps:

- check of positioning of the holes (spacing and end and edge distances);
- evaluation of minimum bolt preloading force ($F_{p.C}$);
- evaluation of design slip resistance at ultimate limit states ($F_{s,Rd}$);
- evaluation of the design bearing resistance ($F_{b,Rd}$);
- evaluation of design ultimate tensile resistance of cover plate net cross-section at holes ($F_{d,u,Rd}$);

Solution

- check of positioning of the holes: with reference to EC3 prescriptions about spacing and edge distances of the holes (Table 13.4) we get:

$p_1 \geq 2.2{\cdot}d_0$	$50\text{ mm} \geq (2.2{\cdot}22)$	$= 48.4\text{ mm}$
$p_1 \leq \min(14t_{\min};200\text{ mm})$	$50\text{ mm} \leq \min(14{\cdot}8;200\text{ mm})$	$= 112.0\text{ mm}$
$p_2 \geq 2.4{\cdot}d_0$	$70\text{ mm} \geq (2.4{\cdot}22)$	$= 52.8\text{ mm}$
$p_2 \leq \min(14t_{\min};200\text{ mm})$	$70\text{ mm} \leq \min(14{\cdot}8;200\text{ mm})$	$= 112.0\text{ mm}$
$e_1 \geq 1.2{\cdot}d_0$	$35\text{ mm} \geq (1.2{\cdot}22)$	$= 26.4\text{ mm}$
$e_1 \leq 40\text{ mm} + 4t_{\min}$	$35\text{ mm} \leq (40 + 4{\cdot}8)$	$= 72.0\text{ mm}$
$e_2 \geq 1.2{\cdot}d_0$	$35\text{ mm} \geq (1.2{\cdot}22)$	$= 26.4\text{ mm}$
$e_2 \leq 40\text{ mm} + 4t_{\min}$	$35\text{ mm} \leq (40 + 4{\cdot}8)$	$= 72.0\text{ mm}$

- evaluation of minimum bolt preloading force (Eq. (13.22b))

$$F_{p,C} = 0.7{\cdot}f_{tb}{\cdot}A_s = 0.7 \times 1000 \times 245{\cdot}10^{-3} = 171.5\text{kN (38.56 kips)}$$

- For class A friction surfaces assume $\mu = 0.5$ for computing design slip resistance (Eq. (13.22a)) and consider two friction surfaces per bolt.

$$F_{s.Rd} = \frac{k_s{\cdot}n{\cdot}\mu}{\gamma_{M3}}{\cdot}F_{p.C} = \frac{1 \times 2 \times 0.5}{1.25} \times 171.5 = 137.2\text{kN (30.84 kips)}$$

Compute design slip resistance of the connection:

$$F_{d,S,Rd} = 4 \cdot F_{s,Rd} = 4 \times 137.2 = 548.8\,\text{kN}\,(123.4\text{ kips})$$

- Design bearing resistance of each plate (Eqs. (13.18), (13.19a,b) and (13.20a,b))
  edge row bolt:

$$\alpha_b = \min\left\{\frac{e_1}{3 \cdot d_0}; \frac{f_{ub}}{f_u}; 1.0\right\} = \min(0.530; 2.778; 1) = 0.530$$

$$k_1 = \min\left\{2.8 \cdot \frac{e_2}{d_0} - 1.7; 2.5\right\} = \min\{2.755; 2.5\} = 2.5$$

$$F_{b,Rd,ext} = \frac{k_1 \cdot \alpha_b \cdot f_u \cdot d \cdot t}{\gamma_{M2}} = \frac{2.5 \times 0.530 \times 360 \times 20 \times 8}{1.25} \cdot 10^{-3} = 61.05\,\text{kN}\,(13.73\,\text{kips})$$

internal row bolt:

$$\alpha_b = \min\left\{\frac{p_1}{3 \cdot d_0} - \frac{1}{4}; \frac{f_{ub}}{f_u}; 1.0\right\} = \min(0.508; 2.778; 1) = 0.508$$

$$k_1 = \min\left\{1.4 \cdot \frac{p_2}{d_0} - 1.7; 2.5\right\} = \min\{2.755; 2.5\} = 2.5$$

$$F_{b,Rd,\text{int}} = \frac{k_1 \cdot \alpha_b \cdot f_u \cdot d \cdot t}{\gamma_{M2}} = \frac{2.5 \times 0.508 \times 360 \times 20 \times 8}{1.25} \cdot 10^{-3} = 58.52\,\text{kN}\,(13.16\,\text{kips})$$

Design bearing resistance of the connection (two plates, two edge holes and two internal holes for each plate)

$$F_{b,Rd} = 2 \times (2 \times 61.05 + 2 \times 58.52) = 478.28\,\text{kN}\,(107.5\,\text{kips})$$

- Cover plates design tensile resistance (Eqs. (5.3a) and (5.3b))
  Cover plate gross area,
  $AA = 8 \times 140 = 1120\text{ mm}^2$

$$N_{pl,Rd} = \frac{A \cdot f_y}{\gamma_{M0}} = \frac{1120 \times 235}{1.00} \cdot 10^{-3} = 263.2\,\text{kN}\,(59.2\text{ kips})$$

Cover plate net area, A_{net}:

$$A_{\text{net}} = (8 \times 140) - 2 \cdot (8 \times 22) = 768\,\text{mm}^2\,(1.19\text{ in.}^2)$$

$$N_{u,Rd} = 0.9 \frac{A_{net} \cdot f_u}{\gamma_{M2}} = 0.9 \times \frac{768 \times 360}{1.25} \cdot 10^{-3} = 199.1\,\text{kN}\,(44.75\,\text{kips})$$

$$F_{d,u,Rd} = 2 \cdot \min\{N_{pl,Rd}; N_{u,Rd}\} = 2 \cdot \min\{263.2; 199.1\} = 2 \times 199.1$$
$$= 398.2\,\text{kN}\,(89.52\,\text{kips})$$

Design shear resistance of the connection is the minimum between design slip resistance (548 kN), design bearing resistance (478.28 kN) and cover plates design tensile resistance (398.2 kN): so it is 398.2 kN.

Example E13.4 Evaluation of the Resistance of a Slip-Critical Connection Subjected to Shear Force According to AISC 360-10

Evaluate shear design resistance, according to AISC 360-10, of connection illustrated in Figure E13.4.1 (all dimensions are in inches). The connection is slip-critical. Bolts have a ¾ in. (19 mm) diameter, ASTM A490, preloaded. Holes have a 13/16 in. (20.6 mm) diameter. Each bolt has two friction surfaces. Elements to be connected and cover plates are made of ASTM A36 steel: $F_y = 36$ ksi (248 MPa), $F_u = 58$ ksi (400 MPa). Friction surfaces are class B: uncoated blast-cleaned steel surfaces.

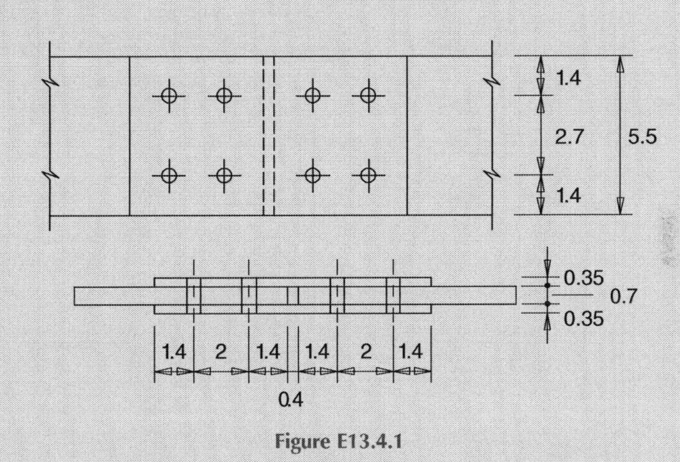

Figure E13.4.1

Check of positioning of the holes

The distance between centres of standard holes shouldn't be less than 2–2/3 times the nominal diameter:

$$d = 3/4 \text{ in.} = 0.75 \text{ in.} (19.1 \text{ mm})$$

$$\min\{2;2.7\}/d = 2/d = 2/0.75 = 2.67 > 2-2/3 = 2.667 \quad \text{OK}$$

Minimum edge distance from centre of standard hole to every edge of connected cart (see Table 13.7a) for 3/4 bolt: 1 in. (25.4 mm)

$$\min(1.4;1.4) = 1.4 \text{ in.} > 1 \text{ in.} \quad \text{OK}$$

Check of design slip resistance

Minimum bolt pretension (from Table 13.9a):

$$T_b = 35 \text{ kips} (155.7 \text{ kN})$$

For class B faying surfaces assume $\mu = 0.5$ for computing bolt design slip resistance (Eq. (13.39)) and consider: two slip planes per bolt, one filler and no tension action applied to the connection.

$$R_n = \mu D_u h_f T_b n_s k_{sc} = 0.5 \times 1.13 \times 1 \times 35 \times 2 \times 1 = 39.55 \text{ kips} (176 \text{ kN})$$

Compute design slip resistance of the connection:
LRFD ($\phi = 1.00$):

$$4 \cdot \phi R_n = 4 \times 1.00 \times 39.55 = 158.2\,\text{kps}\,(703.7\,\text{kN})$$

ASD ($\Omega = 1.50$):

$$4 \cdot R_n / 1.50 = 4 \times 39.55 / 1.50 = 105.5\,\text{kips}\,(469.3\,\text{kN})$$

Check of design shear resistance
Nominal shear strength of a bolt in bearing-type connection (A490 bolts when threads are excluded from shear planes, see Table 13.11):

$$F_{nv} = 84\ \text{ksi}(579\ \text{MPa}).$$

$$\text{Nominal unthreaded bolt area}: A_b = 0.442\ \text{in.}^2\ \left(285\ \text{mm}^2\right)$$

LRFD approach:
Design shear strength of a bolt (two shear planes):

$$\phi R_n = \phi F_{nv} A_b = 0.75 \times 84 \times (2 \times 0.442) = 55.7\,\text{kips}\,(247.8\,\text{kN})$$

Design bearing strength at the bolt hole (deformation at service loads *is* a design consideration):

edge row bolt:

$$\phi \cdot 1.2 l_c t F_u = 0.75 \times 1.2 \times [1.4 - 0.5 \times (13/16)] \times 0.7 \times 58 = 36.31\ \text{kips}\,(161.5\,\text{kN})$$

$$\phi \cdot 2.4 d t F_u = 0.75 \times 2.4 \times (3/4) \times 0.7 \times 58 = 54.81\,\text{kips}\,(243.8\ \text{kN})$$

$$\phi R_n = \min\{\phi \cdot 1.2 l_c t F_u; \phi \cdot 2.4 d t F_u\} = \min\{36.31; 54.81\} = 36.31\ \text{kips}\,(161.5\ \text{kN})$$

internal row bolt:

$$\phi \cdot 1.2 l_c t F_u = 0.75 \times 1.2 \times [2.0 - (13/16)] \times 0.7 \times 58 = 43.39\,\text{kips}\,(193\,\text{kN})$$

$$\phi \cdot 2.4 d t F_u = 0.75 \times 2.4 \times (3/4) \times 0.7 \times 58 = 54.81\,\text{kips}\,(243.8\ \text{kN})$$

$$\phi R_n = \min\{\phi \cdot 1.2 l_c t F_u; \phi \cdot 2.4 d t F_u\} = \min\{43.39; 54.81\} = 43.39\,\text{kips}\,(193\ \text{kN})$$

Design strength for an edge bolt (minimum between shear and bearing strength):

$$\phi R_n = \min\{55.7; 43.39\} = 43.39\,\text{kips}\,(193\,\text{kN})$$

Design strength for an internal bolt (minimum between shear and bearing strength):

$$\phi R_n = \min\{55.7; 36.31\} = 36.31\ \text{kips}\,(161.5\,\text{kN})$$

Design strength of the connection:

$$36.31 \times 2 + 43.39 \times 2 = 159.4\ \text{kips}\,(709\ \text{kN})$$

ASD approach:
Allowable shear strength of a bolt (two shear planes):

$$R_n/\Omega = F_{nv}A_b/\Omega = 84 \times (2 \times 0.442)/2.00 = 37.12\,\text{kips}\,(165.2\,\text{kN})$$

Design bearing strength at the bolt hole (deformation at service loads *is not* a design consideration):

edge row bolt:

$$1.2l_ctF_u/\Omega = 1.2 \times [1.4 - 0.5 \times (13/16)] \times 0.7 \times 58/2.00 = 24.21\,\text{kips}\,(48.4\,\text{kN})$$
$$2.4dtF_u/\Omega = 2.4 \times (3/4) \times 0.7 \times 58/2.00 = 36.54\,\text{kips}\,(162.5\,\text{kN})$$
$$R_n/\Omega = \min\{1.2l_ctF_u/\Omega;2.4dtF_u/\Omega\} = \min\{24.21;36.54\} = 24.21\,\text{kips}\,(107.7\,\text{kN})$$

internal row bolt:

$$1.2l_ctF_u/\Omega = 1.2 \times [2.0 - (13/16)] \times 0.7 \times 58/2.00 = 28.93\,\text{kips}\,(128.7\,\text{kN})$$
$$2.4dtF_u/\Omega = 2.4 \times (3/4) \times 0.7 \times 58/2.00 = 36.54\,\text{kips}\,(162.5\,\text{kN})$$
$$R_n/\Omega = \min\{1.2l_ctF_u/\Omega;2.4dtF_u/\Omega\} = \min\{28.93;36.54\} = 28.93\,\text{kips}\,(128.7\,\text{kN})$$

Design strength for an edge bolt (minimum between shear and bearing strength):

$$R_n/\Omega = \min\{37.12;24.21\} = 24.21\,\text{kips}\,(107.7\,\text{kN})$$

Design strength for an internal bolt (minimum between shear and bearing strength):

$$R_n/\Omega = \min\{37.12;28.93\} = 28.93\,\text{kips}\,(128.7\,\text{kN})$$

Design strength of the connection:

$$24.21 \times 2 + 28.93 \times 2 = 106.3\,\text{kips}\,(472.8\,\text{kN})$$

Resistance of the connection.

LRFD approach:

Design slip resistance:	158.2 kps	(703.7 kN)
Design shear and bearing resistance:	159.4 kips	(709 kN)

ASD approach:

Design slip resistance:	105.5 kps	(469.3 kN)
Design shear and bearing resistance:	106.3 kips	(472.8 kN)

The design resistance of the connection is the minimum resistance between slip resistance and shear and bearing resistance. So in this case we have:

LRFD approach:

$$\text{Design connection resistance}: 158.2 \text{ kps } (703.7 \text{ kN})$$

ASD approach:

$$\text{Design connection resistance}: 105.5 \text{ kips } (469,3 \text{ kN})$$

Design connection resistance means that the connection does not slip at service loads and it does not fail for bearing or bolt shear at ultimate loads.
But the maximum tensile force that can be sustained by the connected elements depends on their tensile resistance and on cover plates' tensile resistance.
Connected elements or cover plates tensile strength
Refer to Chapter 5 for comprehending formulas and symbols.
Connected elements area is equal to cover plate area, so verify just connected elements.

Tensile yielding in gross section:

$$A_g = 5.5 \times 0.7 = 3.85 \text{ in.}^2 (24.8 \text{ cm}^2)$$

$$\text{LFRD: } \phi A_g F_y = 0.90 \times 3.85 \times 36 = 124.7 \text{ kips } (555 \text{ kN})$$

$$\text{ASD: } A_g F_y / \Omega = 3.85 \times 36 / 1.67 = 83 \text{ kips } (369 \text{ kN})$$

Tensile rupture in net section:

$$A_e = [5.5 - 2 \times (13/16 + 1/16)] \times 0.7 = 2.63 \text{ in.}^2 (17 \text{ cm}^2)$$

$$\text{LFRD: } \phi A_e F_u = 0.75 \times 2.63 \times 58 = 113.4 \text{ kips } (509 \text{ kN})$$

$$\text{ASD: } A_e F_u / \Omega = 2.63 \times 58 / 2.00 = 76.3 \text{ kips } (339 \text{ kN})$$

Tensile strength:

$$\text{LFRD: } \min\{124.7; 113.4\} = 113.4 \text{ kips}(509 \text{ kN})$$
$$\text{ASD: } \min\{83; 76.3\} = 76.3 \text{ kips}(339 \text{ kN})$$

So the connection is stronger than the connected plates.

CHAPTER 14
Welded Connections

14.1 Generalities on Welded Connections

Welding is an assembling process that allows us to permanently join two metallic elements causing fusion of the adjoining parts. When comparing welded connections to bolted, nailed or riveted ones, it is apparent that the former are inherently monolithic and are at the same time stiffer and less complicated, allowing more freedom to the designer. These advantages are balanced by the need of additional detailing and fabrication requirements, especially for that which concerns the assurance and verification of the quality of welded joints, in order to prevent potential partial loss of strength or stiffness, or possibly brittle fractures. This is the reason why the welding process should always be performed by qualified welders. Additionally, in the presence of cyclic loads, fatigue design becomes particularly important, both when a large number of cycles is expected ($>10^4$) and when low-cycle fatigue is to be considered. In fact, the weld area, because of the stress concentrations induced by both thermal effects and load path effects, is a critical location for the formation and propagation of cracks.

In welded connections the connected elements are identified as *base material*, while the *weld material*, when applicable, refers to the material that is added to the joint in its liquid state during the welding process. A classification of welding processes can be made from:

- *autogenous processes*: the base metal participates to the formation of the joint by fusion or crystallization with the weld metal, if present. The oldest autogenous welding process, in use for several millennia, is forge welding. In the Bronze Age, people would heat the base material to a cherry red colour and pound it together until bonding occurred. Modern autogenous processes are typically characterized by a combined fusion of both base material and weld material. These processes are classified basing on the specific technique employed to attain sufficient heat input, as well as on the basis of protecting the weld pool, which is the combination of fused materials in the weld region during the welding process. The most common processes are: oxy-acetylene (oxyfuel) welding, arc welding with consumable or non-consumable electrodes, submerged arc welding (SAW), shielded metal arc welding (SMAW), gas metal arc welding (GMAW), also known as metal inert gas welding (MIG), metal active gas welding (MAG), gas tungsten arc welding (GTAW), also known as tungsten inert gas welding (TIG), and electroslag welding (ESW), used mostly for automatic applications for large welds.

Structural Steel Design to Eurocode 3 and AISC Specifications, First Edition. Claudio Bernuzzi and Benedetto Cordova.

- *heterogeneous processes*: in these processes, only the base material is the weld material used at a temperature lower than the melting temperature of the base material. A classic example is provided by soldering or brazing processes.

As a consequence of the metallurgical phenomena (creation of the weld, solidification of the weld pool and thermal effects in the base material surrounding the weld region, known as heat affected zone – HAZ), there can be *defects* in the welded connection. These are classified into *metallurgical* and *geometric defects*.

Defects affect the proper strength performance of the welds; their potential presence must be ascertained in order to avoid potentially dangerous conditions during service. Among the most important metallurgical defects, there are:

- *cracks*: that is the typical discontinuities generated by tearing of the material (Figure 14.1), which can be classified into *hot cracks* or *cold cracks*. During the welding process, in the weld pool there are impurities that segregate in preferential zones, and then solidify at lower temperatures with respect to the base metal. This causes a loss of cohesion of the material, due to shrinkage stresses (which arise during the cooling process), causing in turns the formation of cracks. These are defined as *hot cracks* and are influenced by the carbon content, the presence of impurities in the metal and by shrinkage effects of the weld. The *cold cracks* arise near or even after the conclusion of the cooling process (even within 48 hours from the end of welding process), and are due to the absorption of hydrogen during the formation of the weld pool by both base and weld materials;
- *lamellar tearing* (Figure 14.2): a special family of cracks originated from tensile stresses can be found in the base material, perpendicular to the rolling direction of the material. The main cause of lamellar tearing can be identified in high shrinkage stresses developing during the cooling process, especially when the base material is characterized by large thickness and prevented deformation;
- *inclusions*: that is anomalous regions within the weld due to the presence in the weld pool of materials other than the base and the weld metal. There can be solid inclusions (e.g. slag or tungsten) or gaseous inclusions (gas pockets created by gases trapped within the weld pool).

The most important geometric defects can be listed as follows:

- *excess of weld metal*: this defect takes place when an excessive amount of weld metal is deposited in the weld. This can have deleterious effects due to the potential discontinuities that can be

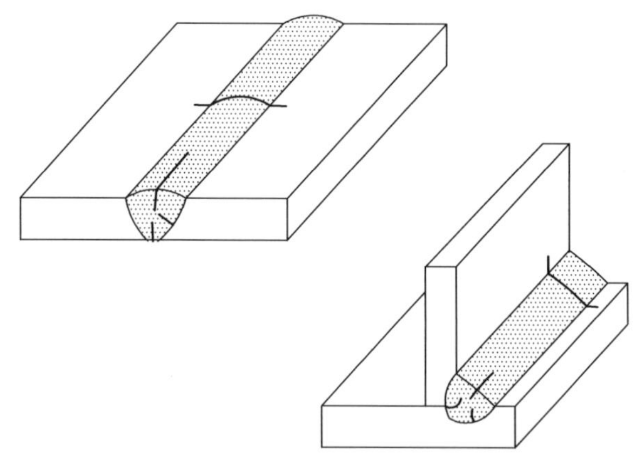

Figure 14.1 Cracks in welds.

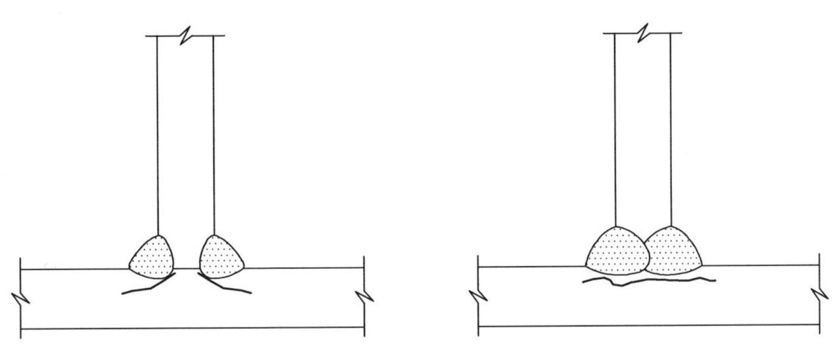

Figure 14.2 Typical cases of lamellar tearing.

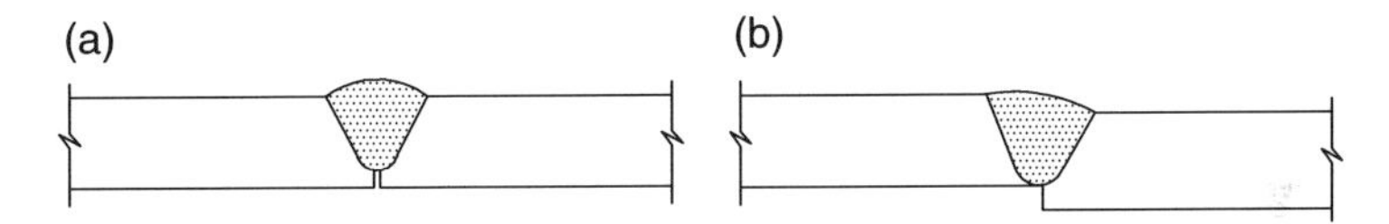

Figure 14.3 Weld defects: (a) lack of penetration and (b) lack of alignment.

created, which in turn can be dangerous for particular service conditions (e.g. fatigue, impact loads, low temperatures);

- *lack of penetration (lack of fusion)*: this defect arises when there are regions in the joint area in which the weld pool has not reached the desired depth, thus creating discontinuities within the welded connection (Figure 14.3a);
- *lack of alignment*: this defect is due to an improper alignment of the connected elements; this can cause a non-negligible change in the geometry of the joined parts (Figure **14.3**b), thus creating an eccentricity that is not accounted for during design.

In most instances, welded connections must be inspected in order to ascertain the presence of defects. In most situations, non-destructive tests (NDTs) are performed, which do not affect the proper performance of the joint in service. Among these testing techniques, there are: visual inspections, dye penetrant testing, magnetic particle testing, ultrasonic testing, testing with radiation imaging systems, radiographic testing and eddy current testing.

In the following paragraph, a concise list of norms and specifications is provided, subdivided by NDT technique used.

14.1.1 European Specifications

14.1.1.1 Visual Testing

- EN 1330-10: Non-destructive Testing – Terminology – Part 10: Terms Used In Visual Testing;
- EN ISO 17637: Non-destructive testing of welds – Visual testing of fusion-welded joints.

14.1.1.2 Dye-Penetrant Testing

- EN 571-1: Non-destructive testing – Penetrant testing – Part 1: General principles;
- EN ISO 3059: Non-destructive testing – Penetrant testing and magnetic particle testing – Viewing conditions;

- EN ISO 3452-2: Non-destructive testing – Penetrant testing – Part 2: Testing of penetrant materials;
- EN ISO 3452-3: Non-destructive testing – Penetrant testing – Part 3: Reference test blocks;
- EN ISO 3452-4: Non-destructive testing – Penetrant testing – Part 4: Equipment;
- EN ISO 3452-5: Non-destructive testing – Penetrant testing – Part 5: Penetrant testing at temperatures higher than 50°C;
- EN ISO 3452-5: Non-destructive testing – Penetrant testing – Part 5: Penetrant testing at temperatures higher than 50°C;
- EN ISO 3452-6: Non-destructive testing – Penetrant testing – Part 6: Penetrant testing at temperatures lower than 10°C;
- EN ISO 3452-6: Non-destructive testing – Penetrant testing – Part 6: Penetrant testing at temperatures lower than 10°C;
- EN ISO 12706: Non-destructive testing – Penetrant testing – Vocabulary;
- EN ISO 23277: Non-destructive testing of welds – Penetrant testing of welds – Acceptance levels;
- EN ISO 10893-4: Non-destructive testing of steel tubes – Part 4: Liquid penetrant inspection of seamless and welded steel tubes for the detection of surface imperfections.

14.1.1.3 Magnetic Particle Testing

- UNI EN 1330-7: Non-destructive testing – Terminology – Part 7: Terms used in magnetic particle testing;
- EN ISO 3059: Non-destructive testing – Penetrant testing and magnetic particle testing – Viewing conditions;
- EN ISO 9934-1: Non-destructive testing – Magnetic particle testing – Part 1: General principles;
- EN ISO 9934-2: Non-destructive testing – Magnetic particle testing – Part 2: Detection media;
- EN ISO 9934-3: Non-destructive testing – Magnetic particle testing – Part 3: Equipment;
- EN ISO 17638: Non-destructive testing of welds – Magnetic particle testing;
- EN ISO 23278: Non-destructive testing of welds – Magnetic particle testing of welds – Acceptance levels;
- EN ISO 10893-1: Non-destructive testing of steel tubes – Part 1: Automated electromagnetic testing of seamless and welded (except submerged arc-welded) steel tubes for the verification of hydraulic leak tightness;
- EN ISO 10893-3: Non-destructive testing of steel tubes – Part 3: Automated full peripheral flux leakage testing of seamless and welded (except submerged arc-welded) ferromagnetic steel tubes for the detection of longitudinal and/or transverse imperfections;
- EN ISO 10893-5: Non-destructive testing of steel tubes – Part 5: Magnetic particle inspection of seamless and welded ferromagnetic steel tubes for the detection of surface imperfections.

14.1.1.4 Radiographic Testing

- EN 444: Non-destructive testing – General principles for radiographic examination of metallic materials by X- and gamma-rays;
- EN 1330-3: Non-destructive testing – Terminology – Part 3: Terms used in industrial radiographic testing;
- EN 1435: Non-destructive examination of welds – Radiographic examination of welded joints;
- EN ISO 10893-6: Non-destructive testing of steel tubes – Part 6: Radiographic testing of the weld seam of welded steel tubes for the detection of imperfections;

- EN ISO 17635: Non-destructive testing of welds – General rules for metallic materials;
- EN 12517-1: Non-destructive testing of welds – Part 1: Evaluation of welded joints in steel, nickel, titanium and their alloys by radiography – Acceptance levels;
- EN 12517-2: Non-destructive testing of welds – Part 2: Evaluation of welded joints in aluminium and its alloys by radiography – Acceptance levels;
- EN 12681: Founding – Radiographic examination;
- EN 13100-2: Non-destructive testing of welded joints in thermoplastics semi-finished products – Part 2: X-ray radiographic testing.

14.1.1.5 Ultrasonic Testing

- EN 583-1: Non-destructive testing – Ultrasonic examination – Part 1: General principles;
- EN 1330-4:2010 Non-destructive testing – Terminology – Part 4: Terms used in ultrasonic testing;
- EN ISO 11666: Non-destructive testing of welds – Ultrasonic testing – Acceptance levels;
- EN ISO 23279: Non-destructive testing of welds – Ultrasonic testing – Characterization of indications in welds;
- EN ISO 17640: Non-destructive testing of welds – Ultrasonic testing – Techniques, testing levels and assessment;
- EN ISO 10893-11: Non-destructive testing of steel tubes – Part 11: Automated ultrasonic testing of the weld seam of welded steel tubes for the detection of longitudinal and/or transverse imperfections;
- EN ISO 10893-8: Non-destructive testing of steel tubes – Part 8: Automated ultrasonic testing of seamless and welded steel tubes for the detection of laminar imperfections;
- EN ISO 22825: Non-destructive testing of welds – Ultrasonic testing – Testing of welds in austenitic steels and nickel-based alloys;
- EN ISO 7963: Non-destructive testing – Ultrasonic testing – Specification for calibration block No. 2.

14.1.1.6 Eddy Current Testing

- EN ISO 15549: Non-destructive testing – Eddy current testing – General principles;
- EN ISO 12718: Non-destructive testing – Eddy current testing – Vocabulary;
- EN 1711: Non-destructive examination of welds – Eddy current examination of welds by complex plane analysis;
- EN ISO 10893-1: Non-destructive testing of steel tubes – Part 1: Automated electromagnetic testing of seamless and welded (except submerged arc-welded) steel tubes for the verification of hydraulic leak tightness;
- EN ISO 10893-2: Non-destructive testing of steel tubes – Part 2: Automated eddy current testing of seamless and welded (except submerged arc-welded) steel tubes for the detection of imperfections.

14.1.2 US Specifications

All structural steel welding requirements are contained in a document set forth by the American Welding Society, AWS D1.1/D1.1M:2010 Structural Welding Code – Steel (2010). In particular, inspection requirements are outlined in Section 14.6 of that document.

14.1.2.1 Visual Inspection

- AWS D1.1/D1.1M:2010 Section 6.9;
- AWS B1.11:2000 Guide for the visual examination of welds.

14.1.2.2 Dye-Penetrant Testing

- AWS D1.1/D1.1M:2010 Section 6, Part C;
- ASTM E165-09 Standard practice for liquid penetrant examination for general industry.

14.1.2.3 Magnetic Particle Testing

- AWS D1.1/D1.1M:2010 Section 6, Part C;
- ASTM E709-08 Standard Guide for Magnetic Particle Testing.

14.1.2.4 Ultrasonic Testing

- AWS D1.1/D1.1M:2010 Section 6, Part F.

14.1.2.5 Radiation Imaging Systems

- AWS D1.1/D1/1M:2010 Section 6, Part G for radiation imaging systems;
- ASTM E1000-98 Standard Guide for Radioscopy.

14.1.2.6 Radiographic Testing

- AWS D1.1/D1.1M:2010 Section 6 Part E;
- ASTM E94-04 Standard Guide for Radiographic Examination;
- ASTM E747-04 Standard Practice for Design, Manufacture and Material Grouping Classification of Wire Image Quality Indicators (IQIs) Used for Radiology;
- ASTM E1032-06 Standard Test Method for Radiographic Examination of Weldments.

14.1.3 Classification of Welded Joints

The load-resisting elements of a welded joint are the welds. Based on the relative position of the elements to be joined, there can be (Figure 14.4): butt joints, edge joints, corner joints, T-joints; L-joints (assimilated to corner joints in US practice), lap joints.

Based on the position of the weld and on the direction of the force to be transferred (Figure 14.5), there can be: longitudinal welds, transverse welds and inclined welds.

Finally, based on the type of weld, there can be: groove welds, fillet welds, slot welds and plug welds. Each type of weld is characterized by its advantages and disadvantages. The vast majority of welds are fillet welds, due to their economy and ease of fabricating, both in the field and in a shop, followed by groove welds. Plug and slot welds are limited in applicability.

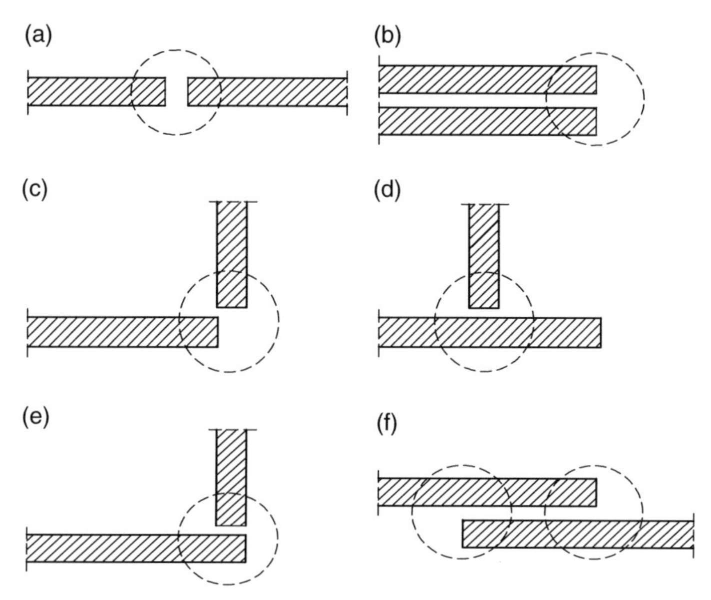

Figure 14.4 Classification based on the relative position of the elements to be joined: (a) butt joint, (b) edge joint, (c) corner joint, (d) T-joint, (e) L-joint and (f) lap joint.

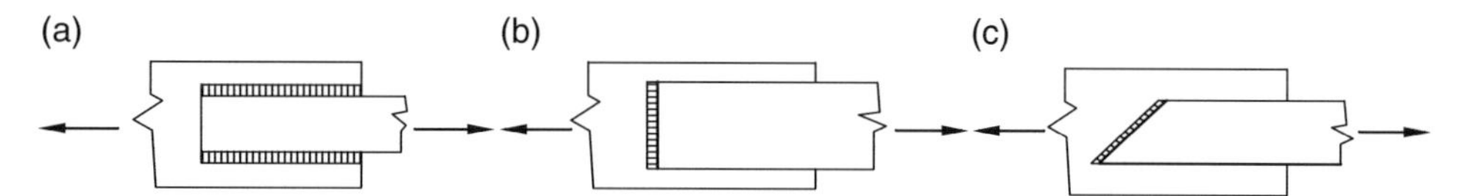

Figure 14.5 Classification based on the location of the weld with respect to the force to be transferred: (a) longitudinal, (b) transverse and (c) inclined.

Groove welds can extend through the entire thickness of the joint, in which case they are called complete joint penetration (CJP) welds, or they can extend only partially between the connecting members, in which case they are called partial joint penetration (PJP) welds. In both circumstances, groove welds require in most instances surface preparation of the connecting elements, which would affect the cost and the effectiveness of the weld. Some examples of surface preparations are shown in Figure 14.6. Fillet welds are obtained by depositing weld material to create what is usually a 45° fillet along the weld line. Normally, no surface preparation such as any of those shown in Figure 14.6 is necessary because the edge conditions resulting from shearing or flame cutting are usually acceptable for the fillet welding procedure. Plug and Slot welds are typically used in combination with fillet welds, in situations where there is no sufficient room to fully develop a fillet weld; other applications include connecting overlapping plates to prevent buckling and construction welding to keep connected members temporarily in place.

14.2 Defects and Potential Problems in Welds

Several factors contribute to the overall quality of a weld, such as the type of electrode used for arc welding, diameter of the electrode, amount of current used, configuration of the weld (horizontal, vertical, overhead), edge preparation, detailing of the weld, distortion of connected elements,

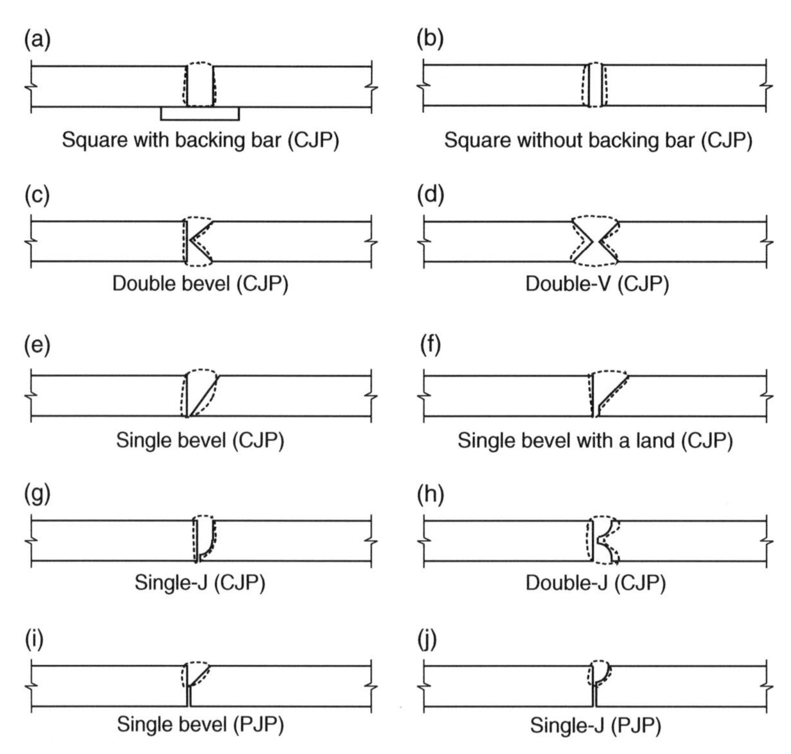

Figure 14.6 Examples of surface preparations for groove welds: (a–h) show CJP welds and (i–j) show PJP welds.

heating of the connected elements and, in no small proportion, ability of the operator. Among the most common defects, there are: lack of fusion, lack of joint penetration, undercutting, slag inclusion, porosity and cracking.

Lack of fusion is due to a poor penetration of the weld into the base metal, typically due to poorly prepared surfaces (presence of mill scale, impurities or other coatings), to excessively rapid passes of the electrode, preventing sufficient heating of the region or to an insufficient amount of current used for welding. Also, welding of two connecting members of considerably different thicknesses can cause this defect, unless the thicker part is properly pre-heated, in order to avoid excessive dispersion of heat through the larger member.

Lack of joint penetration, primarily affecting groove welds, takes place when the molten pool fails to penetrate for the whole depth of a groove, thus creating potential fracture initiation areas. The main causes for this defect relate to the choice of too large electrode, too little current or excessively fast welding passes. Lack of penetration can also be ascribed to a poor choice of weld detail (e.g. surface preparation) with respect to the welding process employed.

Undercutting of a weld happens when too large current is employed, thus digging out a trench along the direction of welding that remains unfilled. This is probably the easiest defect to visually identify and is also the simplest to rectify.

Inclusion of slag represents a discontinuity within the solidified molten pool, which can be the point of initiation of a crack and at the same time takes away from the resisting cross-section of the weld. Slag usually would float to the surface of the molten pool, unless an excessively rapid cooling traps it within. Overhead welds are particularly susceptible to this defect, as slag will float upwards. This defect, as well as the porosity defect, is not detectable by naked eye and thus ultrasonic or radiographic testing is needed to identify it.

Porosity is characterized by the presence of air or gas pockets that are trapped within the molten pool during solidification. This defect is often due to excessively high currents or to the creation of too long an arc during the welding process.

Finally, welds can develop *cracks* due to internal stresses arising during the cooling process. Cracks can be longitudinal or transverse to the weld line and they can also extend from the weld metal into the base metal. The presence of impurities can cause a crack to form when the material is still mostly molten ('hot' cracks); it's important to maintain a heat distribution as uniform as possible, also a slower cooling rate can help preventing these cracks. 'Cold' cracks are due to the presence of hydrogen (a phenomenon that can be likened to the hydrogen embrittlement noticed in high strength fasteners) and are more likely when the weld has a high degree of restraint (i.e. boundary conditions that tend to prevent free shrinkage and reconfiguration of the material). The use of electrodes with low hydrogen content, paired with pre- and post-heating of the base material, can help preventing these cracks. Weld defects become of greatest concern when the welded joint is subject to repeated cycles of loading, as fatigue phenomena (both low- and high-cycle) can substantially reduce the capacity of the connection.

14.3 Stresses in Welded Joints

In order to calculate stresses in welded joints, it is useful to introduce the concept of an *effective area*, which is also called the *effective throat area*. The effective area is calculated as the *effective throat* dimension multiplied by the length of the weld. Depending on the type of weld, this effective throat is calculated differently. For CJP groove welds, assuming that no defects are present, the effective throat dimension can be taken as the thickness of the thinner of the connecting members. For example, in a butt joint with a CJP weld, in presence of a tensile force parallel to the longitudinal axis of the connected members (Figure 14.7), the state of stress can be considered equivalent to that of a continuous member with a cross-section calculated taking the thinner of the two connected members and multiplying it by the length of the weld.

In fillet welds, the effective area is calculated multiplying the length of the fillet by an effective throat dimension that is normally taken as the height of a triangle inscribed within the fillet itself (Figure 14.8). In current practice, the effective throat dimension can be taken as the height of the largest isosceles triangle that can be inscribed within the fillet.

For plug and slot welds, the effective area (typically resisting the external actions through shearing stresses) is given by their nominal area contained in the shear plane, and usually corresponding to the effective diameter of the hole or slot dimensions filled by the weld.

The stresses acting on the effective area can be conventionally indicated using the following symbols (Figure 14.9):

- $\sigma_\perp$, which represents the normal stress, acting perpendicularly to the effective area;
- $\tau_\perp$, which represents the shearing stress in the plane of the effective area, perpendicular to the longitudinal axis of the fillet;
- $\tau_{//}$, which represents the shearing stress in the plane of the effective area, parallel to the longitudinal axis of the fillet;

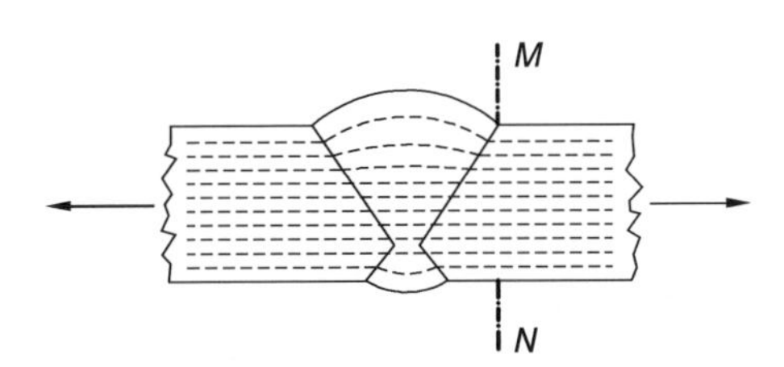

Figure 14.7 Butt joint showing the distribution of tensile stresses through the cross-section.

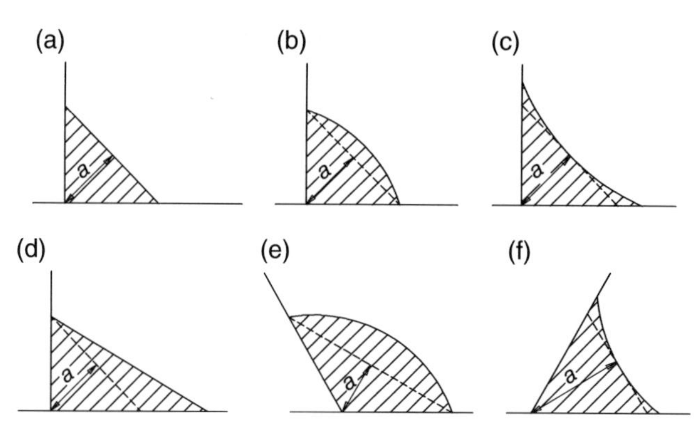

Figure 14.8 Effective throat dimension for various fillet shapes.

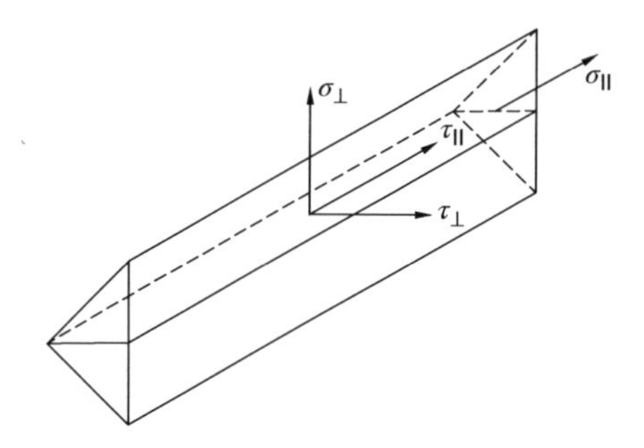

Figure 14.9 State of stress in the effective throat area.

- $\sigma_{//}$, which represents the normal stress perpendicular to the cross-section of the fillet. This stress value is usually neglected, with the notable exception of the case of fatigue checks.

In the following, some examples of typical situations are illustrated for which the relevant stress values are calculated. For the sake of simplicity, the assumption has been made that the stresses on the effective throat area are uniformly distributed.

The effective throat dimension is indicated with a, whereas L, h or b are used, as appropriate, to indicate the length of the fillet. In order to simplify the calculation of the stresses to be used for design, the effective throat area can be rotated onto the horizontal or the vertical plane, depending on whichever is most convenient.

14.3.1 Tension

In the case of a welded joint that is supposed to transmit a tensile force equal to F, the fillet welds can be parallel to the direction of the force (*longitudinal fillets*), perpendicular to the force (*transverse fillets*) or inclined through a generic angle (*inclined fillets*).

Longitudinal fillet welds: With reference to Figure 14.10, if the fillets are parallel to the force (there are a total of four fillets in the figure), the resulting stresses can be calculated directly based on the effective throat area of each fillet in its actual location, or by rotating it onto the horizontal or vertical plane. Shearing stresses are of the $\tau_{//}$ type, the amount of which is given by the following expression:

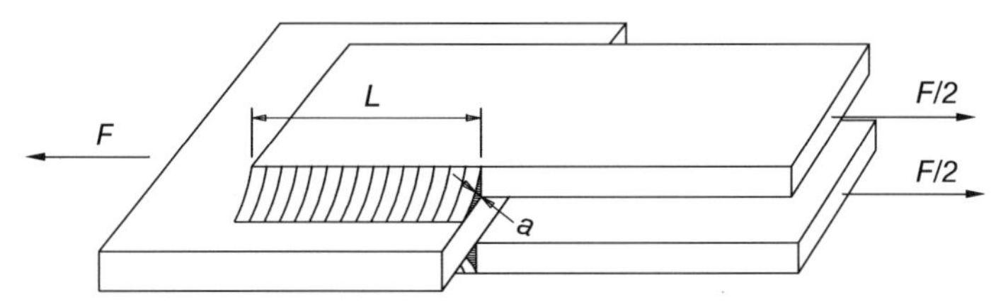

Figure 14.10 Plates connected by longitudinal fillet welds.

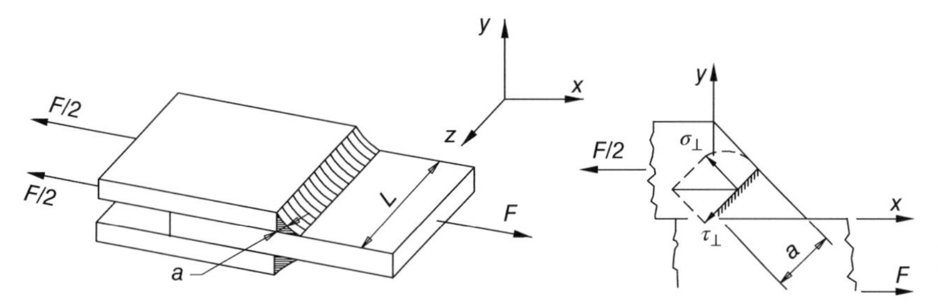

Figure 14.11 Plates connected with transverse fillet welds.

$$\tau_{//} = \frac{F}{4 \cdot L \cdot a} \tag{14.1}$$

Transverse fillet welds: With reference to Figure 14.11, if the (two) fillets are perpendicular to the force, in order to calculate stresses directly on the effective throat area, inclined by 45° with respect to the horizontal (x–z plane), the resulting stress components are:

$$\sigma_{\perp} = \frac{F}{2 \cdot L \cdot a} \cdot \frac{\sqrt{2}}{2} \tag{14.2a}$$

$$\tau_{//} = \frac{F}{2 \cdot L \cdot a} \cdot \frac{\sqrt{2}}{2} \tag{14.2b}$$

In order to simplify the calculation of stresses, the effective throat area can be rotated onto the vertical plane (y–z plane) or onto the horizontal plane (x–z plane).

In the former case, the stresses that develop are perpendicular to the y axis ($\sigma_{\perp}$) given by:

$$\sigma_{\perp} = \frac{F}{2 \cdot L \cdot a} \tag{14.3a}$$

In the latter case, by rotating the effective throat area onto the x–z plane, stresses parallel to the x–axis will develop ($\tau_{\perp}$):

$$\tau_{\perp} = \frac{F}{2 \cdot L \cdot a} \tag{14.3b}$$

Inclined fillets: In the case of two fillets placed obliquely with respect to the direction of the force, in the effective throat area there will be two components of the force; one tangential to the longitudinal axis of the weld ($V = F \cos\theta$) and one perpendicular to it ($N = F \sin\theta$), thus creating a state of stress that is more complicated with respect to the two previous cases.

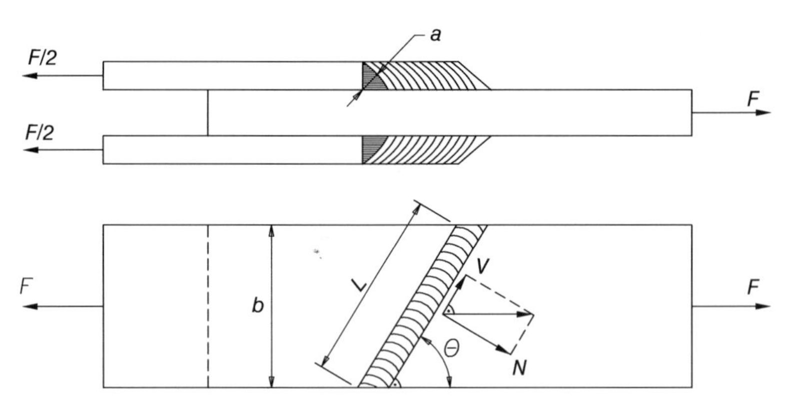

Figure 14.12 Welded connection with inclined fillets.

With reference to Figure 14.12, by rotating the effective throat area onto the horizontal plane, all associated stresses are contained within that plane. In particular, we have:

$$\tau_{\perp} = \frac{F \cdot \sin\theta}{2 \cdot L \cdot a} \tag{14.4a}$$

$$\tau_{//} = \frac{F \cdot \cos\theta}{2 \cdot L \cdot a} \tag{14.4b}$$

If the effective throat area is rotated onto the vertical plane, the state of stress instead becomes:

$$\sigma_{\perp} = \frac{F \cdot \sin\theta}{2 \cdot L \cdot a} \tag{14.5a}$$

$$\tau_{//} = \frac{F \cdot \cos\theta}{2 \cdot L \cdot a} \tag{14.5b}$$

14.3.2 Shear and Flexure

The combination of shear and flexure is very common in welded joints for both residential and industrial use. In the following, reference is made to a welded joint subject to an eccentric shear force F, which generates a bending moment M equal to $F \cdot L_b$.

Longitudinal fillets: The total effective area (Figure 14.13) lies within the vertical plane and consists of two rectangular surfaces, corresponding to the effective throat area of each fillet with throat dimension a and length h.

Rotating the effective throat areas onto the y–z plane, the following stresses develop, associated to shear ($\tau_{//}$) and flexure ($\sigma_{\perp,\max}$), respectively:

$$\tau_{//} = \frac{F}{2 \cdot a \cdot h} \tag{14.6a}$$

$$\sigma_{\perp,\max} = \frac{F \cdot L_b}{W} = \frac{F \cdot L_b \cdot 3}{a \cdot h^2} \tag{14.6b}$$

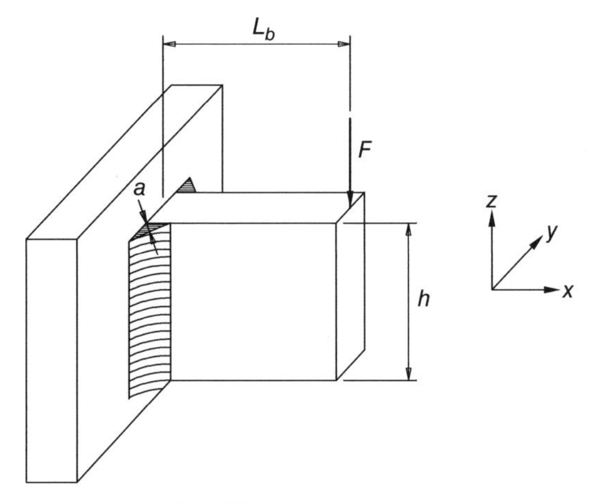

Figure 14.13 Joint in flexure with longitudinal fillet welds.

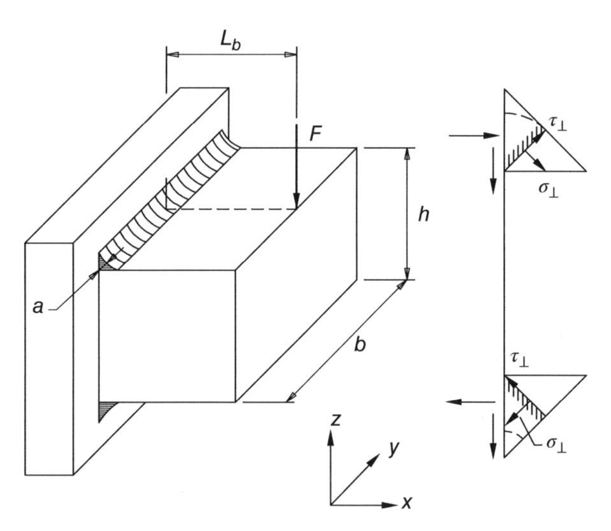

Figure 14.14 Joint in flexure with transverse fillet welds.

Transverse fillets: The total effective area (Figure 14.14) consists of two horizontal cross-sections with effective throat dimension a and length b. The distance between the centroids of the two fillets can be conservatively taken as h (whereas in reality it would be slightly larger than that).

By rotating the effective throat areas onto the y–z plane, the following stresses develop, associated to shear ($\tau_\perp$) and flexure ($\sigma_{\perp,\max}$), respectively:

$$\tau_\perp = \frac{F}{2 \cdot a \cdot b} \tag{14.7a}$$

$$\sigma_{\perp,\max} = \frac{F \cdot L_b}{W} = \frac{F \cdot L_b}{(b \cdot a) \cdot h} \tag{14.7b}$$

Combination of fillets: When connecting I-beams, it is common practice to use combinations of longitudinal and transverse fillet welds (Figure 14.15). If the various parts of the joint have the same stiffness, and the size of the fillet welds are appropriate with respect to the thickness of web

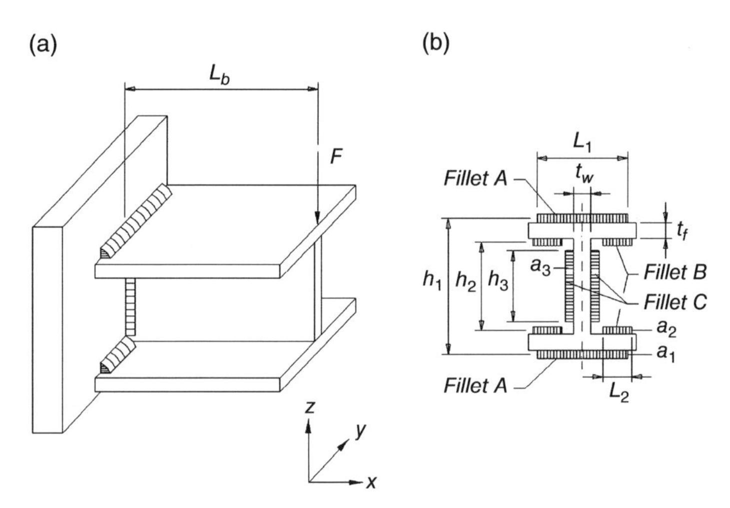

Figure 14.15 Combination of longitudinal and transverse fillet welds.

and flanges of the connecting element, the stress calculation can be performed similarly to that presented previously, considering the mechanical properties of an effective area consisting of both web fillets and flange fillets.

In everyday practice, it is commonplace to simplify calculations by assuming that the shearing force is resisted entirely by the web fillets (fillets C in Figure 14.15), whereas the bending moment are resisted by the flange fillets (fillets A and B). By rotating the effective throat areas onto the y–z plane, the following stresses develop, associated with shear ($\tau_{//}$) and flexure ($\sigma_{\perp,\max}$), respectively:

$$\tau_{//} = \frac{F}{2 \cdot a_3 \cdot L_3} \tag{14.8a}$$

$$\sigma_{\perp,\max} = \frac{F \cdot L_b}{W} = \frac{F \cdot L_b}{(L_1 \cdot a_1 \cdot h_1) + 2(L_2 \cdot a_2 \cdot h_2)} \tag{14.8b}$$

14.3.3 Shear and Torsion

Due to the effect of eccentric actions on a welded joint, in which the line of application of the force and the fillet welds are contained in the same plane, the joint can be subjected to a combination of torsion and shear (Figure 14.16). In the following, reference is made to a joint subjected to a shear force F and a torque equal to $F \cdot e$, in which e is the distance between the line of action of the force and the centroid of the effective areas of the fillet welds.

Transverse fillets: For the case shown in Figure 14.16, in which two fillet welds are provided perpendicular to the line of application of the force, the torsional couple is equilibrated by a couple of forces developing within the fillets that can be estimated as:

$$H = \frac{F \cdot e}{h} \tag{14.9}$$

Corresponding to the force H, a shearing stress $\tau_{//}$ in the fillets develops, equal to:

$$\tau_{//} = \frac{F \cdot e}{h \cdot (a \cdot L)} \tag{14.10}$$

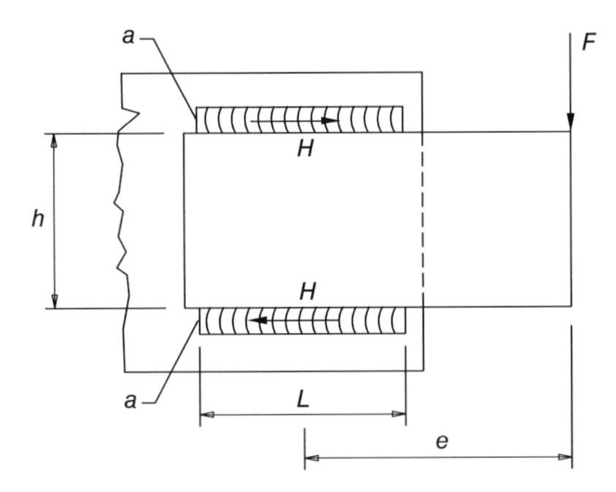

Figure 14.16 Joint under torsion with transverse fillet welds.

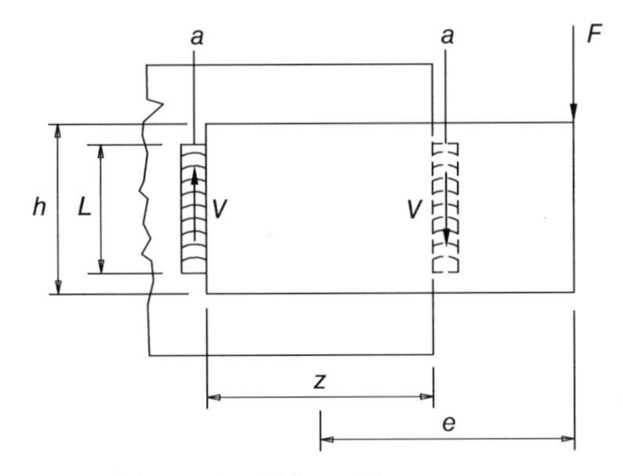

Figure 14.17 Joint under torsion with longitudinal fillet welds.

By rotating the effective throat area onto the horizontal plane, it is possible to calculate the state of stress associated with the shear force F. In particular, the $\tau_{\perp}$ stresses are:

$$\tau_{\perp} = \frac{F}{2 \cdot L \cdot a} \tag{14.11a}$$

Alternatively, by rotating the effective throat area onto the vertical plane (perpendicular to the page), the $\sigma_{\perp}$ stresses associated with F are:

$$\sigma_{\perp} = \frac{F}{2 \cdot L \cdot a} \tag{14.11b}$$

Longitudinal fillets: In the case shown in Figure 14.17, in which two longitudinal fillet welds are provided, the torsional couple is equilibrated by a couple of forces of intensity, V, calculated as:

$$V = \frac{F \cdot e}{z} \tag{14.12}$$

Similar to what was done before, by rotating the effective throat area onto the horizontal plane (or alternatively onto the vertical one), it is possible to calculate the associated state of stress. In particular, the shearing stress, $\tau_{//,1}$, associated with the force V developing as a result of the applied torsion is:

$$\tau_{//,1} = \frac{F \cdot e}{z \cdot (a \cdot L)} \quad (14.13)$$

Corresponding to the applied force F, the resulting shearing stress $\tau_{//,2}$ is:

$$\tau_{//,2} = \frac{F}{2 \cdot (a \cdot L)} \quad (14.14)$$

The overall shearing stress $\tau_{//}$ is obtained as the sum of the previous two components:

$$\tau_{//} = \tau_{//,1} + \tau_{//,2} = \frac{F}{a \cdot L}\left(\frac{e}{z} + \frac{1}{2}\right) \quad (14.15)$$

Combination of fillets: When a joint is made using both transverse and longitudinal fillets, subject to a force F applied with an eccentricity e with respect to the centroid of the fillets (Figure 14.18), it is possible to assume that the total torque T (taking $T = F \cdot e$) is split into two contributions T_1 and T_2, which are resisted by the transverse fillets (1) and by the longitudinal fillets (2), respectively.

If $T_{1,\max}$ and $T_{2,\max}$ represent the resistance developed by the pairs of fillets (1) and (2), respectively, the values for T_1 and T_2 can be calculated as follows:

$$T_1 = T \frac{T_{1,\max}}{T_{1,\max} + T_{2,\max}} \quad (14.16a)$$

$$T_2 = T \frac{T_{2,\max}}{T_{1,\max} + T_{2,\max}} \quad (14.16b)$$

It is possible to substitute terms $T_{1,\max}$ and $T_{2,\max}$ with the corresponding expressions involving the strength of the individual fillets. When doing so, it should be noted that the fillet strength appears both at the numerator and at the denominator of the expressions, thus demonstrating that the distribution of resisting torques depends only on the geometric parameters of the weld. It is thus possible to calculate T_1 and T_2 as a sole function of the geometry of the fillet welds, as follows:

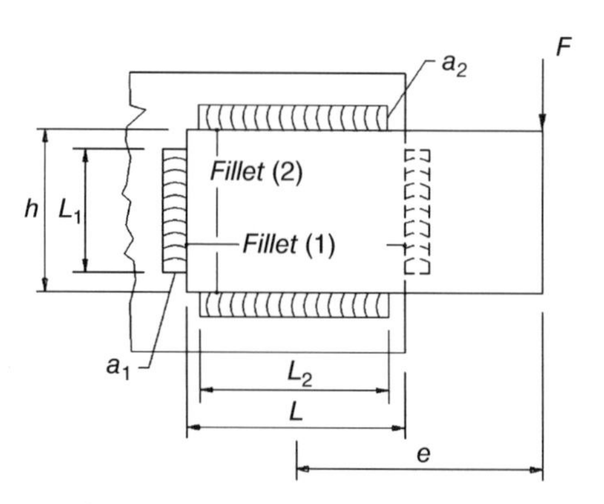

Figure 14.18 Joint under torsion with transverse and longitudinal fillet welds.

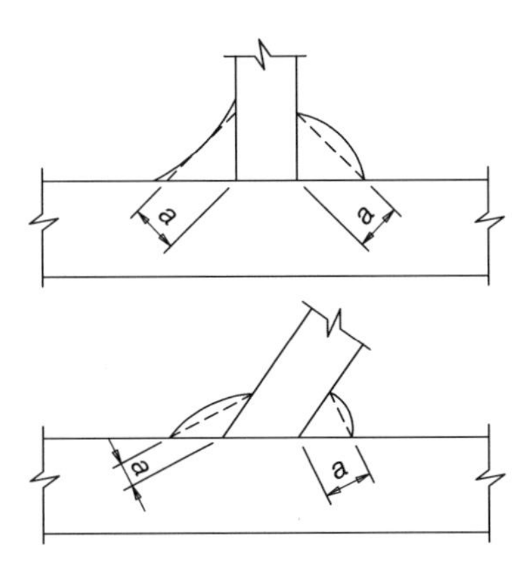

Figure 14.19 Definition of effective throat dimension.

$$T_1 = T\frac{L\cdot(L_1\cdot a_1)}{L\cdot(L_1\cdot a_1)+h\cdot(L_2\cdot a_2)} \quad (14.17a)$$

$$T_2 = T\frac{h\cdot(L_2\cdot a_2)}{L\cdot(L_1\cdot a_1)+h\cdot(L_2\cdot a_2)} \quad (14.17b)$$

The apportioned quote of torque acting on each fillet weld will generate a state of stress, which can be determined following that discussed previously for the transverse fillets (Figure 14.18) or the longitudinal fillets (Figure 14.19), as appropriate.

14.4 Design of Welded Joints

The basic approach that is employed in the design process consists of the transformation of the multidimensional state of stress found in a weld into an equivalent uniaxial state of stress, which can then be compared with a design reference value for the material and this can be appropriately reduced to account for the presence of defects or other considerations.

The methods contained in various specifications have mostly an empirical origin and can be traced back to the seminal work of Van den Eb (1952–1953) who performed experimental tests with the goal of defining the spatial domain for the resistance of fillet welds in terms of the stress contributions $\sigma_\perp$, $\tau_\perp$ and $\tau_{//}$.

14.4.1 Design According to the European Approach

The design strength of a complete penetration butt (CJP) joint and of a T-joint having an effective throat dimension no smaller than the thickness t of the stem of the T, and, in the case of a PJP configuration, having a thickness of the part that is not welded not larger of the smaller of $t/5$ and 3 mm, can be taken to be equal to the design strength of the weaker of the connecting members. This is true if the weld is placed using appropriate electrodes or weld metal that are characterized by yielding and ultimate strengths that are not smaller than those of the base metal.

The design strength of a partial penetration butt (PJP) joint is calculated in a similar way to a fillet weld, using a measure of the penetration that is effectively reached for the effective throat dimension.

For fillet welds, after Eurocode 3, the design strength per unit length $F_{w,Rd}$ can be calculated based on either of the following methods:

- *directional method;*
- *simplified method.*

The *directional method* requires the determination of the state of stress in the effective throat area without rotations, and thus the stresses σ and τ are the normal and shear stresses in the plane of the effective throat area, respectively. This method requires checking the following limit states:

$$\sqrt{\sigma_{\perp}^2 + 3\cdot\left(\tau_{\perp}^2 + \tau_{//}^2\right)} \leq \frac{f_u}{\beta_w \cdot \gamma_{M2}} \tag{14.18a}$$

$$\sigma_{\perp} \leq \frac{0.9 f_u}{\gamma_{M2}} \tag{14.18b}$$

where f_u is the nominal tensile strength of the weakest element in the joint, γ_{M2} is the partial safety factor and β_w is an appropriate correlation coefficient as shown in Table 14.1.

The *simplified method* establishes that the design strength of a fillet weld should be taken, independently on the orientation of the weld, as:

$$F_{w,Rd} = f_{vw,d} \cdot a \tag{14.19}$$

where a is the effective throat dimension (which, as mentioned before, is the height of the generic inscribed triangle, as in Figure 14.19) and $f_{vw,d}$ is the design shear strength of the weld, defined as:

$$f_{vw,d} = \frac{f_u}{\sqrt{3} \cdot \beta_w \cdot \gamma_{M2}} \tag{14.20}$$

where f_u is the nominal tensile strength of the weakest element of the joint, γ_{M2} is the partial safety factor and β_w is an appropriate correlation coefficient as listed in Table 14.1.

A commonly occurring configuration has a plate, or a flange of a beam, welded to an unstiffened flange of an I-beam or other profile (Figure 14.20). In this case, an effective width b_{eff} is used, taken as the portion of the fillet weld that is effectively engaged in the joint.

Table 14.1 Design of welded joints.

Steel grade and reference provisions			Correlation coefficient β_w
EN 10025	**EN 10210**	**EN 10219**	
S 235, S 235 W	S 235 H	S 235 H	0.8
S 275, S 275 N/NL, S 275 M/ML	S 275 H, S 275 NH/NLH	S 275 H, S 275 NH/NLH, S 275 MH/MLH	0.85
S 355, S 355 N/NL, S 355 M/ML, S 355 W	S 355 H, S 355 NH/NLH	S 355 H, S 355 NH/NLH, S 355 MH/MLH	0.9
S 420 N/NL, S 420 M/ML		S 420 MH/MLH	1.0
S 460 N/NL, S 460 M/ML, S 460 Q/QL/QL1	S 460 NH/NLH	S 460 NH/NLH, S 460 MH/MLH	1.0

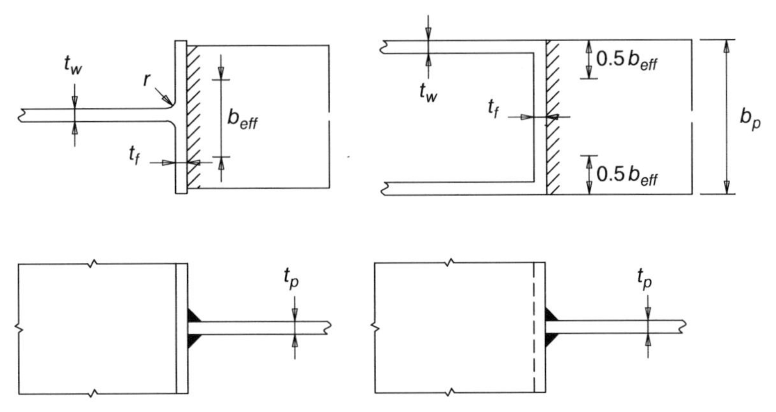

Figure 14.20 Effective width in an unstiffened T-joint.

In the case of a welded joint to an unstiffened I-section, the effective width b_{eff} is given by:

$$b_{\text{eff}} = t_w + 2 \cdot s + 7 \cdot k \cdot t_f \tag{14.21a}$$

with:

$$k = \left(\frac{t_f}{t_p}\right) \cdot \left(\frac{f_{y,f}}{f_{y,p}}\right) \leq 1 \tag{14.21b}$$

in which f_y is the yielding strength, t is the thickness and the subscripts p and f refer to the plate and the flange of the profile, respectively.

The term s is assumed to be equal to the fillet radius at the flange/web juncture (k-zone), and for built-up sections is taken as 1.41 times ($\approx \sqrt{2}$) the effective throat dimension.

In addition to the limit state checks for all members, the following condition must also be verified:

$$b_{\text{eff}} \geq \left(\frac{f_{y,p}}{f_{u,p}}\right) b_p \tag{14.22}$$

in which subscripts y and u refer to yielding and ultimate conditions, respectively, whereas subscript p refers to the plate and b_p is the length of the fillet weld.

In the case of welded joints with other shape cross-sections, such as box or channel sections, if the width of the connecting plate is similar to that of the flange, the effective width b_{eff} can be obtained as follows:

$$b_{\text{eff}} = \left(2t_w + 5t_f\right) \leq 2t_w + 5kt_f \tag{14.23}$$

Even when $b_{\text{eff}} \leq b_p$, the welds connecting the plate to the flange of the profile have to be sized so that the design strength of the plate (calculated as $b \cdot t_p \cdot f_{y,p} / \gamma_{M0}$) can be transmitted, assuming a uniform stress distribution.

In lap splices, the design strength of a fillet weld must be reduced by means of a coefficient β_{Lw} that accounts for the effects of non-uniform stress distribution along the length of the weld.

In lap splices longer than $150a$ (where a is the effective throat dimension expressed in mm), the reduction factor β_{Lw} is assumed to be equal to $\beta_{Lw,1}$, which in turn is calculated as follows:

$$\beta_{Lw,1} = 1.2 - 0.2L_j/(150a) \leq 1.0 \quad (14.24)$$

where L_j is the total length of the splice in the direction of transmitted force.

For fillet welds longer than 1700 mm, which connect transverse stiffeners in built-up sections, the reduction factor β_{Lw} can be taken to be equal to $\beta_{Lw,2}$ calculated as:

$$\beta_{Lw,2} = 1.1 - L_w/17 \text{ with } \beta_{Lw,2} \leq 1.0 \text{ and } \beta_{Lw,2} \geq 0.6 \quad (14.25)$$

where L_w is the weld length expressed in metres.

14.4.2 Design According to the US Practice

As discussed in the general section, ANSI/AISC 360-10 assigns an effective area for groove welds equal to the length of the weld times the effective throat dimension. For a CJP weld, the effective throat dimension is equal to the thickness of the thinner of the connected parts. For a PJP weld, the effective throat dimension is provided in Table 14.2 (from table AISC-J2.1), as a function of the surface preparation (U- or J-groove, 45° or 60° bevel, 60° V-groove), of the welding position (flat, horizontal, vertical, or overhead), and of the welding process used (SMAW, GMAW, FCAW, SAW). In the case of SMAW welding, for all welding positions and a 45° bevel, the effective throat dimension is to be taken as the depth of the groove minus 1/8-in. (3 mm). The same throat dimension is also to be used for vertical and overhead welds using GMAW and FCAW (flux cored arc welding) on a 45° bevel preparation. For all other cases, the effective throat dimension is given by the actual groove depth.

The AISC Specification also allows the accounting for larger effective throat dimensions if the larger values can be demonstrated experimentally, through a qualification procedure.

For the special category of flare groove welds, which occur when welding round bars, or formed profiles (such as Hollow Structural Steel – HSS – sections) by completely filling the groove to form a flat surface, the Specification provides a mean to establish the effective throat dimension for flare welds, as a function of the radius of the joint surface (see Table 14.3, from table AISC-J2.2). A note suggests that in the case of HSS, a dimension equal to twice the thickness t of the shape can be used in place of the radius.

Table 14.2 Effective throat of partial-joint-penetration groove welds (from Table J2.1 of AISC 360-10).

Welding process	Welding position F (flat), H (horizontal), V (vertical), OH (overhead)	Groove type (AWS D1.1/D1.M, Figure 3.3)	Effective throat
Shielded metal arc (SMAW) Gas metal arc (GMAW) Flux cored arc (FCAW)	All	J or V groove 60° V	depth of groove
Submerged arc (SAW)	F	J or V groove 60° bevel or V	
Gas metal arc (GMAW) Flux cored arc (FCAW)	F, H	45° bevel	depth of groove
Shielded metal arc (SMAW)	All	45° bevel	depth of groove minus 1/8 in. (3 mm)
Gas metal arc (GMAW) Flux cored arc (FCAW)	V, OH		

Table 14.3 Effective weld throats of flare groove welds (from Table J2.2 of AISC 360-10).

Welding process	Flare bevel groove	Flare V-groove
GMAW and FCAW-G	(5/6)R	(3/4)R
SMAW and FCAW-S	(5/16)R	(5/8)R
SAW	(5/16)R	(1/2)R

R = radius of joint surface (can be assumed to be 2 t for HSS).

Table 14.4 Minimum effective throat of partial-joint-penetration groove welds (from Table J2.3 of AISC 360-10).

Material thickness of thinner part joined, in. (mm)	Minimum effective throat, in. (mm)
To 1/4 (6) inclusive	1/8 (3)
Over 1/4 (6) to 1/2 (13)	3/16 (5)
Over 1/2 (13) to 3/4 (19)	1/4 (6)
Over 3/4 (19) to 1-1/2 (38)	5/16 (8)
Over 1-1/2 (38) to 2-1/4 (57)	3/8 (10)
Over 2-1/4 (57) to 6 (150)	1/2 (13)
Over 6 (150)	5/8 (16)

For PJP groove welds, the minimum effective throat dimension has to be at least sufficient to transmit the calculated forces. Also, the Specification provides minimum effective throat dimensions for PJP welds, as a function of the material thickness of the smaller of the joined parts (see Table 14.4, from table AISC-J2.3). The minimum throat dimensions range from 1/8 to 5/8-in. (3–16 mm), for plates of thicknesses from less than ¼-in. (6 mm) to over 6-in. (150 mm).

For fillet welds, the Specification defines the effective area as the product of the effective throat dimension of the weld times its effective length. As discussed in Section 14.1, the effective throat dimension is taken as the height of the triangle inscribed within the fillet. A larger value can be used for the effective throat dimension, extending inside the root of the weld, if weld penetration is consistently demonstrated through experimental testing using the same production process and procedure variables.

The Specification also takes into consideration the case of fillet welds placed inside holes or slots: in this case, the effective length of a fillet is measured along a line contained in the plane of the throat, and located at mid-length of the throat dimension. If fillets overlap, filling the hole or slot, the effective area cannot be larger than the nominal cross-sectional area of the hole or slot in the plane of the connected surfaces (faying surface).

A fillet weld has to be of adequate size to carry the calculated forces, and it must follow the minimum leg dimension requirements provided in Table 14.5 (from table AISC-J2.4), which contains the minimum leg size of a fillet weld as a function of the material thickness of the thinner part joined. A single pass of the electrode must be used to deposit at least the prescribed minimum quantity of weld metal in the fillet. It is worth noting that these minimum leg sizes only apply to fillet welds used for strength and do not apply to fillet weld reinforcements, which are commonly used to finish CJP or PJP groove welds, preventing potential surface defects.

The Specification also prescribes maximum sizes for fillet welds: for connecting parts less than ¼-in. (6 mm) in thickness, the fillet cannot be thicker than the material; for connecting parts ¼-in.-thick (6 mm) or thicker, the fillet cannot be thicker than the material minus 1/16-in.

Table 14.5 Minimum size of fillet welds (from Table J2.4 of AISC 360-10).

Material thickness of thinner part joined, in. (mm)	Minimum size of fillet weld, in. (mm)
To 1/4 (6) inclusive	1/8 (3)
Over 1/4 (6) to 1/2 (13)	3/16 (5)
Over 1/2 (13) to 3/4 (19)	1/4 (6)
Over 3/4 (19)	5/16 (8)

(2 mm), unless the weld is specifically designated in the detail drawings to be built out to obtain full-throat thickness.

In terms of *length*, the Specification requires the minimum length of a fillet weld designed for strength to be at least equal to four times the nominal size of the weld leg. If the weld is shorter than that, its effective length is taken as one-fourth of its actual length. A common case is the connection of flat-bar tension members by means of pairs of longitudinal fillet welds: in this case, the Specification requires a minimum length equal to the perpendicular distance between the two fillets in the pair.

For an end-loaded fillet weld, that is a longitudinal fillet weld, parallel to the force, transferring force to the end of a member, the distribution of stresses along the weld length is non-uniform, and difficult to evaluate. When the length of the end-loaded fillet is up to 100 times the size of the fillet weld leg, the effective length can be taken to be equal to the actual length. For lengths above 100 times the leg size, a factor that reduces the effective length of the weld must be used, given by:

$$\beta = 1.2 - 0.002(l/w) \leq 1.0 \tag{14.26}$$

where l is the actual length of the end-loaded weld and w is the size of the weld leg.

Finally, if the weld is longer than 300 times the weld leg size, the effective length of the weld should be taken as $180w$.

The Specification further deals with intermittent fillet welds, that is fillet welds that are not placed continuously, but only along discrete lengths. The length of these fillet welds has to be at least 1-½-in. (38 mm), and in any case no less than four times the leg size of the weld.

For lap splice joints, the minimum amount of lap set by the Specification is five times the thickness of the thinner joined part, and it cannot be less than 1 in. (25 mm). Lap joints connecting members subjected to tension and using only transverse fillet welds to carry the force must be welded along the end of both lapped parts, with the exception of the case in which the deformation of both parts is sufficiently restrained to prevent the opening of a gap at the splice at the maximum load.

The Specification also covers the details of fillet weld termination. In the case of overlapping elements in which the edge of one connected part extends beyond the edge of another connected part that is subject to tensile stress, fillet welds will have to be terminated at a distance from that edge not smaller than the weld leg size. Other special cases are connections in which flexibility of the outstanding elements is required, fillet welds joining transverse stiffeners to plate girders of thickness ¾-in. (19 mm) or less and fillet welds placed on opposite sides of a common plane.

The Specification requires the electrode choice for the weld to be in accordance with what specified by AWS D1.1/D1.1M. Based on the grade and thickness of the base metal, and to some extent on the welding process, the nominal capacity of the electrode, indicated with F_{EXX}, in which XX usually corresponds to the strength in ksi, some common requirements are as follows:

- For grade A36 steels, up to ¾-in. (19 mm) thick, use 60 or 70 ksi (414 or 483 MPa) electrodes;
- For grade A36 steels thicker than ¾-in. (19 mm), A992 steels, A588 steels, A572-Gr.50 steels use 70 ksi (483 MPa) electrodes;
- For grade A913 steels, use 80 ksi (552 MPa) filler material.

The complete requirements for matching filler metal are contained in AWS D1.1/D1.1M.

14.4.2.1 Design Strength

For all types of welds, the *design strength* is defined as the lower value of the base metal strength, calculated according to the limit states of tensile rupture and shear rupture, and of the weld metal strength, calculated for the limit state of rupture. For base metal, the nominal strength is given by:

$$R_n = F_{nBM} A_{BM} \tag{14.27}$$

where F_{nBM} is the nominal stress of the base metal and A_{BM} is the cross-sectional area of the base metal. Similarly, for the weld metal, the nominal strength is given by:

$$R_n = F_{nw} A_{we} \tag{14.28}$$

where F_{nw} is the nominal stress of the weld and A_{we} is the effective area of the weld, calculated as discussed in the previous section.

The Specification distinguishes various cases, based on the type of load and its direction with respect to the weld axis and based on the type of weld being evaluated. All cases are summarized in Table 14.6 (from table AISC J2.5).

For *CJP groove welds*, in the presence of a tension load normal to the axis of the weld, the strength of the joint is controlled by the base metal, and a matching weld metal (filler metal) must be used, as discussed earlier. In the case of a compressive load normal to the axis of the weld, the joint strength is once again controlled by the base metal and the Specification permits the use of a filler metal one strength level lower than the matching filler. In the case of a force parallel to the weld axis, the Specification does not require a weld check and the weld metal used can be the

Table 14.6 Available strength of welded joints, ksi (MPa) (from Table J2.5 – part 1 of AISC 360-10).

Complete-joint-penetration groove welds		
Tension normal to weld axis	Strength of the joint is controlled by the base metal	Matching filler metal shall be used. For T- and Tension corner joints with backing left in place, notch tough filler metal is required. See Section J2.6
Compression Normal to weld axis	Strength of the joint is controlled by the base metal	Filler metal with a strength level equal to or one strength level less than matching filler metal is permitted
Tension or compression Parallel to weld axis	Tension or compression in parts joined parallel to a weld need not be considered in design of welds joining the parts	Filler metal with a strength level equal to or less than matching filler metal is permitted
Shear	Strength of the joint is controlled by the base metal	Matching filler metal by the base metal shall be used

matching filler metal, or a lower strength filler. Finally, for the case of a shear force, the strength of the joint is controlled once again by the base metal, and the weld metal prescribed is the matching filler. These cases are summarized in the Table 14.6.

For *PJP groove welds*, for the case of tensile force normal to the weld axis, the base material should be checked with Eq. (14.27) applying a resistance factor of 0.75, and the base material nominal stress should be taken as the ultimate material strength F_u. The area to be used in Eq. (14.27) is the effective net area of the connecting part. In order to check the weld material, using Eq. (14.28), a resistance factor of 0.80 is to be used. The weld metal nominal stress is taken as 0.60 times the nominal capacity of the matching electrode used. The effective area to be used has been discussed in the previous section. When considering a PJP connecting a column to a base plate, or connecting column splices, the Specification allows us to discount the effects of compressive stress in the design of the weld.

For connections in compression designed to bear members other than columns, the Specification requires us to check the base metal using a resistance factor of 0.90, a base metal stress equal to the yielding stress of the base material F_y and an effective area equal to the gross area of the base metal. For the weld metal check, the resistance factor to be used is 0.80, the weld metal stress is to be taken equal to 0.60 times the nominal capacity of the electrode used and the effective area is the one discussed in the previous section for PJP welds.

For connections in compression of members that are not finished to bear, the base metal checks are the same as the previous case, while for the weld metal check a higher value of the nominal stress can be used, equal to 0.90 times the nominal capacity of the electrode used.

For the case of tension or compression parallel to the weld axis, as it was the case for CJP joints, the Specification allows to ignore the effects of that force in the design of the weld.

For the case of shear force, the base metal check uses a resistance factor of 1.0, a base metal nominal stress equal to 0.60 times the ultimate strength of the base metal F_u and an area equal to the effective net area in shear A_{nv} of the base material. For the weld check, the resistance factor to be used is 0.75, the weld material nominal stress is to be taken as 0.60 times the nominal capacity of the electrode used and the effective area is the one discussed in the previous section for PJP welds. Table 14.7 summarizes these requirements for PJP joints.

For *fillet welds* (including fillets in holes and slots) and skewed T-joints (Table 14.8), in the case of shear force, the base metal check is once again made using a resistance factor of 0.75, a base metal nominal stress equal to 0.60 times the ultimate strength of the base metal F_u and an area equal to the effective net area in shear A_{nv} of the base material. For the weld metal check, the resistance factor to be used is 0.75, the nominal weld stress is calculated as 0.60 times the nominal capacity of the electrode used and the effective area is calculated following the approach discussed in the previous section for fillet welds.

In the case of compressive or tensile force parallel to the weld, as before, the Specification allows the designer to discount the effects of this force on the weld design.

As an alternative, for a linear weld group of *fillet welds* loaded through its centre of gravity, the available strength can also be calculated using a resistance factor of 0.75, and a nominal stress for the weld equal to:

$$F_{nw} = 0.60F_{EXX}\left(1.0 + 0.50\sin^{1.5}\theta\right) \tag{14.29}$$

where θ is the angle of loading measured from the longitudinal axis of the weld (i.e. 0° for a force parallel to the weld, and 90° for a force perpendicular to the weld).

Also, for weld groups that are concentrically loaded and that consist of combinations of longitudinal and transverse fillet welds, the Specification allows us to take the nominal stress for the weld equal to the greater of:

$$R_n = \max\{(R_{nwl} + R_{nwt}),(0.85R_{nwl} + 1.5R_{nwt})\} \tag{14.30}$$

Table 14.7 Available strength of welded joints, ksi (MPa) (from Table J2.5 – part 2 of AISC 360-10).

Load type and direction relative to weld axis	Pertinent metal	ϕ and Ω	Nominal stress (F_{nBM} or F_{nw}) ksi (MPa)	Effective area (A_{BM} or A_{we}) in.^{2} (mm^2)	Required filler metal strength level
Partial-joint-penetration groove welds including flare V-groove and flare bevel groove welds					
Tension normal to weld axis	Base	$\phi = 0.75$ $\Omega = 2.00$	F_u	A_e	Filler metal with a strength level equal to or less than matching filler metal is permitted
	Weld	$\phi = 0.80$ $\Omega = 1.88$	$0.60F_{EXX}$	A_{we}	
Compression column to base plate and column splices designed per Section J1.4(1)	Compressive stress need not be considered in design of welds joining the parts				
Compression connections of members designed to bear other than columns as described in Section J1.4(2)	Base	$\phi = 0.90$ $\Omega = 1.67$	F_y	A_g	
	Weld	$\phi = 0.80$ $\Omega = 1.88$	$0.60F_{EXX}$	A_{we}	
Compression connections not finished-to-bear	Base	$\phi = 0.90$ $\Omega = 1.67$	F_y	A_g	
	Weld	$\phi = 0.80$ $\Omega = 1.88$	$0.90F_{EXX}$	A_{we}	
Tension or compression parallel to weld axis	Tension or compression in parts joined parallel to a weld need not be considered in design of welds joining the parts				
	Base	$\phi = 1.0$ $\Omega = 1.5$	$0.60F_U$	A_{gv}	
	Weld	$\phi = 0.75$ $\Omega = 2.00$	$0.60F_{EXX}$	A_{we}	

A_g = gross area; A_e = effective net area; A_{gv} = gross area subject to shear and A_{we} = PJP weld area (previously discussed).

Table 14.8 Available strength of welded joints, ksi (MPa) (from Table J2.5 – part 3 of AISC 360-10).

Load type and direction relative to weld axis	Pertinent metal	ϕ and Ω	Nominal stress (F_{nBM} or F_{nw}) ksi (MPa)	Effective area (A_{BM} or A_{we}) in.^{2} (mm^2)	Required filler metal strength level
Fillet welds including fillets in holes and slots and skewed T-joints					
Shear	Base	$\phi = 1.0$ $\Omega = 1.5$	$0.60F_U$	A_{gv}	Filler metal with a strength level equal to or less than matching filler metal is permitted
	Weld	$\phi = 0.75$ $\Omega = 2.00$	$0.60F_{EXX}$	A_{we}	
Tension or compression parallel to weld axis	Tension or compression in parts joined parallel to a weld need not be considered in design of welds joining the parts				

A_{gv} = gross area subject to shear and A_{we} = PJP weld area (previously discussed).

Table 14.9 Available strength of welded joints, ksi (MPa) (from ACI 360-10, Table J2.5 – part 4).

Load type and direction relative to weld axis	Pertinent metal	ϕ and Ω	Nominal stress (F_{nBM} or F_{nw}) ksi (MPa)	Effective area (A_{BM} or A_{we}) in.^{2} (mm^2)	Required filler metal strength level
Plug and slot welds					
Shear parallel to faying surface on the effective area	Base	$\phi = 0.75$ $\Omega = 2.00$	$0.60F_U$	A_{nv}	Filler metal with a strength level equal to or less than matching filler metal is permitted
	Weld	$\phi = 0.75$ $\Omega = 2.00$	$0.60F_{EXX}$	A_{we}	

A_{we} = PJP weld area (previously discussed) and A_{nv} = net area subject to shear.

where R_{nwl} is the total nominal strength of longitudinal fillet welds, as calculated above and R_{nwt} is the total nominal strength of transverse fillet welds, calculated as described earlier, but not using the alternate method.

Finally, for *plug* and *slot welds* (Table 14.9), for the case of shear parallel to the faying surface on the surface on which the effective area is calculated, the resistance factor to be used is 0.75, the nominal stress of the weld is taken as 0.60 times the nominal capacity of the electrode used and the effective area is calculated following the considerations that were presented in the previous section for plug and slot welds.

14.5 Joints with Mixed Typologies

Within a joint, it is good practice to avoid the usage of different joining methods, that is welds and bolts, or welds and rivets. Simultaneous use of multiple joining techniques is allowed, as long as at least one of these is capable of carrying the whole force demand.

EC3 allows one exception to this rule: in slip critical joints, where slip is considered to be an ultimate limit state (category C), the shearing force to be transmitted can be shared among pre-tensioned bolts and welds, as long as the fasteners are installed after the welds have been placed.

ANSI/AISC360-10 also allows an exception to this rule. Bolts installed in standard or short slots holes (with the slot perpendicular to the direction of loading) are permitted to share load with longitudinal fillet welds, with the limitation of taking the available strength of the bolts no greater than 50% of the available bearing strength in the connection. Also, in retrofits and modifications of a bolted slip-critical connection, the additional welds can be sized to resist just the additional loads, using the pre-tensioned bolts to resist their original design load.

14.6 Worked Examples

Example E14.1 Welded Connection According to EC3 for a Tension Member

Verify in accordance with EC3, the welded connection in Figure E14.1.1 between a plate 250 × 20 mm (9.84 × 0.787 in.) in tension and a column flange, realized by one fillet weld orthogonal to the force axes.

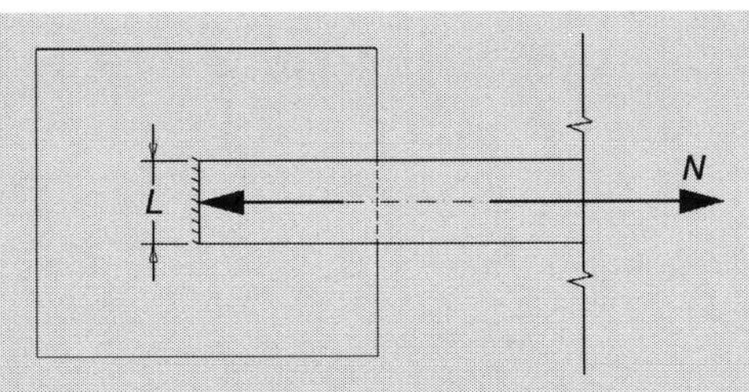

Figure E14.1.1

Applied tension force:	N = 900 kN (202.3 kips)
Steel:	S275 $f_u = 430$ N/mm^2 (62.4 ksi)
	$\beta_w = 0.85$
Fillet length:	$l = 250$ mm (9.84 in.)
Fillet side:	$d = 20$ mm (0.787 in.)

(1) EC3 directional method
Compute fillet throat dimension, a:

$$a = d/\sqrt{2} = 20/\sqrt{2} = 14\,\text{mm}\ (0.551\,\text{in.})$$

$$\sigma_\perp = \tau_\perp = \frac{N}{l \cdot a} \cdot \frac{\sqrt{2}}{2} = \frac{900 \cdot 10^3}{250 \times 14} \cdot \frac{\sqrt{2}}{2} = 181.8\,\text{N/mm}^2$$

The verification formulas are:

$$\sqrt{\sigma_\perp^2 + 3\left(\tau_\perp^2 + \tau_\parallel^2\right)} = \sqrt{181.8^2 + 3 \times (181.8^2 + 0^2)} = 363.6\,\text{N/mm}^2 \leq$$

$$\leq \frac{f_u}{\beta_w \cdot \gamma_{M2}} = \frac{430}{0.85 \times 1.25} = 404.7\,\text{N/mm}^2\ \text{OK}\ (52.7\,\text{ksi} \leq 58.7\,\text{ksi})$$

$$\sigma_\perp = 181.8\,\text{N/mm}^2 \leq \frac{0.9 f_u}{\gamma_{M2}} = \frac{0.9 \times 430}{1.25} = 309.6\,\text{N/mm}^2\ \text{OK}\ (26.4\,\text{ksi} \leq 44.9\,\text{ksi})$$

(2) Simplified EC3 method
The verification formula is:

$$F_{w,Ed} \leq F_{w,Rd} = \frac{f_u \cdot a}{\sqrt{3} \cdot \beta_w \cdot \gamma_{M2}}$$

$$F_{w,Ed} = N/l = \left(900 \cdot 10^3\right)/250 = 3600\,\text{N/mm}\ (20.56\,\text{kips/in.})$$

$$F_{w,Rd} = \frac{f_u \cdot a}{\sqrt{3} \cdot \beta_w \cdot \gamma_{M2}} = \frac{430 \times 14}{\sqrt{3} \times 0.85 \times 1.25} = 3271\,\text{N/mm} < 3600\,\text{N/mm}$$

NOT VERIFIED (18.68 kips/in. < 20.56 kips/in.)

As can be noted, in this case the EC3 simplified method is more conservative than the directional method and the weld results are verified with the second method but not with the first one.

Example E14.2 Welded Connection According to EC3 for a Member in Bending and Shear

Verify in accordance with the EC3 provisions, the welded connection between a UPN 240 profile in bending, connected to a gusset plate by two fillet welds of same length (Figure E14.2.1).

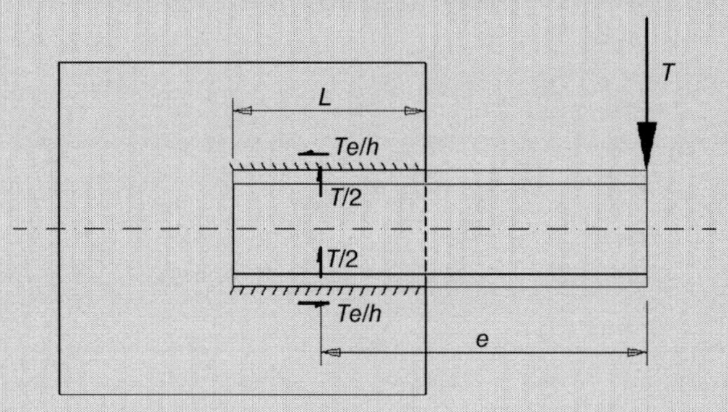

Figure E14.2.1

Applied load: T = 90 kN(20.2 kips)

Load eccentricity: e = 600 mm (23.6 in.)

Steel: S275 $f_u = 430$ N/mm^2(62.4 ksi)

$\beta_w = 0.85$

Use two fillet welds of side length of 10 mm (0.394 in.) and length of 200 mm (7.87 in.) each.

Forces acting on each fillet:

(a) Force parallel to fillet axes (h is profile depth):

$$T_1 = \frac{T \cdot e}{h} = \frac{90 \times 600}{240} = 225\,\text{kN}\,(50.6\,\text{kips})$$

(b) Force orthogonal to fillet axes:

$$T_2 = T/2 = 90/2 = 45\,\text{kN}\,(10.1\,\text{kips})$$

(1) EC3 directional method

Compute fillet throat dimension, a:

$$a = d/\sqrt{2} = 10/\sqrt{2} = 7\,\text{mm}\,(0.276\,\text{in.})$$

$$\sigma_\perp = \tau_\perp = \frac{T/2}{l \cdot a} \cdot \frac{\sqrt{2}}{2} = \frac{(90/2) \cdot 10^3}{200 \times 7} \cdot \frac{\sqrt{2}}{2} = 22.7\,\text{N/mm}^2(3.29\,\text{ksi})$$

$$\tau_\parallel = \frac{T_1}{a \cdot l} = \frac{225 \cdot 10^3}{7 \times 200} = 160.7\,\text{N/mm}^2(23.3\,\text{ksi})$$

The verification formulas are:

$$\sqrt{\sigma_\perp^2 + 3\left(\tau_\perp^2 + \tau_\parallel^2\right)} = \sqrt{22.7^2 + 3 \times (22.7^2 + 160.7^2)} = 282\,\text{N/mm}^2$$

$$\leq \frac{f_u}{\beta_w \cdot \gamma_{M2}} = \frac{430}{0.85 \times 1.25} = 404.7\,\text{N/mm}^2\ \text{OK}\,(40.9\,\text{ksi} \leq 58.7\,\text{ksi})$$

$$\sigma_\perp = 22.7\,\text{N/mm}^2 \leq \frac{0.9 f_u}{\gamma_{M2}} = \frac{0.9 \times 430}{1.25} = 309.6\,\text{N/mm}^2\ \text{OK}\,(3.29\,\text{ksi} \leq 44.9\,\text{ksi})$$

$$\text{Stress rate}: 282/404.7 = 0.70$$

(2) Simplified EC3 method
The verification formula is:

$$F_{w,Ed} \leq F_{w,Rd} = \frac{f_u \cdot a}{\sqrt{3} \cdot \beta_w \cdot \gamma_{M2}}$$

$$F_{w,Ed} = \sqrt{\left(\frac{T_1}{l}\right)^2 + \left(\frac{T_2}{l}\right)^2} = \sqrt{\left(\frac{225 \cdot 10^3}{200}\right)^2 + \left(\frac{45 \cdot 10^3}{200}\right)^2} = 1147\,\text{N/mm}\,(6.55\,\text{kips/in.})$$

$$F_{w,Rd} = \frac{f_u \cdot a}{\sqrt{3} \cdot \beta_w \cdot \gamma_{M2}} = \frac{430 \times 7}{\sqrt{3} \times 0.85 \times 1.25} = 1636\,\text{N/mm} > 1147\,\text{N/mm}$$

$$\text{OK}\,(9.34\,\text{kips/in.} > 6.55\,\text{kips/in.})$$

$$\text{Stress rate}: 1147/1636 = 0.70$$

As can be noted, stress rate is the same with both methods (= 0.70).

CHAPTER 15

Connections

15.1 Introduction

The selection of the proper connections for mono-dimensional members is an extremely important phase of each design. The choice of bolted connections, welded connections or connections with some components bolted and others welded has to be done considering not only the structural performance but also economic factors associated with shop execution as well as site assemblage, especially with reference to the costs of the manpower and of the impact on the building erection schedule. The use of either bolting or welding has certain advantages and disadvantages. Bolting requires either the punching or drilling of holes in all the plies of members to join. As introduced in Chapter 13, these holes may be standard size, oversized, short-slotted or long-slotted depending on the type of connection. It is not unusual to have one ply of material prepared with a standard hole while another ply of the connection is prepared with a slotted hole in order to allow for easier and faster erection of the structural framing. However, the welding process requires a greater level of skill than installing the bolts. Welding eliminates the need for punching or drilling the plies of members to be connected to each other. It is preferable to avoid site welding owing to the fact that shop welding guarantees a better quality level as well as higher structural performance.

Preliminary to the description of the most common type of connections, it should be noted that they can be distinguished in articulated connections and joints on the basis of the effects produced by the relative displacements between the members to be connected. All structures in fact move to some extent and these movements may be permanent and irreversible or short-term and possibly reversible. The effects can be non-negligible in terms of the behaviour of the structure, the performances of components and sub-systems during its lifetime. With reference to this aspect, it is possible to distinguish in:

- *articulated connections*: which allow, under normal service conditions, relative movements between the connected members in elastic range, without causing any plasticization of them as well as of the required devices (bolts, web, plate, angles, etc.). These connections can be distinguished in pin joints, bearing joints or joints in synthetic material. Articulated connections were frequently used up to the beginning of this century: structural design strictly followed the elastic theory and the constraint conditions, on which calculations were based, were complied with as faithfully as possible. When plastic theory was developed and it became clear that each equilibrated calculation model was in favour of safety, provided that localized failures and

Structural Steel Design to Eurocode 3 and AISC Specifications, First Edition. Claudio Bernuzzi and Benedetto Cordova.

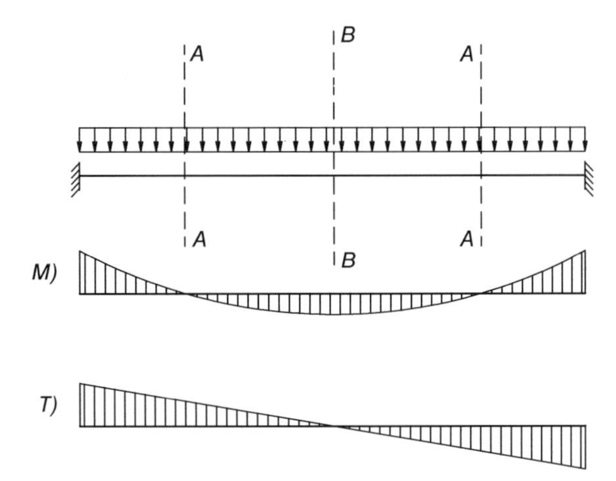

Figure 15.1 Example of cross-sections where locate intermediate partial-strength connections.

buckling phenomena did not take place, the importance of constraint compliance with their model was retrenched.

- *joints*: which do not allow any relative displacements in elastic range but provide the values of design displacements or rotations via the spreading of plasticity. In this case, adequate ductility of material is required associated with an appropriate choice of the connection details necessary to avoid brittle failures and instability phenomena.

Furthermore, depending on the resistance of the connections, which has to be compared with the one of the connected members, following types of connection can be distinguished:

- *partial strength connection*: if the connection is weaker than the connected members;
- *full strength connection*: when failure occurs in one of the connected members, before than in the connection.

The position of the connections depends, in some cases, on the internal forces and bending moment distribution on the members intended to connect. As an example, reference can be made to the beam in Figure 15.1, which is supposed to have quite a long span, for example $L = 20$ m (65.6 ft), greater than the normal limit of transportability. If connections are supposed to be located in the cross-sections (A), where no moment acts, they must be able to transfer only shear force while, if it is preferable only a connection at the beam midspan (B), connection design has to be developed considering the need to transfer the sole bending moments. In both cases, partial strength connections should be used guaranteeing the transfer of the sole design shear (A) or of the sole design moment (B).

15.2 Articulated Connections

Nowadays, articulated connections are mainly used in truss members, typically for bridge supports and for frames supporting machinery or moving equipment.

The cost of making a pin joint is quite high because of the machining required for the pin and its holes and also because of difficulties in assembly. Furthermore, pins are used in special architectural features where relative rotation occurs between the members being connected.

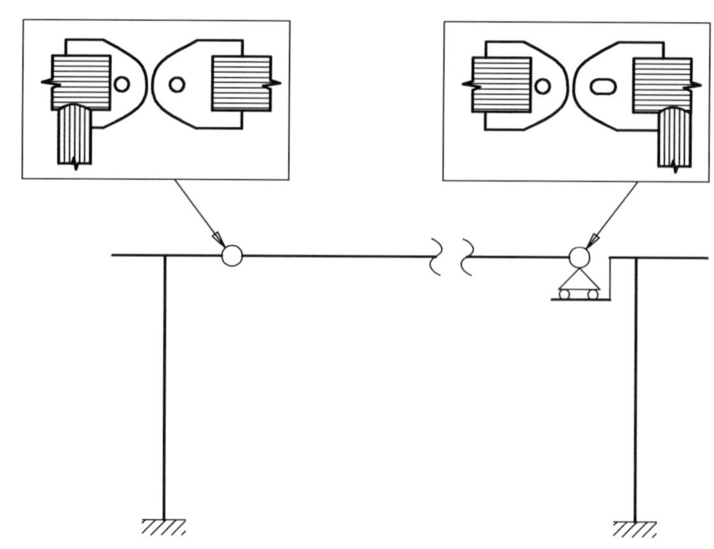

Figure 15.2 Typical connection detail for a pinned simply supported beam.

An aspect sometimes critical for connections is the required performance associated with the practical realization of the kinematic mechanism considered in the design phase, especially with reference to the detailing associated with hinges and simple supports. As presented in Section 15.5.1, in addition to the checks for geometry and resistance, checks are also necessary to verify the compatibility between the required displacement and the one guaranteed by the chosen detailing.

The movements of a structure are not in themselves detrimental. Problems may arise where movements are restrained, either by the way in which the structure is connected to the ground, or by surrounding elements such as claddings, adjacent buildings or other fixed or more rigid items. If no adequate attention has been paid in the design phase to such movements and to the associated forces and moments, it is possible that they will lead to, or contribute towards, deterioration in one or more elements. Deterioration in this context could range from cracking or disturbance of the finishes on a building to buckling or failure of primary structural elements due to large forces developed through inadvertent restraint. As an example, the portal frame in Figure 15.2 can be considered, which is characterized by a roof beam that is pinned to the column at one end and simply supported at the other beam end.

A plate is attached at both beam ends, which has a hole to insert a pin to realize the connection with the column. In case of hinge restraints, coupled plates with a circular hole of equal diameter are attached to the column. In a similar way, in correspondence of the other beam end, connection to the column is realized via coupled plates but has a slotted hole to allow for horizontal displacements. The slot should have an appropriate length in order to hamper the contact between the pin and the hole of the column plate, hence avoiding the transfer of horizontal forces to the vertical member, which induces shear forces and bending moments on the column.

15.2.1 Pinned Connections

A pinned connection is generally composed of steel coupled plates (brackets) welded at the ends of the elements to be connected, suitably stiffened to contain the local effects related to the concentration of forces, and drilled to accommodate the pin, thereby allowing the kinematics expected in the design phase (Figure 15.3).

Figure 15.3 Examples of pinned connections.

The members to connect must be arranged so as to avoid any possible eccentricity in the transmission of the load and, at the same time, must be characterized by an adequate size to ensure that the force is transmitted via connection without any significant stress concentration. Furthermore, when the use conditions might lead to an accidental extraction of the pins, these must be blocked with appropriate safety devices (plugs anti-release).

Usually, shear force and bending moment acting on the pin are calculated assuming that the brackets behave as simple supports and considering the reaction forces uniformly distributed along the length of contact of each part.

15.2.2 Articulated Bearing Connections

Articulated bearing connections are frequently used as supports for bridges and can easily be obtained by means of a direct contact between the metal surfaces of the elements in the bond competitors. Usually, these connections can be divided in two types on the basis of the contact surfaces:

- *connections with contact between two surfaces of which at least one is curved*: contact point can be obtained via a ball bearing and linear contact is typically due to an interposed cylindrical round, which is presented in Figure 15.4a,b. Appropriate guides or wedges must be provided to control surface rolling and to transmit possible transverse forces;
- *connections with concentrated contact between a plate and a plate with a knife* (Figure 15.4c,d): which are frequently used for small span bridges and often for the supports of important crane runway beams. Excessive contact pressure between the plates must be avoided in order not to deform the surfaces.

The evaluation of the state of stress is based on conventional Hertz formulas. Design is developed considering, as limit for the strength the stress value f_{lim}, which has to be evaluated in accordance with the considered Code of practice. This value is obtained from the tension limit strength multiplied by a factor significantly greater than unity, accounting for the benefits associated with the tri-axial state of stress in correspondence of the joint, which has a beneficial confinement effect.

In cases of linear contact, that is contact due to a cylindrical hinge (cases a– d of Figure 15.5), it is required that:

$$\sigma \leq 4 f_{\text{lim}} \tag{15.1}$$

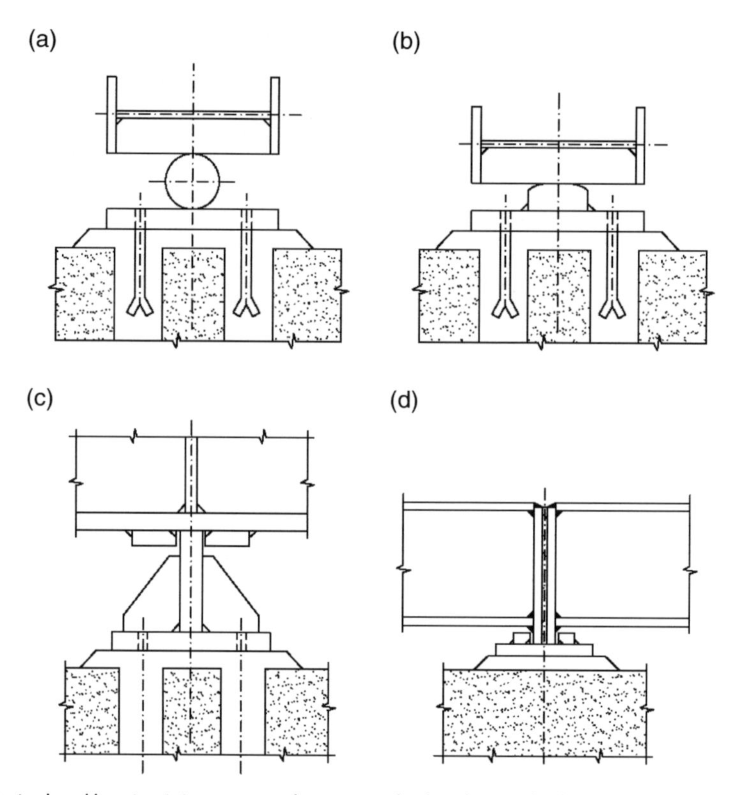

Figure 15.4 Articulated bearing joints: contact between cylindrical (a) and spherical (b) surface, contact plate (c) and knife contact plate (d).

If cylindrical hinge has a length b, with reference to the symbols in Figure 15.5, assuming that F is the transferred force, the design stress is evaluated as:

case (a) with $\frac{r_2}{r_1} \geq 2$:

$$\sigma = \sqrt{0.18 \cdot E \cdot F \frac{r_2 - r_1}{r_1 \cdot r_2 \cdot b}} \tag{15.2a}$$

case (b):

$$\sigma = \sqrt{\frac{0.18 \cdot E \cdot F}{r \cdot b}} \tag{15.2b}$$

case (c):

$$\sigma = \sqrt{\frac{0.20 \cdot E \cdot F}{2r \cdot b}} \tag{15.2c}$$

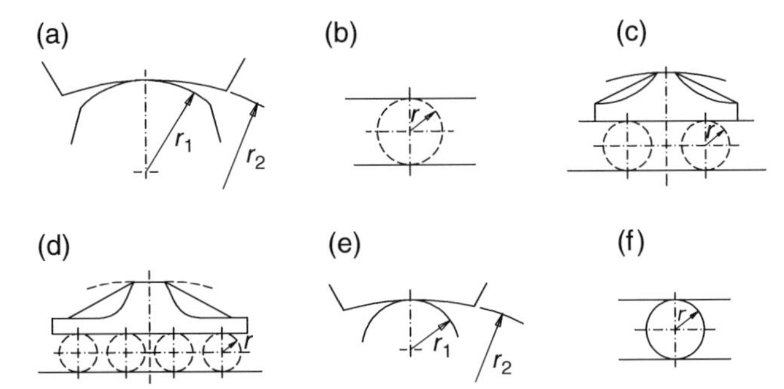

Figure 15.5 Most common types of contact surfaces: linear contact via cylindrical pin (a–d) and via spherical ball (e) and (f).

case (d) with term n identifying the number of the rollers:

$$\sigma = \sqrt{\frac{0.24 \cdot E \cdot F}{n \cdot r \cdot b}} \tag{15.2d}$$

In case of punctual contact, that is contact due to a spherical hinge (cases e and f in Figure 15.5), it is required that:

$$\sigma \leq 5.5 f_{\mathrm{lim}} \tag{15.3}$$

Design stress σ is evaluated as:

- in case (e) of Figure 15.5:

$$\sigma = \sqrt[3]{\frac{0.06 E^2 F (r_2 - r_1)^2}{r_2^2 \cdot r_1^2}} \tag{15.4a}$$

- in case (f) of Figure 15.5:

$$\sigma = \sqrt[3]{\frac{0.06 E^2 F}{r^2}} \tag{15.4b}$$

15.3 Splices

As already introduced in Chapter 3, steel buildings have a skeleton frame, which is site-erected by connecting mono-dimensional members. In addition to problems associated with the choice of the types of connections, also strictly depending on the connected members (secondary beam, girders, diagonal bracing, etc.) sometimes connections are necessary to realize a member of great length, splitting it into parts of a length suitable for transport by tracks (as already mentioned in Section 15.1). Typically, for columns in buildings as well as for trussed beams it can be convenient to site join single components via splice connections.

There are many ways of making splices. For example, traditional cover plates may be used for full load transfer or just for continuity; welds or bolts may be chosen as fasteners. Herein, reference is made to the following types of splices, which are the most commonly used:

- *beam splices*: for example connections between horizontal members, generally under bending moment and shear force;
- *column splices*: for example connections between vertical members, generally under compression and shear forces and bending moment.

It should be noted that for these types of connections, as well as for all the other components, it is preferable to have shop welding instead of site welding in order to guarantee a better quality of the product as well as to reduce the costs associated with the execution of connections.

15.3.1 Beam Splices

As for other types of connection, beam splices can be full or partial strength connections. In the latter case, it can be convenient to place the joint in correspondence of zones interested by suitable internal forces and moments, as already discussed with reference to the example of Figure 15.1. Some typical beam splices are illustrated in Figure 15.6:

(a) connection with extended end plate shop welded to beam end and site bolted;
(b) connection with cover plates site bolted to beam flanges, to transfer the bending moment and/or to the beam web to transfer the shear force;
(c) connection with welded cover plates for the flanges and for the web. These plates can be site welded at both beam ends or, more conveniently, shop welded at one beam end and site welded to the other beam;
(d) butt welded connection, which requires appropriate surface preparations (as discussed with reference to Figure 15.7).

It should be noted that the lacking of the appropriate thickness and/or width of the cover plates makes the splice a partial strength connection.

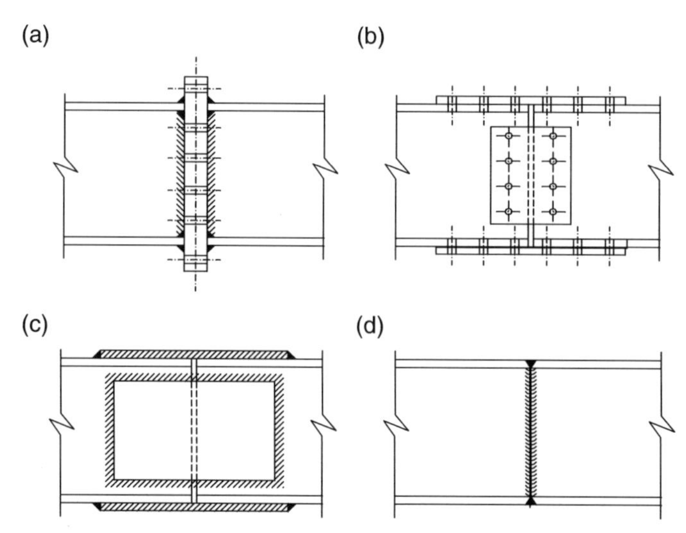

Figure 15.6 Typical splice for beams.

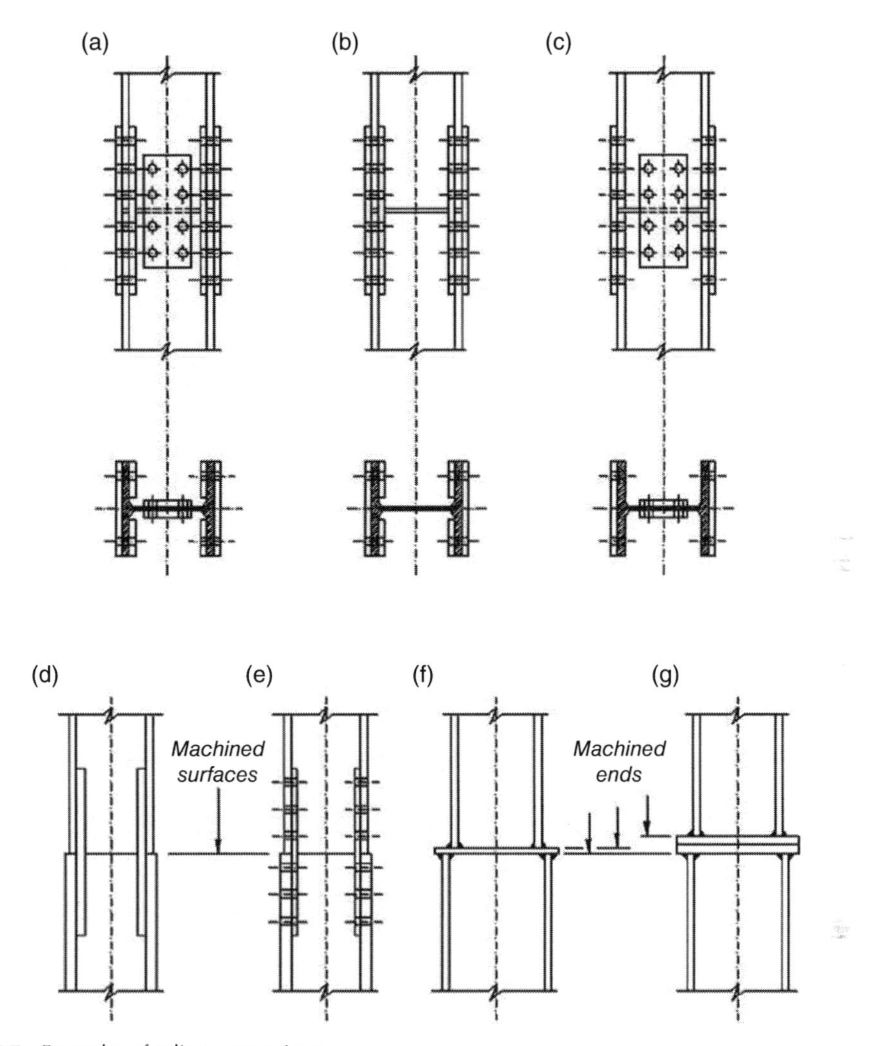

Figure 15.7 Examples of splice connections.

15.3.2 Column Splices

Columns in framed systems are generally subjected to compression or to compression, shear and bending moments, hence resulting in predominant compression stresses. As a consequence, from a theoretical point of view, no splice connection is required, since the compression force is transferred by direct bearing. However, due to the presence of geometric imperfections (lack of straightness of the column) as well as of unavoidable erection eccentricities and to the fact that even carefully machined surfaces will never assure full contact, column splices have necessarily to be realized.

Even when the column is subject to pure compression and full contact in bearing is assumed, all the appropriate design verifications are necessary, as required by the Codes of practice. The location of the splice should be selected so that any adverse effect on column stability is avoided, that is the distance of the connection from the floor level should be kept as low as possible. A limit of 0.25 times the storey height is usually accepted. If this requirement cannot be fulfilled, account should be taken of the bending moment due to deformations induced by member imperfections.

More significant values of the bending resistance may be required in splices when columns are subject to primary bending moments, as in frame models assuming hinges at, or outside, the column outer face, as shown in the discussion on joint modelling (see Section 15.5). In addition, in columns acting as chords of cantilever bracing trusses, tensile forces may arise (uplift) in some load conditions(uplift), which are transmitted by splices.

Typical compression column splices suitable for use in simple frames are shown in Figure 15.7 and in particular it is possible to recognize:

(a) splice connection with coupled cover plates site bolted to each column flange and to the column web;
(b) splice connection with coupled cover plates site bolted to the sole column flanges;
(c) splice connection with coupled cover plates site bolted to the column web and a single cover plate per each column flange, site bolted to the outside faces of the column flanges in order to reduce the plan area occupied by the splice;
(d) splice connection with fillet welds and internal welded cover plates;
(e) splice connection with light cover plates, which are bolted to the column flanges;
(f) splice connection with an interposed plate, which is welded to both columns (generally shop welded to one column and site welded to the other column);
(g) splice connection with interposed plates shop welded to the column ends and site bolted to each other.

When a bolted solution is adopted (a–c in Figure 15.7), forces are transmitted through the cover plates and distributed among the connecting plates in proportion to the stress resultant in the cross-sectional elements, for example for simple compression in proportion to the areas of the flanges and of the web. Differences in column flange thickness may be accommodated by the use of packs. These may be positioned preferably on the outside faces of the flanges.

Splice connection types (d)–(g) rely on direct bearing. When the surfaces of the end cross-sections of the two column shapes are sawn and considered to be flat, and squareness between these surfaces and the member axis is guaranteed, the axial force may be assumed to be transmitted by bearing. Both fillet welds (d) or light cover plates (e) are provided to resist possible secondary shear force and bending moment when the upper and lower columns differ in serial size. Plates are flattened by presses in the range of thicknesses up to 50 mm (1.97 in.), and machined by planning for thicknesses greater than 100 mm (3.94 in.). For intermediate thicknesses either working process may be selected. In case of significant variation of cross-sectional dimensions, as in the arrangement of type (f), the plate must be checked for bending resistance. A possible conservative model assumes the plate as a cantilever of a height equal to the width and clamped to the upper column flange. The axial force, which is transmitted between the corresponding column flanges, is applied as an external load at the mean plane of the flange of the lower column.

For larger differences in column size, a short vertical stiffener has to be located directly below the flange(s) of the upper column to directly assist in transferring the locally high force. If full penetration welding connections realize these splices, the verification of the welded connections can be omitted in most practical cases. Where the cross-section of the members connected via splices vary considerably, a separating plate is required to realize the splice, the design of which is more complex if subject to bending.

In Figure 15.8 some typical splice solutions are presented for sensible cross-section variations:

(a) A splice with a shop welded plate on the top of the lower column, which results very useful for the direct positioning and the site welding of the upper column. Web stiffeners are always

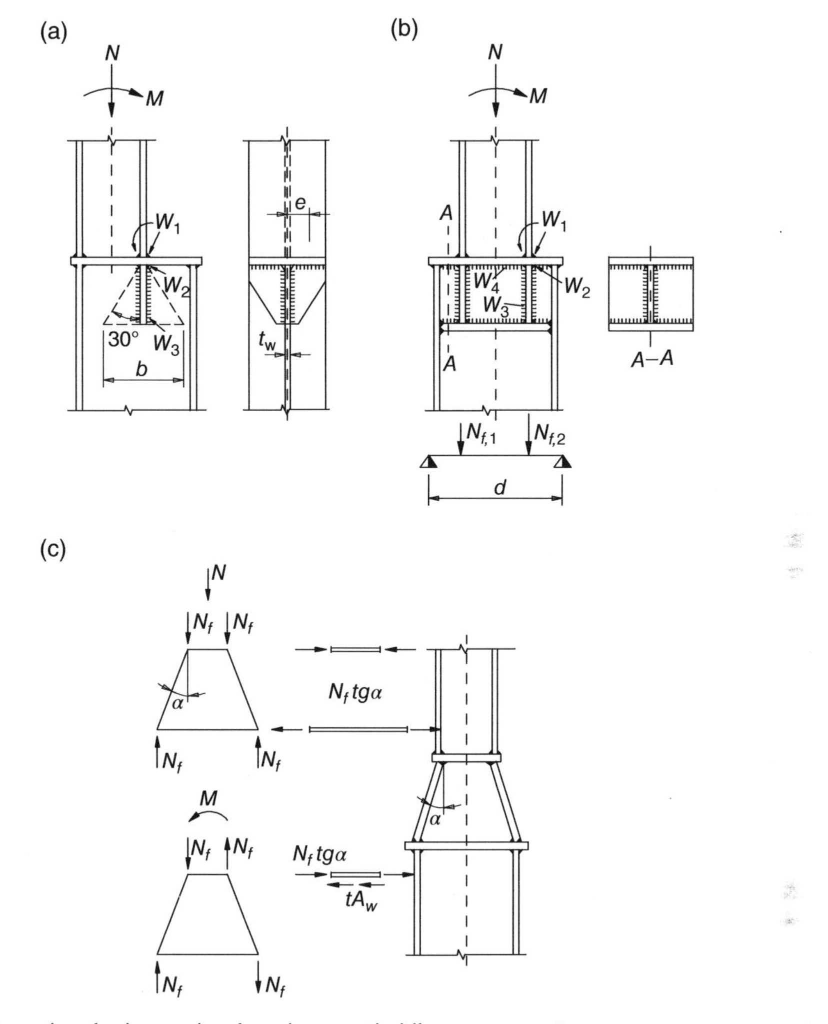

Figure 15.8 Examples of column splice for columns with different cross-sections.

required because they receive the load transmitted by the upper flange and transfer it to the lower web, which in turn redistributes it in the underlying part. Due to the differences between column axis, an additional bending moment also has to be transferred via the splice;

(b) A splice with longitudinal and transverse shop welded stiffeners at the top of the lower column. The bottom end of the upper column is site welded to the plate. This splice is recommended when loads acting on the column are high and it is hence convenient to maintain the longitudinal axis of the column coincident in order to avoid additional bending moments. A simplified model related to a stocky I-shaped section loaded by two concentrated forces in correspondence of the flanges of the upper column can be adopted. Height of the beam is the distance between the horizontal splice plates, which are considered the beam flanges.

(c) A tapered splice. The size of the columns is different in this case too and the splice is tapered: at the end of each column a horizontal plate is welded and diagonal plates are placed to connect the column flanges. A design approach based on the use of the strut-and-tie model can be adopted, very similar to the one adopted for the design of isolated concrete stocky foundations.

15.4 End Joints

Several types of end joints, for example joints connecting the end of elements with their longitudinal axis not parallel, can be considered, and in the following reference is made to:

- beam-to-column joints;
- beam-to-beam joints;
- bracing connections;
- base-plate connections;
- beam-to-concrete wall connections.

15.4.1 Beam-to-Column Connections

Beam-to-column joints can be realized by connecting the beam flanges and/or the beam web, depending on the required joint performance. In Figure 15.9 some common types of beam-to-column joints are presented, which refer to a beam attached to the column flange but can also be adopted if the web column is considered. In particular, the following solutions are frequently adopted:

(a) *web angle (cleat) connection*: a couple of angles is site bolted to the beam web and to the column flange;
(b) *fin plate connection*: a plate, parallel to column web, is shop welded to the column flange and site bolted to the beam web. This joint is also one of the few arrangements suitable for use with rectangular or square hollow section columns as no bolting to the column is necessary;
(c) *header plate connection (partial depth end plate)*: a plate, parallel to the column flange, with a depth lower than the one of the web is shop welded to the beam web and site bolted to the column;
(d) *flush end plate (full depth end plate)*: a plate having the depth of the beam and parallel to the column flange, is shop welded both to the beam flanges and to the beam web and site bolted to the column. Furthermore, when high joint performance is required, it can be convenient to extend the plate beyond the tension flange of the beam in order to allow the positioning of an additional bolt row external to the beam (*extended end plate*).

Both solutions (a) and (b) provide some allowance for tolerance (through the clearance in the beam web holes) on member length. Type (b) connection permits beams to be lifted in from one side. Furthermore, types (c) and (d) connections require a more strict control of the beam length and of the squareness of the cross-section at the end of the beam. The flush end plate scheme of solution (d) is sometimes preferred to the part depth end plate of solution (c) in order to reduce the chances of damage during transportation. Partial depth endplates should not normally be less than about 0.6 times the beam depth or the end torsional restraint to the beam may be reduced. All the presented types of connections can improved by adopting column panel stiffeners, in correspondence of the beam flange, which are necessary to avoid local weakness zone of the joint.

Additional joint solutions frequently used in steel structures practice are also proposed in the next part of the chapter (Section 15.5) where joints are presented with reference to their guaranteed performances.

15.4.2 Beam-to-Beam Connections

Floor decks in buildings are usually supported by grids of secondary beams and main girders. Most common types of these connections, which generally have a very limited degree of flexural resistance, are reported in Figure 15.10, where it is possible to recognize:

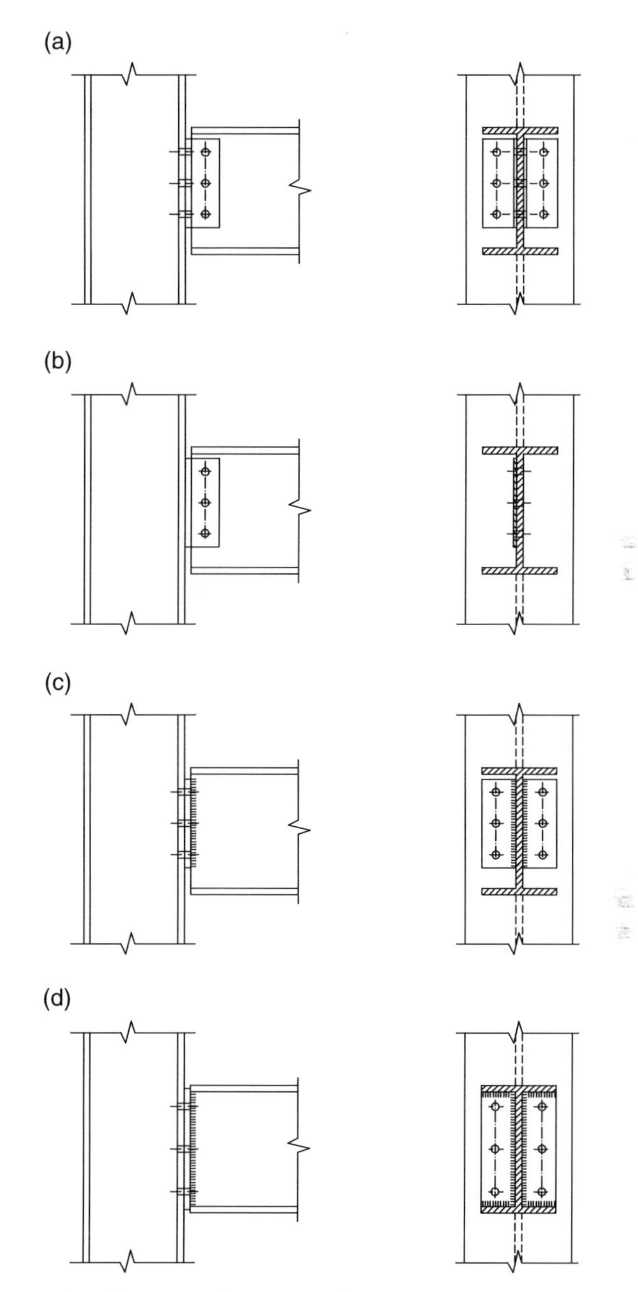

Figure 15.9 Typical examples of beam-to-column connections.

(a) *Web cleat bolted connection*: angles are site bolted to the web of the secondary beam and of the main girder. It should be noted that the top flange of the secondary beam, which supports the floor system, is located at a lower level than the corresponding one of the primary beam. This solution is convenient if the upper part of the girder can be embedded in the concrete floor or enclosed in partitions. Furthermore, it is good practice to place the angles as close as possible to the upper flange of the girder in order to minimize cracking of the concrete floor slab due to the beam rotation.

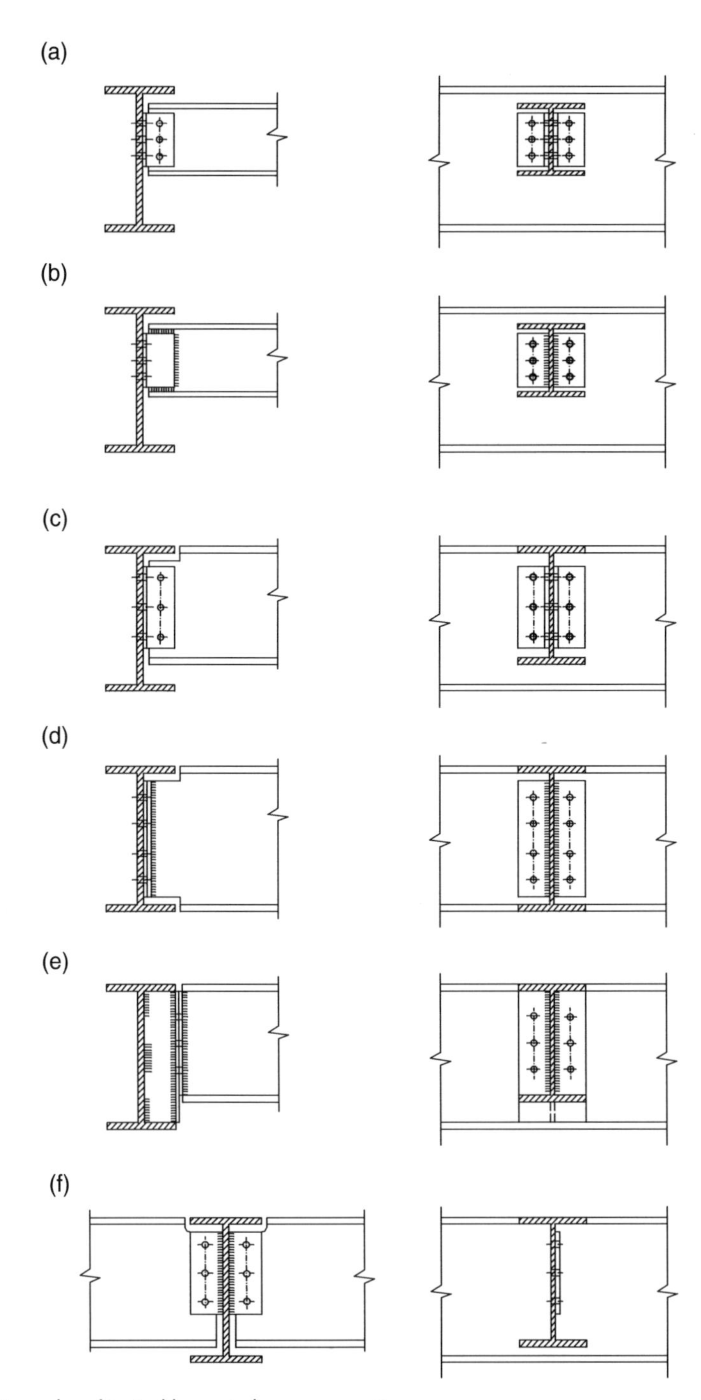

Figure 15.10 Examples of typical beam-to-beam connections.

(b) *Web cleated bolted-welded connection*: angles are shop welded to the web of the secondary beam and site bolted to the web of the main girder. As for the connection type (a), also in this case the top flanges of the secondary beam and the girder are placed at different levels.

(c) *Web cleated connection with coped secondary beam (single notched beam-to-beam connection)*: this solution, where angles can be bolted to the web of the beams or welded to the secondary beam and bolted to the main girders, differs from previous solutions (a) and (b) because in this case, the beam and girder top flanges meet at the same level. The coped beam is thus locally weakened and appropriate checks are required, as discussed next.

(d) *Flush end plate connection*: this connection is used when the same cross-section is adopted for secondary beam and main girder. Also in this case a coped beam is used as secondary member but both beam flanges are removed (double notched beam-to-beam connection). Flush end plate is shop welded to the web of the secondary beam and site bolted to the web of the main girder.

(e) *Stiffened flush end plate connection*: a flush end plate is shop welded to the secondary beam and site bolted to a tee stiffener shop welded to the girder. This connection is in many cases very expensive due to the costs associated with the shop welding techniques.

(f) *Fin plate connection*: this connection can be considered a variant to web angles connections. A single plate is shop welded to the primary beam end and site bolted to the secondary beam. A fin plate connection is particularly simple both to fabricate and to erect, but it requires careful design as it has to function as a hinge.

Connection types (d) and (e) possess some predictable stiffness and strength, but in routine design they are modelled as pins. In particular, there is a need to decide where the 'hinge' is located as explained in Section ***15.5.1***. Furthermore, it should be noted that types (a) and (c), which make use of web cleats bolted to both the girder and the beam, were extensively used in the past. Type (b) with the cleats bolted to the girder and welded to the beam, and types (d) and (e) where a flush end plate is adopted, may cause lack-of-fit during erection due to the dimensional tolerances.

When a beam is coped, as in connection type (c) or (d), it should be verified that no failure occurs at the beam section that has been weakened (block shear failure mechanism). Particularly in these types of connection, it should be noted that the presence of bolt holes often weakens the holed component(s) of the member as well as the connection plate. Failure may occur locally there by bearing or plate punching, or in an overall mode along a path whose position is determined by the hole location and by the actions transferred by the plate, such as that considered in Chapter 5 for tension members connected via staggered holes.

Generally, if the plate proportions are appropriate to avoid instability, it results conservative for the strength limit against general yield and satisfactory to design against local fracture, which is generally brittle.

The actual failure stress distributions in bolted connection plates are both uncertain and complicated. Plate sections may be subjected to simultaneous normal and shear stress, as in the case of the splice plates shown in Figure 15.11a,b. These may be designed conservatively against general yield by using the shear and bending stresses determined by elastic analysis of the gross cross-section in the combined yield Von Mises criterion of Eqs. (1.1) and (1.3), and against fracture by using the stresses determined by elastic analyses of the net section.

Furthermore, block failure may occur in some connection plates as shown in Figure 15.11c,d and it may be assumed that the total resistance is provided partly by the tensile resistance across one section of the failure path, and partly by the shear resistance along another section of the failure path. This assumption implies considerable redistribution from the elastic stress distribution, which is likely to be very different from uniform.

15.4.3 Bracing Connections

Connections within the bracing members or between the bracing members and the main framing components transfer generally forces between a number of differently oriented members.

In Figure 15.12 typical connections for horizontal bracings (i.e. floor bracings and roofs bracings) are proposed. As can be noted, diagonal elements are usually bolted to a plate fixed to the

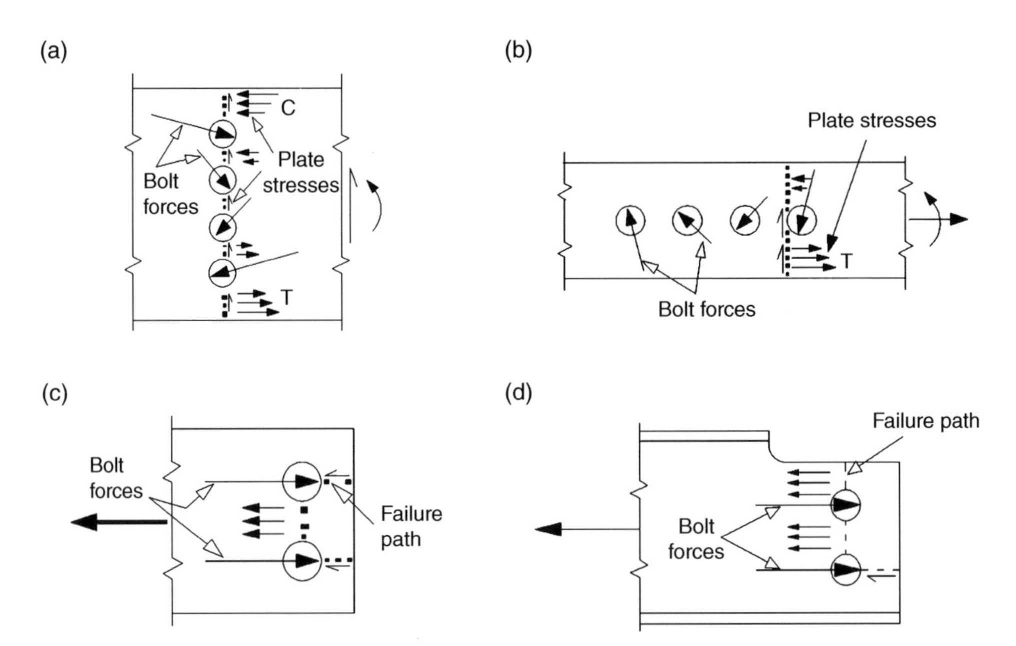

Figure 15.11 Stress distribution in bolted plates in shear and tension: (a) shear and bending, (b) tension and bending, (c) block tearing in plate and (d) block tearing in a coped beam.

primary beams. The location of these plates has to be defined considering the possible interferences with structural and non-structural components. In absence of limitation due possible interactions with the slab as well as the roof, connections can be realized at the top flange level, as in solution (a) and (c) in Figure 15.12. Otherwise, different solutions have to be considered, attaching the plate to the beam or to the bottom flange of the beam. Usually, bracing members are realized with slender profiles, owing to the fact that in service conditions, the floor slab directly transfers horizontal forces to the vertical bracing systems. The removal of these bracings after the erection of the skeleton frame should result non-economical and for this reason, they are usually embedded in the concrete slab. Regarding vertical bracings, Figure 15.13 proposes typical details used for cross bracing and k-bracings.

In case of cross brace, type (a) connection attaches the bracing via a plate shop welded to the column and bolted to the bracing (diagonal and horizontal) members. Type (b) is an internal bracing connection. In case of k-bracing connections, types (c) and (b) combine both functions by making the beams part of the bracing system.

With reference to the most common cross-bracing types that can be designed considering active both tension and compression members, diagonal members should be connected to each other at the intersection point, as shown in Figure 15.14. This internal connection, realized in many cases via one bolt, is usually enough to restraint efficiently the overall stability of the compression diagonal, reducing its effective length.

15.4.4 Column Bases

As for beam-to-column joints, column bases also have to be correctly classified in order to select the appropriate model to use in structural analysis. It is worth mentioning that the relationships between the moment and the rotation is significantly influenced by the level of the axial load acting at the base joint location, which increases significantly base joint performance in terms of both rotational stiffness and moment resistance.

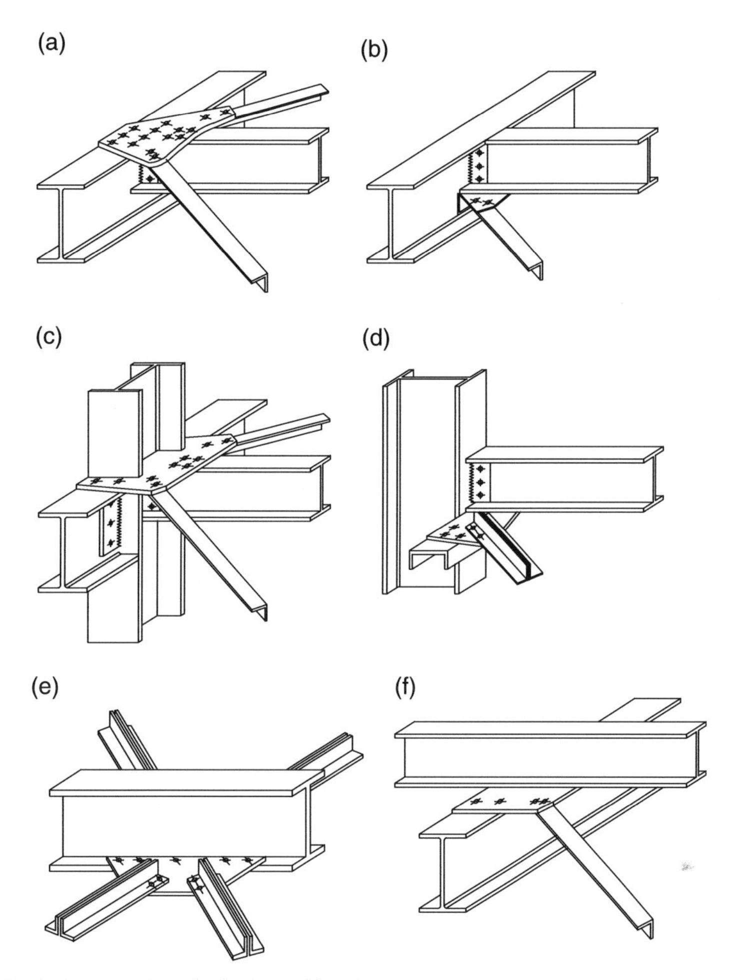

Figure 15.12 Typical connections for horizontal bracings.

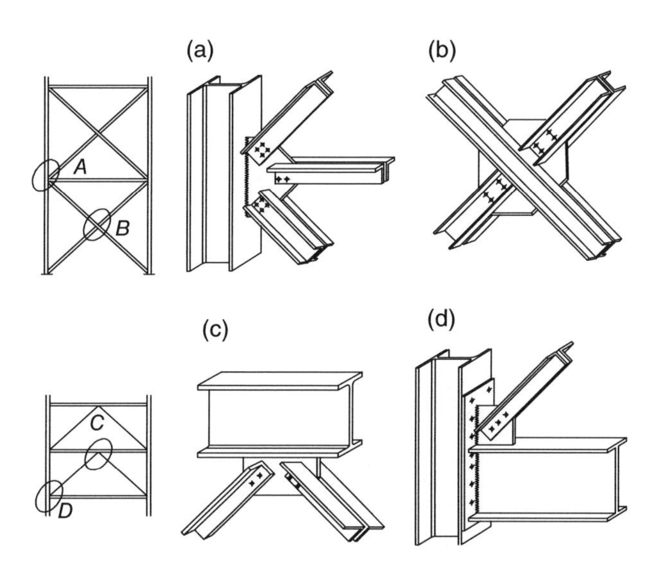

Figure 15.13 Typical connections for vertical bracings.

Figure 15.14 Example of cross bracing with an internal connection.

A column base connection always consists of a plate welded to the foot of the column and bolted down to the foundations (Figure 15.15). In a few cases, a second steel plate, usually rather thicker, can be incorporated into the top of the foundation, helping both to locate the base of the column accurately and in spreading the load into the weaker (concrete or masonry) foundation material.

The plate is always attached to the column by means of shop fillet welds. However, if the column carries only compression loads, direct bearing may be assumed into design, provided that the contact surfaces are machined or can be considered to be flat. Machining is generally omitted when the transfer loads are relatively small. No verification of the welds is then required.

Baseplate connections in simple frames are typically subjected to the sole axial load; they are generally modelled as ideal pins, and are designed to transfer concentrated force (compression or tension). As it results from the base connections in Figure 15.15a,b anchor bolts are required only for the erection phase, making possible to regulate the height of the column base. A combination of axial and shear force acts on the column base, usually when the column is part of the bracing system (c). In tension connections, the baseplate thickness is often governed by the bending moments produced by the holding down bolts to transfer the acting load to foundation. In some cases, the use of stiffeners can be required (d), which significantly increases the fabrication content and therefore the cost of the column base as compared with the cases (a) and (b).

With reference to concrete foundation in case of moderate tension forces, the bolts holding down are usually cast into the foundation (Figure 15.16).

Hooked anchor bolts allow clearance for their positioning and can contribute to transfer tension force by friction between steel and concrete. As to the failure mode, owing to the importance of a ductile failure mode with anchor bolt yielding, it is recommended to use an appropriate steel grade in order to avoid brittle failure. Furthermore, when high tensile forces should be transferred to the foundation, which is frequent in high-rise and tall buildings, it is necessary to provide appropriate anchorage to the bolts. For example, threaded bolts may be used in conjunction with channel sections embedded in the concrete (hammer head anchor bolts) as in Figure 15.17b or with a washer plate as in Figure 15.16c.

In case of high shear forces, appropriate devices can be used in order to transfer these forces by contact between the devices and the concrete. As an example, Figure 15.18 can be considered, which proposes:

solution (a) with a device obtained by shop welding at the bottom of the base plate a short length of profile identical to that of the column and filled in the concrete after the base plate positioning;
solution (b) with some appropriate stiffened plates shop welded to the bottom face of the base-plate.

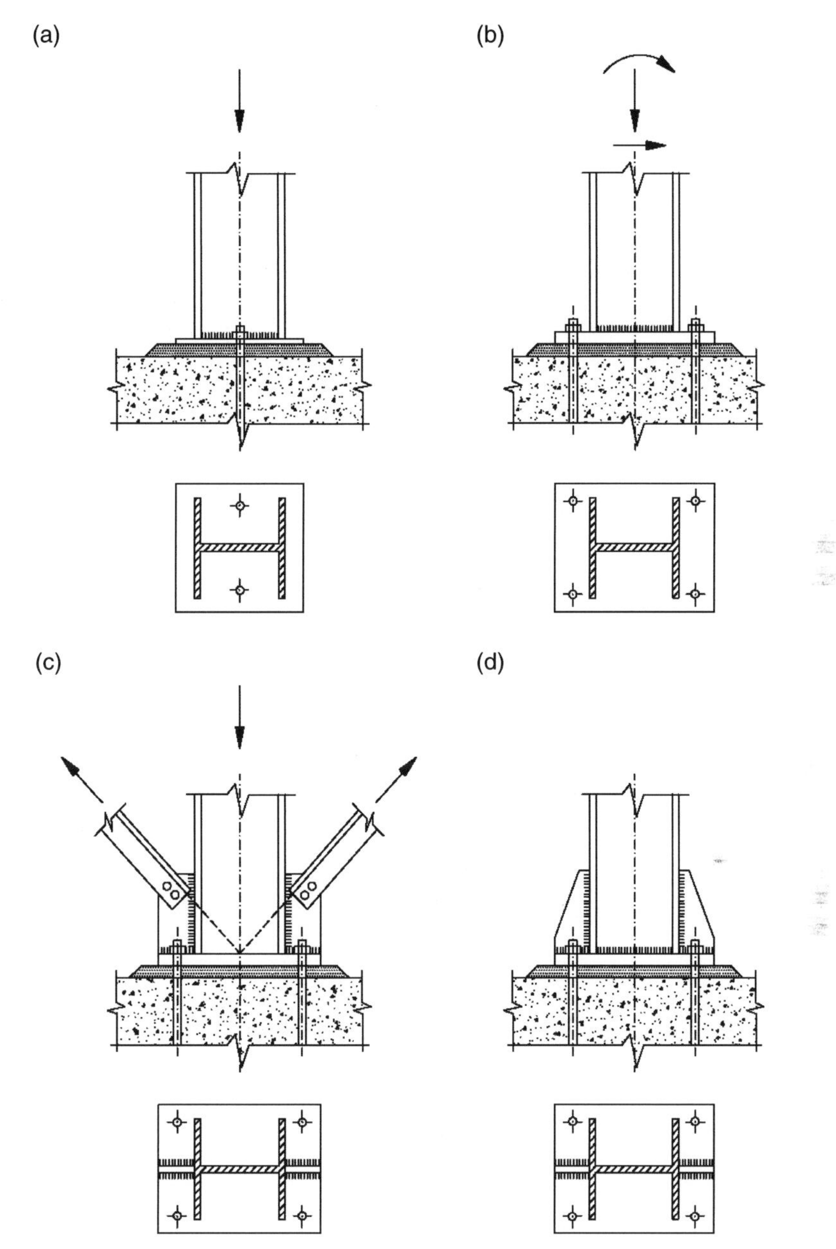

Figure 15.15 Typical column bases for simple frames.

15.4.5 Beam-to-Concrete Wall Connection

In steel buildings for residence and commercial destinations, frequently concrete cores, which are used to sustain stairs and uplifts, brace efficiently the skeleton frame; that is transfer to foundations the horizontal forces. Furthermore, concrete shear walls can be used to increase the frame performance to lateral loads as well as to improve seismic resistance of the structures. In all these cases, the steel structure resisting gravity loads is combined with a concrete core resisting horizontal forces.

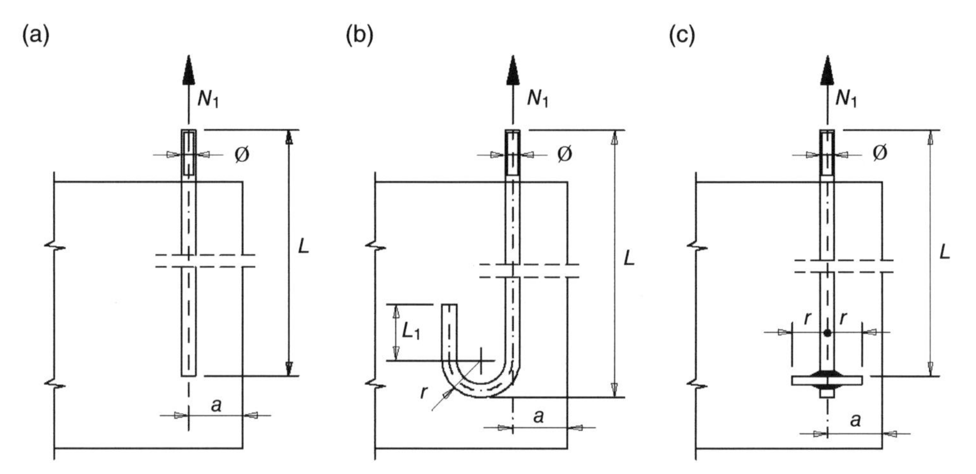

Figure 15.16 Examples of anchor bolts embedded in the concrete of foundation: simple anchor bolt (a) hooked anchor bolt (b) and anchor bolt with a washer plate (c).

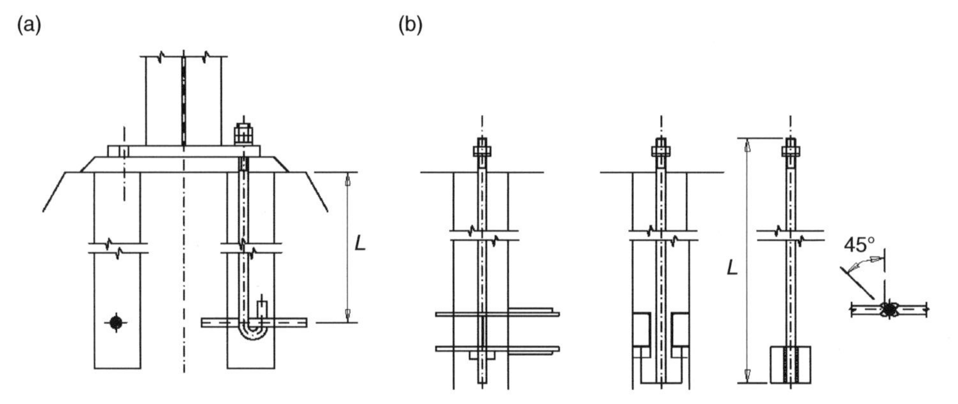

Figure 15.17 Typical solutions to anchor the foundation when bolts are in presence of high values of the tension force: hooked anchor bolts (a) and hammer head anchor bolts (b).

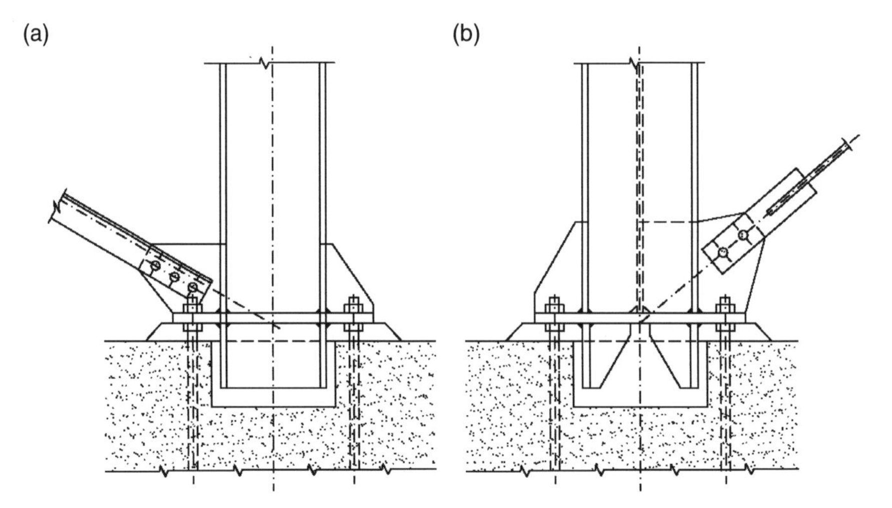

Figure 15.18 Examples of appropriate device to transfer shear load via direct contact steel-concrete.

Particular attention must be paid to the connection systems between these two materials. The two systems are built with dimensional tolerances of a different order of magnitude: millimetres in case of steel components and centimetres for concrete cores and walls. Special care is required to account for the relative sequence of erection of the concrete and steel components, the method of construction of the core (influencing significantly concrete tolerances) and the feasibility of compensating for misalignments. Furthermore, it should be noted that the details in the concrete wall must be suitably designed to disperse connection forces safely. In particular, joint detailing is especially important when deep beams are required to transmit high vertical loads.

The most common connections between steel beam and concrete cores (or concrete walls) are illustrated in Figure 15.19. The connection of the steel beam can be made via different steel details and in particular:

(a) fin plate site welded to the adjustment tool and bolted to the beam web;
(b) fin plate site welded to the steel plate encased in the concrete core and site welded to the beam web;
(c) fin plate site welded to the steel plate encased in the concrete core and site bolted to the beam web;
(d) rigid block and web angles site welded to the steel plate encased in the concrete core;
(e) fin plate encased in the concrete core and site bolted to the beam web;
(f) fin plate bolted via appropriate fasteners to the concrete core and to the beam web.

Connection type (a) with a pocket in the wall is convenient for simplicity of adjustment, but complex in terms of core erection: types from (b) to (f), where part of the connection is encased in the core wall during concrete pouring, may be preferable.

Reinforcing bars (rebars) and/or headed studs can also be used to anchor the connection in the concrete components. Full penetration welds are preferred when rebars are connected directly to

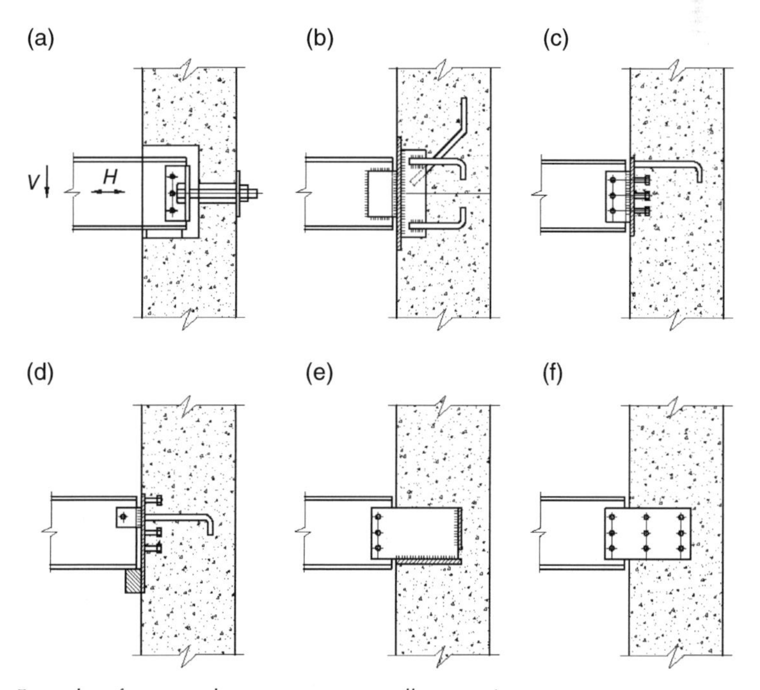

Figure 15.19 Examples of common beam-to-concrete wall connections.

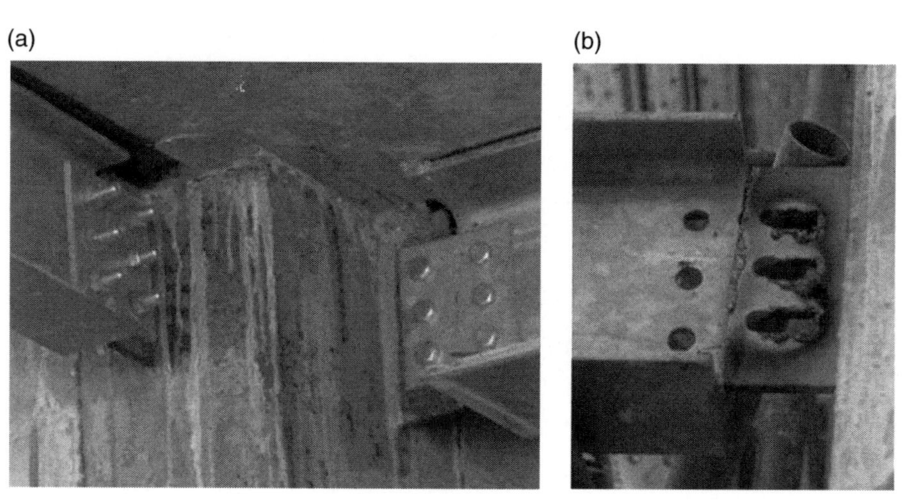

Figure 15.20 Examples of beam-to-concrete wall connection (a) and site slots of the thin plate to assemble the connection with geometrical tolerances that are too large (b).

the steel plate, so that eccentricity of the force with respect to the welds is avoided or significantly reduced.

Checking of the various components within the connection should be conducted in a consistent manner, ensuring that the principles of connection design, for example that the assumed distribution of forces satisfies equilibrium, are observed.

It could eventually occur that the tolerances associated with shop working as well as site assembling are not completely respected. As a consequence, problems could arise during the construction phase, hampering a correct erection of the skeleton frame. Site attempts to assembly members out of tolerances are very dangerous and have to be avoided. Figure 15.20a presents some examples of typical steel beam-concrete wall connections via steel plate site bolted to the beam. Due to the excessive working tolerances in few cases it was not possible to assembly the beam in the fin plate due to the non-correspondence of the holes. As a consequence normal holes of thin plate were site modified by slotting (Figure 15.20b) via oxyhydrogen flame but this technique has to be absolutely avoided.

15.5 Joint Modelling

It can be noted that, although the terms connections and joints are often regarded as having the same meaning, their definitions are slightly different:

- term *connection* identifies the location at which two or more elements meet. For design purposes it is the assembly of the basic components required to represent the behaviour during the transfer of the relevant internal forces and moments through steel members. Term connection identifies a set of components such as plates, cleats, bolts and welds that actually join the members together. With reference to the Figure 15.21, connection is realized by top and seat angles bolted to the beam and the column flanges;
- term *joint* identifies the zone where two or more members are interconnected. For design purposes it is the assembly of all the basic components required to represent the behaviour during the transfer of the relevant internal forces and moments between the connected members.

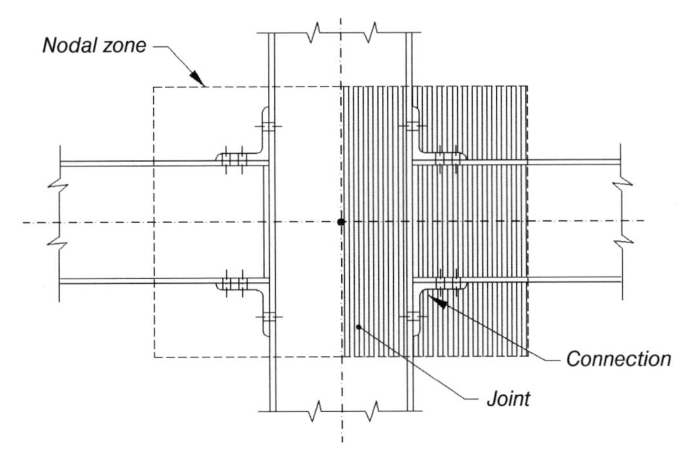

Figure 15.21 Terms and definitions.

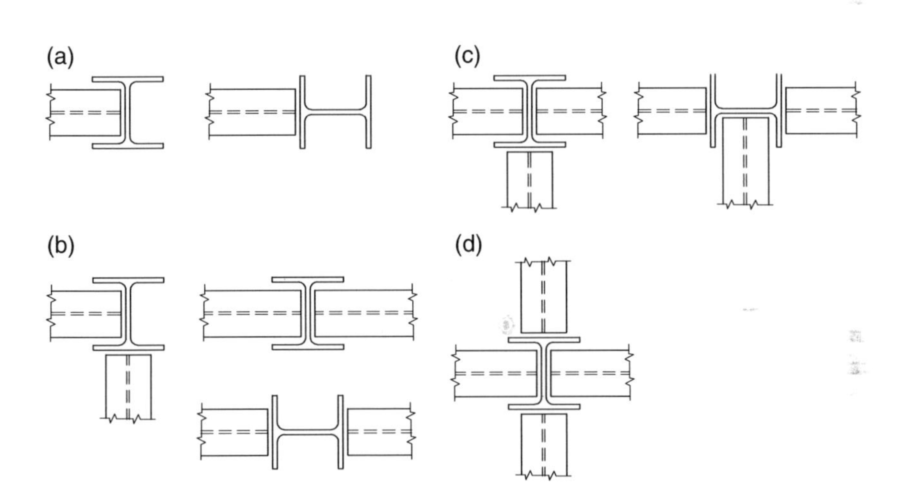

Figure 15.22 Classification of the nodes on the basis of the connected members.

A beam-to-column joint consists of a web panel and either one connection (single sided joint configuration) or two connections (double sided joint configuration).

Furthermore, *the nodal zone* is the zone where interactions among joints occur, that is where one or more joints and connections are located. From a practical point of view, it is defined as the point at which the axes of two or more interconnected structural elements converge.

It should be noted that nodes can be classified also on the basis of the number of the connected beams. With reference to Figure 15.22, it is possible to identify a:

(a) *one way node*, where a sole beam is connected to the column;
(b) *two way node*, where two beams are connected to the column;
(c) *three way node*, where three beams are connected to the column;
(d) *four way node*, where four beams are connected to the column.

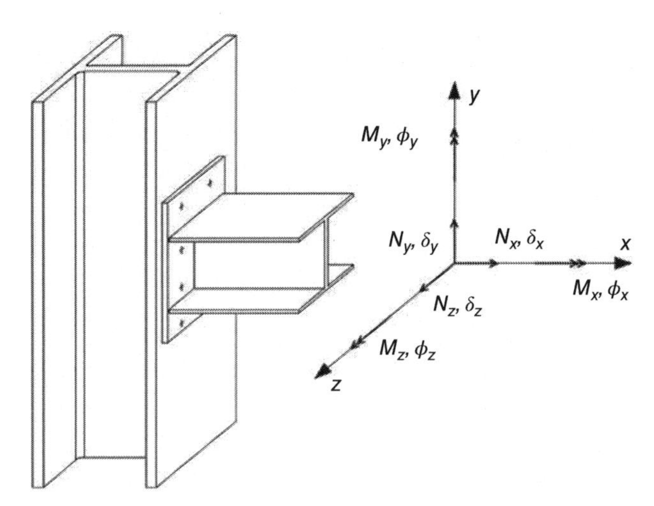

Figure 15.23 Stress and deformations in a typical beam-to-column joint.

It is worth mentioning that all the approaches developed for the design of members cannot be used to design connections, owing to the fact that concentrate forces are transferred via zones of limited extension. The approach to the description of the joint behaviour is still considerably complex. Construction forms typical of framed structures and, in particular, of multi-storey buildings, allow for the introduction of practical assumptions, whereby the problem can be made simpler. The state of deformation produced by beam-to-column joint interaction is by its own nature complex and involves significant local distortions. Being difficult to represent it accurately for the purpose of frame analysis, it is, however, convenient to use a behaviour representation in global terms and describe the stress state and the deformation of the beam-to-column joint with reference to the six components of the stress resultant (N_x, N_y, N_z, M_x, M_y, M_z) and the associated components of the overall deformation (δ_x, δ_y, δ_z, ϕ_x, ϕ_y, ϕ_z) shown in Figure 15.23.

The joint response to the force components corresponding to the two shear forces N_y and N_z, and the twisting action M_x, exhibits a negligible deformability when compared to that of the members, if these are, as is usual, made of open sections. Besides, as the axial deformation δ_x is insignificant in terms of structural response, joint behaviour may be represented by the in-plane and out-of-plane moment rotation relations (M_x–ϕ_x and M_y–ϕ_y, respectively). The stiffness of the floors in their own planes is usually large enough in order that also the latter deformation component can be overlooked in the analysis. Ultimately, the joint response can be described through the sole equation governing its in-plane rotational behaviour. Experimental results show that the interaction between the different stress components is modest. This relation, briefly referred to as the moment-rotation relationship (M–ϕ), is assumed to express the joint performances.

All the components of the connection, as well as the part of the mono-dimensional connected members, significantly influence the overall response of the beam-to-column joint. As an example, if the one-way joint in Figure 15.24 is considered, its deformation can be considered as the sum of the following distortion components:

- shear and bending deformation of the column web panel zone (b);
- bending of the column flange (c). Furthermore, the compressed beam flange can be subjected to local buckling at the joint location;
- deformation of the connecting elements (d), that is plate and bolts.

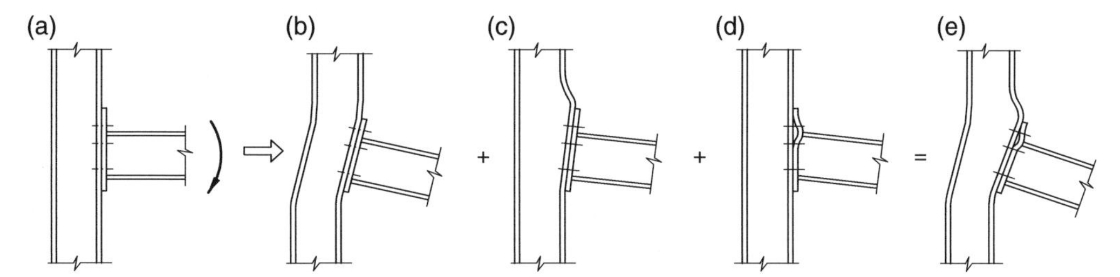

Figure 15.24 A typical external beam-to-column joint and the main contribution to its response.

As to the characterization of the beam-to-column joint performance, as for other types of joint, reference has to be made to the relationship between the moment (M) at the beam end and the rotation (Φ) in the plane of frame, which is considered as the relative rotation between the beam and the column (Figure 15.25), that is the one obtained as the difference between the beam (ϑ_b) and the column (ϑ_c) rotation. With reference to the typical moment–rotation (M–Φ) relationship for connection under monotonic loading presented in Figure 15.25, it can be noted that:

- initially, the response is approximately linear (*elastic branch*) and joint response is associated with the sole value of the rotational stiffness C_i, until the elastic bending moment M_e is achieved. Experimental studies developed since last decades pointed out that in this phase higher values of the rotational stiffness can be obtained preloading the bolts;
- in the *post-elastic branch*, joint response is characterized by a rotational stiffness C_{red}, significantly lower than the one of the elastic branch, mainly due to the local spread of plasticity in the connection components as well as non-elastic phenomena. The end of this phase corresponds to the achievement of the plastic moment of the joint, M_p;
- eventually, an additional phase (*strain-hardening branch*), in which the M–Φ relationship has a very moderate slope with a low value of the associated rotational stiffness, C_p, until the load carrying capacity of the joint under flexure (M_u) is achieved.

Furthermore, it can be noted that the stiffness in the un-loading branches, C_{unl}, is practically constant, independent on the level of bending moment reached in the loading phase.

The ability to approximate at least the key parameters of the moment-rotation joint curve is an essential prerequisite to allow for a safe design by using semi-continuous frame model. Mainly, in last four decades several extensive studies have been developed on this topic all over the world in order to develop suitable approaches to predict joint behaviour. Nowadays, the main tools available for this purpose are;

- *experimental tests*;
- *mathematical expressions*;
- *finite element models*;
- *theoretical models.*

Experimental tests on joint specimens allow for direct evaluation of the moment-rotation joint response with the limitation that the results are referred to the sole tested geometry and to the sole characteristics of the materials composing the specimen. These tests are quite expensive and require very refined resources. As an example, in Figure 15.26 the specimen of a top-and-seat cleated connection with a very rigid column is presented together with some instruments of the measuring systems. Different inductive transducers need to be placed in order to capture the different contributions to the connection rotation. In particular:

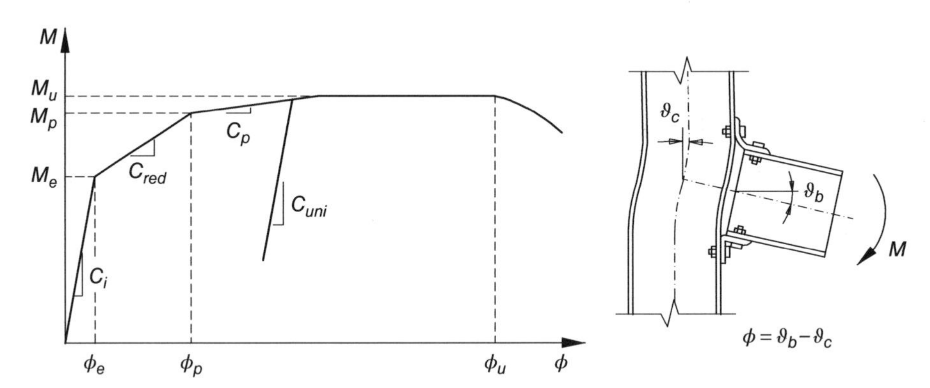

Figure 15.25 Typical moment-rotation (M–Φ) relationship for a beam-to-column joint.

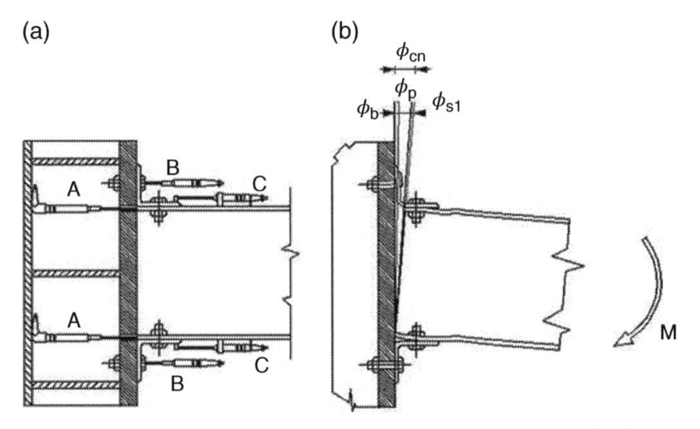

Figure 15.26 Moment-rotation determined via the experimental approach: (a) details of the measuring system for the connection and contribution to joint rotation (b).

- transducers A to evaluate the overall joint rotation, ϕ_{cn};
- transducers B to evaluate the contribution due to bolt elongation, ϕ_b;
- transducers C to evaluate the rotation due to the slip between the leg angle and the beam flanges (ϕ_{sl}).

The contribution due to the angle rotation, ϕ_p, can be obtained by subtracting bolt and slip contribution to the overall joint rotation, as:

$$\phi_p = \phi_{cn} - \phi_b - \phi_{sl} \tag{15.5}$$

Mathematical expressions used to predict joint response have the considerable advantage of being extremely simple and immediately implementable in structural analysis programs; however, they fail to notice the changes in the behaviour of the same joint as a function of the geometrical and strength characteristics of the single components. Hence, their reliability does not appear fully satisfactory, because they were calibrated basing on a necessarily small number of tests.

Finite element models are too complex by far to be adopted in design practice, despite having the advantage of allowing us to model joint geometry accurately. Moreover, a few modelling problems, which are important to simulate the bolted joint behaviour and, in particular, those aspects

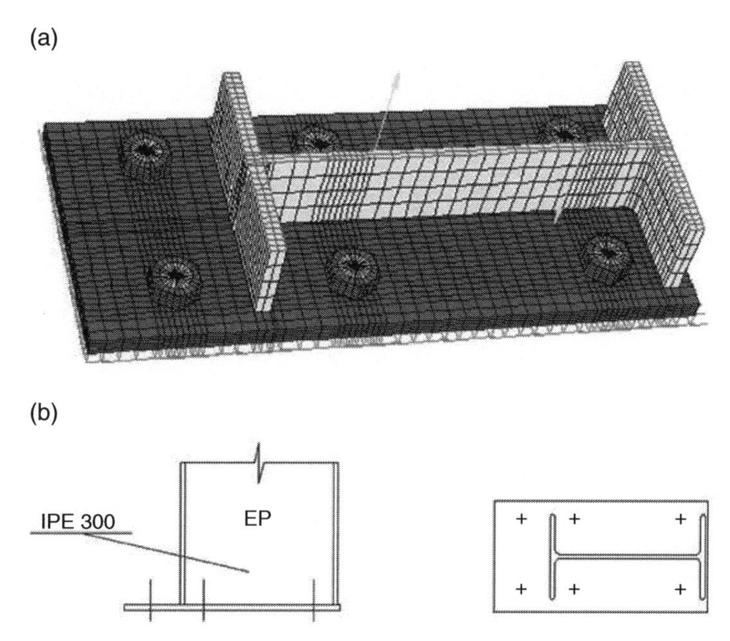

Figure 15.27 The finite element model (a) to appraise the moment-rotation curve an extended end plate connection (b).

concerning the plate-bolt interaction and the contact between the plate elements, have not been completely solved so far. As an example, Figure 15.27 reports the mesh of the finite element model of an extended end plate connection.

Theoretical models, also identified in literature as physical-mechanical models, appear to be the most suitable ones for using at a design stage, as they are simple enough and, at the same time, allow to clearly relate the characteristics of the joint overall response (or of one of its components) to the local mechanisms of behaviour. Besides, methods based on mechanical models possess generally a remarkable capability to adapt themselves to the wide range of node configurations that may occur in practice for the same type of connection. The model of the node can be conceived as a collection of models representative of the single parts brought together either in series or in parallel, thus recognizing that the overall behaviour is the result of the distortion of the single components. The most popular theoretical prediction method is the *component approach*: each component of the joint is simulated via an axial spring with an elastic-perfectly plastic behaviour, for which the elastic stiffness as well as the resistance is calculated via suitable equations based on the geometry as well as on mechanical properties. These potential forces are converted to the actual forces by considering equilibrium and compatibility conditions. The moment capacity of the connection is then calculated making reference to the centre of compression.

As an example, Figure 15.28 presents the spring system used to simulate the response of a typical external beam-to-column joint, realized by means of an extended end plate connection. As to the key components necessary to reproduce adequately the connection response, reference has been made at least to:

(1) column web panel in shear;
(2) column web in compression;
(3) beam flange in compression;
(4) bolts in tension;

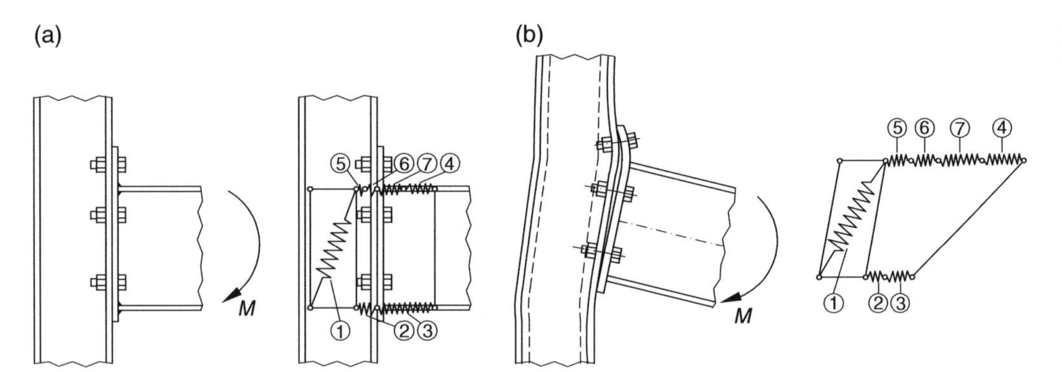

Figure 15.28 The component approach: the undeformed (a) and the deformed (b) joint configuration.

(5) column web in tension;
(6) flange column in bending;
(7) end plate in bending.

Three zones can be distinguished, which differ from each other for the internal state of stress: part of the connection in tension (components 4–7), part of the connection in shear (component 1) and part of the connection in compression (component 2 and 3).

15.5.1 Simple Connections

In steel design practice, the simple frame model is frequently adopted and hence beam-to-column connections must be designed to prevent the transmission of significant bending moments with a very limited rotational stiffness. Their purpose is to transfer the load from the supported members into the supporting members in such a way that essentially only direct forces are involved, for example vertical shear in a beam-to-column or beam-to-beam connection, axial tension or compression in a lattice girder chord splice, column base or column splice connection. They may, therefore, only be used in situations where appropriate bracing systems are located, guaranteeing that joints assumed to function as pins have adequate structural performances. Popular arrangements include lattice girders and bracing systems or connections between beams and columns in rectangular frames in which lateral loadings are resisted by stiff systems of shear walls, cores or braced bays. Hinge connections can be realized using different connection details and in Figure 15.29 some of them are presented:

(a) *fin plate connection*: a thin plate is shop welded to the column and side bolted to the beam web;
(b) *web cleat connection*: a couple of angles is site bolted both to the web (or flange) of the column and to the web of the beam;
(c) *flush end plate connection*: a flush end plate is shop welded to the beam end and site bolted to the column;
(d) *web and seat cleat connection*: a couple of angles is site bolted to the beam web as in case (a) but an additional seat angle is placed in correspondence of the bottom beam flange to facilitate the bolting of the beam to the column. In some instances, an additional angle is bolted to the top beam flange and to the column flange to provide restraints to lateral stability of the top flange of the beam;
(e) *fin plate for the connection of tubular columns.* A fin plate is shop welded to the columns and site bolted to the beam, avoiding the direct connection to the columns via bolts;

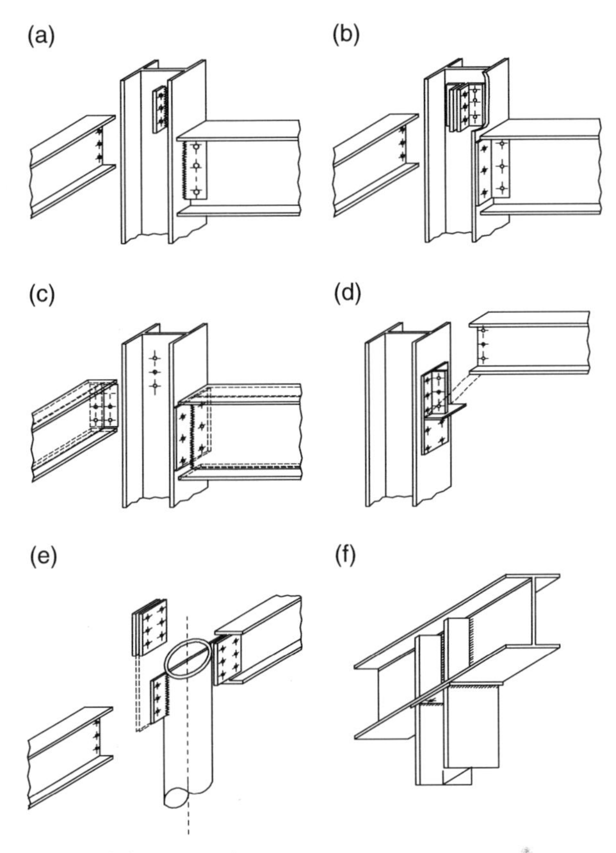

Figure 15.29 Examples of simple beam-to-column joints.

(f) *simple support to guarantee the beam continuity.* Suitable stiffeners are necessary at the joint location in order to avoid premature failure due to the transfer of elevate reactions. The connection is obtained via a shop welded plate at the end of the column, which is site bolted to the beam.

If connections are designed as hinges, at least two bolts must always be used. The use of a single bolt, despite allowing a better match between the actual joint response and the hinged model considered into design, is particularly sensitive to its possible defectiveness. In addition, it is not able to transfer the limited value of bending moments resulting from the eccentricity associated with the connection details. To better understand this aspect, the beam in Figure 15.30 can be considered, which belongs to simple braced frames and, as a consequence, it can be designed as a simply supported beam. As to the model for structural analysis, reference should be made to three schemes, differing in the position of their ideal hinges.

Scheme 1: it is assumed that the hinges are located at the intersection between the longitudinal axes of beams and columns. The design of the beam is developed considering a beam span of length L, slightly greater than the actual beam length $(L - 2a)$, due to the presence of the columns. As to the connections, on the basis of the theoretical calculation model, they have to be designed in order to transfer, in addition to the shear end reaction (R_i) also a very modest value of bending moment, due to the distance between the position of the ideal hinge and the one of the bolt rows. By defining with T and M the internal shear force and the bending moment,

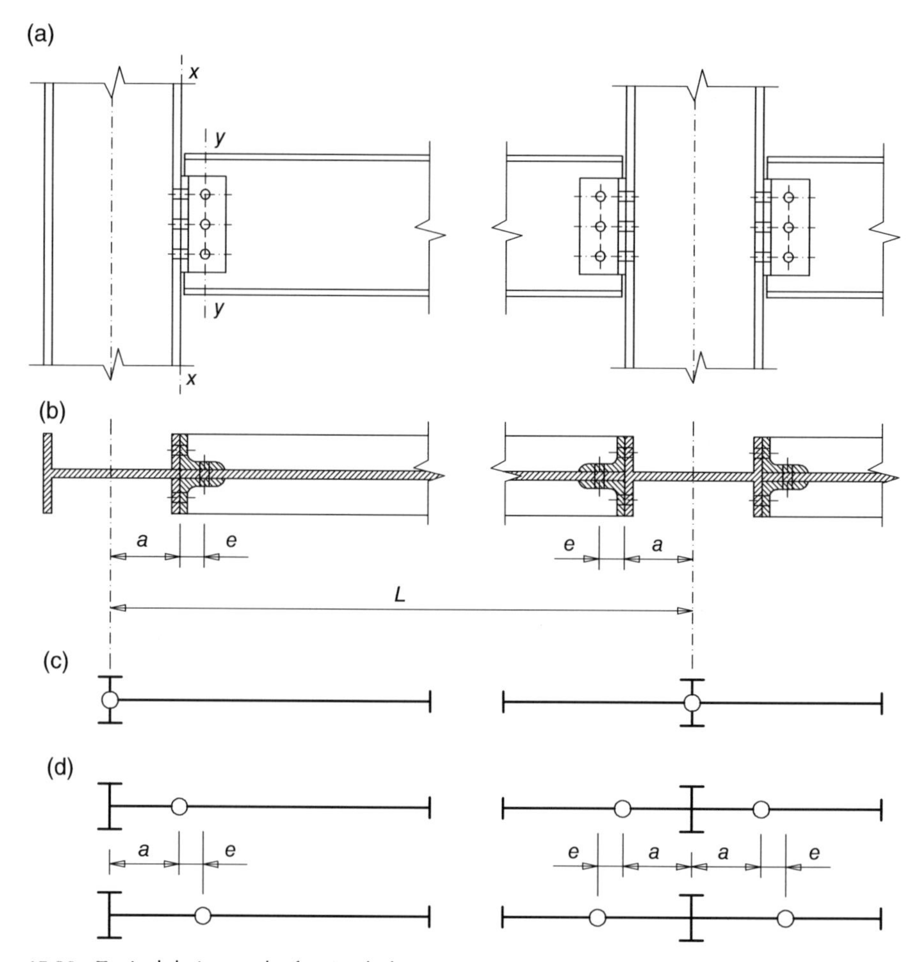

Figure 15.30 Typical design modes for simple frames.

respectively, and assuming p as the uniform distributed load on the beam, design must be based on the following data:

Section x–x, in correspondence with the outstand face of the column:

$$T_{x-x}=\frac{p\cdot L}{2}-p\cdot a=R_i-p\cdot a\approx R_i \tag{15.6a}$$

$$M_{x-x}=R_i\cdot a-\frac{p\cdot a^2}{2}\approx R_i\cdot a \tag{15.6b}$$

Section y–y, in correspondence with the bolts row on the beam web:

$$T_{y-y}=R_i-p\cdot(a+e)\approx R_i \tag{15.7a}$$

$$M_{y-y}=R_i\cdot(a+e)-\frac{p\cdot(a+e)^2}{2}\approx R_i\cdot(a+e) \tag{15.7b}$$

Bolts in section x–x are subjected to shear and tension forces, while bolts in section y–y transfer bending moment via a shear force mechanism.

Scheme 2: it is assumed that hinges are located in correspondence of the angle legs connected to the column flange (section x–x), that is reference is made to the actual beam span length $(L - 2a)$, as shown in Figure 15.30c. It can be noted that:

Section x–x, which is in correspondence of the outstanding face of the column, is interested by the sole shear force associated with the beam end reaction.
Section y–y, which is in correspondence of the bolt row on the beam web, presents bolts subjected also to a shear force contribution due to bending moment acting on the section:

$$M_{y-y} = R_i \cdot e - \frac{p \cdot e^2}{2} \approx R_i \cdot e \tag{15.8}$$

In accordance with this scheme, columns are subjected to compression and bending moment due to the hinge eccentricity, the value of which can be approximated, from the safe side, as $R_i \cdot a$.

Scheme 3: it is assumed (Figure 15.30d) that hinges are located in correspondence of the row of the bolts connected to the beam web (section y–y) and the beam span results reduced with reference to one of the previous schemes.

Section x–x in correspondence of the outstanding face of the column a bending moment due to the hinge eccentricity $(R_i \cdot e)$ acting in addition to the shear force R_i.
Section y–y in correspondence of the bolts row on the beam web, which transfers a shear force R_i.

Simple connections must be able to transfer only a very limited value of bending moment, in order to guarantee the full compatibility between the structure and the design model. An adequate rotational capacity must be guaranteed by an appropriate selection of the connection details, which has to be properly taken into account when designing the beam. Table 15.1 is referred to some common design schemes for isolated beams and reports the values of the required rotation (determined neglecting the influence of the shear deformability) to be compared with the rotational capacity of the connection.

Table 15.1 Value of the design rotation required for some common types of loaded beams.

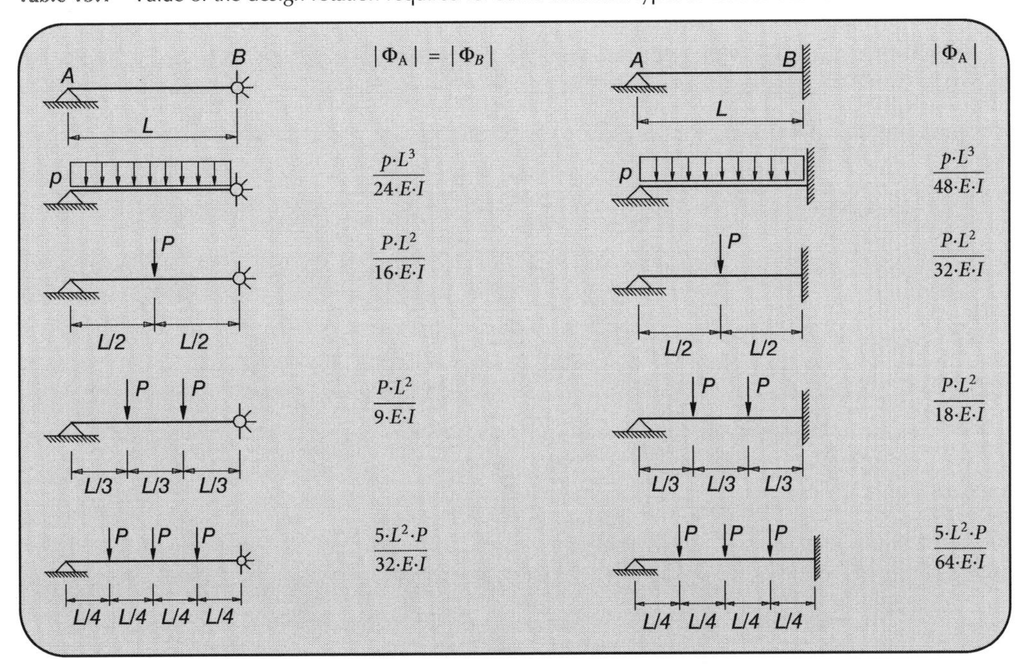

Scheme (A–B, span L)	$\lvert \Phi_A \rvert = \lvert \Phi_B \rvert$	Scheme (A–B, span L)	$\lvert \Phi_A \rvert$
p (uniform load)	$\frac{p \cdot L^3}{24 \cdot E \cdot I}$	p (uniform load)	$\frac{p \cdot L^3}{48 \cdot E \cdot I}$
P; L/2, L/2	$\frac{P \cdot L^2}{16 \cdot E \cdot I}$	P; L/2, L/2	$\frac{P \cdot L^2}{32 \cdot E \cdot I}$
P, P; L/3, L/3, L/3	$\frac{P \cdot L^2}{9 \cdot E \cdot I}$	P, P; L/3, L/3, L/3	$\frac{P \cdot L^2}{18 \cdot E \cdot I}$
P, P, P; L/4, L/4, L/4, L/4	$\frac{5 \cdot L^2 \cdot P}{32 \cdot E \cdot I}$	P, P, P; L/4, L/4, L/4, L/4	$\frac{5 \cdot L^2 \cdot P}{64 \cdot E \cdot I}$

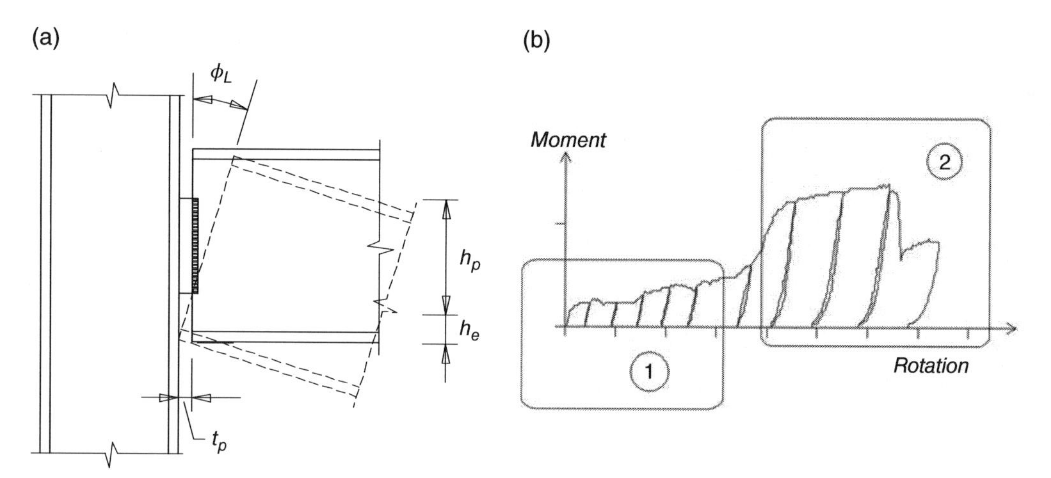

Figure 15.31 Header plate connection: (a) contact between the beam and the column flanges and (b) moment-rotation joint curve before the contact (zone 1) and after the contact (zone 2).

As bending moments developed in the joint, the bolts and the welds are subjected to tension forces in addition to shear forces. Premature failure of those elements, which exhibit a brittle failure and which are more heavily loaded in reality than in the calculation model, therefore has to be strictly avoided. To enable rotation without increasing the bending moment that develops into the joint too much, contact between the lower beam flange and the supporting member has to be strictly avoided. So, it is imperative that the height h_p of the plate is less than that of the supported beam web. If such a contact takes place, a compression force develops at the place of contact; it is equilibrated by tension forces in the bolts and a significant bending moment develops, as can be noted from Figure 15.31. The level of rotation at which the contact occurs is obviously dependent on the geometrical characteristics of the beam and of the header plate, but also on the actual deformations of the joint components. A simple criterion that designers could apply, before any calculation, is to check whether the risk of contact may be disregarded. In particular, the following rough assumptions are made:

- the supporting element remains un-deformed;
- the centre of rotation of the beam is located in correspondence of the lower extremity of the header plate.

On the basis of such assumptions, a safe estimation (i.e. a lower bound) of the joint limit rotation (ϕ_L) is:

$$\phi_L = \frac{t_p}{h_e} \tag{15.9}$$

where t_p is the thickness of the plate.

In order to increase ϕ_L it is enough to reduce the distance h_c; that is to weld the plate close to the beam bottom flange. In this case, a joint has adequate ductility if failure is due to the plasticity on the header plate, that is if this collapse occurs before failures of bolts as well as of the welding, which are typically brittle failure modes.

15.5.2 Rigid Joints

As previously introduced, if the rigid joint frame model is adopted, no relative rotation is expected between the beam and the column and the joint detailing has to be adequate to transfer the

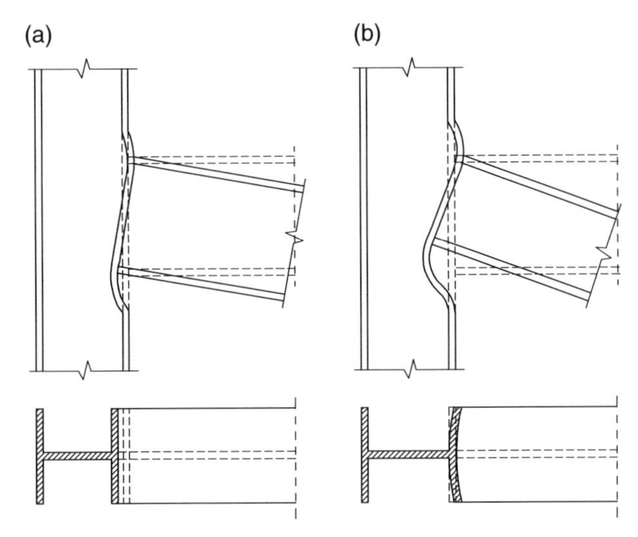

Figure 15.32 Contributions to the deformation of an external rigid joint.

bending moment acting at the beam end. Rigid joints are also identified with the term *moment-resisting joints* and special rules have to be followed when they are used in seismic zones.

The most natural solution to realize a rigid joint is the one that requires the full welding directly of end of the beam directly to the outside face of the column flange. This node should be very weak with reference to three failure modes: failure of the web column in shear (Figure 15.32a) for crushing and/or buckling, failure for local bending of column flange (in tension and compression) and failure for crushing or buckling of the compression flange of the beam (Figure 15.32b). This last failure mode is not relevant if the beam section is compact or, even if not, having column flange thickness that is not too much greater than beam flange thickness.

All these modes can be achieved for a value of the load significantly lower than the one associated with plastic bending beam resistance and, usually, the adoption of appropriate stiffeners is strongly recommended to increase joint performance, which are generally shop welded at the column web panel. A quantitative evaluation of the resistance of the column flange in bending and column web in shear can be found in AISC 360-10 in paragraphs J10.1 and J10.6.

(a) *Column flange local bending:*
This can happen for the tension force transmitted by upper beam flange or compression force transmitted by lower beam flange (Figure 15.32a).
The *available strength* is evaluated as:

LRFD approach	ASD approach
ϕR_n with $\phi = 0.90$	R_n/Ω with $\Omega = 1.67$

Term R_n represents the nominal strength and shall be determined as follows:

$$R_n = 6.25 F_{yf} \cdot t_f^2 \tag{15.10}$$

where F_{yf} is the specified minimum yield stress of the column flange and t_f is the thickness of column flange.

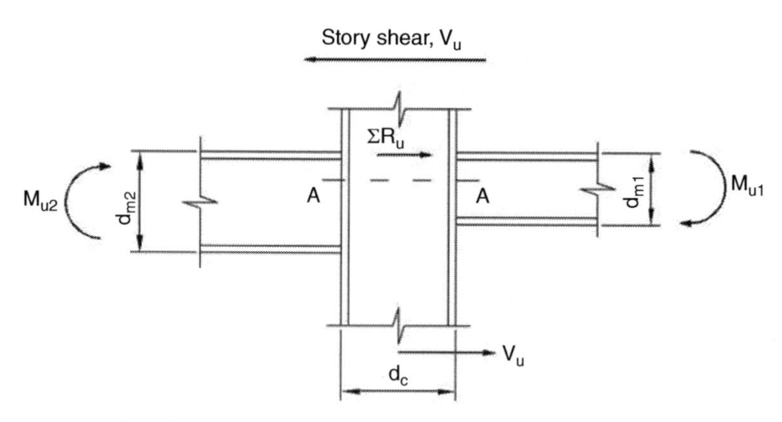

Figure 15.33 Forces in the column panel zone.

(b) *Column web panel zone shear*
The column web area delimited by connected beam(s) (panel zone) is subjected to shear stress due to column shear and beam moments (Figure 15.33). This total web shear $\sum R_u$ for load and resistance factor design (LRFD) or $\sum R_a$ for allowable stress design (ASD) can be evaluated, with reference to the LRFD forces, as:

LRFD approach	ASD approach
$$\sum R_u = \frac{M_{u1}}{d_{m1}} + \frac{M_{u2}}{d_{m2}} - V_u \quad (15.11a)$$	$$\sum R_a = \frac{M_{a1}}{d_{m1}} + \frac{M_{a2}}{d_{m2}} - V_a \quad (15.11b)$$
where M_{u1}, M_{u2} are factored moments and d_{m1}, d_{m2} are the distance between beam flange centroids.	where M_{a1}, M_{a2} are nominal moments and d_{m1}, d_{m2} are the distance between beam flange centroids.

In order to avoid reinforcing web panel, adding stiffeners, for example the following expressions have to be satisfied:

LRFD approach	ASD approach
$$\sum R_u \le \phi R_n \quad (15.12a)$$	$$\sum R_a \le R_n/\Omega \quad (15.12a)$$
where ϕR_n ($\phi = 0.90$) is the *available strength* of the web panel zone for the shear failure mode.	where R_n/Ω ($\Omega = 1.67$) is the *available strength* of the web panel zone for the shear failure mode.

R_n is the nominal strength and can be determined as follows (not considering web plastic resistance):

(1) for $P_r \le 0.4\ P_c$:

$$R_n = 0.60 F_y \cdot d_c \cdot t_w \quad (15.13a)$$

(2) for $P_r > 0.4\ P_c$:

$$R_n = 0.60 F_y \cdot d_c \cdot t_w \left(1.4 - \frac{P_r}{P_c}\right) \quad (15.13b)$$

where d_c is the column depth, t_w is the column web thickness, P_r is the requited axial strength (according to LRFD or ASD loading combinations), $P_c = 0.60\ P_y$ and P_y is the column axial yield strength ($P_y = F_y A_g$).

As an example, in Figure 15.34 typical common solutions for rigid beam-to-column joints are presented. In particular, note the:

(a) fully welded connection of an external node at the roof level. Horizontal stiffeners are shop welded to the column in correspondence of the bottom beam flange;

(b) bolted knee-connection of an external node at the roof level. Both ends of beam and column are completed with shop welded stiffeners. Furthermore, the internal flange and the web of the column have been shop removed in order to allow a quick site assemblage via traditional bolting technique;

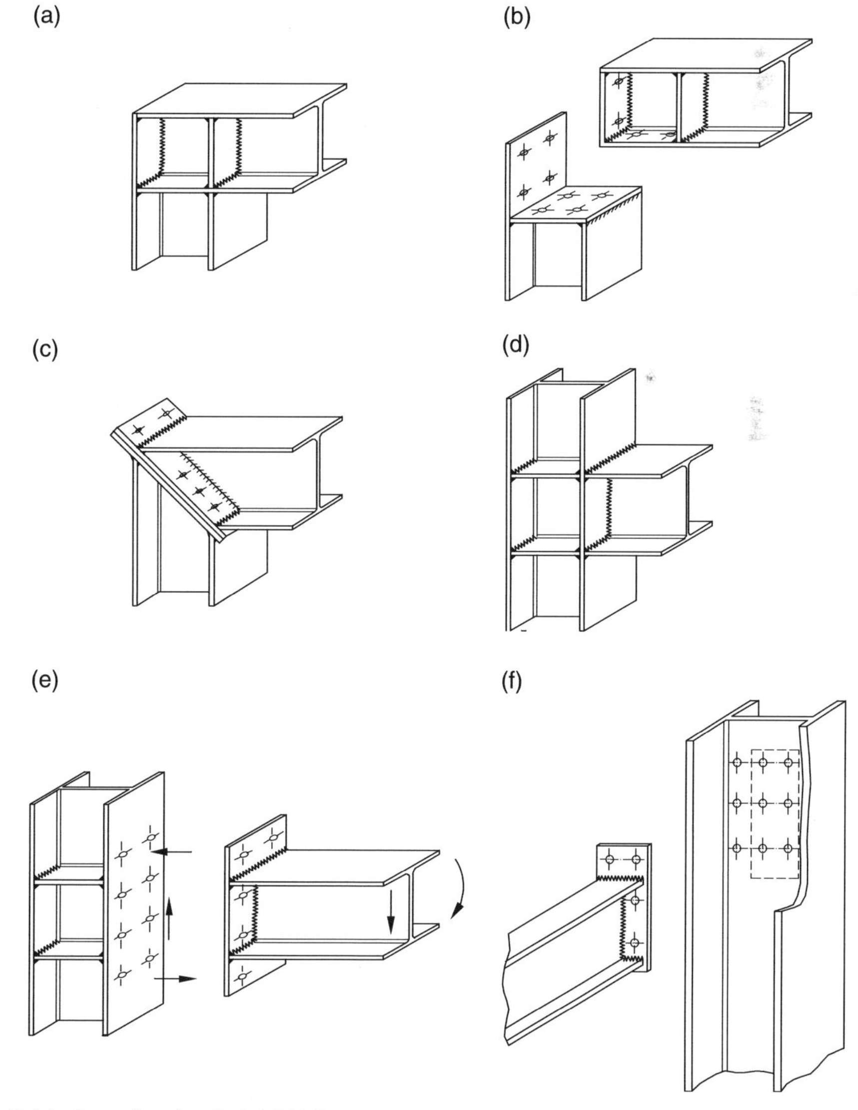

Figure 15.34 Examples of typical rigid joints.

(c) knee-connection of an external node at the roof level with external end plate shop welded to the member ends. In this case the diagonal cutting of the beam end is required;
(d) welded T-connection to join a floor beam to an external column. Stiffeners have been placed in correspondence of both the top and the bottom flange of the beam;
(e) bolted end plate connection to join a floor beam to the flange of an internal column;
(f) bolted end plate connection to join a floor beam to the web of an internal column.

15.5.3 Semi-Rigid Joints

As previously introduced in Section 3.4, the models of hinged and rigid joints, which have been extensively used in the past, have a quite limited application nowadays, owing to the need to account for the actual joint response. Classification in the past was mainly based on the connection joint components, independent of the mechanical properties and this could lead, in some cases, to unsafe design. As an example, the results of a research carried out at the University of Trento (I) on beam-to-column joints are presented in Figure 15.35, where the non-dimensional moment-rotation curves, which have been evaluated with reference to a beam spam of 6 m (19.7 ft), are presented for some of the tested joints. As to the joint classification in the figure, reference is made to the EC3 criteria. It can be noted that EPBC and EPC connections, which are traditionally

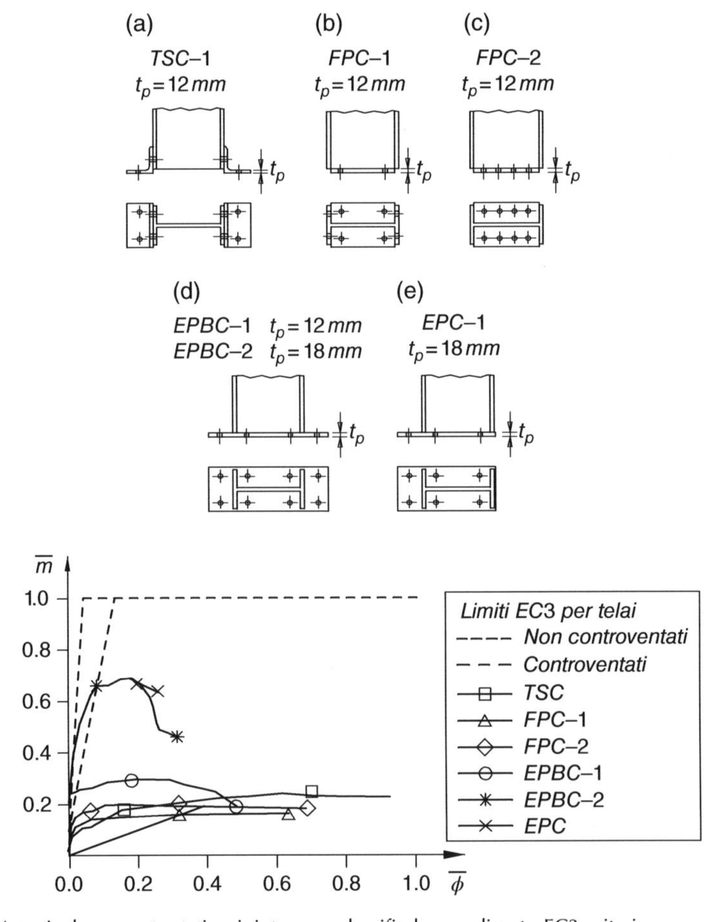

Figure 15.35 A typical moment-rotation joint curve classified according to EC3 criteria.

considered rigid connections owing to the presence of the extended end plate, have an actual semi-rigid behaviour. Increasing the thickness of the plate from 12 mm (0.47 in.) for EPBC-1 to 18 mm (0.71 in.) for EPBC-2 and EPC-1, bending load carrying capacity increases too, but the response remains in the semi-rigid domain. On the other hand, joints traditionally considered as hinges, as top-and-seat angle connection (TSC-1) and flush end plate connection (FPC-1) show a semi-rigid behaviour, despite the limited value of the bending capacity, slightly lower than 25% of the plastic moment of the beam, that is the limit to be classified as partial-strength connections.

Correct design procedure for steel frames implies that the appropriate design models for structural analysis have to be selected and, in case of semi-continuous frames, each joint should be considered as semi-rigid, that is a rotational spring should be used to simulate its response.

15.5.3.1 Plastic Analysis Applied to Semi-Continuous and to Rigid Frames

In many cases, three-dimensional framed systems are regular both in plant and in elevation. The use of plastic analysis approaches lead hence directly to the evaluation of the ultimate load (or ultimate load multiplier) by simple hand calculations, as shown with reference to the frames with semi-rigid joints presented in Figure 15.36, with n_c bays (each of them of span L_b) and n_p storeys (each of them at the level h_i with respect to the foundation planes). The considered load condition is very common in the steel design practice: each beam is loaded by a uniform load (q) and a concentrated horizontal load (F_{Hi}) is applied to each story as a fraction, via an appropriate multiplier β, of the resulting vertical load applied to the storey:

$$F_{Hi} = \beta \cdot q \cdot L_b \cdot n_c \tag{15.14}$$

Semi-continuity has been taken into account by considering the flexural resistance of beam-to-column joints ($M_{j,btc}$) and of base-plate connections ($M_{j,b}$). It has been assumed that $M_{j,btc} \le M_b$, where M_b is the plastic bending resistance of the beam and $M_{j,b} \le M_c$, where M_c is the plastic bending resistance of the columns.

It should be noted that the approach herein proposed can also be directly applied to rigid frames, assuming, in the proposed equations, $M_{j,btc} = 0.5\ M_b$ and $M_{j,b} = M_c$.

The plastic analysis theory can be used and, in particular, reference is herein made to the kinematic mechanism method, which is based on the upper-bound theorem of plastic analysis (*a load computed on the basis of an assumed mechanism will always be greater than, or at best equal to, the true ultimate load*). The ultimate plastic load (or equivalently the ultimate plastic multiplier) can

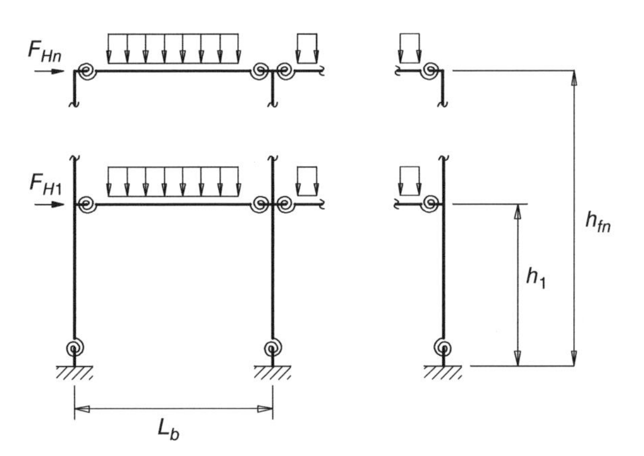

Figure 15.36 A regular semi-continuous planar frame.

be evaluated directly on the basis of hand calculations. Defining q_u as the uniform beam load associated with a full plastic mechanism (or equivalently α_u) the ultimate load multiplier associated to the reference uniform load q_s (with $q_u = \alpha_u \cdot q_s$) we can obtain:

- Beam mechanism (Figure 15.37a):

$$q_u = \frac{8\left(M_{j,btc} + M_b\right)}{L_b^2} \tag{15.15a}$$

$$\alpha_u = \frac{8\left(M_{j,btc} + M_b\right)}{q_s L_b^2} \tag{15.15b}$$

- Panel mechanism (Figure 15.37b):

$$q_u = \frac{2M_{j,btc} \cdot n_c \cdot n_p + M_{j,b}\left(n_c + 1\right)}{\beta \cdot n_c \cdot L_b \sum_{i=1}^{n_p} h_i} \tag{15.16a}$$

$$\alpha_u = \frac{2M_{j,btc} \cdot n_c \cdot n_p + M_{j,b}\left(n_c + 1\right)}{\beta \cdot n_c \cdot L_b \cdot q_s \sum_{i=1}^{n_p} h_i} \tag{15.16b}$$

- Mixed mechanism (Figure 15.37c):

$$q_u = \frac{M_{j,btc} \cdot n_c \cdot n_p + M_{j,b}\left(n_c + 1\right) + 2 \cdot M_b \cdot n_c \cdot n_p}{\beta \cdot n_c \cdot L_b \sum_{i=1}^{n_p} h_i + 0{,}25 \cdot n_c \cdot n_p \cdot L_b^2} \tag{15.17a}$$

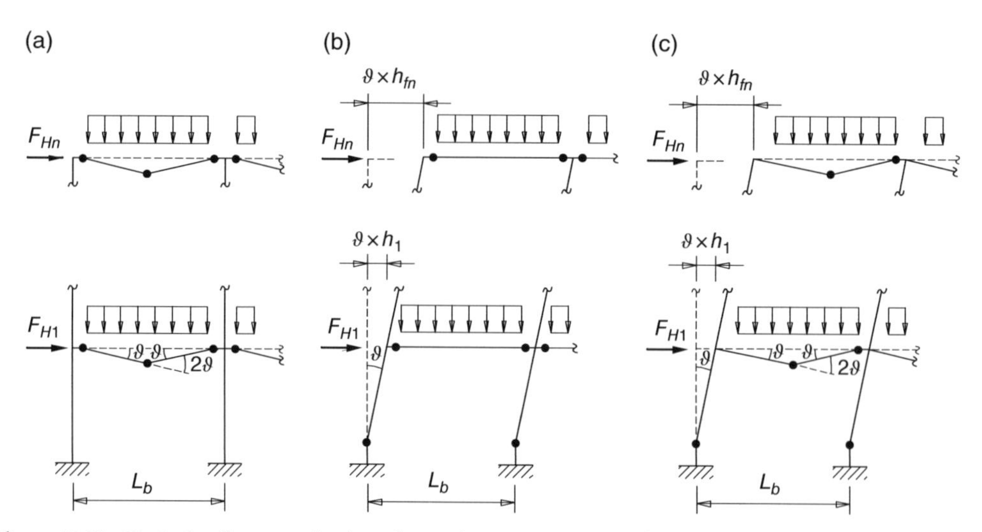

Figure 15.37 Typical collapse mechanisms for regular semi-continuous frames: (a) beam mechanism, (b) panel mechanism and (c) mixed mechanism.

$$\alpha_u = \frac{M_{j,btc} \cdot n_c \cdot n_p + M_{j,b}(n_c + 1) + 2 \cdot M_b \cdot n_c \cdot n_p}{q_s \left[\beta \cdot n_c \cdot L_b \sum_{i=1}^{n_p} h_i + 0,25 \cdot n_c \cdot n_p \cdot L_b^2 \right]} \tag{15.17b}$$

As in case of hinged joints, for semi-rigid and rigid joints it is also necessary to guarantee an appropriate level of rotational capacity, owing to the fact that a complete plastic mechanism is generally activated only if relative rotations can occur at the plastic hinge locations.

As an example of the importance of the rotational capacity of the plastic hinge, the isolated beam with semi-rigid joints presented in Figure 15.38 can be considered. The static method to evaluate the ultimate load is applied, which is based on the lower-bound theorem of the plastic analysis (*a load computed on the basis of an assumed distribution of internal forces and bending moments with the applied loading and where member resistance is not exceeded is less than, or at best equal to the true ultimate load*). Joint bending resistance ($M_{pl,j}$) is supposed to be equal to half of the one of the beam ($M_{pl,b}$), that is partial strength joints with $M_{pl,j} = M_{pl,b}$ are used.

As in the example of fixed-end beams (see Section 3.6.1) and also in case of semi-rigid joints, the first two plastic hinges, which are at the joint location, are activated simultaneously for a uniform load p_e evaluated as:

$$p_e = \frac{12 \cdot M_{pl,j}}{L^2} = \frac{6 \cdot M_{pl,b}}{L^2} \tag{15.18}$$

Also in this case, the beam does not collapse and the additional load Δp (Figure 15.39) can be increased until another plastic hinge is activated (Figure 15.40).

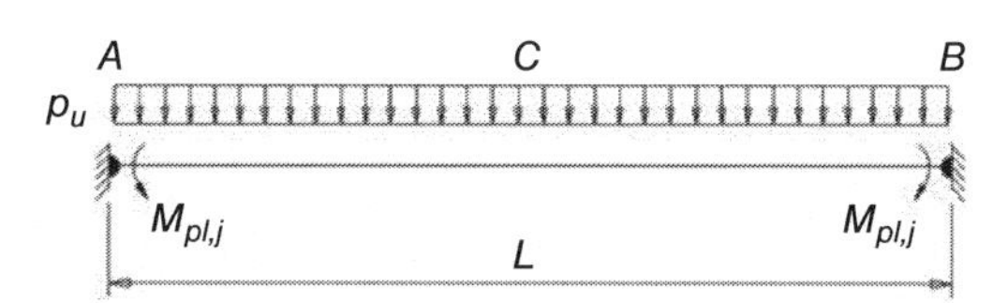

Figure 15.38 Isolated beam with semi-rigid joints.

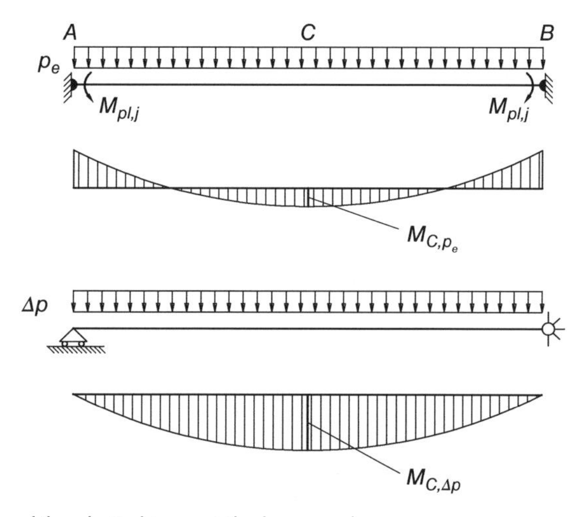

Figure 15.39 Activation of the plastic hinges at the beam ends.

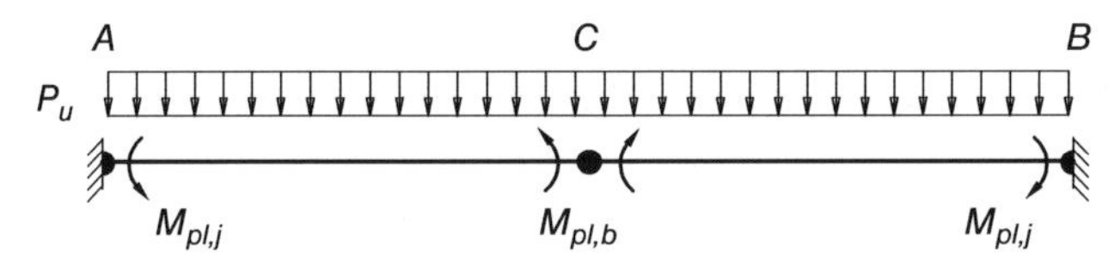

Figure 15.40 Collapse mechanism.

This value of Δp corresponds to the formation of the third plastic hinge, which transforms the structure in a mechanism and hence is named Δp_u, obtained by the condition:

$$\frac{M_{pl,b}}{4}+\frac{\Delta p_u \cdot L^2}{8}=M_{pl,b} \tag{15.19}$$

It should be noted that it immediately evaluates the values of the rotation required to the joint to activate the plastic mechanism $\Delta\phi_u$, which is:

$$\Delta\phi_u=\frac{\Delta p_u \cdot L^3}{24 \cdot E \cdot I}=\frac{\frac{6M_{pl,b}}{L^2}L^3}{24 \cdot E \cdot I}=\frac{M_{pl,b} \cdot L^3}{4 \cdot E \cdot I}=\frac{M_{pl,j} \cdot L^3}{8 \cdot E \cdot I} \tag{15.20}$$

It is of fundamental importance that the required rotation $\Delta\phi_u$ can be reached by the selected type of semi-continuous joints avoiding brittle and premature failure.

15.6 Joint Standardization

In a typical braced multi-storey frame, joints represent about the 5% of total steel weight, but the 30% or more of the cost of steel structure. Starting from this consideration, BCSA (British Constructional Steelwork Association) and SCI (Steel Construction Institute) in the UK have developed standard connections in the field of simple joints as well as moment connections.

Joint standardization reduces the number of connection types and promotes the use of standard components for fittings: M20 8.8 bolts fully threaded whenever possible, with holes generally 22 mm in diameter, punched or drilled, spaced at standard values, end plates and fin plates 10 or 12 mm of thickness, and so on.

Using standardized components improves availability, leads to a material cost reduction and reduces time for buying, storage and handling. Furthermore, on the side of design, using standardized joints guarantees that every joint has a good reliability because it has been computed in advance by very skilled engineers.

BCSA standardization is available in the so called 'green books', several publications issued by BCSA and SCI (see the Bibliography in Appendix B). These standard joints have been computed according to EC3 (according to BS 5950 formerly) but, unfortunately, they use United Kingdom's UB and UC profiles only, excluding European shapes such as the HE and IPE series.

In Table 15.2 an example of joint standardization is shown.

A similar effort has been performed by AISC. In AISC Steel Construction Manual, detailed procedures for any kind of joint are listed, with a lot of practical tables to help dimensioning the connections (Table 15.3).

Unfortunately analogous effort has not been performed up to now in developing EC3. The Code states principles and rules, but no one has prepared construction manuals starting from it.

Joint standardization, especially for moment connections, is really important also for design of steel structures in seismic areas. Moment connections in frames designed for areas with high seismicity actually have to exhibit two important properties:

Table 15.2 Excerpt from *Joint in Steel Construction: Simple Joints to Eurocode 3*, by BCSA-SCI, Publication P358 (2011).

Beam S275
End plates S275
Bolts M20 8.8

Partial depth end plates, ordinary or flow drill bolts
200 × 12 OR 150 × 10 mm end plate

		Un-notched		Single notch		Double notch		Fitting (end plate)				Fillet weld	Min support thickness		Tying	
Beam size ($V_{Rd,beam}$)	Bolt rows, n_1	Shear resist, V_{Rd} (kN)	Critical Check	Shear resist, V_{Rd} (kN)	Maximum length, l_n (mm)	Shear resist, V_{Rd} (kN)	Maximum length, l_n (mm)	Width, b_p (mm)	Thck, t_p (mm)	Height, h_p (mm)	Gauge, p_3 (mm)	Leg, s (mm)	S275	S355	Resist, $N_{Rd,u}$ (kN)	Critical check
610 × 229 × 140	7	902	4	902	208	776	100	200	12	500	140	8	4.8	4.2	554	11
(1300 kN)	6	776	4	776	312	776	100	200	12	430	140	8	4.9	4.2	477	11
610 × 229 × 125	7	819	4	819	201	697	100	200	12	500	140	8	4.4	3.8	548	11
(1170 kN)	6	705	4	705	305	697	100	200	12	430	140	8	4.4	3.9	471	11
610 × 229 × 113	7	764	4	764	193	643	100	200	12	500	140	8	4.1	3.6	544	11
(1000 kN)	6	657	4	657	296	643	100	200	12	430	140	8	4.1	3.6	468	11
610 × 229 × 101	7	750	4	750	182	624	100	200	12	500	140	8	4.0	3.5	525	11
(1060 kN)	6	645	4	645	284	624	100	200	12	430	140	8	4.0	3.5	462	11

Table 15.3 End plate connections according to the *AISC Steel Construction Manual*.

W44		**Table 10-4 bolted/welded shear end-plate vonnections**					**3 in./4 in. bolts 12 rows**	
Bolt and end-plate available strength (kips)								
ASTM design	**Thread condition**	**Hole type**	**End-plate thickness (in.)**					
			1/4		**5/16**		**3/8**	
			ASD	**LRFD**	**ASD**	**LRFD**	**ASD**	**LRFD**
A325/F1852	N	—	197	295	246	369	254	382
	X	—	197	295	246	369	295	443
	SC Class A	STD	177	266	177	266	177	266
		OVS	128	192	128	192	128	192
		SSLT	151	226	151	226	151	226
	SC Class B	STD	197	295	246	369	253	380
		OVS	183	274	183	274	183	274
		SSLT	195	293	215	323	215	323
A490	N	—	197	295	246	369	295	443
	X	—	197	295	246	369	295	443
	SC Class A	STD	197	295	221	332	221	332
		OVS	160	240	160	240	160	240
		SSLT	188	282	188	282	188	282
	SC Class B	STD	197	295	246	369	295	443
		OVS	196	294	229	343	229	343
		SSLT	195	293	244	366	269	403

Weld and beam available strength (kips)				Support available Strength per inch Thickness (kips/ft)	
70 ksi weld size (in.)	Minimum beam web thickness (in.)	R_n/Ω kips ASD	ϕR_n kips LRFD	ASD	LRFD
3/16	0.286	196	293	1400	2110
¼	0.381	260	390		
5/16	0.476	324	486		
3/8	0.571	387	581		

STD = standard holes; OVS = oversized holes; SSLT = short-slotted holes transverse to direction of load

N = threads included; X = threads excluded; SC = slip critical

End-plate: $F_y = 36$ ksi, $F_u = 58$ ksi; Beam: $F_y = 50$ ksi, $F_u = 65$ ksi

(1) they must be more resistant than the beam they connect;

(2) they must maintain the greater part of their moment resistance with node rotation prescribed by Code, without showing brittle behaviour.

To achieve both targets it is necessary to test experimentally different typologies and verify their behaviour with special attention to their ductility. Calculations cannot guarantee the ductility of a joint type: sometimes after changing just a detail a joint has shown a more ductile behaviour. AISC has tested few typologies of moment connections (prequalified connections) to be used in seismic frames and has developed a specific standard design procedure to use for them, which is contained in document AISC 358-10 'Prequalified Connections for Special and Intermediate Steel Moment Frames for Seismic Applications'. In the document, the scope is clearly declared:

> *The connections contained in this Standard are prequalified to meet the requirements in the AISC Seismic Provisions only when designed and constructed in accordance with the*

> *requirements of this Standard. Nothing in this Standard shall preclude the use of connection types contained herein outside the indicated limitations, nor the use of other connection types, when satisfactory evidence of qualification in accordance with the AISC Seismic Provisions are presented to the authority having jurisdiction.*

In other words, the engineer can adopt different connections, but he has to demonstrate that such connections meet AISC requirements and this is not an easy job.

No similar prequalification of joints to be used for frames in seismic zones exists in Eurocode 3 up to now. So the choice of a joint that shows the correct behaviour under seismic actions is not a straightforward task.

CHAPTER 16
Built-Up Compression Members

16.1 Introduction

Individual members may be combined in a quite great variety of ways to produce a more efficient compound cross-section member. The main advantage in the use of built-up members is the high value of the load-carrying capacity that can be achieved by combining suitably very slender isolated members. Furthermore, resistance can significantly exceed the sum of the axial resistances of the component members, which can, as a result, be significantly limited by instability phenomenon in the range of standard products.

16.2 Behaviour of Compound Struts

It can be noted that built-up compression members are composed by isolated members (chords) appropriately connected in a non-continuous way. For the sake of simplicity, chords can be compared with the flanges of I- or H-shaped hot-rolled profiles, the web of which, in built-up members, is realized by means of lacings, battens or plates. From a practical point of view, struts can be classified on the basis of the distance between the centroids of the chords (h_0) and of the radius of gyration of the chord along the axis where the element is compounded (i_1). In particular it is possible to distinguish:

- *strut with distant chords*, if $h_0 > 6i_1$, such as laced struts and struts with batten plates, typically used as columns, that is to sustain vertical compression axial load;
- *struts with close chords*, if $h_0 < 3i_1$, such as buttoned struts (also named closely spaced built-up members), which are typically used for the chords of both trusses or for the struts in case of elevated axial loads.

With reference to the type of connection between the chords, it is possible to distinguish:

- *Laced members* (Figure 16.1), where lacing members are interested by axial forces and each chord can be considered as a simple strut, with a buckling length equal to the joint spacing. Shear deformability mainly depends on the axial deformability of the lattice members. Some typical laces struts are presented in Figure 16.1c, which differ for the type of panel components. Most commonly used solutions are presented by N-type (a) and V-type laced panels (b) one,

Structural Steel Design to Eurocode 3 and AISC Specifications, First Edition. Claudio Bernuzzi and Benedetto Cordova.

where both diagonal and batten members can be subjected to compression or tension force. Solution (c) is obtained from (b) by inserting battens in order to reduce the effective length of the chord while solutions (d) and (e) present, in the same panel two laced members: one under tension and one under compression.

- *Struts with batten plates* (Figure 16.2), composed by chords rigidly connected by batten plates. Chords are compressed and bent, with a linear distribution of the bending moments null at the middle of each panel. Moment distribution can be approximated with reference to the undeformed situation, while the vertical load buckling effects are considered balanced by equal and opposite axial, forces in the chords. These struts have a typical Vierendeel beam behaviour.
- *Buttoned struts* (Figure 16.3), in which the single chord is compressed and bent with a bending moment distribution that cannot be approximated as linear. In this case the overall lateral

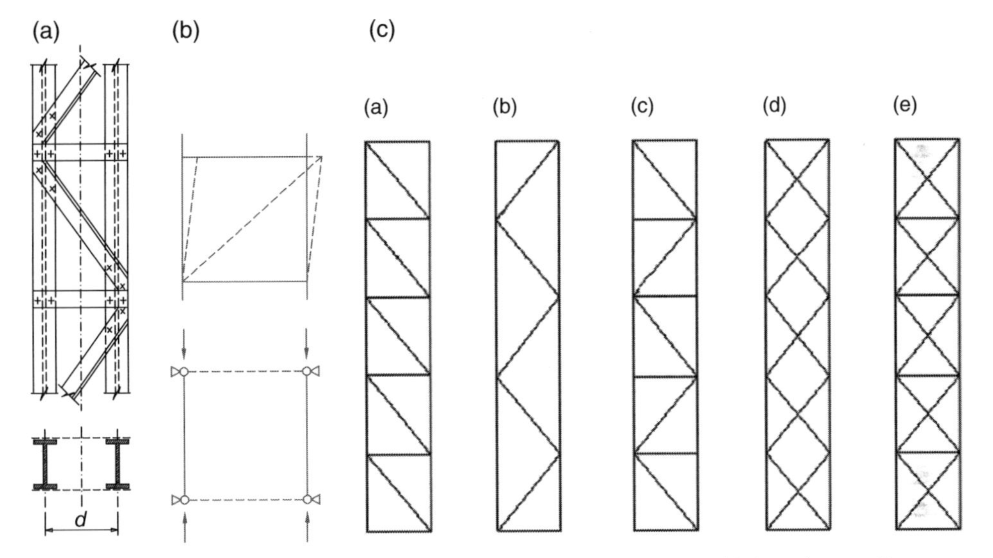

Figure 16.1 Typical arrangements for laced struts (a), the associated design model (b) and types of lacing panels (c).

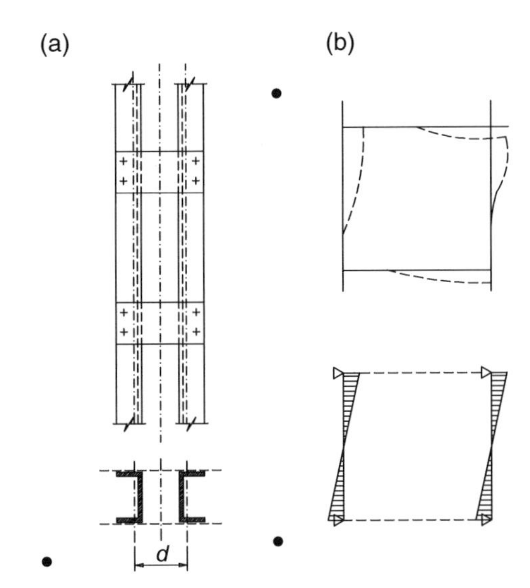

Figure 16.2 Typical arrangements for struts with batten plates (a) and the associated design model (b).

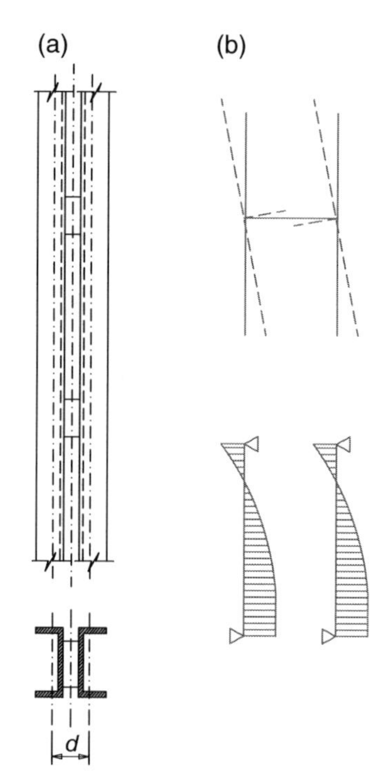

Figure 16.3 Typical arrangements for buttoned struts (a) and the associated design model (b).

bending contribution becomes significant if compared with the local one of the single chord so that bending moment values must be computed with reference to the deformed configuration (second order analysis). Shear deformability depends mainly on the flexural deformability of the chords and on bolt slippage when bolted joints are used.

Load carrying capacity of built-up members is strictly influenced by several aspects, such as overall and local response of members and connection behaviour.

Overall response of built-up members depends significantly on the deformability due to bending and shear, which can significantly influence lateral deflection of the compound members for the presence of initial geometrical imperfections. Bending deformability depends on the value of the moment of inertia of the compound struts while shear deformability is mainly affected by the performances of lacings and battens and by the deformability of the connections.

Local behaviour of each chord and of the other strut components has to be verified in accordance with appropriate criteria as for isolated members.

Connections between the constituent members, which must be able to absorb any sliding action between the profiles forming the cross-section can be characterized by a significant deformability able to increase significantly the overall and lateral deflection of the member and thus the destabilizing effect of vertical loads increases, too. Furthermore, it should be noted that connections represent a critical aspect of the compound struts, due to the fact that an excessive deformability decreases significantly the load carrying capacity. Two types of connections can be distinguished:

- connection with static function if able to resist sliding force between the isolated members;
- connection with kinematic function if able to prevent the buckling (local) of the isolated member in the weakest direction.

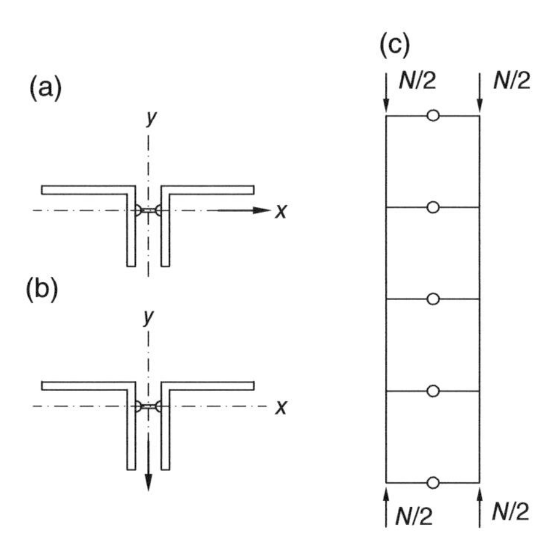

Figure 16.4 Examples of connections able to absorb slippage force.

As an example, the case presented in Figure 16.4 can be considered with connections resisting to sliding forces. The two angles behave as single isolated members for bending in direction *y* (about the *x*-axis) and hence a connection with kinematic function can be adequate. Otherwise, when bending in direction *x* (about the *y*-axis) is considered (a), the section must behave as a built-up member (c) and hence a static function is required. If the angle legs are equal, there is no reason for statically connecting the two bars, when their buckling lengths are equal in both planes. It might be economical to connect the angles statically, but only if their legs are not equal or if buckling lengths in the two planes differ. However, a suitable composition can equalize the slenderness values in the two planes. For the same reason, it is always convenient to compose a section of two channels when buckling lengths in the two planes are equal.

Furthermore, it is worth mentioning that the influence of local behaviour on the overall performance of a strut is difficult to quantify. It is preferable to ignore interaction between local and overall response but, at the same time, to define dimensional limitations to built-up member geometry that, if complied with, guarantees that the overall behaviour of the strut is practically independent on local behaviour of any single chord.

With reference to isolated members, as previously introduced in Chapter 6, when shear deformability is considered, the elastic critical load $N_{cr,id}$ can be defined on the basis of the Eulerian load, N_{cr}, which is evaluated considering only the flexural contribution, as:

$$N_{cr,id} = \frac{N_{cr}}{1 + \frac{\chi_T}{G \cdot A} \cdot N_{cr}} = \frac{1}{\frac{1}{N_{cr}} + \frac{\chi_T}{G \cdot A}} = \frac{\pi^2 \cdot E \cdot A}{\lambda_{eq}{}^2} \tag{16.1}$$

where χ_T is the shear factor of the cross-section of area A and E and G are the Young's and the shear modulus, respectively.

For isolated members, the equivalent slenderness, λ_{eq}, can be evaluated as:

$$\lambda_{eq} = \sqrt{\lambda^2 + \frac{\chi_T \cdot \pi^2 \cdot E}{G}} \tag{16.2}$$

A similar approach can be adopted also in the design of strut members and slenderness λ_{eq} depends strictly on the type of struts as well as on the panel geometry, as more clearly explained in the following section.

Recent design approaches for struts are based on a structural analysis accounting directly for second order effects. The model of a pinned-end strut is proposed, with a length L and an initial sinusoidal imperfection characterized by the maximum amplitude e_0. As already considered for isolated compression member with geometrical imperfections, deflection of the midspan v due to an axial load N acting on the imperfect element can be approximated as:

$$v = \frac{e_0}{1 - \dfrac{N}{N_{cr,id}}} \tag{16.3}$$

where $N_{cr,id}$ is the critical load of the strut, which can be expressed as a function of the elastic critical load, N_{cr}, and of the panel shear stiffness, S_v, as:

$$N_{cr,id} = \frac{1}{\dfrac{1}{N_{cr}} + \dfrac{1}{S_v}} \tag{16.4}$$

With reference to the midspan cross-section (Figure 16.5), the maximum bending moment M acting can be expressed, considering second order effects, as:

$$M = \frac{e_0 \cdot N}{1 - \dfrac{N}{N_{cr,id}}} \tag{16.5}$$

As a consequence, the maximum axial load on the chord, N_f, can be evaluated directly on the basis of equilibrium equation as:

$$N_f = \frac{N}{2} + \frac{M}{h_0} = \frac{N}{2} \cdot \left(1 + \frac{\dfrac{2 \cdot e_0}{h_0}}{1 - \dfrac{N}{N_{cr,id}}} \right) \tag{16.6}$$

A very important aspect of built-up design regards the prevention of instability, which has to be developed by taking adequately into account the presence of concentrated connections between the chords. Two different buckling modes have to be explicitly considered into design:

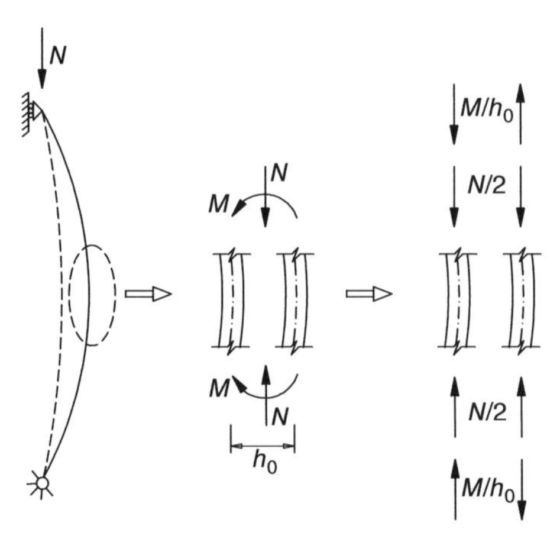

Figure 16.5 Second order effects on the strut and axial load on the chords at the midspan.

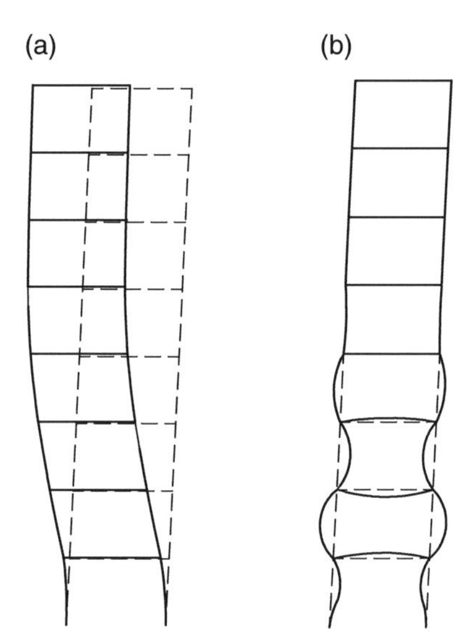

Figure 16.6 Typical failure modes for a battened column: (a) overall buckling and (b) local buckling of the chord between the panels.

- *overall buckling mode*, affecting built-up members in its whole length (Figure 16.6a) and significantly influenced by the loading condition as well as by the member end restraints;
- *local buckling* mode, affecting the chord between each panel (Figure 16.6b), with an effective length depending on the end panel restraints and a buckling mode depending on the chord cross-section geometry.

In addition to the checks on the components realizing the struts, load carrying capacity depends significantly on the stability and hence the evaluation of the shear stiffness (S_v) of the panel is required as preliminary to the verification. In the following, two common cases are considered as examples of the procedure to determine S_v. Several studies carried out in the past demonstrated that a design based on these equations shows an adequate safety level if at least four panels form the strut.

16.2.1 Laced Compound Struts

The built-up laced column in Figure 16.7 is characterized by an equal N-type panel, the shear deformability of which influences the overall compression response. With reference to each panel, shear displacements are due to lengthening of the diagonal lacing and to the shortening of the batten.

As to the first contribution, lacing length is $L_d = a/\sin\phi$ and tension force is $N_d = T/\cos\phi$. Elongation Δ of the diagonal element can be evaluated as:

$$\Delta = \varepsilon \cdot L_d = \frac{N_d}{E \cdot A_d} \cdot L_d \tag{16.7}$$

Lacing elongation can be re-written as:

$$\Delta = \frac{T}{\cos\phi} \cdot \frac{1}{EA_d} \cdot \frac{a}{sen\phi} = \frac{T}{EA_d} \cdot \frac{a}{sen\phi \cdot \cos\phi} \tag{16.8}$$

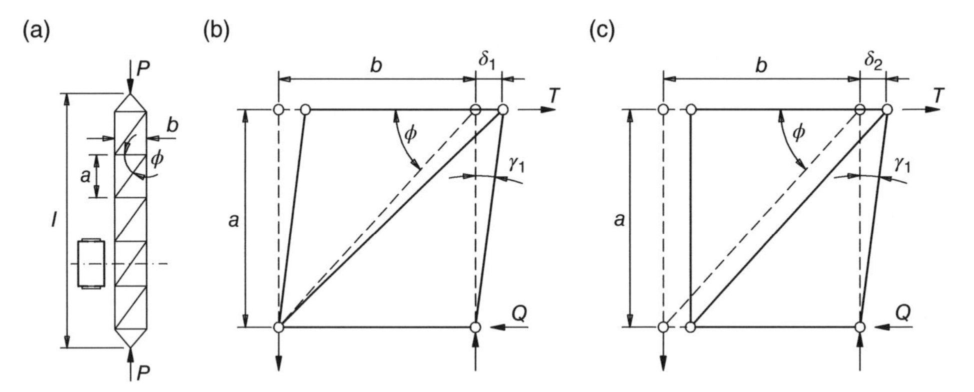

Figure 16.7 Example of built-up laced member (a): deformed shape of the panel due to deformation of the diagonal (b) and of the batten (c).

The corresponding lateral displacement δ_1 of the panel, assuming that displacements are small, is:

$$\delta_1 = \frac{\Delta}{\cos\phi} = \frac{T}{EA_b} \cdot \frac{a}{sen\phi \cdot \cos^2\phi} \tag{16.9a}$$

Shortening of the batten, subjected to the external horizontal force T (Figure 16.7c) is:

$$\delta_2 = \frac{Tb}{EA_b} \tag{16.9b}$$

By the sum of δ_1 and δ_2, the total angular displacement produced by the horizontal force T applied to the panel, γ, is:

$$\gamma = \frac{\delta_1 + \delta_2}{a} = T\left(\frac{1}{EA_d sen\phi \cos^2\phi} + \frac{b}{aEA_b}\right) \tag{16.10}$$

As a consequence, shear stiffness, S_v, can be directly evaluated from the relationship $\gamma = \frac{T}{S_v}$, hence resulting in:

$$\frac{1}{S_V} = \frac{1}{EA_d sen\phi \cos^2\phi} + \frac{b}{aEA_b} \tag{16.11}$$

On the basis of the previous Eq. (16.4), substituting the expression of the shear stiffness, it can be obtained:

$$N_{cr,id} = \frac{1}{\frac{1}{N_{cr}} + \left(\frac{1}{EA_d \ \cos\phi sen^2\phi} + \frac{b}{aEA_b}\right)} \tag{16.12}$$

Influence of the shear stiffness could be also taken into account considering $N_{cr,id} = \frac{\pi^2 EI}{k^2_{\beta,eq} L^2}$ and introducing an appropriate effective length factor $k_{\beta,eq}$, defined as:

$$k_{\beta,eq} = \sqrt{1 + \pi^2 \frac{EI}{L^2}\left(\frac{1}{EA_d sen\phi \cos^2\phi} + \frac{b}{aEA_b}\right)} \tag{16.13a}$$

Considering that $\cos\phi = \frac{b}{L_d}$, $\sin\phi = \frac{a}{L_d}$, and using $\pi^2 \approx 10$, term $k_{\beta,eq}$ can be alternatively defined as:

$$k_{\beta,eq} = \sqrt{1 + \pi^2 \frac{I}{L^2} \frac{1}{b^2 a} \left(\frac{L_d^3}{A_d} + \frac{b^3}{A_d} \right)} \tag{16.13b}$$

In both Eqs. (16.12) and (16.13a), the function $f(\phi) = \sin\phi \cos^2\phi$ is contained, which assumes maximum value approximately for $\phi = 35°$. Moreover, if the angle ϕ ranges between 30° and 45° the value of $f(\phi)$ is very similar to $f(\phi = 35°)$, which guarantees the maximum efficiency of the built-up lattice member.

16.2.2 Battened Compound Struts

In the case of a compound strut only made of batten plates, as the one in Figure 16.8a, the chords are subjected to bending, shear and axial load while battens are mainly affected by shear and bending moments. For these members, by using the same approach already adopted for lattice built-up members, reference has to be made to an internal panel, as the one delimited by the sections m–n and m_1–n_1.

Shear stiffness of the panel can be evaluated on the basis of the horizontal displacements δ due to the following contributions:

(1) flexural deformation of the chords ($\delta_{F,\text{cor}}$) (as in Figure 16.8b);
(2) flexural deformation of the batten ($\delta_{F,\text{cal}}$) (as in Figure 16.8c);
(3) shear deformation of the batten ($\delta_{T,\text{cal}}$).

Battened members are internally, statically indeterminate structures but the evaluation of internal forces and moments is usually carried out by assuming that each panel is connected with the other via hinges, considering that the deflection of the chords has a point of inflection at sections m–n and m_1–n_1.

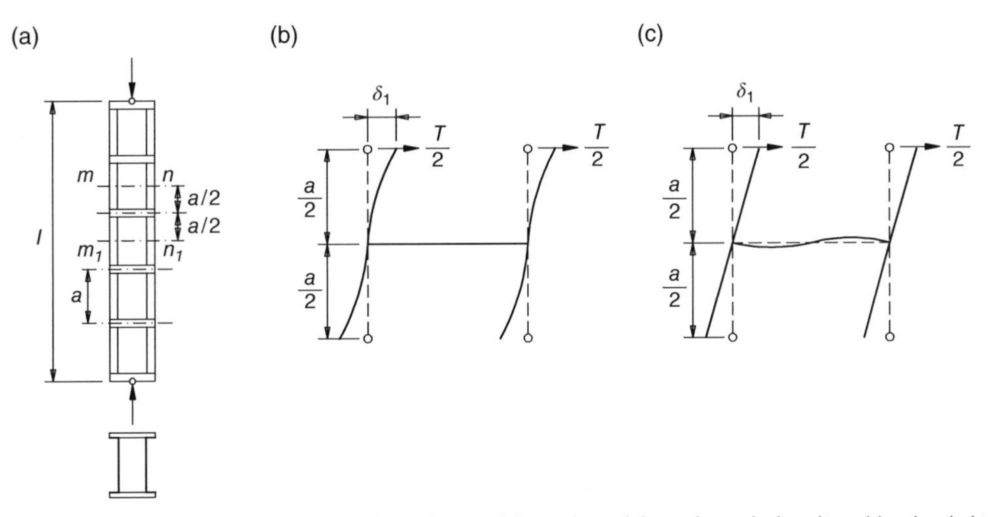

Figure 16.8 Example of a built-up battened member (a): deformed panel shape due to the bending of the chords (b) and due to the bending of the batten (c).

The total horizontal displacement δ due to a horizontal force, T, applied to the end of the considered panel can be obtained as:

$$\delta_{TOT} = \delta_{F,\text{cor}} + \delta_{F,\text{cal}} + \delta_{T,\text{cal}} \tag{16.14}$$

where subscripts F and T are related to the deformation associated with flexure and shear, respectively, while *cor* and *cal* are related to the chord and to the batten, respectively.

Flexure of the Chords Term δ_1, which is related to the displacement of the end of one of the two chords, is evaluated considering a cantilever beam subjected to a lateral force equal to $T/2$:

$$\delta = \frac{T}{2}\left(\frac{a}{2}\right)^3 \frac{1}{3EI_{cor}} = \frac{Ta^3}{48EI_{cor}} \tag{16.15a}$$

$$\delta_{F,cor} = 2\delta_1 = 2\frac{Ta^3}{48EI_{cor}} = \frac{Ta^3}{24EI_{cor}} \tag{16.15b}$$

Flexure of the Batten The total displacement δ_2 of the batten in bending is due to the rotation θ at the chord-to-batten node associated with the bending moment. For each chord, each of the two cantilever beams of length equal to $a/2$ are subjected to a moment of $\frac{T}{2}\left(\frac{a}{2}\right)$ and hence the total moment at each end of the batten is $\frac{Ta}{2}$ (given by $2\left[\frac{T}{2}\left(\frac{a}{2}\right)\right]$). As a consequence it is possible to evaluate rotation and the associated displacement, respectively as:

$$\theta = \frac{Ta}{2}\cdot\frac{1}{3EI_{cal}}\cdot\frac{b}{2} = \frac{Tab}{12EI_{cal}} \tag{16.16a}$$

$$\delta_{F,cal} = 2\theta\frac{a}{2} = \frac{Ta^2b}{12EI_{cal}} \tag{16.16b}$$

Shear Deformation of the Batten The displacement $\delta_{T,cal}$ due to the shear deformation of the batten is evaluated with reference to the model of a beam under a constant shear load of $\frac{T\cdot a}{b}$. Considering the shear strain $\gamma = \frac{\chi_T Ta}{bGA_{cal}}$, where χ_T is the shear factor of the batten, the associated displacement is:

$$\delta_{T,cal} = 2\gamma\frac{a}{2} = \frac{\chi_T}{A_{cal}G}\cdot\frac{Ta^2}{b} \tag{16.17}$$

Shear stiffness of the battened panel, S_V, which depends on the values of these displacements, can be determined as:

$$\frac{1}{S_V} = \frac{\gamma_{TOT}}{T} = \frac{\delta_{F,cor} + \delta_{F,cal} + \delta_{T,cal}}{aT} = \frac{a^2}{24EI_{cor}} + \frac{ab}{12EI_{cor}} + \frac{\chi_T a}{bA_{cal}G} \tag{16.18}$$

Finally, the elastic critical load for a battened struts with hinged ends, $N_{cr,id}$, can be expressed as:

$$N_{cr,id} = \frac{\pi^2 EI}{L^2}\left[\frac{1}{1+\pi^2\frac{EI}{L^2}\left(\frac{a^2}{24EI_{cor}}+\frac{a\cdot b}{12EI_{cal}}+\frac{\chi_T\cdot a}{bA_{cal}G}\right)}\right] \quad (16.19)$$

where I is the moment of inertia of the compound strut.

Reference can be made to an appropriate effective length factor, $k_{\beta,eq}$, which for struts with battened plates is defined as:

$$k_{\beta,eq} = \sqrt{1+\pi^2\frac{EI}{L^2}\left(\frac{a^2}{24EI_{cor}}+\frac{ab}{12EI_{cal}}+\frac{\chi_T a}{bA_{cal}G}\right)} \quad (16.20)$$

Furthermore, neglecting the shear deformability of the battens, Eq. (16.20) simplifies to:

$$k_{\beta,eq} = \sqrt{1+\pi^2\frac{EI}{L^2}\frac{a^2}{24EI_{cor}}} \quad (16.21)$$

Alternatively, design can refer to the overall equivalent slenderness, λ_{eq}, defined as:

$$\lambda_{eq} = \frac{\beta_{eq}L}{r} = \sqrt{\left(\frac{L}{r}\right)^2+\pi^2\frac{I}{24I_{cor}}\frac{a^2}{r^2}} \quad (16.22a)$$

where L is the strut length and r is the radius of gyration, defined as $r=\sqrt{\frac{I}{2A_{cor}}}$.

Due to $\frac{I}{r^2} = 2A_{cor}$ and $I_{cor} = A_{cor}\, r^2_{cor}$, Eq. (16.22a) can be re-written as:

$$\lambda_{eq} = \frac{\beta_{eq}L}{r} = \sqrt{\left(\frac{L}{r}\right)^2+\frac{\pi^2}{12}\left(\frac{a}{r_{cor}}\right)^2} \quad (16.22b)$$

It can be noted that the slenderness of the strut depends strictly on the overall slenderness of the chords rigidly connected (L/r) to each other in the compound member and on the local slenderness of the isolated chords delimited by two contiguous battens (a/r_{cor}).

16.3 Design in Accordance with the European Approach

EC3 in its general part 1-1 deals with the verification on compound struts. In particular, reference is made to uniform built-up compression members with hinged ends that are laterally supported. For these struts, a bow imperfection e_0 has to be considered, never lower than $L/500$, where L is the strut length ($e_0 \geq L/500$). Furthermore, the elastic deformations of lacings or battenings can be considered due to a continuous (smeared) shear stiffness S_V of the column. The proposed approach for uniform built-up compression members can be applied if the lacings or battens compound consist of equal panels with parallel chords and the minimum number of panels in a member is three. If these assumptions are fulfilled, the structure is considered regular and it is hence possible to smear the discrete structure to a continuum.

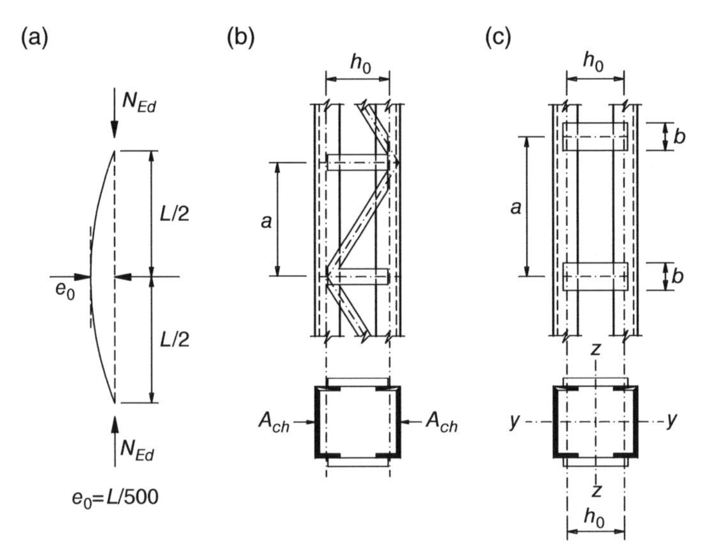

Figure 16.9 Design model (a) for uniform built-up columns with lacings (b) and battenings (c).

It should be noted that the procedure can be applied also if chords are laced or battened themselves in the perpendicular plane.

With reference to the symbols of Figure 16.9, if N_{Ed} and M_{Ed} are the design axial load and the design maximum moment in the middle of the built-up member considering second order effects, respectively, for a member with two identical chords their design force $N_{ch,Ed}$ is given by the expression:

$$N_{ch,Ed} = 0.5 \cdot N_{Ed} + \frac{M_{Ed} \cdot h_0 \cdot A_{ch}}{2 \cdot I_{eff}} \tag{16.23}$$

where h_0 is the distance between the centroids of chords, A_{ch} is the cross-sectional area of one chord and I_{eff} is the effective moment of inertia of the built-up member and M_{Ed} can be obtained as:

$$M_{Ed} = \frac{N_{Ed} \cdot e_0 + M_{Ed}^I}{1 - \dfrac{N_{Ed}}{N_{cr}} - \dfrac{N_{Ed}}{S_v}} \tag{16.24}$$

where term $M^I{}_{Ed}$ is the design value of the maximum moment (if present) in the middle of the built-up member without second-order effects, S_v is the shear stiffness of the lacings or battened panel (Figure 16.9) and term N_{cr} is the effective critical force of the built-up member that is evaluated considering the sole flexural stiffness of the built-up column as:

$$N_{cr} = \frac{\pi^2 \cdot E \cdot I_{\text{eff}}}{L^2} \tag{16.25}$$

where E is the Young modulus, I_{eff} is the effective moment of inertia of the built-up member and L is the effective length.

The checks for the elements connecting the chords to each other have to be performed for the end panel taking account of the shear force in the built-up member V_{Ed}, defined as:

$$V_{Ed} = \frac{\pi \cdot M_{Ed}}{L} \tag{16.26}$$

16.3.1 Laced Compression Members

The chords and diagonal lacings subject to compression should be designed for buckling and secondary moments may be neglected.

The effective moment of inertia of laced built-up member, I_{eff}, is given by:

$$I_{eff} = 0.5 \cdot h_0^2 \cdot A_{ch} \tag{16.27}$$

where A_{ch} is the cross-sectional area of one chord.

As to constructional details, single lacing systems on opposite faces of the built-up member with two parallel laced planes should be corresponding systems as shown in Figure 16.10a, arranged so that one is the shadow of the other. When the single lacing systems on opposite faces of a built-up member with two parallel laced planes are mutually opposed in direction as shown in Figure 16.10b, the resulting torsional effects in the member should be taken into account. Furthermore, tie panels should be provided at the ends of lacing systems, at points where the lacing is interrupted and at joints with other members.

The values of the shear stiffness for the most common cases are reported in Figure 16.11.

16.3.2 Battened Compression Members

The effective moment of inertia of battened built-up members is:

$$I_{eff} = 0.5 \cdot h_0^2 \cdot A_{ch} + 2 \cdot \mu \cdot I_{ch} \tag{16.28}$$

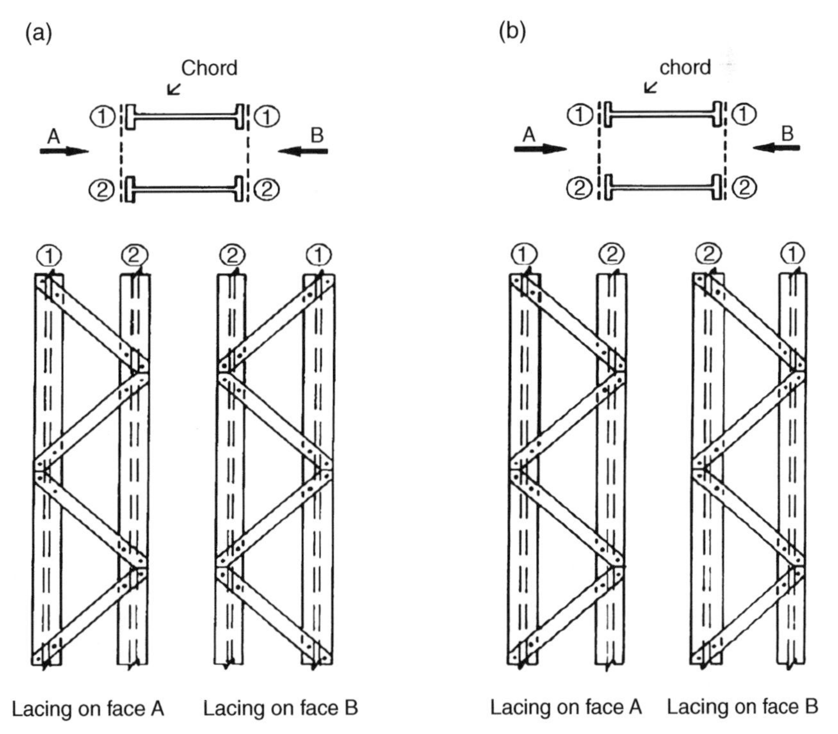

Figure 16.10 Single lacing system on opposite faces of a built-up member with two parallel laced planes: (a) corresponding lacing system (recommended system) and (b) mutually opposed lacing system (not recommended).

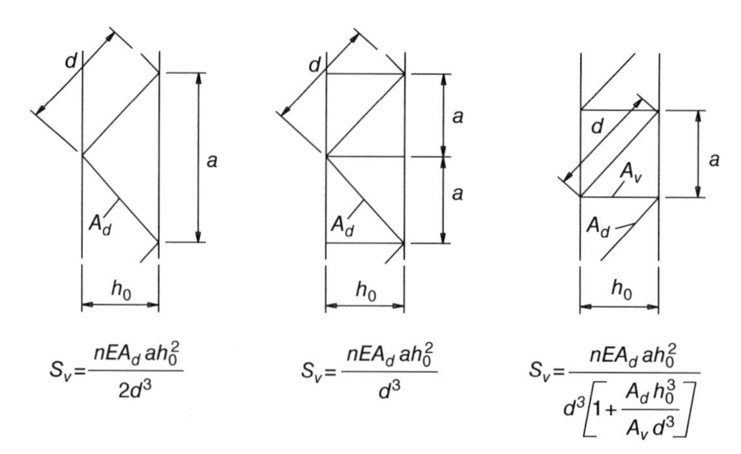

Figure 16.11 Shear stiffness of lacings of built-up members.

Table 16.1 Efficiency factor μ.

Criterion	Efficiency factor μ
$\lambda \geq 150$	$\mu = 0$
$75 < \lambda < 150$	$\mu = 2 - \lambda/75$
$\lambda \leq 75$	$\mu = 1.0$
$\lambda = \dfrac{L}{\sqrt{\dfrac{0.5 \cdot h_0^2 \cdot A_{ch} + 2 \cdot I_{ch}}{2 \cdot A_{ch}}}}$	

where I_{ch} is the in plane moment of inertia of one chord and μ is the efficiency factor to be computed Table 16.1.

Batten plates as well as their connections have to be verified to the distribution of forces and bending moments presented in Figure 16.12, which are considered when applied on the plane panel.

The shear stiffness, S_v, is defined as:

$$S_v = \frac{24 \cdot E \cdot I_{ch}}{a^2 \cdot \left[1 + \dfrac{2 \cdot I_{ch}}{n \cdot I_b} \cdot \dfrac{h_0}{a}\right]} \leq \frac{2 \cdot \pi^2 \cdot E \cdot I_{ch}}{a^2} \tag{16.29}$$

where I_b is the in plane moment of inertia of one batten.

16.3.3 Closely Spaced Built-Up Members

Built-up compression members with chords in contact or closely spaced and connected, with reference to Figure 16.14, through packing plates (a) or star battened angle members connected by pairs of battens in two perpendicular planes (b) should be checked for buckling as a single integral member ignoring the effect of the shear stiffness ($S_V = \infty$), when the conditions in Table 16.2 are met.

In the case of unequal-leg angles (Figure 16.13b), buckling about the y–y axis may be verified with:

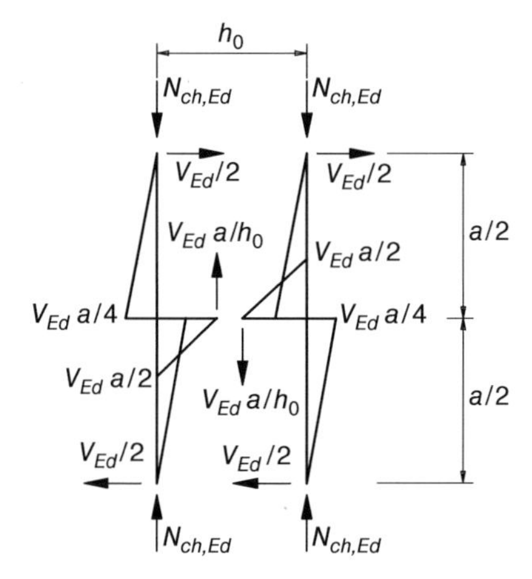

Figure 16.12 Moments and forces in an end panel of a battened built-up member.

Table 16.2 Maximum spacing for interconnections in closely spaced built-up or star battened angle members.

Type of built-up member	Maximum spacing between interconnections [a]
Members according to Figure 16.13a connected by bolts or welds	15 i_{min}
Members according to Figure 16.13b connected by a pair of battens	70 i_{min}

[a] Centre-to-centre distance of interconnections i_{min} is the minimum radius of gyration of one chord or one angle.

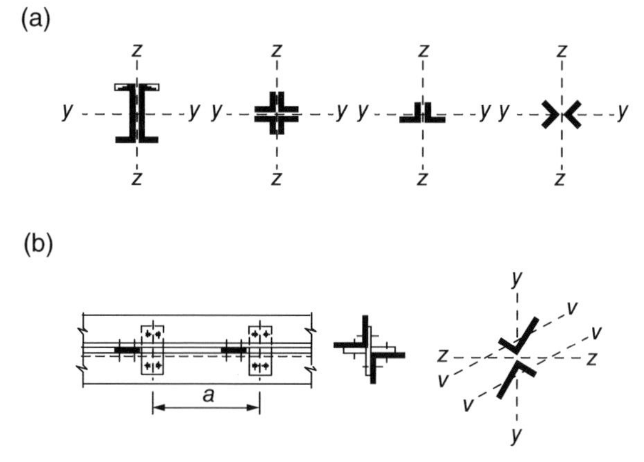

Figure 16.13 Closely spaced built-up members (a) and star-battened angle members (b).

$$i_y = \frac{i_0}{1.15} \tag{16.30}$$

where i_0 is the minimum radius of gyration of the built-up member.

EC3 does not treat the case of closely spaced built-up members connected (by bolts or welds) at a distance >15 i_{min}. In common practice such built-up members are actually connected typically at a distance of 50 i_{min}. Other Codes, like, for example the British Standard BS 5950-1:2000 (no more used at the present), addressed such a case, prescribing how to design the built-up member, which can be a single integral member provided that an equivalent slenderness λ_{eq} is used, computed as:

$$\lambda_{eq} = \sqrt{\lambda_m^2 + \lambda_c^2} \tag{16.31}$$

where λ_m is the slenderness of the whole member and λ_c is the slenderness of the single profile, computed using its minimum radius of gyration, using the distance between adjacent interconnections as length. Its value should not exceed 50.

It worth mentioning that some Codes, like the AISC 360-10 and the already mentioned BS 5950, treat the case of laced and battened built-up compression members with the same previously explained criteria of an equivalent (increased) slenderness, as shown in the next paragraph.

16.4 Design in Accordance with the US Approach

Design according to the provisions for *load and resistance factor design* (LRFD) satisfies the requirements of AISC Specification when the *design compressive strength* $\phi_c P_n$ of each structural component equals or exceeds the *required compressive strength* P_u determined on the basis of the LRFD load combinations. Design has to be performed in accordance with the following equation:

$$P_u \leq \phi_c P_n \tag{16.32}$$

where ϕ_c is the *compressive resistance factor* ($\phi_c = 0.90$) and P_n represents the *nominal compressive strength*.

Design according to the provisions for *allowable strength design* (ASD) satisfies the requirements of AISC Specification when the *allowable compressive strength* P_n/Ω_c of each structural component equals or exceeds the *required compressive strength* P_a determined on the basis of the ASD load combinations. Design has to be performed in accordance with the following equation:

$$P_a \leq P_n/\Omega_c \tag{16.33}$$

where Ω_c is the *compressive safety factor* ($\Omega_c = 1.67$).

The nominal compressive strength P_n is determined as:

$$P_n = F_{cr} A_g \tag{16.34}$$

AISC 360-10 specifications treat the case of built-up members composed of two shapes (angles, channels, etc.) either (i) closely spaced and interconnected by bolts or welds (Figure 16.13) or (ii) put at greater distance with at least one open side interconnected by perforated cover plates (Figure 16.14b) or lacing with tie plates (Figure 16.14a). The end connection has to be welded or connected by means of pretensioned bolts. Intermediate connections can be bolted snug-tight or connected with pretensioned bolts or welds. In the first case the effective length increases because of the greater shear deformability.

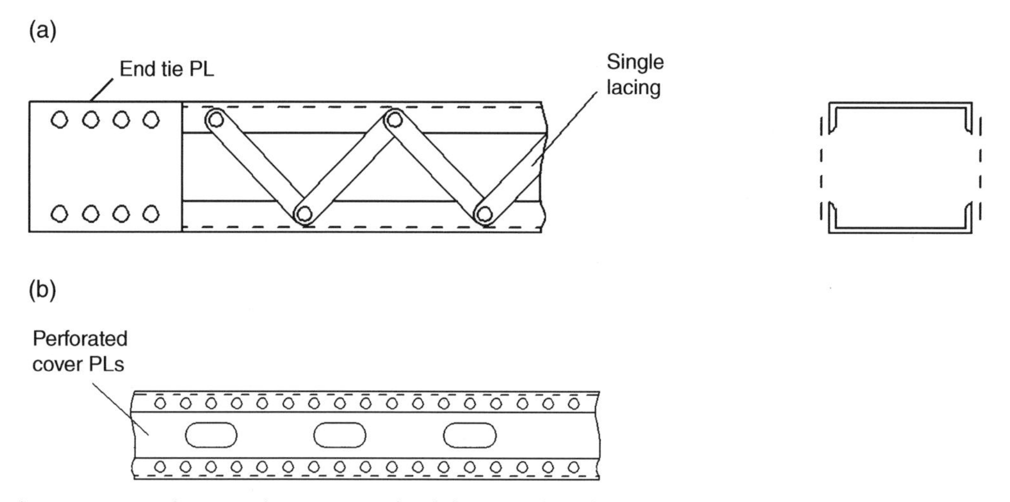

Figure 16.14 Built-up members connected with lacing and tie plates (a) and with perforated cover plates (b).

The *critical stress* F_{cr} must be determined according to the following equation:

$$F_{cr} = \left(\frac{F_{cry} + F_{crz}}{2H}\right)\left[1 - \sqrt{1 - \frac{4F_{cry}F_{crz}H}{\left(F_{cry} + F_{crz}\right)^2}}\right] \tag{16.35}$$

where F_{cry} is taken as F_{cr} from Eqs. (6.32) to (6.33) (see Section 6.2.3), using for KL/r the value:

(1) For intermediate connectors that are bolted snug-tight:

$$\frac{KL}{r} = \left(\frac{KL}{r}\right)_m = \sqrt{\left(\frac{KL}{r}\right)_o^2 + \left(\frac{a}{r_i}\right)^2} \tag{16.36}$$

(2) For intermediate connectors that are welded or connected by means of pretensioned bolts:

When $\dfrac{a}{r_i} \leq 40$:

$$\left(\frac{KL}{r}\right)_m = \left(\frac{KL}{r}\right)_o \tag{16.37}$$

When $\dfrac{a}{r_i} > 40$:

$$\left(\frac{KL}{r}\right)_m = \sqrt{\left(\frac{KL}{r}\right)_o^2 + \left(\frac{K_i a}{r_i}\right)^2} \tag{16.38}$$

where $\left(\dfrac{KL}{r}\right)_m$ is the modified slenderness ratio of built-up members and $\left(\dfrac{KL}{r}\right)_o$ is the slenderness ratio of built-up members acting as a unit in the buckling direction being considered, K_i is a numerical coefficient (0.50 for angles back-to-back, 0.75 for channels back-to-back and 0.86 for all other cases), a is the distance between adjacent connectors and r_i is the minimum radius of gyration of individual component.

16.5 Worked Examples

Example E16.1 Battened Built-Up Member according to EC3

Verify, according to EC3, a battened built-up member composed of 2 UPN 240 as chords, connected by battens realized with 340 × 150 × 12 plates (13.8 × 5.91 × 0.472 in.) put at a distance of a = 1200 mm (47 in.). The distance between centroids of 2 UPN is h_0 = 355 mm (14 in.). The built-up member is 20 m (65.6 ft) long. The effective length in the plane of battens is 20 m. In the orthogonal plane the built-up member is restrained so that flexural instability shall not be considered. The design axial load is 800 kN (180 kips).

Chords	2 UPN 240
Steel	S275 f_y = 275 MPa (40 ksi)
Area of 1 chord	A_{ch} = 42.3 cm^2 (6.56 in.^{2})
Moment of inertia about the y–y axis	I_y = 3599 cm^4 (86.5 in.^{4})
Moment of inertia about the z–z axis	$I_z = I_{ch}$ = 274 cm^4 (6.58 in.^{4})
Radius of inertia about the y–y axis	i_y = 9.22 cm (3.63 in.)
Radius of inertia about the z–z axis	i_z = 2.42 cm (0.953 in.)
Distance between chord centroids	h_0 = 355 mm (14 in.)
Distance between battens	a = 1200 mm (47 in.)
Battens	350 × 150 × 12 plate (13.8 × 5.91 × 0.472 in.)
Built-up member length	L = 2000 cm (65.6 ft)
Design axial load	N_{Ed} = 800 kN (180 kips)

Verify each chord instability between two battens:
Moment of inertia of the whole member:

$$I_1 = 0.5h_0^2 A_{ch} + 2I_{ch} = 0.5 \times 35.5^2 \times 42.3 + 2 \times 274 = 27\ 202\ \text{cm}^4\left(654\ \text{in.}^4\right)$$

Radius of inertia of the whole section:

$$i_0 = \sqrt{\frac{I_1}{2A_{ch}}} = \sqrt{\frac{27\ 202}{2 \times 42.3}} = 17.93\ \text{cm}\ (7.06\ \text{in.})$$

Critical length of built-up member in the plane containing battens:

$$\lambda = \frac{L_{cr}}{i_0} = \frac{2000}{17.93} = 112;\ 75 \le \lambda \le 150\ \text{then}:\ \mu = 2 - \frac{\lambda}{75} = 2 - \frac{112}{75} = 0.51$$

Effective moment of inertia:

$$I_{eff} = 0.5h_0^2 A_{ch} + 2\mu I_{ch} = 0.5 \times 35.5^2 \times 42.3 + 2 \times 0.51 \times 274 = 26\ 934\ \text{cm}^4\left(647\ \text{in.}^4\right)$$

Radius of inertia of the whole section computed using effective moment of inertia:

$$i_{0,eff} = \sqrt{\frac{I_{eff}}{2A_{ch}}} = \sqrt{\frac{26\ 934}{2 \times 42.3}} = 17.84\text{cm}\ (7.02\ \text{in.})$$

Moment of inertia of batten: $I_b = \frac{1}{12} \times 1.2 \times 15^3 = 337.5\ \text{cm}^4\left(8.11\ \text{in.}^4\right)$

Shear stiffness:

$$S_v = \frac{24EI_{ch}}{a^2\left(1+\frac{2I_{ch}}{nI_b}\cdot\frac{h_o}{a}\right)} = \frac{24\times 21\ 000\times 274}{120^2\times\left(1+\frac{2\times 274}{2\times 337.5}\times\frac{35.5}{120}\right)} = 7733\,\text{kN}$$

$$\leq \frac{2\pi^2 EI_{ch}}{a^2} = \frac{2\times\pi^2\times 21\ 000\times 274}{120^2} = 7887\,\text{kN}\ (1738\,\text{kips} \leq 1773\,\text{kips})$$

Let's consider: $e_0 = \frac{L_{cr}}{500} = \frac{2000}{500} = 4\,\text{cm}$ (1.58 in.) (eccentricity).

Built-up member critical load:

$$N_{cr} = \pi^2 EI_{eff}/L^2 = 3.14^2\times 21\ 000\times 26\ 934/2000^2 = 1396\,\text{kN}\ (314\,\text{kips})$$

Second order moment:

$$M_{Ed} = \frac{N_{Ed}e_0}{1-\frac{N_{Ed}}{N_{cr}}-\frac{N_{Ed}}{S_V}} = \frac{800\times 4}{1-\frac{800}{1396}-\frac{800}{7733}} = \frac{3200}{1-0.573-0.103} = 9892\,\text{kN cm}\ (73\,\text{kip-ft})$$

Maximum axial load in a chord considering second order effects:

$$N_{ch,Ed} = 0.5N_{Ed} + \frac{M_{Ed}h_0A_{ch}}{2I_{eff}} = 0.5\times 800 + \frac{9892\times 35.5\times 42.3}{2\times 26\ 934} = 400 + 275 = 675\,\text{kN}\ (151.7\,\text{kips})$$

Verify a chord under compression load $N_{ch,Ed}$, using the distance between chord as unbraced length and using the minimum radius of inertia.

$$\lambda_1 = 93.9\varepsilon = 93.9\sqrt{\frac{235}{f_y}} = 93.9\sqrt{\frac{235}{275}} = 86.8;\ \lambda = \frac{a}{i_z} = \frac{120}{2.42} = 50$$

$$\bar{\lambda} = \lambda/\lambda_1 = 50/86.8 = 0.576;\ \Phi = 0.5\left[1+\alpha(\bar{\lambda}-0.2)+\bar{\lambda}^2\right] = 0.5\times\left[1+0.49\times(0.576-0.2)+0.576^2\right] = 0.758$$

$$\chi = \frac{1}{\Phi+\sqrt{\Phi^2-\bar{\lambda}^2}} = \frac{1}{0.758+\sqrt{0.758^2-0.576^2}} = 0.800$$

Compute design compressive strength of a single chord $N_{ch,Rd}$ and compare it with maximum compression load $N_{ch,Ed}$.

$$N_{ch,Rd} = \chi\frac{A_{ch}f_y}{\gamma_{M1}} = 0.800\times\frac{42.3\times 27.50}{1.00} = 931\,\text{kN} > N_{ch,Ed} = 675\,\text{kN OK}$$

(209 kips > 152 kips)

(Stress ratio: 675/931 = 0.73).

Verify the whole built-up member (two chords) for buckling in the plane of battens

Use the already computed radius of inertia of the whole section, $i_{0,eff}$.

$$\lambda = \frac{L_{cr}}{i_{0,eff}} = \frac{2000}{17.84} = 112; \ \bar{\lambda} = \lambda/\lambda_1 = 112/86.8 = 1.29$$

$$\Phi = 0.5\left[1 + \alpha(\bar{\lambda} - 0.2) + \bar{\lambda}^2\right] = 0.5 \times \left[1 + 0.49(1.29 - 0.2) + 1.29^2\right] = 1.599$$

$$\chi = \frac{1}{\Phi + \sqrt{\Phi^2 - \bar{\lambda}^2}} = \frac{1}{1.599 + \sqrt{1.599^2 - 1.29^2}} = 0.393$$

$$N_{Rd} = \chi \frac{2A_{ch}f_y}{\gamma_{M1}} = 0.393 \times \frac{2 \times 42.3 \times 27.50}{1.00} = 914\,\text{kN} > N_{Ed} = 800\,\text{kN}$$

Stress rate: 800/914 = 0.88.

In this case, the verification of the whole built-up member for buckling on the length of 20 m is dominant on the verification of a single chord for buckling on the distance between battens. In the verification of the whole member the effective moment of inertia has been used.

Verify the whole built-up member (two chords) for buckling in the plane orthogonal to that of the battens.

We hypothesize that built-up member is braced in this plane, so no check is needed.

Batten check:

Shear and moment verification (see Figure 16.13). Max shear:

$$V_{Ed} = \pi \frac{M_{Ed}}{L} = 3.14 \times \frac{9877}{2000} = 15.5\,\text{kN}\,(3.49\,\text{kips})$$

Max shear and bending moment in the batten:

$$V_{cal,Ed} = \frac{V_{Ed} \cdot a}{h_0} = \frac{15.5 \times 120}{35.5} = 52.4\,\text{kN}\,(11.8\,\text{kips})$$

$$M_{cal,Ed} = \frac{V_{Ed} \cdot a}{2} = \frac{15.5 \times 120}{2} = 930\,\text{kN cm}\,(6.86\,\text{kip-ft})$$

Shear area: $A_v = 15 \times 1.2 = 18\,\text{cm}^2 (2.79\,\text{in.}^2)$

Shear strength:

$$V_{c,Rd} = V_{pl,Rd} = \frac{A_v f_y}{\sqrt{3} \cdot \gamma_{M0}} = \frac{18 \times 27.50}{\sqrt{3} \times 1.05} = 272.2\,\text{kN}\,(61.2\,\text{kips})$$

Shear check: $V_{cal,Ed}/V_{c,Rd} = 52.4/272.2 = 0.19 < 1$ OK

Being shear stress ratio <0.50 resisting moment shall not be reduced:

$$W_{cal,el} = \frac{1}{6} \times 1.2 \times 15^2 = 45\,\text{cm}^3$$

$$M_{c,Rd} = \frac{W_{cal,el} f}{\gamma_{M0}} = \frac{45 \times 27.50}{1.00} = 1238\,\text{kN cm} > M_{cal,Ed} = 930\,\text{kN cm OK}$$
$$(9.13\,\text{kip-ft} > 6.86\,\text{kip-ft})$$

Chord verification:

Check for compression and bending moment (Figure 16.13).

Maximum moment: $M_{ch,Ed} = \frac{V_{Ed} \cdot a}{4} = \frac{15.5 \times 120}{4} = 465\,\text{kN cm}\,(3.43\,\text{kip-ft})$

Maximum axial load: $N_{ch,Ed} = 675\,\text{kN}\,(152\,\text{kips})$.

(Near connections of the built-up member the increment of axial load due to imperfection is actually minimal and axial load should be something more than 0.5 N_{Ed} = 400 kN (90 kips). Here, to be on the safe side, use the maximum value.)

$$A_{ch} = 42.3\,\text{cm}^2;\ W_{pl,z} = 75.7\,\text{cm}^3$$

Compression strength:

$$N_{pl,Rd} = \frac{A_{ch} f_y}{\gamma_{M0}} = \frac{42.3 \times 27.50}{1.00} = 1163\,\text{kN}\,(262\,\text{kips})$$

$$N_{ch,Ed} = 675\,\text{kN} > \frac{h_w t_w f_y}{\gamma_{M0}} = \frac{(24 - 2 \times 1.3) \times 0.95 \times 27.50}{1.00} = 559\,\text{kN}\,(152\,\text{kips} > 126\,\text{kips})$$

Influence of axial load on bending strength shall be then considered.

$$a_1 = \frac{A_{web}}{A_{ch}} = \frac{(24 - 2 \times 1.3) \times 0.95}{42.3} = 0.481$$

$$M_{pl,z,Rd} = \frac{W_{pl,z} f_y}{\gamma_{M0}} = \frac{75.7 \times 27.50}{1.00} = 2082\,\text{kN cm}\,(15.4\,\text{kip} - \text{ft})$$

$$M_{N,z,Rd} = \left[1 - \left(\frac{\frac{N_{Ed}}{N_{pl,Rd}} - a_1}{1 - a_1}\right)^2\right] M_{pl,z,Rd} = \left[1 - \left(\frac{\frac{675}{1108} - 0.481}{1 - 0.481}\right)^2\right] \times 2082 = 1955\,\text{kN cm}$$

$$(14.4\,\text{kip} - \text{ft})$$

Check:

$$\frac{N_{ch,Ed}}{N_{pl,Rd}} + \frac{M_{ch,Ed}}{M_{N,z,Rd}} = \frac{675}{1108} + \frac{465}{1955} = 0.61 + 0.24 = 0.85 < 1\ \text{OK}$$

Example E16.2 Battened Built-Up Member According to EC3

Verify, according to EC3, a battened built-up member composed of 2 UPN 240 as chords, like Example E16.1, but with battens put at a distance of a = 1800 mm (70.9 in.). The design axial load is 750 kN (169 kips). All other parameters are equal to those in Example E16.1.

Chords	2 UPN 240
Steel	S275 f_y = 275 MPa (40 ksi)
Area of 1 chord	A_{ch} = 42.3 cm^2 (6.56 in.^{2})
Moment of inertia about the y–y axis	I_y = 3599 cm^4 (86.5 in.^{4})
Moment of inertia about the z–z axis	$I_z = I_{ch}$ = 274 cm^4 (6.58 in.^{4})
Radius of inertia about the y–y axis	i_y = 9.22 cm (3.63 in.)
Radius of inertia about the z–z axis	i_z = 2.42 cm (0.953 in.)
Distance between chord centroids	h_0 = 355 mm (14 in.)
Distance between battens	a = 1800 mm (70.9 in.)
Battens	350 × 150 × 12 plate (13.8 × 5.91 × 0.472 in.)
Built-up member length	L = 2000 cm (65.6 ft)
Design axial load	N_{Ed} = 750 kN (169 kips)

Verify each chord instability between two battens:
Moment of inertia of the whole member:

$$I_1 = 0.5h_0^2 A_{ch} + 2I_{ch} = 0.5 \times 35.5^2 \times 42.3 + 2 \times 274 = 27\ 202\,\text{cm}^4\left(654\,\text{in.}^4\right)$$

Radius of inertia of the whole section:

$$i_0 = \sqrt{\frac{I_1}{2A_{ch}}} = \sqrt{\frac{27\,202}{2 \times 42.3}} = 1793\,\text{cm}\,(7.06\,\text{in.})$$

Critical length of built-up member in the plane containing battens:

$$\lambda = \frac{L_{cr}}{i_0} = \frac{2000}{17.93} = 112;\ 75 \le \lambda \le 150\ \text{then}:\ \mu = 2 - \frac{\lambda}{75} = 2 - \frac{112}{75} = 0.51$$

Effective moment of inertia:

$$I_{eff} = 0.5h_0^2 A_{ch} + 2\mu I_{ch} = 0.5 \times 35.5^2 \times 42.3 + 2 \times 0.51 \times 274 = 26934\,\text{cm}^4\left(647\,\text{in.}^4\right)$$

Radius of inertia of whole section computed using effective moment of inertia:

$$i_{0,eff} = \sqrt{\frac{I_{eff}}{2A_{ch}}} = \sqrt{\frac{26\,934}{2 \times 42.3}} = 17.84\,\text{cm}\,(7.02\,\text{in.})$$

Moment of inertia of batten: $I_b = \frac{1}{12} \times 1.2 \times 15^3 = 337.5\,\text{cm}^4\left(8.11\,\text{in.}^4\right)$.
Shear stiffness:

$$S_v = \frac{24EI_{ch}}{a^2\left(1 + \frac{2I_{ch}}{nI_b}\cdot\frac{h_o}{a}\right)} = \frac{24 \times 21000 \times 274}{180^2 \times \left(1 + \frac{2 \times 274}{2 \times 337.5} \times \frac{35.5}{180}\right)} = 3674\,\text{kN}$$

$$\le \frac{2\pi^2 EI_{ch}}{a^2} = \frac{2 \times \pi^2 \times 21000 \times 274}{180^2} = 3506\,\text{kN; Assume } S_v = 3506\,\text{kN}\,(788\,\text{kips})$$

Let's consider: $e_0 = \frac{L_{cr}}{500} = \frac{2000}{500} = 4\,\text{cm}\,(1.58\,\text{in.})$ (eccentricity).
Built-up member critical load:

$$N_{cr} = \pi^2 EI_{eff}/L^2 = 3.14^2 \times 21\,000 \times 26\,934/2000^2 = 1396\,\text{kN}\,(314\,\text{kips})$$

Second order moment:

$$M_{Ed} = \frac{N_{Ed}e_0}{1 - \frac{N_{Ed}}{N_{cr}} - \frac{N_{Ed}}{S_V}} = \frac{750 \times 4}{1 - \frac{750}{1396} - \frac{750}{3506}} = 12\ 056\,\text{kN cm}\,(88.9\,\text{kip-ft})$$

Maximum axial load in a chord considering second order effects:

$$N_{ch,Ed} = 0.5N_{Ed} + \frac{M_{Ed}h_0A_{ch}}{2I_{eff}} = 0.5 \times 750 + \frac{12\ 056 \times 35.5 \times 42.3}{2 \times 26\ 934} = 375 + 336 = 711\,\text{kN}\,(160\,\text{kips})$$

Verify a chord under compression load $N_{ch,Ed}$ using as unbraced length the distance between chord and using minimum radius of inertia.

$$\lambda_1 = 93.9\varepsilon = 93.9\sqrt{\frac{235}{f_y}} = 93.9\sqrt{\frac{235}{275}} = 86.8;\ \lambda = \frac{a}{i_z} = \frac{180}{2.42} = 74$$

$$\bar{\lambda} = \lambda/\lambda_1 = 74/86.8 = 0.853;\ \Phi = 0.5\left[1 + \alpha(\bar{\lambda} - 0.2) + \bar{\lambda}^2\right] = 0.5 \times \left[1 + 0.49 \times (0.853 - 0.2) + 0.853^2\right] = 1.024$$

$$\chi = \frac{1}{\Phi + \sqrt{\Phi^2 - \bar{\lambda}^2}} = \frac{1}{1.024 + \sqrt{1.024^2 - 0.853^2}} = 0.629$$

Compute the design compressive strength of a single chord $N_{ch,Rd}$ and compare it with maximum compression load $N_{ch,Ed}$.

$$N_{ch,Rd} = \chi\frac{A_{ch}f_y}{\gamma_{M1}} = 0.629 \times \frac{42.3 \times 27.50}{1.00} = 732\,\text{kN} > N_{ch,Ed} = 711\,\text{kN OK}$$

$$(\text{Stress ratio}: 711/732 = 0{,}97).$$

Verify the whole built-up member (two chords) for buckling in the plane of battens:
Use the already computed radius of inertia of the whole section, i_0.

$$\lambda = \frac{L_{cr}}{i_{0,eff}} = \frac{2000}{17.84} = 112;\ \bar{\lambda} = \lambda/\lambda_1 = 112/86.8 = 1.29$$

$$\Phi = 0.5\left[1 + \alpha(\bar{\lambda} - 0.2) + \bar{\lambda}^2\right] = 0.5 \times \left[1 + 0.49(1.29 - 0.2) + 1.29^2\right] = 1.599$$

$$\chi = \frac{1}{\Phi + \sqrt{\Phi^2 - \bar{\lambda}^2}} = \frac{1}{1.599 + \sqrt{1.599^2 - 1.29^2}} = 0.393$$

$$N_{Rd} = \chi\frac{2A_{ch}f_y}{\gamma_{M1}} = 0.393 \times \frac{2 \times 42.3 \times 27.50}{1.00} = 914\,\text{kN} > N_{Ed} = 750\,\text{kN}$$

$$\text{Stress ratio}: 750/914 = 0{,}82.$$

In this case, different from Example E16.1, the verification of a single chord for buckling on the distance between battens is dominant on the verification of the whole built-up member for buckling on the length of 20 m.

Other verifications are equivalent to those in Example E16.1.

In Table E16.2.1 a comparison between stress ratios, for Examples E16.1 and E16.2, is shown.

Table E16.2.1 Stress ratios for Examples E16.1 and E16.2.

Example	Chord local buckling (between two battens)	Built-up member global buckling (over the entire member length)	Dominant buckling mode
E16.1	0.73	0.88	Global
E16.2	0.97	0.82	Local

It can be outlined that, by increasing batten distance from 1200 to 1800 mm, and as a consequence increasing slenderness of a single chord between two battens from 50 to 74, the dimensioning buckling mode switches from the global buckling of the built-up member to the buckling of a chord between two battens.

Example E16.3 Closely-Spaced Built-Up Member According to AISC

Compute according to AISC 360-10 the compressive strength of a closely-spaced built-up member composed by 2 L 4 × 4 × 3/8 (L102 × 102 × 9.5) as chords, 16 ft (4.88 m) length, connected at a distance of $a = 39$ in. (991 mm). The separation between the two angles is 0.5 in. (12.7 mm). Effective length in the plane of connectors is 16 ft. In the orthogonal plane the built-up member is restrained at the middle, so the effective length is 8 ft ($K = 0.5$).

	L 4 × 4 × 3/8 (from AISC Manual Tables 1-7)
Steel	ASTM A36 $F_y = 36$ ksi (248 MPa)
Area of 1 L	$A_s = 2.86$ in.^{2} (18.5 cm^2)
Radius of inertia about the y–y axis	$r_y = 1.23$ in. (3.12 cm)
Radius of inertia about the z–z axis	$r_i = 0.779$ in. (1.98 cm)
Distance of centroid	$\bar{y} = 1.13$ in. (2.87 cm)
Distance between connectors	$a = 39$ in. (991 mm.)
	2 L 4 × 4 × 3/8 (from AISC Manual Tables 1-15)
Area of 2 L	$A_s = 5.71$ in.^{2} (37.0 cm^2)
Thickness of connectors	$d = 3/8$ in. (9.5 mm)
	$\bar{r}_0 = 2.38$ in. (6.05 cm)
	$H = 0.843$

Classify angle L 4 × 4 × 3/8
(See Table 4.4a).

$$b/t = \frac{4}{3/8} = 10.7 < 0.45\sqrt{\frac{E}{F_y}} = 0.45 \times \sqrt{\frac{29\,000}{36}} = 12.8 \;\rightarrow\; \text{The profile is non-slender}$$

Slenderness in the plane orthogonal to connectors.

$$\frac{KL}{r_y} = \frac{0.5 \times 16 \cdot 12}{1.23} = 78$$

Slenderness in the plane of connectors.
Radius of inertia of the whole section:

$$r = \sqrt{r_y^2 + (\bar{y} + d/2)^2} = \sqrt{1.23^2 + (1.13 + (3/8)/2)^2} = 1.80\,\text{in.}\ (4{,}57\,\text{cm})$$

This value has been here computed but it can be found in the AISC Manual Tables 1-15.

(a) intermediate connectors are bolted snug-tight:

$$\left(\frac{KL}{r}\right)_m = \sqrt{\left(\frac{KL}{r}\right)_o^2 + \left(\frac{a}{r_i}\right)^2} = \sqrt{\left(\frac{1\times 16\cdot 12}{1.80}\right)^2 + \left(\frac{39}{0.779}\right)^2} = 118 \text{ governs}$$

(b) intermediate connectors are welded or connected by means of pretensioned bolts:

$$a/r_i = 39/0.779 = 50 > 40$$

$$\left(\frac{KL}{r}\right)_m = \sqrt{\left(\frac{KL}{r}\right)_o^2 + \left(\frac{K_i a}{r_i}\right)^2} = \sqrt{\left(\frac{1\times 16\cdot 12}{1.80}\right)^2 + \left(\frac{0.5\times 39}{0.779}\right)^2} = 110 \text{ governs}$$

$$F_{crz} = \frac{GJ}{A_g \bar{r}_0^2} = \frac{11\,200\times(2\times 0.141)}{5.71\times 2.38^2} = 97.6 \text{ ksi } (673 \text{ MPa})$$

(a) intermediate connectors are bolted snug-tight:

$$\left(\frac{KL}{r}\right)_m = 118 \le 4.71\sqrt{\frac{E}{F_y}} = 4.71\sqrt{\frac{29\,000}{36}} = 134$$

$$F_e = \frac{\pi^2 E}{\left(\frac{kL}{r}\right)^2} = \frac{3.14^2\times 29\,000}{118^2} = 20.6 \text{ ksi } (142 \text{ MPa})$$

$$F_{cry} = \left[0.658^{\left(\frac{F_y}{F_e}\right)}\right]F_y = \left[0.658^{\left(\frac{36}{20.6}\right)}\right]\times 36 = 17.3 \text{ ksi } (119.2 \text{ MPa})$$

$$F_{cr} = \left(\frac{F_{cry}+F_{crz}}{2H}\right)\left[1-\sqrt{1-\frac{4F_{cry}F_{crz}H}{(F_{cry}+F_{crz})^2}}\right] = \left(\frac{17.3+97.6}{2\times 0.843}\right)\left[1-\sqrt{1-\frac{4\times 17.3\times 97.6\times 0.843}{(17.3+97.6)^2}}\right]$$

$$= 16.8 \text{ ksi governs } (115.8 \text{ MPa})$$

(b) intermediate connectors are welded or connected by means of pretensioned bolts:

$$\left(\frac{KL}{r}\right)_m = 110 \le 4.71\sqrt{\frac{E}{F_y}} = 4.71\sqrt{\frac{29\,000}{36}} = 134$$

$$F_e = \frac{\pi^2 E}{\left(\frac{kL}{r}\right)^2} = \frac{3.14^2\times 29\,000}{110^2} = 23.6 \text{ ksi } (162.7 \text{ MPa})$$

$$F_{cry} = \left[0.658^{\left(\frac{F_y}{F_e}\right)}\right]F_y = \left[0.658^{\left(\frac{36}{23.6}\right)}\right]\times 36 = 19.0 \text{ ksi } (131 \text{ MPa})$$

$$F_{cr} = \left(\frac{F_{cry}+F_{crz}}{2H}\right)\left[1-\sqrt{1-\frac{4F_{cry}F_{crz}H}{(F_{cry}+F_{crz})^2}}\right] = \left(\frac{19.0+97.6}{2\times 0.843}\right)\left[1-\sqrt{1-\frac{4\times 19.0\times 97.6\times 0.843}{(19.0+97.6)^2}}\right]$$

$$= 18.3 \text{ ksi governs } (126.2 \text{ MPa})$$

(a) intermediate connectors are bolted snug-tight:
Design compressive strength:

$$\phi_c F_{cr} A_s = 0.90 \times 16.8 \times 5.71 = 86.3\,\text{kips}\ (384\,\text{kN})$$

Allowable compressive strength:

$$F_{cr} A_s / \Omega_c = 16.8 \times 5.71 / 1.67 = 57.4\,\text{kips}\ (255\,\text{kN})$$

(b) intermediate connectors are welded or connected by means of pretensioned bolts:
Design compressive strength:

$$\phi_c F_{cr} A_s = 0.90 \times 18.3 \times 5.71 = 94.0\,\text{kips}\ (418\,\text{kN})$$

Allowable compressive strength:

$$F_{cr} A_s / \Omega_c = 18.3 \times 5.71 / 1.67 = 62.6\,\text{kips}\ (278\,\text{kN})$$

As can be seen, welded connectors or connected by means of pretensioned bolts instead of snug-tight bolts, allow us to increase the compressive strength of about 8–9%.

Appendix A

Conversion Factors

	To convert	To	Multiply by	To convert	To	Multiply by
Lengths	in.	mm	25.4	mm	in.	0.0394
	ft	m	0.3048	m	ft	3.281
Areas	in.^{2}	mm^2	645	mm^2	in.^{2}	0.00155
	in.^{2}	cm^2	6.45	cm^2	in.^{2}	0.155
	ft^2	m^2	0.093	m^2	ft^2	10.764
Forces	lb	N	4.448	N	lb	0.225
	kips	kN	4.448	kN	kips	0.225
Moments	kip-ft	kNm	1.356	kNm	kip-ft	0.0685
Stresses	psi	N/m^2	6895	N/m^2	psi	0.0001450
	ksi	MPa (= N/mm^2)	6.895	MPa (= N/mm^2)	ksi	0.1450
Uniform loads	kip/ft	kN/m	14.59	kN/m	kip/ft	0.06852
	psf	N/m^2	47.88	N/m^2	psf	0.02089
	kip/ft^2	kN/m^2	47.88	kN/m^2	kip/ft^2	0.02089
Temperature	°F	°C	(°F − 32) × (5/9) = °C	—		
	°C	°F	°C × (9/5) + 32 = °F	—		

Structural Steel Design to Eurocode 3 and AISC Specifications, First Edition. Claudio Bernuzzi and Benedetto Cordova.

Appendix B
References and Standards

B.1 Most Relevant Standards For European Design

The main standards for the design and construction of steel buildings are in the following listed in the following groups:

Reference for structural design (see Section B.1.1);
Reference for the materials and technical delivery conditions (see Section B.1.2);
Reference for products and tolerances (see Section B.1.3);
Reference for material tests (see Section B.1.4);
Reference for mechanical fasteners (see Section B.1.5);
Reference for welding (see Section B.1.6);
Reference for protection (see Section B.1.7).

Codes and standards listed here are, in many cases, indicated with the year of issue, valid at the writing phase of this book. Please refer to the updated version, if available.

B.1.1 Reference for Structural Design

- EN 1990 – Eurocode 0: Basis of structural design;
- EN 1991 – Eurocode 1: Actions on structures;
- EN 1992 – Eurocode 2: Design of concrete structures;
- EN 1993 – Eurocode 3: Design of steel structures;
- EN 1994 – Eurocode 4: Design of composite steel and concrete structures;
- EN 1995 – Eurocode 5: Design of timber structures;
- EN 1996 – Eurocode 6: Design of masonry structures;
- EN 1997 – Eurocode 7: Geotechnical design;
- EN 1998 – Eurocode 8: Design of structures for earthquake resistance;
- EN 1999 – Eurocode 9: Design of aluminium structures.
- EN 1993-1-1: Eurocode 3: Design of steel structures – Part 1-1: General rules and rules for buildings;

- EN 1993-1-2: Eurocode 3: Design of steel structures – Part 1-2: General rules – Structural fire design;
- EN 1993-1-3: Eurocode 3: Design of steel structures – Part 1-3: General rules – Supplementary rules for cold-formed members and sheeting;
- EN 1993-1-4: Eurocode 3: Design of steel structures – Part 1-4: General rules – Supplementary rules for stainless steels;
- EN 1993-1-5: Eurocode 3: Design of steel structures – Part 1-5: Plated structural elements;
- EN 1993-1-6: Eurocode 3: Design of steel structures – Part 1-6: Strength and Stability of Shell Structures;
- EN 1993-1-7: Eurocode 3: Design of steel structures – Part 1-7: Plated structures subject to out of plane loading;
- EN 1993-1-8: Eurocode 3: Design of steel structures – Part 1-8: Design of joints;
- EN 1993-1-9: Eurocode 3: Design of steel structures – Part 1-9: Fatigue;
- EN 1993-1-10: Eurocode 3: Design of steel structures – Part 1-10: Material toughness and through-thickness properties;
- EN 1993-1-11: Eurocode 3: Design of steel structures – Part 1-11: Design of structures with tension components;
- EN 1993-1-12: Eurocode 3: Design of steel structures – Part 1-12: Additional rules for the extension of EN 1993 up to steel grades S 700;
- EN 1993-2: Eurocode 3: Design of steel structures – Part 2: Steel Bridges;
- EN 1993-3-1: Eurocode 3: Design of steel structures – Part 3-1: Towers, masts and chimneys – Towers and masts;
- EN 1993-3-2: Eurocode 3: Design of steel structures – Part 3-2: Towers, masts and chimneys – Chimneys;
- EN 1993-4-1: Eurocode 3: Design of steel structures – Part 4-1: Silos;
- EN 1993-4-2: Eurocode 3: Design of steel structures – Part 4-2: Tanks;
- EN 1993-4-3: Eurocode 3: Design of steel structures – Part 4-3: Pipelines;
- EN 1993-5: Eurocode 3: Design of steel structures – Part 5: Piling;
- EN 1993-6: Eurocode 3: Design of steel structures – Part 6: Crane supporting structures.

B.1.2 Standards for Materials and Technical Delivery Conditions

EN ISO 643: Steels – Micrographic determination of the apparent grain size.

EN 10025-1: Hot rolled products of structural steels – Part 1: General technical delivery conditions.

EN 10025-2: Hot rolled products of structural steels – Part 2: Technical delivery conditions for non-alloy structural steels.

EN 10025-3: Hot rolled products of structural steels – Part 3: Technical delivery conditions for normalized/normalized rolled weldable fine grain structural steels.

EN 10025-4: Hot rolled products of structural steels – Part 4: Technical delivery conditions for thermomechanical rolled weldable fine grain structural steels.

EN 10025-5: Hot rolled products of structural steels – Part 5: Technical delivery conditions for structural steels with improved atmospheric corrosion resistance.

EN 10025-6: Hot rolled products of structural steels – Part 6: Technical delivery conditions for flat products of high yield strength structural steels in the quenched and tempered condition.

EN 10027-1: Designation systems for steels – Part 1: Steel names.

EN 10027-2: Designation systems for steels – Part 2: Numerical system.

EN 10149-1: Hot-rolled flat products made of high yield strength steels for cold forming – Part 1: General delivery conditions.

EN 10149-2: Hot-rolled flat products made of high yield strength steels for cold forming – Part 2: Delivery conditions for thermomechanically rolled steels.
EN 10149-3: Hot-rolled flat products made of high yield strength steels for cold forming – Part 3: Delivery conditions for normalized or normalized rolled steels.
EN 10162: Cold rolled steel sections – Technical delivery conditions – Dimensional and cross-sectional tolerances.
EN 10164: Steel products with improved deformation properties perpendicular to the surface of the product – Technical delivery conditions.
EN 10268: Cold rolled steel flat products with high yield strength for cold forming – Technical delivery conditions.

- *Hollow sections*
  EN 10210-1: Hot finished structural hollow sections of non-alloy and fine grain steels – Part 1: Technical delivery conditions.
  EN 10219-1: Cold formed welded structural hollow sections of non-alloy and fine grain steels – Part 1: Technical delivery conditions.
- *Strip and flat products*
  EN 10346: Continuously hot-dip coated steel flat products – Technical delivery conditions.
  EN 10268: Cold rolled steel flat products with high yield strength for cold forming – Technical delivery conditions.

B.1.3 Products and Tolerances

EN 1090-1: Execution of steel structures and aluminium structures – Part 1: Requirements for conformity assessment of structural components.
EN 1090-2: Execution of steel structures and aluminium structures – Part 2: Technical requirements for steel structures.
EN 10204: Metallic products – Types of inspection documents.
EN 10024: Hot rolled taper flange I sections – Tolerances on shape and dimensions.
EN 10034: Structural steel I and H sections – Tolerances on shape and dimensions.
EN 10055: Hot rolled steel equal flange tees with radiused root and toes – Dimensions and tolerances on shape and dimensions.
EN 10056-1: Structural steel equal and unequal leg angles – Part 1: Dimensions.
EN 10056-2: Structural steel equal and unequal leg angles – Part 2: Tolerances on shape and dimensions.
EN 10058: Hot rolled flat steel bars for general purposes – Dimensions and tolerances on shape and dimensions.
EN 10059: Hot rolled square steel bars for general purposes – Dimensions and tolerances on shape and dimensions.
EN 10060: Hot rolled round steel bars for general purposes – Dimensions and tolerances on shape and dimensions.
EN 10061: Hot rolled hexagon steel bars for general purposes – Dimensions and tolerances on shape and dimensions.
EN 10279: Hot rolled steel channels – Tolerances on shape, dimensions and mass.

- *Hollow sections*
  EN 10219-2: Cold formed welded structural hollow sections of non-alloy and fine grain steels – Part 2: Tolerances, dimensions and sectional properties.
  EN 10210-2: Hot finished structural hollow sections of non-alloy and fine grain steels – Part 2: Tolerances, dimensions and sectional properties.

EN 10278:2002: Dimensions and tolerances of bright steel products.

- *Flat products*

EN 10278: Dimensions and tolerances of bright steel products.

R-UNI EN 508-1.

EN 508-1: Roofing products from metal sheet – Specification for self-supporting products of steel, aluminium or stainless steel sheet – Part 1: Steel

EN 10143: Continuously hot-dip coated steel sheet and strip – Tolerances on dimensions and shape.

EN 14782: Self-supporting metal sheet for roofing, external cladding and internal lining – Product specification and requirements.

EN 14509: Self-supporting double skin metal faced insulating panels – Factory made products – Specifications.

B.1.4 Material Tests

EN ISO 9015-1: Destructive tests on welds in metallic materials – Hardness testing – Part 1: Hardness test on arc welded joints.

EN ISO 6892-1: Metallic materials – Tensile testing – Part 1: Method of test at room temperature.

EN ISO 7500-1: Metallic materials – Verification of static uniaxial testing machines – Part 1: Tension/compression testing machines – Verification and calibration of the force-measuring system.

EN ISO 376: Metallic materials – Calibration of force-proving instruments used for the verification of uniaxial testing machines.

EN ISO 9513: Metallic materials – Calibration of extensometers used in uniaxial testing.

EN ISO 148-1: Metallic materials – Charpy pendulum impact test – Part 1: Test method.

EN ISO 148-2: Metallic materials – Charpy pendulum impact test – Part 2: Verification of testing machines.

EN ISO 148-3: Metallic materials – Charpy pendulum impact test – Part 3: Preparation and characterization of Charpy V-notch test pieces for indirect verification of pendulum impact machines.

UNI EN ISO 18265: EN ISO 18265: Metallic materials -- Conversion of hardness values.

B.1.5 Mechanical Fasteners

EN ISO 898-1: Mechanical properties of fasteners made of carbon steel and alloy steel – Part 1: Bolts, screws and studs with specified property classes – Coarse thread and fine pitch thread.

EN ISO 898-5: Mechanical properties of fasteners made of carbon steel and alloy steel – Part 5: Set screws and similar threaded fasteners not under tensile stresses.

EN ISO 898-6: Mechanical properties of fasteners – Part 6: Nuts with specified proof load values – Fine pitch thread.

EN ISO 1478: Tapping screws thread.

EN ISO 1479: Hexagon head tapping screws.

EN ISO 2702: Heat-treated steel tapping screws – Mechanical properties.

EN ISO 4014: Hexagon head bolts – Product grades A and B.

EN ISO 4016: Hexagon head bolts – Product grade C

EN ISO 4017: Hexagon head screws – Product grades A and B.

EN ISO 4018: Hexagon head screws – Product grade C.

EN ISO 7049: Cross-recessed pan head tapping screws.
EN ISO 7089: Plain washers – Normal series – Product grade A.
EN ISO 7090: Plain washers, chamfered – Normal series – Product grade A.
EN ISO 7091: Plain washers – Normal series – Product grade C.
EN ISO 10684: Fasteners – Hot dip galvanized coatings.
EN 14399-1: High-strength structural bolting assemblies for preloading – Part 1: General requirements.
EN 14399-2: High-strength structural bolting assemblies for preloading – Part 2: Suitability test for preloading.
EN 14399-3: High-strength structural bolting assemblies for preloading – Part 3: System HR – Hexagon bolt and nut assemblies.
EN 14399-4: High-strength structural bolting assemblies for preloading – Part 4: System HV – Hexagon bolt and nut assemblies
EN 14399-5: High-strength structural bolting assemblies for preloading – Part 5: Plain washers
EN 14399-6: High-strength structural bolting assemblies for preloading – Part 6: Plain chamfered washers.
EN 15048-1: Non-preloaded structural bolting assemblies – Part 1: General requirements.
EN 15048-2: Non-preloaded structural bolting assemblies – Part 2: Suitability test.
EN 20898-2: Mechanical properties of fasteners – Part 2: Nuts with specified proof load values – Coarse thread.

B.1.6 Welding

B.1.6.1 Welding Processes

EN 1011-1: Welding – Recommendations for welding of metallic materials – Part 1: General guidance for arc welding.
EN 1011-2: Welding – Recommendations for welding of metallic materials – Part 2: Arc welding of ferritic steels.
EN 1011-3: Welding – Recommendations for welding of metallic materials – Part 3: Arc welding of stainless steels.
EN ISO 4063: Welding and allied processes – Nomenclature of processes and reference numbers.
EN ISO 9692-1: Welding and allied processes – Recommendations for joint preparation – Part 1: Manual metal-arc welding, gas-shielded metal-arc welding, gas welding, TIG welding and beam welding of steels.

B.1.6.2 Welding Consumables

EN ISO 14171: Welding consumables – Solid wire electrodes, tubular cored electrodes and electrode/flux combinations for submerged arc welding of non-alloy and fine grain steels – Classification.

B.1.7 Protection

EN ISO 12944-1: Paints and varnishes – Corrosion protection of steel structures by protective paint systems – Part 1: General introduction.
EN ISO 12944-2: Paints and varnishes – Corrosion protection of steel structures by protective paint systems – Part 2: Classification of environments.

EN ISO 12944-3: Paints and varnishes – Corrosion protection of steel structures by protective paint systems – Part 3: Design considerations.
EN ISO 12944-4: Paints and varnishes – Corrosion protection of steel structures by protective paint systems – Part 4: Types of surface and surface preparation.
EN ISO 12944-5: Paints and varnishes – Corrosion protection of steel structures by protective paint systems – Part 5: Protective paint systems.
EN ISO 12944-6: Paints and varnishes – Corrosion protection of steel structures by protective paint systems – Part 6: Laboratory performance test methods.
EN ISO 12944-7: Paints and varnishes – Corrosion protection of steel structures by protective paint systems – Part 7: Execution and supervision of paint work.
EN ISO 12944-8: Paints and varnishes – Corrosion protection of steel structures by protective paint systems – Part 8: Development of specifications for new work and maintenance.
EN ISO 8501-1: Preparation of steel substrates before application of paints and related products – Visual assessment of surface cleanliness – Part 1: Rust grades and preparation grades of uncoated steel substrates and of steel substrates after overall removal of previous coatings
EN ISO 8501-2: Preparation of steel substrates before application of paints and related products – Visual assessment of surface cleanliness – Part 2: Preparation grades of previously coated steel substrates after localized removal of previous coatings.
EN ISO 8501-3: Preparation of steel substrates before application of paints and related products – Visual assessment of surface cleanliness – Part 3: Preparation grades of welds, edges and other areas with surface imperfections.
EN ISO 8501-4: Preparation of steel substrates before application of paints and related products – Visual assessment of surface cleanliness – Part 4: Initial surface conditions, preparation grades and flash rust grades in connection with high-pressure water jetting.
EN ISO 8503-1: Preparation of steel substrates before application of paints and related products – Surface roughness characteristics of blast-cleaned steel substrates – Part 1: Specifications and definitions for ISO surface profile comparators for the assessment of abrasive blast-cleaned surfaces.
EN ISO 8503-2: Preparation of steel substrates before application of paints and related products – Surface roughness characteristics of blast-cleaned steel substrates – Part 2: Method for the grading of surface profile of abrasive blast-cleaned steel – Comparator procedure.
EN ISO 8503-3: Preparation of steel substrates before application of paints and related products – Surface roughness characteristics of blast-cleaned steel substrates – Part 3: Method for the calibration of ISO surface profile comparators and for the determination of surface profile – Focusing microscope procedure.
EN ISO 8503-4: Preparation of steel substrates before application of paints and related products – Surface roughness characteristics of blast-cleaned steel substrates – Part 4: Method for the calibration of ISO surface profile comparators and for the determination of surface profile – Stylus instrument procedure.
EN ISO 8503-5: Preparation of steel substrates before application of paints and related products – Surface roughness characteristics of blast-cleaned steel substrates – Part 5: Replica tape method for the determination of the surface profile.
EN ISO 1461 : Hot dip galvanized coatings on fabricated iron and steel articles – Specifications and test methods.
EN ISO 14713-1: Zinc coatings – Guidelines and recommendations for the protection against corrosion of iron and steel in structures – Part 1: General principles of design and corrosion resistance.
EN ISO 14713-2: Zinc coatings – Guidelines and recommendations for the protection against corrosion of iron and steel in structures – Part 2: Hot dip galvanizing.

B.2 Most Relevant Standards for United States Design

B.2.1 Reference for Structural Design

ANSI/AISC 360-10: Specification for Structural Steel Buildings
ANSI/AISC 341-10: Seismic Provisions for Structural Steel Buildings
ANSI/AISC 358-10: Prequalified Connections for Special and Intermediate Steel Moment Frames for Seismic Applications
ASCE/SEI 7-10: Minimum Design Loads for Buildings and Other Structures.
ANSI/AISC 303-10: Code of Standard Practice for Steel Buildings and Bridges.
Research Council on Structural Connections (RCSC) – Specification for Structural Joints Using High-Strength Bolts.

B.2.2 Standards for Materials and Technical Delivery Conditions

ASTM A6/A6M – 14: Standard Specification for General Requirements for Rolled Structural Steel Bars, Plates, Shapes and Sheet Piling.
ASTM A992/A992M – 11: Standard Specification for Structural Steel Shapes.
ASTM A572/A572M-13a: Standard Specification for High-Strength Low-Alloy Columbium-Vanadium Structural Steel.
ASTM A913/A913M-14a: Standard Specification for High-Strength Low-Alloy Steel Shapes of Structural Quality, Produced by Quenching and Self-Tempering Process (QST).
ASTM A588/A588M-10 Standard Specification for High-Strength Low-Alloy Structural Steel, up to 50 ksi (345 MPa) Minimum Yield Point, with Atmospheric Corrosion Resistance.
ASTM A242/A242M-13: Standard Specification for High-Strength Low-Alloy Structural Steel.
ASTM A36/A36M-12: Standard Specification for Carbon Structural Steel.
ASTM A529/A529M-05: Standard Specification for High-Strength Carbon-Manganese Steel of Structural Quality.

- *Hollow sections*
  ASTM A500/A500M-13: Standard Specification for Cold-Formed Welded and Seamless Carbon Steel Structural Tubing in Rounds and Shapes.
  ASTM A501-07: Standard Specification for Hot-Formed Welded and Seamless Carbon Steel Structural Tubing.
  ASTM A550-06: Standard Specification for Ferrocolumbium.
  ASTM A847/A847M-14: Standard Specification for Cold-Formed Welded and Seamless High-Strength, Low-Alloy Structural Tubing with Improved Atmospheric Corrosion Resistance.
  ASTM A618/A618M-04: Standard Specification for Hot-Formed Welded and Seamless High-Strength Low-Alloy Structural Tubing.
  ASTM A53/A53M-12: Standard Specification for Pipe, Steel, Black and Hot-Dipped, Zinc-Coated, Welded and Seamless.
- *Strip and flat products*
  ASTM A514/A514M-14: Standard Specification for High-Yield-Strength, Quenched and Tempered Alloy Steel Plate, Suitable for Welding.
  ASTM A852/A852M – 01: Standard Specification for Quenched and Tempered Low Alloy Structural Steel Plate with 70 ksi (485 MPa) Minimum Yield Strength to 4 in. (100 mm) Thick.
  ASTM A606/A606M-09a: Standard Specification for Steel, Sheet and Strip, High-Strength, Low-Alloy, Hot-Rolled and Cold-Rolled, with Improved Atmospheric Corrosion Resistance.

ASTM A1011/A1011M-14: Standard Specification for Steel, Sheet and Strip, Hot-Rolled, Carbon, Structural, High-Strength Low-Alloy, High-Strength Low-Alloy with Improved Formability and Ultra-High Strength.

B.2.3 Material Tests

ASTM A673/A673M: standard specification for sampling procedure for impact testing of structural steel.

B.2.4 Mechanical Fasteners

ASTM A307 – Standard Specification for Carbon Steel Bolts, Studs and Threaded Rod 60 000 PSI Tensile Strength
ASTM A325 – Standard Specification for Structural Bolts, Steel, Heat Treated, 120/105 ksi Minimum Tensile Strength
ASTM A325M – Standard Specification for Structural Bolts, Steel, Heat Treated 830 MPa Minimum Tensile Strength (Metric)
ASTM A354 – Standard Specification for Quenched and Tempered Alloy Steel Bolts, Studs and Other Externally Threaded Fasteners
ASTM A449 – Standard Specification for Hex Cap Screws, Bolts and Studs, Steel, Heat Treated, 120/105/90 ksi Minimum Tensile Strength, General Use
ASTM A490 – Standard Specification for Structural Bolts, Alloy Steel, Heat Treated, 150 ksi Minimum Tensile Strength
ASTM A490M – Standard Specification for High-Strength Steel Bolts, Classes 10.9 and 10.9.3, for Structural Steel Joints (Metric)
ASTM F1852 – Standard Specification for "Twist Off" Type Tension Control Structural Bolt/Nut/Washer Assemblies, Steel, Heat Treated, 120/105 ksi Minimum Tensile Strength
ASTM F2280 – Standard Specification for "Twist Off" Type Tension Control Structural Bolt/Nut/Washer Assemblies, Steel, Heat Treated, 150 ksi Minimum Tensile Strength
ASTM F959 – Standard Specification for Compressible-Washer Type Direct Tension Indicators for Use with Structural Fasteners
ASTM F436-11: Standard Specification for Hardened Steel Washers.
ASTM F1136: Standard Specification for Zinc/Aluminium Corrosion Protective Coatings for Fasteners
ASTM A563-07a: Standard Specification for Carbon and Alloy Steel Nuts.
ASTM F1554-07ae1: Standard Specification for Anchor Bolts, Steel, 36, 55 and 105-ksi Yield Strength.
ASTM A502 – 03: Standard Specification for Rivets, Steel, Structural.

B.2.5 Welding

B.2.5.1 Welding Processes

Aws D1.1/D1.1m Structural Welding Code – Steel

B.2.5.2 Welding Consumables

AWS A5.1/A5.1M – Specification for Carbon Steel Electrodes for Shielded Metal Arc Welding
AWS A5.5 Low-Alloy Steel Electrodes for Shielded Metal Arc Welding

AWS A5.17/A5.17M – Specification for Carbon Steel Electrodes and Fluxes for Submerged Arc Welding
AWS A5.18 Carbon Steel Electrodes and Rods for Gas Shielded Arc Welding
AWS A5.20/A5.20M – Carbon Steel Electrodes for Flux Cored Arc Welding.
AWS A5.23 Low-Alloy Steel Electrodes and Fluxes for Submerged Arc Welding.
AWS A5.25 Carbon and Low-Alloy Steel Electrodes and Fluxes for Electroslag Welding
AWS A5.26 Carbon and Low-Alloy Steel Electrodes for Electrogas Welding.
AWS A5.28 Low-Alloy Steel Electrodes and Rods for Gas Shielded Arc Welding.
AWS A5.29 Low-Alloy Steel Electrodes for Flux Cored Arc Welding.
AWS A5.32 Specification for Welding Shielding Gases.

B.2.6 Protection

SSPC SP2 – SSPC Surface Preparation Specification No. 2, Hand Tool Cleaning.
SSPC SP6 – SSPC Surface Preparation Specification No. 6, Commercial Blast Cleaning.

B.3 Essential bibliography

In the following, the main references used are listed, on which this volume has been based.

AAVV (2005) *Steel Designer's Manual*, (eds B. Davison and G.W. Owens), The Steel Construction Institute, Blackwell Science Ltd, Oxford, UK.
AAVV-ECCS (2006) *Rules for Member Stability in EN 1993-1-1, Background Documentation and Design Guidelines*, European Convention for Constructional Steelwork.
AAVV-ECCS n. 123 (2008), *Worked Examples According to EN 1993-1-3 Eurocode 3, Part 1-3*, European Convention for Constructional Steelwork.
Ballio, G. and Mazzolani, F.M. (1983) *Theory and Design of Steel Structuers*, Taylor & Francis.
Chen, W.F. (ed.) (1997) *Handbook of Structural Engineering*, CRC Press.
Dowling, P.J., Harding, J.E. and Bjorhovde, R. (eds) (1992) *Constructional Steel Design: an International Guide*, Elsevier Applied Science.
Faella, C., Piluso V. and Rizzano, G. (2000), *Structural Steel Semirigid Connections*, CRC Press.
Gardner, L. and Nethercot, D.A. (2005) *Designers' Guide to EN 1993-1-1 – Eurocode 3: Design of Steel Structures General Rules and Rules for Buildings*, Thomas Telford.
Ghersi, A., Landolofo, R. and Mazzolani, F.M. (2002) *Design of Metallic Cold-Formed Thin-Walled Members*, Spon Press.
Johansson, B., Maquoi, R., Sedlacek, G., Müller, C., and Beg D. (2007) Commentary and Worked Examples to EN 1993-1-5. Plated Structural Elements. Joint Report Prepared under the JRC – ECCS cooperation agreement for the evolution of Eurocode 3 (programme of CEN / TC 250).
Rodhes, J. (1991) *Design of Cold Formed Steel Members*, Elsevier Applied Science.
Sedlacek, G., Feldmann, M., Kühn, B., Tschickardt, D., Höhler, S., Müller, C., Hensen, W., Stranghöner, N., Dahl, W., Langenberg, P., Münstermann, S., Brozetti, J., Raoul, J., Pope, R., and Bijlaard, F. (1993) Commentary and Worked Examples to EN 1993-1-10. Material Toughness and Through Thickness Properties and other Toughness Oriented Rules in EN 1993. Joint Report Prepared under the JRC – ECCS cooperation agreement for the evolution of Eurocode 3 (programme of CEN/TC 250).
Simoes da Silva, L., Simoes, R., and Gervasio, H. (2010) *Design of Steel Structure- Eurocode 3: Design of Steel Structures – Part 1-1- General Rules and Rules for Building*, Ernst Sohn, A Wiley Company.
Trahair, N.S., Bradford, M.A., Nethercot, D.A. and Gardner, L. (2007) *The Behaviour and Design of Steel Structures to EC3*, Taylor & Francis Group.

In the following, a list of some websites is proposed that specialize in steel structures and from which free software is available.

http://www.access-steel.com
http://www.bauforumstahl.de/

http://dicata.ing.unibs.it/gelfi
http://ceeserver.cee.cornell.edu/tp26
http://www.ce.jhu.edu/bschafer
http://www.constructalia.com/it_IT/tools/catherramientas.jsp
http://www.construiracier.fr/
http://www.cticm.com/
http://eurocodes.jrc.ec.europa.eu/home.php
http://www.infosteel.be/
http://www.ruukki.com/
http://www.sbi.se/default_en.asp
http://www.steel-ncci.co.uk/
http://www.steel-sci.org/
http://www.steelconstruct.com

Index

Note: Pages number in *italics* and **Bold** denotes figures and tables

Printed and bound by CPI Group (UK) Ltd, Croydon, CR0 4YY

17/04/2025

14658897-0001